Aquaculture Systems and Engineering

Aquaculture Systems and Engineering

Editor: Olando Martin

www.callistoreference.com

Callisto Reference,
118-35 Queens Blvd., Suite 400,
Forest Hills, NY 11375, USA

Visit us on the World Wide Web at:
www.callistoreference.com

ISBN: 978-1-63239-974-8 (Hardback)

Trademark Notice: Registered trademark of products or corporate names are used only for explanation and identification without intent to infringe.

Cataloging-in-Publication Data

Aquaculture systems and engineering / edited by Olando Martin.
    p. cm.
Includes bibliographical references and index.
ISBN 978-1-63239-974-8
1. Aquaculture. 2. Aquacultural engineering. I. Martin, Olando.
SH135 .A68 2018
639.8--dc23

TABLE OF CONTENTS

Preface..IX

Chapter 1   **Augmenting the Post-Transplantation Growth and Survivorship of Juvenile**
               Scleractinian Corals via Nutritional Enhancement...1
               Tai Chong Toh, Chin Soon Lionel Ng, Jia Wei Kassler Peh, Kok Ben Toh and
               Loke Ming Chou

Chapter 2   **Model-Based Assessment of Estuary Ecosystem Health using the Latent Health**
               Factor Index, with Application to the Richibucto Estuary..10
               Grace S. Chiu, Margaret A. Wu and Lin Lu

Chapter 3   **Fish Oil Supplementation Alters the Plasma Lipidomic Profile and Increases**
               Long-Chain PUFAs of Phospholipids and Triglycerides in Healthy Subjects.............................22
               Inger Ottestad, Sahar Hassani, Grethe I. Borge, Achim Kohler, Gjermund Vogt,
               Tuulia Hyötyäinen, Matej Orešiĉ, Kirsti W. Brønner, Kirsten B. Holven,
               Stine M. Ulven and Mari C. W. Myhrstad

Chapter 4   **Preference Alters Consumptive Effects of Predators: Top- Down Effects of**
               a Native Crab on a System of Native and Introduced Prey...33
               Emily W. Grason and Benjamin G. Miner

Chapter 5   **Optimisation of Mesh Enclosures for Nursery Rearing of Juvenile Sea Cucumbers**............................39
               Steven W. Purcell and Natacha S. Agudo

Chapter 6   **Climate Change, Precipitation and Impacts on an Estuarine Refuge from Disease**.............49
               Jeffrey Levinton, Michael Doall, David Ralston, Adam Starke and Bassem Allam

Chapter 7   **A New Cryptic Species of South American Freshwater Pufferfish of the Genus**
               ***Colomesus* (Tetraodontidae), based on both Morphology and DNA Data**.............................57
               Cesar R. L. Amaral, Paulo M. Brito, Dayse A. Silva and Elizeu F. Carvalho

Chapter 8   **Spatial Variation in the Population Structure and Reproductive Biology of**
               ***Rimicaris hybisae* (Caridea: Alvinocarididae) at Hydrothermal Vents on the**
               Mid-Cayman Spreading Centre...72
               Verity Nye, Jonathan T. Copley and Paul A. Tyler

Chapter 9   **Capturing Ecosystem Services, Stakeholders' Preferences and Trade-Offs in**
               Coastal Aquaculture Decisions: A Bayesian Belief Network Application..............................87
               Laetitia Helene Marie Schmitt and Cecile Brugere

Chapter 10   **Spatial Variability of Benthic-Pelagic Coupling in an Estuary**
               Ecosystem: Consequences for Microphytobenthos Resuspension
               Phenomenon..99
               Martin Ubertini, Sébastien Lefebvre, Aline Gangnery, Karine Grangeré,
               Romain Le Gendre and Francis Orvain

Chapter 11   **Assessing the Health of the U.S. West Coast with a Regional-Scale Application of the Ocean Health Index**..**116**
Benjamin S. Halpern, Catherine Longo, Courtney Scarborough, Darren Hardy, Benjamin D. Best, Scott C. Doney, Steven K. Katona, Karen L. McLeod, Andrew A. Rosenberg and Jameal F. Samhouri

Chapter 12   **Fish Oil Replacement in Current Aquaculture Feed: Is Cholesterol a Hidden Treasure for Fish Nutrition?**..**132**
Fernando Norambuena, Michael Lewis, Noor Khalidah Abdul Hamid, Karen Hermon, John A. Donald and Giovanni M. Turchini

Chapter 13   **Effect of Fish Oil Supplementation on Fasting Vascular Endothelial Function in Humans: A Meta-Analysis of Randomized Controlled Trials**..**140**
Wei Xin, Wei Wei and Xiaoying Li

Chapter 14   **Whole Transcriptome Profiling of Successful Immune Response to *Vibrio* Infections in the Oyster *Crassostrea gigas* by Digital Gene Expression Analysis**..................**150**
Julien de Lorgeril, Reda Zenagui, Rafael D. Rosa, David Piquemal and Evelyne Bachère

Chapter 15   **Using Temporal Sampling to Improve Attribution of Source Populations for Invasive Species**..**161**
Sharyn J. Goldstien, Graeme J. Inglis, David R. Schiel and Neil J. Gemmell

Chapter 16   **A Nonluminescent and Highly Virulent *Vibrio harveyi* Strain is Associated with "Bacterial White Tail Disease" of *Litopenaeus vannamei* Shrimp**..............................**171**
Junfang Zhou, Wenhong Fang, Xianle Yang, Shuai Zhou, Linlin Hu, Xincang Li, Xinyong Qi, Hang Su and Layue Xie

Chapter 17   **Experts and Novices use the Same Factors–But Differently–To Evaluate Pearl Quality**..**177**
Yusuke Tani, Takehiro Nagai, Kowa Koida, Michiteru Kitazaki and Shigeki Nakauchi

Chapter 18   **Larval Dispersal Modeling of Pearl Oyster *Pinctada margaritifera* following Realistic Environmental and Biological Forcing in Ahe Atoll Lagoon**..................................**184**
Yoann Thomas, Franck Dumas and Serge Andréfouët

Chapter 19   **A Regional-Scale Ocean Health Index for Brazil**..**199**
Cristiane T. Elfes, Catherine Longo, Benjamin S. Halpern, Darren Hardy, Courtney Scarborough, Benjamin D. Best, Tiago Pinheiro and Guilherme F. Dutra

Chapter 20   **Survival, Growth and Reproduction of Cryopreserved Larvae from a Marine Invertebrate, the Pacific Oyster (*Crassostrea gigas*)**..**210**
Marc Suquet, Catherine Labbé, Sophie Puyo, Christian Mingant, Benjamin Quittet, Myrina Boulais, Isabelle Queau, Dominique Ratiskol, Blandine Diss and Pierrick Haffra

Chapter 21   **Spatial and Temporal Dynamics of Mass Mortalities in Oysters is Influenced
by Energetic Reserves and Food Quality**...216
Fabrice Pernet, Franck Lagarde, Nicolas Jeannée, Gaetan Daigle, Jean Barret,
Patrik Le Gall, Claudie Quere and Emmanuelle Roque D'orbcastel

Permissions

List of Contributors

Index

Preface

The cultivation of fishes, algae and other aquatic organisms is known as aquaculture. It includes fish farming, algaculture, oyster farming, etc. The methods included in aquaculture mainly are aquaponics and integrated multitrophic aquaculture. The need for sustainable practices in the field of aquaculture increases as the levels of marine pollution rises. This book includes some of the vital pieces of work being conducted across the world, on various topics related to aquaculture. Scientists and students actively engaged in this field will find it full of crucial and unexplored concepts. It will help the readers in keeping pace with the rapid changes in this field.

Various studies have approached the subject by analyzing it with a single perspective, but the present book provides diverse methodologies and techniques to address this field. This book contains theories and applications needed for understanding the subject from different perspectives. The aim is to keep the readers informed about the progresses in the field; therefore, the contributions were carefully examined to compile novel researches by specialists from across the globe.

Indeed, the job of the editor is the most crucial and challenging in compiling all chapters into a single book. In the end, I would extend my sincere thanks to the chapter authors for their profound work. I am also thankful for the support provided by my family and colleagues during the compilation of this book.

Editor

Augmenting the Post-Transplantation Growth and Survivorship of Juvenile Scleractinian Corals via Nutritional Enhancement

Tai Chong Toh*, Chin Soon Lionel Ng, Jia Wei Kassler Peh, Kok Ben Toh, Loke Ming Chou

Reef Ecology Laboratory, Department of Biological Sciences, National University of Singapore, Singapore, Singapore

Abstract

Size-dependant mortality influences the recolonization success of juvenile corals transplanted for reef restoration and assisting juvenile corals attain a refuge size would thus improve post-transplantation survivorship. To explore colony size augmentation strategies, recruits of the scleractinian coral *Pocillopora damicornis* were fed with live *Artemia salina* nauplii twice a week for 24 weeks in an *ex situ* coral nursery. Fed recruits grew significantly faster than unfed ones, with corals in the 3600, 1800, 600 and 0 (control) nauplii/L groups exhibiting volumetric growth rates of 10.65 ± 1.46, 4.69 ± 0.9, 3.64 ± 0.55 and 1.18 ± 0.37 mm^3/week, respectively. Corals supplied with the highest density of nauplii increased their ecological volume by more than 74 times their initial size, achieving a mean final volume of 248.38 ± 33.44 mm^3. The benefits of feeding were apparent even after transplantation to the reef. The corals in the 3600, 1800, 600 and 0 nauplii/L groups grew to final sizes of 4875 ± 260 mm^3, 2036 ± 627 mm^3, 1066 ± 70 mm^3 and 512 ± 116 mm^3, respectively. The fed corals had significantly higher survival rates than the unfed ones after transplantation (63%, 59%, 56% and 38% for the 3600, 1800, 600 and 0 nauplii/L treatments respectively). Additionally, cost-effectiveness analysis revealed that the costs per unit volumetric growth were drastically reduced with increasing feed densities. Corals fed with the highest density of nauplii were the most cost-effective (US\$0.02/mm^3), and were more than 12 times cheaper than the controls. This study demonstrated that nutrition enhancement can augment coral growth and post-transplantation survival, and is a biologically and economically viable option that can be used to supplement existing coral mariculture procedures and enhance reef restoration outcomes.

Editor: Brian Gratwicke, Smithsonian's National Zoological Park, United States of America

Funding: Funding for the research was provided by Wildlife Reserves Singapore Conservation Fund awarded to TCT. TCT was supported by the National University of Singapore Research Scholarship and the SingHaiyi Scholarship. The funders had no role in study design, data collection and analysis, decision to publish, or preparation of the manuscript.

Competing Interests: The authors have declared that no competing interests exist.

* E-mail: taichong.toh@gmail.com

Introduction

The global decline of coral reefs and the loss of associated ecological services have necessitated immediate intervention measures to reverse their further deterioration [1,2]. Active coral reef restoration initiatives have increasingly been incorporated into coastal management frameworks to supplement existing measures of rehabilitating impacted reefs [3,4]. Of the myriad techniques which have been developed, coral transplantation remains one of the most widely used, largely due to its ability to promote rapid colonization of the reefs and its ease of application [3,5]. The potential for generating large quantities of coral material via the "coral gardening" approach [6] for eventual transplantation to degraded reefs led to a greater emphasis on coral mariculture techniques. Asexual propagation techniques such as fragmentation allow coral material to be generated easily [3], but the drawbacks of this approach include a lack of genetic diversity of the clonal fragments and susceptibility of the donor colonies to stress arising from the fragmentation process, hence impeding large-scale production [7,8,9]. Recent developments have enabled the use of sexually derived coral juveniles as material for transplantation onto degraded reefs [10,11]. As scleractinian corals are highly fecund, this ensures that large numbers of genetically diverse coral propagules would be generated. While direct artificial seeding of coral larvae onto reefs can enhance initial recruitment [12], early post-settlement mortality of the recruits is exceedingly high due to competition by fouling communites and predation [13].

The use of *ex situ* coral mariculture in reef restoration can improve coral post-settlement survivorship. The rearing conditions can be carefully monitored and regulated to minimize the impacts of disturbances from fouling communities, temperature fluctuations and predator infestations by allowing the timely introduction of mitigative measures [14,15]. In spite of these benefits, the cost of setting up and operating *ex situ* mariculture facilities can be very expensive [4]. For instance, the cost of maintaining juvenile coral culture in the Philippines for six months constitutes 42.9% of the total project expenditure [11] and this inevitably increases with labour costs [16]. Unfortunately, such detailed financial estimates are rarely reported in the existing scientific literature due to the complexities involved and the rigorous efforts required to provide a reliable estimate. Cost-effectiveness analyses of cost-per-coral reveal clearly that as mortality rate increases, so does the cost of each colony [4]. Given that the highest mortality rates occur during the early developmental phases of the coral life cycle [13], augmenting the survivorship of juvenile corals would improve cost-effectiveness and increase the availability of source material for transplantation.

Size is an important determinant of survivorship in scleractinian corals and thus affects the rate of establishment of coral transplants on degraded reefs [17]. Smaller colonies tend to be more vulnerable since the refuge size required for surviving injuries arising from predation and incidental grazing is not yet attained [15,18]. Increasing coral colony size prior to transplantation is thus advantageous for enhancing post-transplantation growth, survivorship and promoting sexual maturity – factors which are essential for the maintenance of a viable coral community [17,19,20].

Scleractinian corals exhibit substantial inter- and intra-specific variations in growth rates [19,21], and one potential approach to promote rapid colony growth is to facilitate colony fusion [17]. However tissue resorption and somatic germ-cell parasitism may instead retard colony growth [22,23,24]. Another strategy involves enhancing the autotrophic and heterotrophic modes of coral nutrition by adjusting the conditions in *ex situ* mariculture prior to transplantation. Various studies have demonstrated that photo-synthetic and feeding rates could be increased by the manipulation of light intensity, flow rate and nutrient levels [25,26,27]. Although information on the effects of these manipulations on long-term coral growth rates is limited, the effects of nutritional enhancement are remarkably consistent for coral species from the families Faviidae, Acroporidae and Pocilloporidae. Compared to non-live feeds, live feeds were particularly useful for inducing faster coral growth [28], as were increments in *ex situ* feeding densities [28,29]. With heterotrophy in scleractinian corals commencing as early as two to seven days post-settlement [30,31], enhancing nutrition in the early stages should be explored as this would assist coral juveniles in attaining a size refuge as early as possible and reduce mortality.

The present study aims to evaluate the feasibility of nutritional enhancement as a strategy to improve the post-transplantation growth and survivorship of juveniles of the scleractinian coral *Pocillopora damicornis*. We hypothesize that the growth and survivorship of fed corals would be augmented both during the *ex situ* mariculture phase and after transplantation to the reef. To assess the economic viability of this approach for both coral mariculture and reef restoration efforts, we also determined the cost estimates for the study and examined the cost-effectiveness of *ex situ* nutritional enhancement. The findings of this study will facilitate planning of future coral mariculture and reef restoration initiatives.

Materials and Methods

Study Species and Planulae Collection

Pocillopora damicornis (Linnaeus, 1758) is a hermaphroditic scleractinian coral commonly found inhabiting shallow coastal areas within the Indo-Pacific region [32]. The reproductive and feeding biology of this coral has been well-studied [31,33,34] and it has been used extensively as a model species for developmental studies [17]. *Pocillopora damicornis* is also highly fecund and broods monthly [35], making it a popular candidate for propagation for the aquarium trade and reef restoration. This research was conducted with permission from Singapore National Parks Board (permit number NP/RP13-016), and no permit is required for collecting coral propagules in Singapore. Ten donor colonies of *P. damicornis* were collected from the fringing reef off Kusu Island, Singapore (1°13′25′′N, 103°51′38′′E) two days before the new moon in July 2012. Only colonies spaced 5 m apart and at least 20 cm in diameter were collected to ensure that they were sexually

mature and to minimize the chances of collecting identical genets [36]. The colonies were then transported to the Tropical Marine Science Institute on St. John's Island, Singapore (1°13′44′′N, 103°50′73′′E) and maintained in aerated outdoor aquaria (190×100×40 cm) with flow-through filtered seawater [9], which functioned as an *ex situ* coral nursery.

Biologically conditioned 'plugs', made of plastic wall plugs embedded in cement hemispheres (40 mm diameter), were fabricated and used as settlement substrates [15]. These allowed the *P. damicornis* recruits to be handled easily and facilitated their eventual transplantation onto the reef. One day before the new moon, all donor colonies were transferred and isolated in polyethylene planulation tanks (45 cm×30 cm×30 cm) with flow-through filtered seawater. Five centimetres below the rim of each tank, an outlet (3 cm diameter) was created to ensure water exchange. It was also covered with 100 μm plankton mesh (Sefar Pte. Ltd., Singapore) which helped to retain the coral planulae within the tanks. The tanks were filled with approximately 8 cm of sand which the conditioned plugs were inserted into, leaving only their hemispherical surfaces exposed for planulae settlement. Each plug was monitored daily for newly settled recruits, and plugs with at least three recruits were removed from the tank and maintained in the outdoor aquaria. In this study, all the colonies planulated within one to six days after the new moon, and the planulae were observed to settle within a day after planulation.

Feeding Regime in *ex situ* Coral Nursery

A total of 288 plugs with live juvenile corals were used for this study. On each plug, one primary polyp which had settled at least 10 mm away from the rest of the recruits was identified, measured and tagged by mapping the coral's position on the plug. This served to reduce the chances of colony fusion which would affect growth rates [17]. The plugs were randomly assigned among 16 holding tanks, each tank corresponding to one of the four feeding densities (0, 600, 1800 and 3600 nauplii/L following Petersen et al. (2008); $n = 4$ tanks). In each replicate tank, 18 plugs spaced 5 cm apart were secured on an elevated PVC frame. All plugs were maintained in the outdoor aquaria for one week before the start of the feeding regime [28].

The juvenile corals were fed with cultured day-old *Artemia salina* (approximately 400 μm; Bio-Marine Inc., California, U.S.A.), wherein each nauplius provided around 9.77 μcal [37], for 4 hours (between 12:00 to 16:00) twice every week for 24 weeks (from August 2012 to February 2013). During each feeding session, all the plugs were transferred to 10 L polyethylene feeding tanks containing 6 L of filtered sea water with gentle aeration. The corresponding volume of nauplii stock solution was added to make up the required densities for each treatment tank. The positions of the feeding tanks were randomised during each feeding session to minimize potential spatial influences on heterotrophic rates. After feeding, the plugs were gently flushed with filtered seawater to remove any remaining nauplii, and subsequently transferred back to the holding tanks. Fouling macroalgae were physically removed twice a week as these would otherwise rapidly overgrow the coral juveniles and compromise colony health [15].

The survivorship and growth – length (*l*), width (*w*) and height (*h*) – of the 18 tagged coral juveniles in each replicate tank were measured using vernier calipers every four weeks and the ecological volume of each coral was estimated following the calculation for right cylindrical volumes, $V = \pi r^2 h$, where $r = (l+w)/4$ [38]. Weekly radial and volumetric growth were calculated by dividing the respective differences in colony radii and ecological volumes at the start of the *ex situ* feeding regime and at the end of the *ex situ* feeding regime (24 weeks). The data obtained for all the surviving corals in each replicate tank was then averaged. The mean daily temperature and light irradiance (Onset Computer Corporation Inc., Massachusetts, U.S.A.) in the aquaria were $29 \pm 0.01°C$ ($n = 168$ days) and 128.7 ± 18.1 Lux ($n = 168$ days) respectively.

Transplantation and Monitoring

After 24 weeks, eight plugs with live corals were randomly selected from each holding tank to be transplanted back to the donor reef at Kusu Island. Four limestone outcrops (approximately 3.5 m in diameter and 2.5 m in height) that were at least 5 m apart were identified for the transplantation of the juvenile corals. Four sets of eight holes were then drilled on each outcrop and each replicate treatment was randomly assigned to one set of holes, such that the corals belonging to the same replicate holding tank were transplanted on the same outcrop ($n = 4$ outcrops). The plugs were inserted into the holes and stabilized using two-part marine epoxy [11].

The survivorship and colony dimensions of the tagged coral juveniles on each plug were recorded every four weeks for 24 weeks (from February to August 2013). Weekly radial and volumetric growth were calculated by dividing the respective differences in colony radii and ecological volumes at the start of the ex situ feeding regime and at the end of the entire study with the duration of the entire study (48 weeks). The data obtained for all the surviving corals in each replicate outcrop was then averaged. The mean daily temperature (Onset Computer Corporation Inc., Massachusetts, U.S.A.) in the transplant site was $29.9 \pm 0.07°C$ (24 readings per day, $n = 168$ days).

Cost Analysis

Cost-estimates were tabulated for each of the five phases of this study: (1) Collection of source materials and establishment of coral culture, (2) Maintenance and ex situ monitoring, (3) Feeding, (4) Transplantation and (5) In situ monitoring, and further itemized into equipment costs, labour costs and boat trips following Edwards et al. (2010) and Villanueva et al. (2012). The cost per coral produced before and after transplantation to the reef were then calculated. In addition, the cost per unit volumetric growth of each treatment group for both the ex situ feeding and post-transplantation phases was also estimated based on the total production costs, the mean weekly volumetric growth rates, duration and the number of tagged colonies alive at the end of each phase.

Statistical Analysis

Data for the final ecological volume, weekly radial and volumetric growth rates were first tested for homogeneity of variances using Levene's test and normality using Shapiro-Wilk test, followed by one-factor ANOVA with Tukey's HSD (Honestly Significant Difference) post-hoc test for all possible pairwise comparisons. As the variances for post-transplantation volumetric growth rates were heterogenous and not normally distributed, a non-parametric Kruskall-Wallis analysis was used. Subsequent pairwise comparisons were analysed using Mann-Whitney U test. These analyses were computed using SPSS v 17.0 (SPSS Inc). Data for the survivorship was analyzed using Cox Proportional-Hazards regression model and logrank test (R 2.14.2), using the independent factors initial colony radius, treatment and the interaction between radius and treatment for analysis. The model that best explained the trend was then selected using Akaike Information Criteria (AIC).

Results

Growth of Pocillopora damicornis Juveniles in ex situ Feeding Phase

The initial mean colony volume of the coral juveniles (approximately 3.5 mm³) did not differ significantly among the treatment groups ($F_{3,12} = 0.687$, $p = 0.557$). The mean colony volume across the treatments increased monotonically over the ex situ feeding phase of the study (Fig. 1a; Fig. 2). Juvenile corals in the 3600 nauplii/L treatment group grew by more than 74 times their initial sizes and attained a mean final ecological volume of 248.38 ± 33.44 mm³ (mean $\pm$ S.E.; 4.03 ± 0.18 mm radius). The final volumes of the colonies in the 1800, 600 and 0 nauplii/L were 111.66 ± 20.8 mm³ (34 times the initial volume; 3.63 ± 0.25 mm radius), 87.18 ± 12.91 mm³ (24 times the initial volume; 2.78 ± 0.12 mm radius) and 30.65 ± 8.65 mm³ (8 times the initial volume; 2.13 ± 0.05 mm radius), respectively.

The weekly radial growth rates of the colonies (Fig. 1b) significantly differed among treatments ($F_{3,12} = 30.8$, $p < 0.001$). Colonies in the 3600 and 1800 nauplii/L treatment groups grew at rates of 0.13 ± 0.008 mm/week (mean $\pm$ S.E.) and 0.11 ± 0.009 mm/week respectively, and were significantly faster than those in the 600 nauplii/L (0.08 ± 0.005 mm/week, $p < 0.001$) and control (0.05 ± 0.002 mm/week, $p < 0.001$) groups. Weekly volumetric growth rates (Fig. 1c) were also significantly different among treatments ($F_{3,12} = 19.2$, $p < 0.001$) with the colonies in the 3600 nauplii/L treatments growing significantly faster (10.65 ± 1.46 mm³/week) than colonies in the 1800 nauplii/L (4.69 ± 0.9 mm³/week, $p = 0.003$), 600 nauplii/L (3.64 ± 0.55 mm³/week, $p = 0.001$) and control (1.18 ± 0.37 mm³/week, $p < 0.001$) groups.

Growth of Pocillopora damicornis Juveniles after Transplantation

The mean colony sizes of all juvenile corals continued to increase steadily after transplantation to the reef (Fig. 2; Fig. 3a), with the colonies in the 3600 nauplii/L treatment group exhibiting the largest increase in size (1534 times the initial size at the start of the study). Final mean colony volumes for the 0, 600 1800 and 3600 nauplii/L groups were 512 ± 116 mm³ (mean $\pm$ S.E.; 137 times the initial volume; 5.03 ± 0.49 mm radius), 1066 ± 70 mm³ (284 times the initial volume; 6.35 ± 0.14 mm radius), 2036 ± 627 mm³ (486 times the initial volume; 7.25 ± 0.80 mm radius) and 4875 ± 260 mm³ (10.5 ± 0.29 mm radius), respectively.

Weekly radial growth rates (Fig. 3b) differed among the treatment groups ($F_{3,12} = 26.05$, $p < 0.001$). Colonies in the 3600 nauplii/L group (mean $\pm$ S.E.; 0.198 ± 0.005 mm/week) had significantly faster growth rates than the colonies in the 1800 nauplii/L (0.128 ± 0.016 mm/week; $p = 0.001$), 600 nauplii/L (0.109 ± 0.003; $p < 0.001$) and the control (0.082 ± 0.01 mm/week; $p < 0.001$) groups. A significant difference between the 1800 nauplii/L and control groups was also present ($p < 0.05$). The weekly volumetric growth rates (Fig. 3b) were also significantly different ($p = 0.006$), displaying a similar trend as that of the radial growth rates. The mean volumetric growth rates (Fig. 3c) were 101.5 ± 5.4 mm³/week, 42.3 ± 13.1 mm³/week, 22.1 ± 1.5 mm³/week and 10.6 ± 2.4 mm³/week for the 3600, 1800, 600 and 0 nauplii/L groups respectively.

Survivorship of Juvenile Pocillopora damicornis in ex situ Feeding Phase and after Transplantation

In the ex situ feeding phase (Fig. 4a), there were no significant differences in survivorship across treatments (logrank test = 1.22,

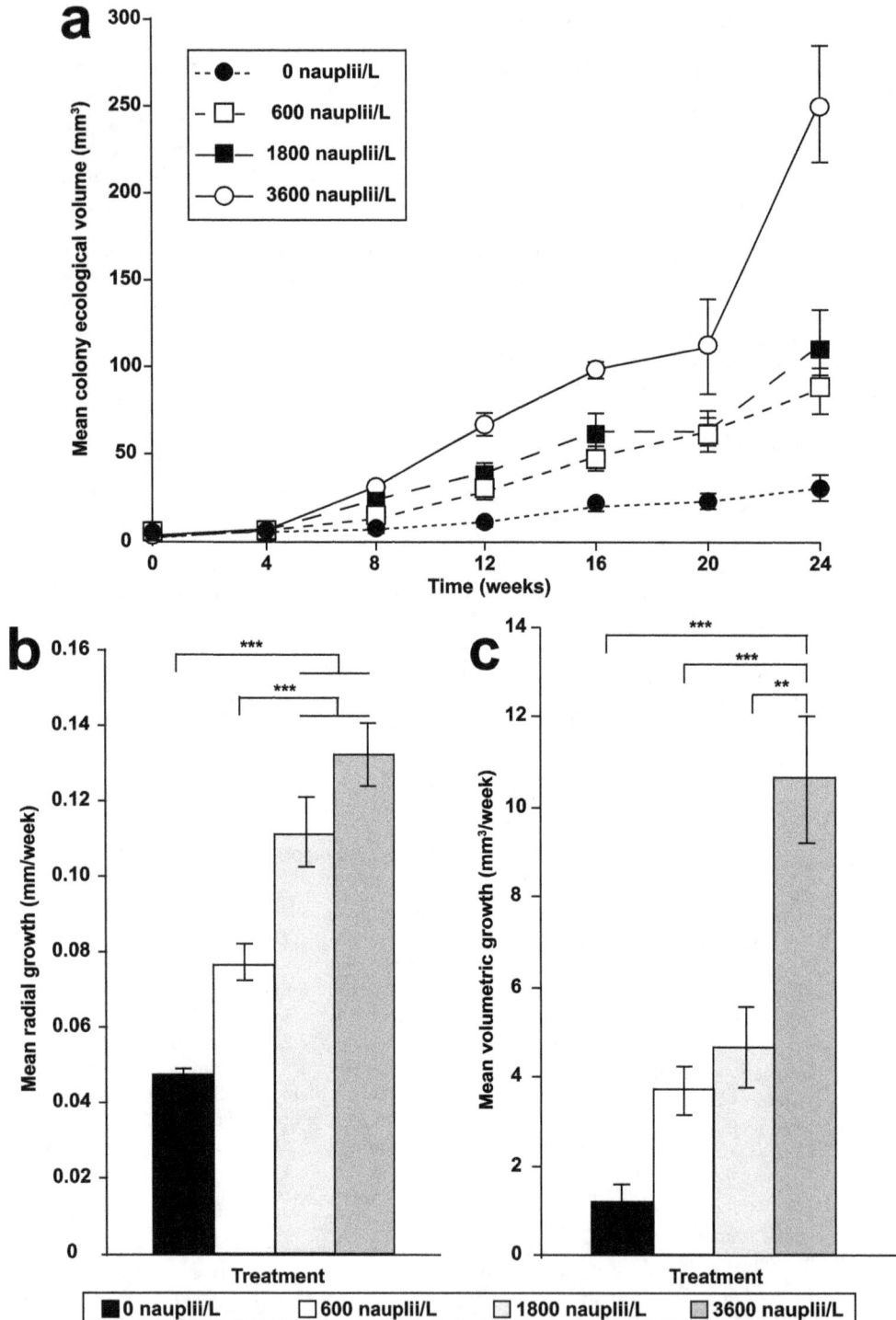

Figure 1. Growth of *Pocillopora damicornis* juveniles over a 24-week *ex situ* feeding regime. Graphs show the (a) mean ecological volumes, (b) mean weekly radial and (c) volumetric growth rates (± S.E.) of the corals in the 0 (control), 600, 1800 and 3600 nauplii/L treatment groups. The symbols *, **, and *** denote statistical significance at $p = 0.05$, $p = 0.01$, $p = 0.001$ respectively.

$d.f. = 1$, $p = 0.27$). Survival rates of the *P. damicornis* juveniles in the control, 600, 1800 and 3600 nauplii/L groups at the end of 24 weeks were 45%, 54%, 58% and 47% respectively, and the overall survival was 51%. Corals in the control, 600, 1800 and 3600 nauplii/L groups had post-transplantation survival rates of 38%, 56%, 59% and 63% respectively (overall survival of 54%), and these were significantly different across treatments (Fig. 4b).

Post-hoc pairwise comparisons revealed that the corals in the 3600 and 1800 nauplii/L groups ($p = 0.016$ and $p = 0.044$, respectively) had significantly higher survival rates than the control. The difference in survivorship was accounted for by both the initial radius prior to transplantation (logrank test = 6.86, A.I.C. value = 535, $d.f. = 1$, $p = 0.009$) and treatment (logrank

Figure 2. Growth of *Pocillopora damicornis* juveniles throughout the study. *Pocillopora damicornis* juveniles in the 0 (control), 600, 1800 and 3600 nauplii/L treatment groups (a, c, e, g) after the 24-week *ex situ* feeding regime, and (b, d, f, h) 24 weeks after transplantation to the reef. Scale bar = 1 cm, arrows indicate the positions of the corals.

test = 6.26, A.I.C. value = 536, *d.f.* = 1, *p* = 0.012), and both factors were highly correlated (r = 0.6).

Cost Analysis

The total cost for producing 288 coral plugs in the *ex situ* feeding phase and transplanting 128 corals to the reef was an estimated US$10467 (see Table S1 for detailed cost estimates). Over 40% was attributed to the cost of establishing the coral culture, which included the harvesting of donor colonies, setting up of culture tanks and the collection of planulae (Table 1). 34.3% of the total costs arose from transplanting and subsequent monitoring of the coral transplants, while feeding and maintenance of the coral juveniles contributed the remaining 9.6% and 7.2% respectively.

The cost of propagating 288 corals was estimated at US$20.90/coral. Upon taking into account the mean survival rate of 51% at the end of the *ex situ* feeding phase, the cost per coral was US$40.98 (Table 1). The cost of each transplanted coral was estimated at US$81.78. With a 54% mean survival rate 24 weeks

after transplantation, the cost per coral was US$151.44 (Table 1). In the *ex situ* feeding phase, the cost per unit growth decreased with increasing feeding densities (Table 2), making the 3600 nauplii/L treatment group the most cost-effective. The cost per unit volumetric growth was US$0.18/mm³, which was more than seven times cheaper than that of the control group. A similar trend was observed for the corals after transplantation – the cost per unit volumetric growth for the 3600 nauplii/L treatment was US$0.02/mm³, which was more than 12 times cheaper than the control treatments.

Discussion

Scleractinian corals supplement up to 35% of their daily metabolic requirements with a wide range of items such as dissolved organic matter, suspended particulate matter and zooplankton [29,39,40]. While corals reared in ex situ systems are routinely supplied with zooplankton, microalgae and commercial dry food [28], those fed with live zooplankton – a highly nutritious feed – consistently grow faster [28,39]. The use of *Artemia* nauplii as coral feed in this study significantly augmented the growth of *P. damicornis* juveniles. Coral volumetric growth rates increased by up to 9 times with the addition of higher densities of *Artemia* nauplii, leading to final ecological volumes that were 2.9 to 8.8 times greater than those in the control groups after 24 weeks (Fig. 1). These results were comparable to work by Petersen et al. (2008), who reported that *Acropora tenuis* juveniles fed with 3750 *Artemia* nauplii/L and *Favia fragum* juveniles fed with 300 nauplii/L respectively grew eight and five times larger than those in the control group. Since juvenile coral growth was proportionate to feed densities and the growth rates did not slow down even at 3600 nauplii/L, further increment of feeding densities and frequency would likely augment coral growth further. Additionally, as heterotrophy is known to play an important role in mitigating effects during stress events such as coral bleaching [41], introducing live feed during the early life stages can assist juvenile corals in attaining the required refuge size faster and cope with the effects of acute environmental stress.

Although *ex situ* mariculture can help to enhance the survivorship of coral fragments (>98%) [42,43] and sexual propagules (60–75%) [15,16,28] by providing a conducive environment for the coral material to grow, the facilities are usually expensive to run [6], inadvertently placing limits on the duration of rearing as well as the potential for any improvements to survivorship [28]. As survivorship increases with colony size [44], it is important to explore ways of accelerating the growth of juvenile corals in the least possible time. In this study, coral survivorship did not improve substantially despite significant increases in growth, as was consistent with that observed by Petersen et al. (2008). However, at 51%, the mean survival rate across treatments were more than four times higher than if juvenile corals of the same size class were to be transplanted to the field [17], underscoring the usefulness of feeding corals in *ex situ* mariculture to optimise restoration outcomes.

Twenty-four weeks after transplantation, the juvenile coral transplants were 1.5 to 2.1 times larger than their initial sizes (Fig. 3). This corroborated with other studies wherein 6-months-old and 18-months-old branching juvenile corals grew 1.5 to five times their initial diameters six months after transplantation [10,17]. More importantly, the growth rates of fed corals remained consistently higher than those of the unfed corals even after transplantation to the reef, suggesting the possibility that benefits obtained from the *ex situ* feeding regime will continue even after feeding has stopped.

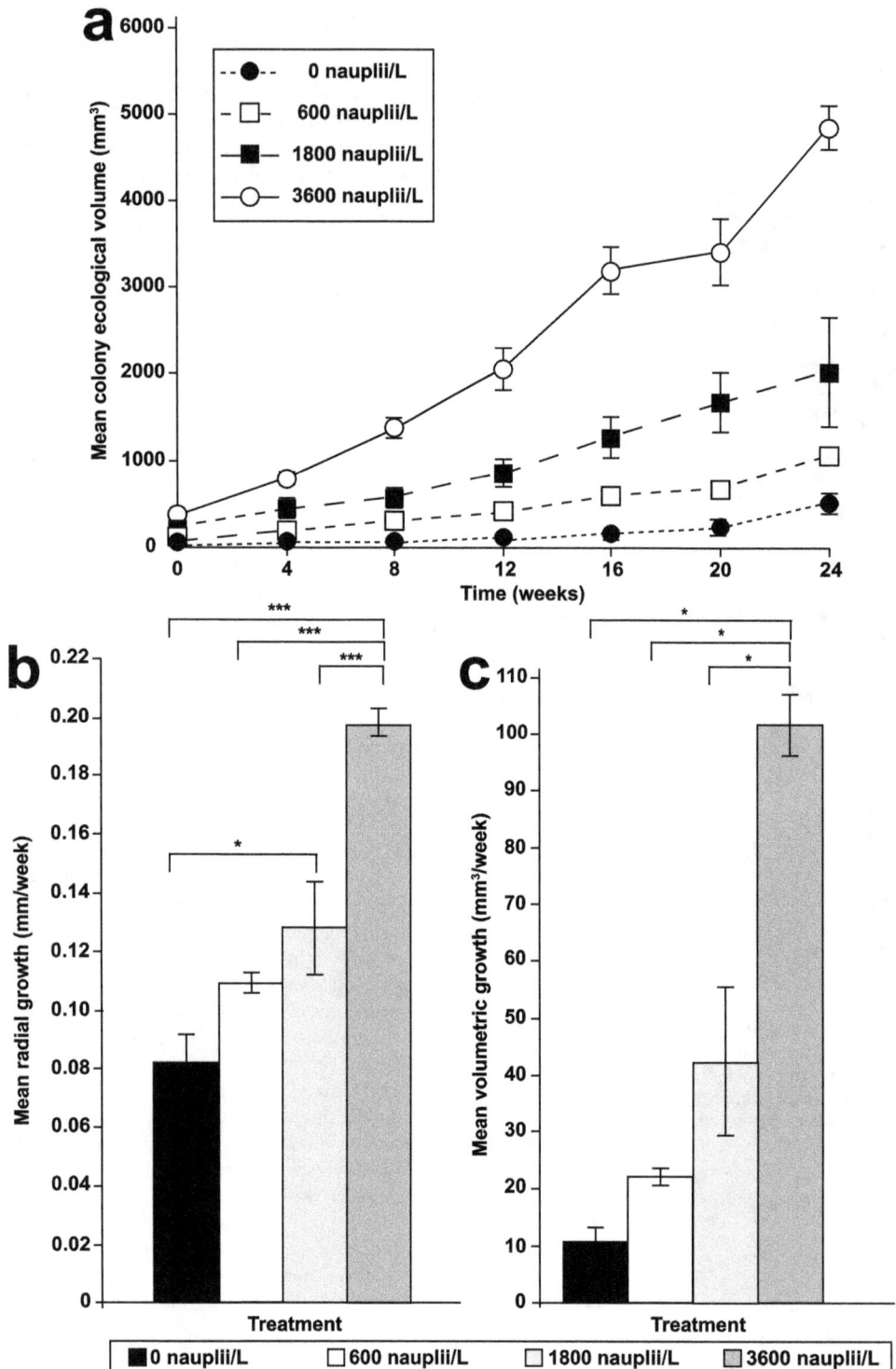

Figure 3. Growth of *Pocillopora damicornis* juveniles over 24 weeks after transplantation to the reef. Graphs show the (a) mean ecological volumes, (b) mean weekly radial and (c) volumetric growth rates (± S.E.) of juvenile *Pocillopora damicornis* in the 0 (control), 600, 1800 and 3600 *Artemia* nauplii/L treatment groups. The symbols *, **, and *** denote statistical significance at $p = 0.05$, $p = 0.01$, $p = 0.001$ respectively.

Interestingly, the enhancement in growth from the *ex situ* feeding regime improved the post-transplantation survivorship of the juvenile corals. Both size and feeding regimes were able to account for the survivorship patterns observed, supporting the observations of size-dependant mortality in scleractinian corals [17]. Since nutrition enhancement was a direct causative agent of the coral growth and both the effects of size and feeding regime on survivorship were highly correlated, it exerted a concomitant effect of augmenting post-transplantation survival. Clearly, size was a key determinant of post-transplantation survival. However, the average post-transplantation mortality rate of all *P. damicornis* juveniles in this study (46%) was higher than that reported from

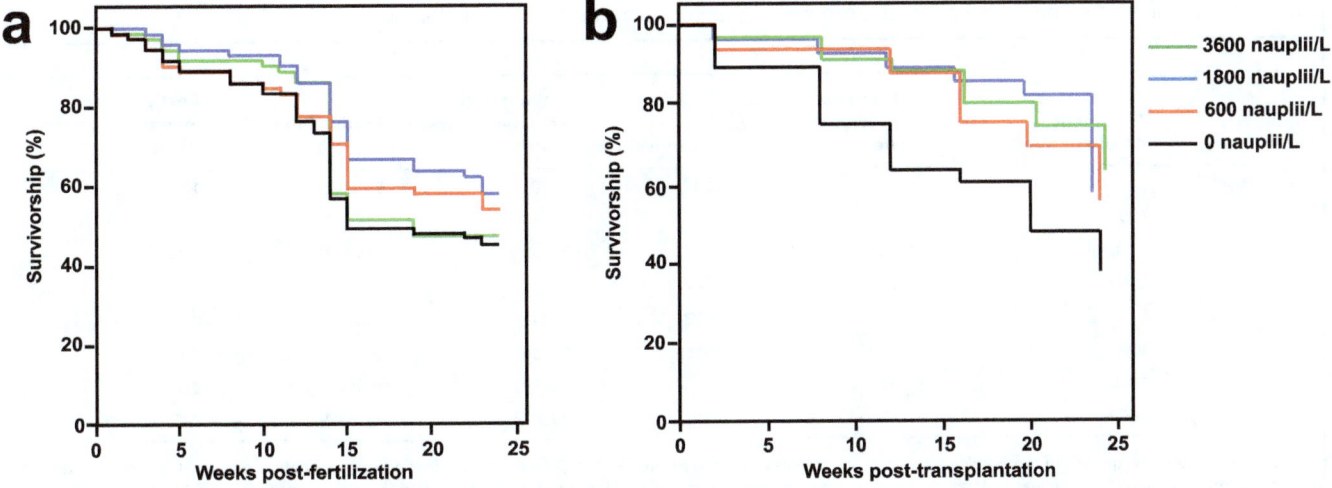

Figure 4. Survivorship of *Pocillopora damicornis* juveniles. Survival curves of *Pocillopora damicornis* juveniles in the 0 (control), 600, 1800 and 3600 nauplii/L groups (a) in the *ex situ* feeding phase (24 weeks, *n* = 72) and (b) after transplantation (24 weeks, *n* = 32).

other studies (11–34%) [10,11,17], likely due to the high sediment levels in Singapore waters, which have been estimated to limit scleractinian recruitment to two individuals m^{-2} [45]. As was observed during monthly visits to the study site, most juvenile colonies were smothered by fine particulate sediment, with obvious damage to the coral tissue. Post-transplantation survivorship can thus be expected to be lower in areas experiencing chronic sedimentation such as Singapore. It is clearly advantageous to boost the survival chances of juvenile corals by implementing an *ex situ* nutritional enhancement regime to increase colony size prior to transplantation.

While nutritional enhancement confers significant ecological advantages to juvenile corals in *ex situ* mariculture, the process should still be thoroughly assessed and reviewed to boost its

economic viability. In the current study, nutritional enhancement constituted only 9% of total production costs. Of this amount, 99% was attributed to the labour required for transferring the corals from the holding tanks to the feeding tanks. Such costs can be reduced further in commercial mariculture systems where the corals do not need to be transferred elsewhere for feeding. The results also showed that corals supplied with the highest density of feed (3600 nauplii/L) attained ecological volumes close to that of the corals in the control group at the end of the 24-week feeding phase, in as early as eight weeks. This corresponds to a one-third reduction in *ex situ* rearing time and translates to significant reductions in operational costs. Additionally, the cost-effectiveness of the method was apparent as the cost per unit volumetric growth of the corals fed with 3600 nauplii/L was more than seven and

Table 1. Summary of the cost estimates.

Phase	Subcategory	Cost (US$)	Percentage of total cost (%)
1. Establishment of coral culture		**4261.69**	**40.7**
2. Maintenance		**757.22**	**7.2**
3. Feeding regime		**1000.9**	**9.6**
	3.1 Control treatment	0	0
	3.2 600 nauplii/L	333.27	3.2
	3.3 1800 nauplii/L	333.58	3.2
	3.4 3600 nauplii/L	334.05	3.2
4. Transplantation		**860.24**	**8.2**
5. *In situ* monitoring		**3587.70**	**34.3**
Grand Total		10467.75	
Ex situ production cost for 288 coral plugs		6019.81	
Ex situ production cost per coral		20.90	
Cost per coral (51% survival)		40.98	
Cost per coral transplanted (128 coral plugs)		81.78	
Cost (54% survival)		151.44	

Summary of the cost estimates of producing 288 plugs with live *Pocillopora damicornis* juveniles under four ex situ feeding regimes (0, 600, 1800, 3600 nauplii/L) for 24 weeks, followed by the transplantation of 128 coral plugs and subsequent monitoring for 24 weeks. Mean survival rates across the treatments were used for the calculation of cost effectiveness at the end of each phase. Costs were estimated in Singapore Dollars (S$) prior to conversion to US$ at the rate of S$ 1.26 = US$ 1.

Table 2. Estimated cost per unit volumetric growth of the *Pocillopora damicornis* colonies.

Phase	Treatment density (nauplii/L)	Mean volumetric growth rates (mm³/week)	Survival (%)	Estimated total volumetric growth (mm³)	Production cost (US$)	Cost per unit volumetric growth (US$/mm³)
Ex situ feeding	0	1.18	45	3670	5018.91	1.37
	600	3.64	54	13586	6351.99	0.47
	1800	4.69	58	18802	6353.11	0.34
	3600	10.65	47	34598	6355.11	0.18
Transplantation	0	10.6	38	24748	9466.85	0.38
	600	22.1	56	76038	10799.93	0.14
	1800	42.3	59	153335	10801.17	0.07
	3600	101.5	63	392878	10803.05	0.03

Cost per unit volumetric growth of the *Pocillopora damicornis* colonies after the *ex situ* feeding (24 weeks, $n = 288$) and transplantation phase (24 weeks, $n = 128$). Total volumetric growth for each phase was estimated based on the mean weekly volumetric growth rates, duration and the number of live tagged colonies at the end of each phase. Production cost for each treatment group was calculated based on the cost estimates for the entire study.

twelve times cheaper than the controls in both the *ex situ* rearing and post-transplantation phases, respectively. However, it must be noted that directly comparing project costs among localities leads to inaccuracies. For example, costs per coral can be as low as US$11 in the Philippines [11] to as high as US$151 in Singapore (this study), mainly due to differences in manpower and equipment costs – labour costs differed by almost six-fold while the cost of boat hire differed by nearly ten-fold. Exploring other options such as recruiting volunteers to reduce labour costs [10] or increasing production for economies of scale [16] would help to improve cost-effectiveness.

The current study showed that supplying live *Artemia salina* nauplii as coral feed enhanced juvenile coral growth rates and survivorship in both the *ex situ* nursery phase as well as six months after they had been transplanted to a reef. These findings are important, because even though sexually-derived corals are increasingly used as material for reef restoration [10,11], the high mortality rates of the juvenile propagules is often a stumbling block in such projects. Since long rearing periods are infeasible due to high operational costs, nutritional enhancement may be considered as a means of reducing the time and cost required for the coral material to be reared in mariculture facilities. The approach is simple, cost-effective, and harbours the potential for large-scale application.

References

1. Bridge TC, Hughes TP, Guinotte JM, Bongaerts P (2013) Call to protect all coral reefs. Nat Clim Chang 3(6): 528–530.
2. Graham NA, Bellwood DR, Cinner JE, Hughes TP, Norström AV, et al. (2013). Managing resilience to reverse phase shifts in coral reefs. Front Ecol Environ 11(10): 541–548.
3. Rinkevich B (1995) Restoration strategies for coral reefs damaged by recreational activities: the use of sexual and asexual recruits. Restoration Ecology 3(4): 241–251.
4. Edwards AJ (2010) Reef Rehabilitation Manual. Coral Reef Targeted Research & Capacity Building for Management Program, St Lucia, Australia. 166 p.
5. Edwards AJ, Clark S (1999) Coral transplantation: a useful management tool or misguided meddling? Mar Pollut Bull 37(8): 474–487.
6. Shafir S, Van Rijn J, Rinkevich B (2006) Steps in the construction of underwater coral nursery, an essential component in reef restoration acts. Mar Biol 149(3): 679–687.
7. Yap HT, Alvarez RM, Custodio III HM, Dizon RM (1998) Physiological and ecological aspects of coral transplantation. J Exp Mar Biol Ecol 229(1): 69–84.
8. Shearer TL, Porto I, Zubillaga AL (2009) Restoration of coral populations in light of genetic diversity estimates. Coral Reefs 28(3): 727–733.
9. Toh TC, Guest J, Chou LM (2012) Coral larval rearing in Singapore: Observations on spawning timing, larval development and settlement of two

common scleractinian coral species. In Tan KS, editor. Contributions to Marine Science. National University of Singapore, Republic of Singapore. 81–87.
10. Omori M, Iwao K, Tamura M (2008) Growth of transplanted Acropora tenuis 2 years after egg culture. Coral Reefs 27(1): 165–165.
11. Villanueva RD, Baria MVB, dela Cruz DW (2012). Growth and survivorship of juvenile corals outplanted to degraded reef areas in Bolinao-Anda Reef Complex, Philippines. Mar Biol Res 8(9): 877–884.
12. Heyward AJ, Smith LD, Rees M, Field SN (2002) Enhancement of coral recruitment by in situ mass culture of coral larvae. Mar Ecol Prog Ser 230: 113–118.
13. Guest JR, Heyward AJ, Omori M, Iwao K, Morse A, et al. (2010) Rearing coral larvae for reef rehabilitation. In: Edwards AJ editor. Reef Rehabilitation Manual. Coral Reef Targeted Research & Capacity Building for Management Program, St Lucia, Australia. 73–92.
14. Forsman ZH, Rinkevich B, Hunter CL (2006) Investigating fragment size for culturing reef-building corals (*Porites lobata* and *P. compressa*) in ex situ nurseries. Aquaculture 261(1): 89–97.
15. Toh TC, Ng CSL, Guest J, Chou LM (2013). Grazers improve the health of scleractinian coral juveniles in *ex situ* mariculture. Aquaculture 414–415: 288–293.

Supporting Information

Table S1 Detailed cost estimates. Cost estimates of producing 288 plugs with live *Pocillopora damicornis* juveniles under four ex situ feeding regimes (0, 600, 1800, 3600 nauplii/L) for 24 weeks, followed by the transplantation of 128 coral plugs and subsequent monitoring for 24 weeks.

Acknowledgments

We would like to thank the staff and students of the NUS Reef Ecology Laboratory and the Tropical Marine Science Institute, for their administrative and logistical support. We would also like to acknowledge K.Y. Chong, A.T.K. Yee, X. Giam, J.R. Guest and A.J. Underwood for their valuable suggestions, and S.K.G. Lo for fabricating the settlement substrates used in this study. The comments provided by four anonymous reviewers greatly enhanced the manuscript. This study was part of T.C. Toh's Ph.D. dissertation work.

Author Contributions

Conceived and designed the experiments: TCT. Performed the experiments: TCT JWKP CSLN. Analyzed the data: TCT JWKP CSLN KBT. Contributed reagents/materials/analysis tools: TCT JWKP CSLN KBT LMC. Wrote the paper: TCT JWKP CSLN KBT LMC.

16. Nakamura R, Ando W, Yamamoto H, Kitano M, Sato A, et al. (2011) Corals mass-cultured from eggs and transplanted as juveniles to their native, remote coral reef. Mar Ecol Prog Ser 436: 161–168.

17. Raymundo LR, Maypa AP (2004) Getting bigger faster: Mediation of size-specific mortality via fusion in juvenile coral transplants. Ecol Appl 14(1): 281–295.

18. Wood R (1993) Nutrients, predation and the history of reef-building. Palaios: 526–543.

19. Hughes TP (1984) Population dynamics based on individual size rather than age: a general model with a reef coral example. Am Nat: 778–795.

20. Wallace CC (1985) Reproduction, recruitment and fragmentation in nine sympatric species of the coral genus Acropora. Mar Biol. 88: 21–233.

21. Bak RPM, Engel MS (1979) Distribution, abundance and survival of juvenile hermatypic corals (Scleractinia) and the importance of life history strategies in the parent coral community. Mar Biol 54(4): 341–352.

22. Buss LW (1982) Somatic cell parasitism and the evolution of somatic tissue compatibility. Proc Nat Acad of Sci U S A 79: 5337–5341.

23. Rinkevich B, Weissman IL (1992) Chimeras vs. genetically homogeneous individuals: potential fitness costs and benefits. Oikos 63: 119–124.

24. Pancer Z, Gershon H, Rinkevich B (1995) Coexistence and possible parasitism of somatic and germ cell lines in chimeras of the colonial urochordate Botryllus schlosseri. Biol Bull 189: 106–112.

25. Sebens KP, Witting J, Helmuth B (1997) Effects of water flow and branch spacing on particle capture by the reef coral Madracis mirabilis (Duchassaing and Michelotti). J Exp Mar Biol Ecol 211(1): 1–28.

26. Marubini F, Barnett H, Langdon C, Atkinson MJ (2001) Dependence of calcification on light and carbonate ion concentration for the hermatypic coral Porites compressa. Mar Ecol Prog Ser 220: 153–162.

27. Hii YS, Soo CL, Liew HC (2009) Feeding of scleractinian coral, Galaxea fascicularis, on Artemia salina nauplii in captivity. Aquac Int 17(4): 363–376.

28. Petersen D, Wietheger A, Laterveer M (2008) Influence of different food sources on the initial development of sexual recruits of reefbuilding corals in aquaculture. Aquaculture 277(3): 174–178.

29. Ferrier-Pagès C, Witting J, Tambutté E, Sebens KP (2003) Effect of natural zooplankton feeding on the tissue and skeletal growth of the scleractinian coral Stylophora pistillata. Coral Reefs 22(3): 229–240.

30. Cumbo VR, Fan T-Y, Edmunds PJ. (2012) Scleractinian corals capture zooplankton within days of settlement and metamorphosis. Coral Reefs: 31: 1155.

31. Toh TC, Peh JWK, Chou LM (2013). Early onset of zooplanktivory in equatorial reef coral recruits. Mar Biodivers 43(3): 177–178.

32. Veron JEN (2000) Corals of the world. Townsville: Australian Institute of Marine Science.

33. Harii S, Kayanne H, Takigawa H, Hayashibara T, Yamamoto M (2002) Larval survivorship, competency periods and settlement of two brooding corals, Heliopora coerulea and Pocillopora damicornis. Mar Biol 141(1): 39–46.

34. Toh TC, Peh JWK, Chou LM (2013). Heterotrophy in recruits of the scleractinian coral Pocillopora damicornis. Mar Freshw Behav Physiol 46(5): 313–320.

35. Chou LM, Quek ST (1993) Planulation in the scleractinian coral Pocillopora damicornis in Singapore waters. In: Proceedings of the 7th International Coral Reef Symposium. Mangilao (GU): University of Guam: p. 500.

36. Harriott VJ (1983) Reproductive seasonality, settlement, and post-settlement mortality of Pocillopora damicornis (Linnaeus), at Lizard Island, Great Barrier Reef. Coral Reefs 2: 151–157.

37. Benijts F, Vanvoorden E, Sorgeloos P (1976) Changes in the biochemical composition of the early larval stages of the brine shrimp, Artemia salina L. In Proceedings of the 10th European Symposium on Marine Biology Volume 1: p. 1–9.

38. Levy G, Shaish L, Haim A, Rinkevich B (2010) Mid-water rope nursery–Testing design and performance of a novel reef restoration instrument. Ecol Eng 36(4): 560–569.

39. Sorokin YI (1973) On the feeding of some scleractinian corals with bacteria and dissolved organic matter. Limnol Oceanogr 18(3): 380–385.

40. Anthony K (1999) Coral suspension feeding on fine particulate matter. J Exp Mar Biol Ecol 232(1): 85–106.

41. Houlbrèque F, Ferrier-Pagès C (2009) Heterotrophy in tropical scleractinian corals. Biol Rev Camb Philos Soc 84(1): 1–17.

42. Shaish L, Levy G, Katzir G, Rinkevich B (2010) Employing a highly fragmented, weedy coral species in reef restoration. Ecol Eng 36(10): 1424–1432.

43. Ng CSL, Ng SZ, Chou LM (2012) Does an ex situ coral nursery facilitate reef restoration in Singapore's waters? In Tan KS, editor. Contributions to Marine Science. National University of Singapore, Republic of Singapore. 95–100.

44. Van Moorsel GWNM (1985) Disturbance and growth of juvenile corals (Agaricia humilis and Agaricia agaricites, Scleractinia) in natural habitats on the reef of Curacao. Mar Ecol Prog Ser 24: 99–112.

45. Dikou A, Van Woesik R (2006) Survival under chronic stress from sediment load: spatial patterns of hard coral communities in the southern islands of Singapore. Mar pollut bull 52(11): 1340–1354.

Model-Based Assessment of Estuary Ecosystem Health Using the Latent Health Factor Index, with Application to the Richibucto Estuary

Grace S. Chiu[1]*, Margaret A. Wu[2], Lin Lu[3]

1 CSIRO Mathematics, Informatics and Statistics, Commonwealth Scientific and Industrial Research Organisation (CSIRO), Canberra, Australian Capital Territory, Australia, **2** Business Methods Survey Division, Statistics Canada, Ottawa, Ontario, Canada, **3** McGregor GeoScience, Bedford, Nova Scotia, Canada

Abstract

The ability to quantitatively assess ecological health is of great interest to those tasked with monitoring and conserving ecosystems. For decades, biomonitoring research and policies have relied on multimetric health indices of various forms. Although indices are numbers, many are constructed based on qualitative procedures, thus limiting the quantitative rigor of the practical interpretations of such indices. The statistical modeling approach to construct the latent health factor index (LHFI) was recently developed. With ecological data that otherwise are used to construct conventional multimetric indices, the LHFI framework expresses such data in a rigorous quantitative model, integrating qualitative features of ecosystem health and preconceived ecological relationships among such features. This hierarchical modeling approach allows unified statistical inference of health for observed sites (along with prediction of health for partially observed sites, if desired) and of the relevance of ecological drivers, all accompanied by formal uncertainty statements from a single, integrated analysis. Thus far, the LHFI approach has been demonstrated and validated in a freshwater context. We adapt this approach to modeling estuarine health, and illustrate it on the previously unassessed system in Richibucto in New Brunswick, Canada, where active oyster farming is a potential stressor through its effects on sediment properties. Field data correspond to health metrics that constitute the popular AZTI marine biotic index and the infaunal trophic index, as well as abiotic predictors preconceived to influence biota. Our paper is the first to construct a scientifically sensible model that rigorously identifies the collective explanatory capacity of salinity, distance downstream, channel depth, and silt–clay content–all regarded a priori as qualitatively important abiotic drivers–towards site health in the Richibucto ecosystem. This suggests the potential effectiveness of the LHFI approach for assessing not only freshwater systems but aquatic ecosystems in general.

Editor: Rodolfo Paranhos, Instituto de Biologia, Brazil

Funding: The research of this paper was conducted through research assistantships to MAW at the University of Waterloo funded by an NSERC Strategic Project Grant (http://www.nsercpartnerships.ca/FundingPrograms-ProgrammeDeSubventions/SPG-SPS-eng.asp) to Prof. J. Grant (Dalhousie University) subcontracted to GSC. Funding agency NSERC had no role in study design, data collection and analysis, decision to publish, or preparation of the manuscript.

* E-mail: grace.chiu@csiro.au

Introduction

Assessment of the "health" of an ecosystem is often of great importance to those interested in the monitoring and conservation of ecosystems. Health is a complex concept often involving many diverse factors, and therefore is not straightforward to quantify. A popular method to estimate ecosystem health is through one or more multimetric indices, each of which is a scalar collapsed from several indicator variables of health, or metrics. Often, ecosystem health metrics are measures of faunal abundance and diversity. For aquatic ecosystems, these biotic metrics typically focus on benthic populations because they are useful indicators of underlying health conditions [1,2]. For example, the AZTI marine biotic index (AMBI) [3] is a quantitative measure of health for an estuarine ecosystem based on the sample counts of categorized benthos. Its popularity is evident from its use across the globe, including Africa [4], Asia [5], Europe [6], North America [7], and South America [8].

AMBI and other common multimetric indices, e.g., infaunal trophic index (ITI) [9], estuarine biotic integrity index [10], benthic response index [11], benthic quality index [12], infaunal quality index [13], have the main appeal that they are conceptually simple and thus easily interpretable. They also contain a high amount of biological content from subject-matter scientists being involved at all stages of the design of the index. Yet, the construction and mathematical formulation of many such indices can involve a substantial amount of investigator-specific definitions that are qualitative in nature. Consequently, rigorous evaluation of index reliability and other quantitative aspects is difficult with conventional indices: for example, detecting relationships between health and environmental or impact-related covariates such as water depth or urbanization; and formally assessing the uncertainty in these estimates of health. Recent multistep approaches towards addressing such concerns (e.g., [11,14]) do not address propagation of uncertainty from one step to another, thereby resulting in inference that is less reliable than that from an integrated statistical methodology. Chiu and Guttorp [8] proposed the SHIPSL approach, a statistically enhanced method to construct multimetric indices. Dobbie and Dail [16] compared SHIPSL with other stream health index approaches through a simulation study and showed SHIPSL to have the most favorable statistical properties. Nonetheless, SHIPSL and conventional multimetric approaches share unresolved issues such as being space- and/or time-specific, and the need for follow-up analyses to determine its relationship with nonfaunal (abiotic) variables in method evaluation or policy-making contexts.

Recently, Chiu et al. [17] devised the latent health factor index (LHFI), a novel statistical model-based ecological index aimed to retain the advantages of conventional multimetric indices while addressing some of their shortcomings. In [17], the LHFI modeling methodology was demonstrated and validated on freshwater ecosystems. Through M.W.'s master's studies [18], we adapted this approach to assess an estuarine ecosystem, utilizing the dataset collected by Lu et al. [19] in the previously unassessed Richibucto estuary in the Canadian province of New Brunswick. The LHFI approach involves a multilevel analysis of covariance generalized linear mixed-effects (regression) model (e.g., [20]), or ANOCOVA GLMM: instead of being treated as *measures* of health, metrics are regarded as *indicators* of underlying health conditions. Thus, metrics are regressed as response variables upon a latent health quantity (latent since it is not directly observable) which is site-specific, forming the main level of the regression; health in turn can be regressed upon available drivers/covariates, such as environmental (e.g., salinity, silt–clay content) and impact-related (e.g., urbanization) variables, forming the optional sublevel in the model hierarchy.

With data on metrics and covariates, latent health can be estimated as a scalar, so that interpretability is retained; the estimated quantity is the value of the index. Additionally, the effect of drivers on health can be evaluated in a single integrated statistical framework. Regressing abundance/richness metrics directly on drivers is common in the literature (e.g. [21,22]). Yet, with latent health additionally sandwiched between metrics and drivers, the LHFI regression hierarchy naturally expresses the abstract notion of health as a quantitative parameter, thus integrating the formal quantification of health with attributing it to drivers. Importantly, statistical modeling is what directly produces the health index under the integrated LHFI framework, as opposed to being employed merely to select relevant metrics before index construction (e.g., in [10]) or to evaluate the resulting index (e.g., in [23]). Thus, the LHFI is much more rigorous than conventional indices, as its definition utilizes universal modeling practices for the definition of the index; its hierarchical modeling framework also allows comprehensive statistical inference without the need for sequential analyses through which the propagation of uncertainty is lost from one analysis to the next. As well, the approach provides a predictive framework under which interpolation of health for a new site can be carried out in a cost effective yet rigorous fashion. Specifically, once an appropriate LHFI model has been identified for a set of existing sites, prediction a posteriori can be accomplished simply with covariate values observed at this new site, thus bypassing the expensive benthic taxonomic laboratory procedures that are required to gather the metric data as required by conventional indices. These desirable properties are gained without sacrificing scientific integrity in the form of subject-matter expertise, which can be involved in the identification of biologically relevant metrics and covariates to form the LHFI. It is also straightforward to use the LHFI framework to handle data that have certain types of spatial and/or

temporal features, thus resolving the space/time-specific issue of other indices.

Recently in [24], LHFI principles were integrated with formal point-referenced spatial modeling [25] to formulate the hierarchical relationship among four levels of quantities: (i) ordinal health metrics each on a five-point scale from "poor" to "excellent," (ii) latent continuous quantities that determine the ordinal metrics, (iii) latent health, and (iv) geographical/environmental covariates. This formulation illustrates the type of unified statistical inference that can be drawn from such an LHFI-based approach for assessing biotic integrity of river basins in Colorado, USA. In contrast, directly modeling the quantitative health indicators based upon which ordinal metrics are defined [17] can avoid the loss of information due to mapping quantitative health metrics to a coarse ordinal scale. This was the approach for our preliminary models, but they had a major limitation: estuarine health was statistically attributable to separate subsets of ecologically important drivers, but when these subsets were integrated into a single LHFI model, all but one driver came out statistically significant. In the following sections, we first discuss our preliminary estuarine LHFI models and main findings (crux of M.W.'s studies [18]). We then proceed to build on these models by considering two possible extensions: (a) a nontrivial covariance structure, and (b) additional level(s) to the regression hierarchy based upon the known associations among various drivers.

Methods

Constructing Estuarine LHFIs

Our data (Appendix S1) were collected by Lu et al. [19] in the Richibucto estuary at 18 sites (Figure 1) who used these data to investigate the relationship between soft-bottom macrobenthic communities and environmental variables. Macrofaunal data–88 species for the estuary–were recorded from 2–3 grab sample replicates per site collected between September (Sites 1–3 and 9–18) and October (Sites 4–8) in 2006. Many dominant species were polychaetes, oligochaetes, amphipods, gastropods, and bivalves (the top five dominant species for each site appear in Table 2 in [19]). Observed alongside benthic fauna were abiotic properties of the estuary (Table 1 in [20]). They included *depth*, the distance (m) from the water surface to the estuary bed at the location of the site from which grab samples were obtained; *water temperature* (°C) and *salinity* (parts per thousand), both measured from a single in situ water sample obtained at the site; SC, the *fraction of silt–clay* (grains of size $<63\mu m$); *median grain size* of sediment; *sorting* (a unitless measure of variability of grain size); and *organic content* (%). The latter four variables were recorded by extracting two subsamples from each grab replicate, then pooling all subsamples for sediment assay.

Sites 2, 4–7, and 14 in the estuary were closest to active oyster farms [19]. Oyster farming activity is perceived to impact site health through its direct influence on sediment properties, although different biotic indicators were reported to show different types of association with proximity to oyster farms [19]. For example, relative to all 18 sites, macrobenthic faunal abundance was moderate for Sites 4–7 and 14 but high for Site 2, while Shannon's diversity [26,27] is relatively even among all sites aside from a slight upward trend with increasing distance from the upper channel instead of from an oyster farm. Even when Lu et al. [19] considered the abundance of various dominant species as a suite of separate indicators, they saw no obvious association between these latter indicators and oyster farm location. Shannon's index has limitations including ambiguity in its interpretation [26,28]; the same is true for other nonmodel-based

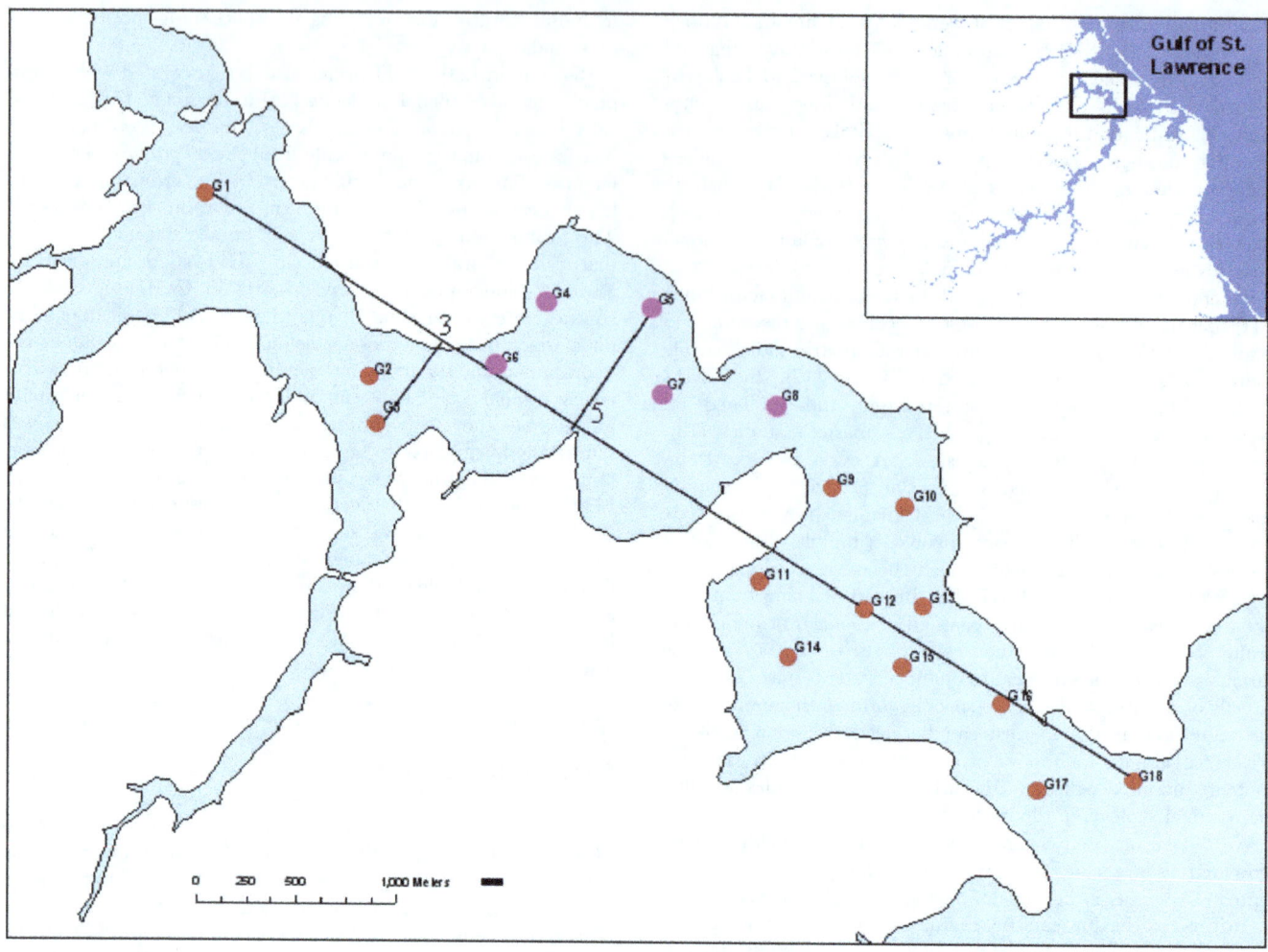

Figure 1. Map of Richibucto Estuary. The 18 monitored sites are shown in red/pink and labeled "G" followed by the site number. Red sites were sampled in September and pink ones, in October. Straight lines illustrate the method for calculating *distance downstream* (DD) for Sites 3 and 5.

indicators such as ones based on single species. This motivated us to build LHFI models for Richibucto based on indicator metrics (Tables 1–2) used to construct the AMBI and ITI. Specifically, AMBI and ITI metrics are popular estuarine ecosystem health indices, being better tailored for estuaries than the generic indicators of abundance, richness, and diversity; and they are more comprehensive than indicators based on single species. However, biotic health indicators alone do not explicitly reveal the collective impact on overall health from abiotic variables: benthic fauna in Richibucto are believed to be related to organic

Table 1. Metrics based on the definition of AMBI, used to construct LHFIs for the Richibucto estuary.

Metric Number	AMBI Abundance[a] Metric	Preconceived Association with Health
1	species (including specialist carnivores and some deposit-feeding tubicolous polychaetes) very sensitive to organic enrichment and present under unpolluted conditions	+
2	species (including suspension feeders, less selective carnivores and scavengers) indifferent to enrichment, always present in low densities with nonsignificant variations with time	±[b]
3	species tolerant to excess organic matter enrichment (including surface deposit-feeding species, e.g., tubicolous spionids)	−
4	second-order opportunistic species; mainly small-sized polychaetes: subsurface deposit-feeders, e.g., cirratulids	−
5	first-order opportunistic species: deposit-feeders, which proliferate in reduced sediments	−

[a]Organisms with the specified characteristics, given all benthic organisms in the grab sample.
[b]Neither clearly positive nor clearly negative.

enrichment (plausibly affected by oyster farming activity), freshwater input (salinity gradient), variability of sediment particle size, water temperature, and topography (channel and water depth), as well as their interactions [19]. To this end, we considered two sets of preliminary LHFI models. The first included only metrics from AMBI (denoted by LHFI-A), and the second, metrics from both AMBI and ITI (denoted by LHFI-A-I).

Identifying Drivers of Estuary Health

For each of LHFI-A and LHFI-A-I, we investigated which and how covariates might influence site health as reflected by biotic metrics. As discussed in [17], a thorough understanding of the relationship between covariates and health is key to rigorous yet cost effective interpolation of site health. Indeed, interpolated biotic conditions would be unreliable when the LHFI model includes weakly predictive abiotic covariates, such as an environmental gradient that exhibits little change across the study area. On the other hand, an LHFI model with good predictive power could prove to be an enormous asset to biologists and policy makers for biomonitoring purposes.

To this end, we implemented preliminary LHFI-A and LHFI-A-I models with different combinations of the covariates listed in the previous subsection, as well as two additional candidates: *month* (September or October) and DD, the *distance downstream* (km). DD is measured by extending a straight line from the western-most site (Site 1) to the eastern-most site (Site 18), then defining DD for any site as the distance between the site's perpendicular projection onto the straight line and Site 1 (Figure 1, Table 3). An alternative covariate to DD would be two-dimensional spatial coordinates of sites. Though, as shown in Figure 1, the study sites roughly align diagonally across a small geographical domain of approximately (4km)×(23km). Thus, we expect little loss of information through collapsing the two-dimensional coordinates into DD. In fact, this single spatial covariate can avoid collinearity between the spatial dimensions.

These preliminary models were considered in a Bayesian statistical framework, as follows. For LHFI-A, AMBI metrics are abundances of five disjoint taxonomic groups. We denote the metrics by Y. Due to the difference in the preconceived direction of their association with health (Table 1), we split the metrics into two groups: $s =$ "$-$" for Metrics 3–5 (negatively related to health), and $s =$ "$+$" for the remaining metrics. In the LHFI model, each member of Group s is modeled as a multinomial random variable. The link function for the GLMM is a generalized logit for $s =$ "$+$", and an inverted generalized logit for $s =$ "$-$". Thus, large metric values for $s =$ "$-$" and "$+$" reflect, respectively, poor health and otherwise. More precisely, let $Y_{ijk\ell s}$ denote the value of

the $j(s)th$ metric (j nested in the sth group) for the kth replicate grab sample at the $i(\ell)th$ site (i nested in the ℓth month for $\ell = 9, 10$). Let $N_{ik\ell}$ be the total number of benthic organisms in the kth replicate sample at the $i(\ell)th$ site, and $p_{ij\ell s}$ be the unknown probability that a random organism from the $i(\ell)th$ site belongs to the $j(s)th$ taxonomic group. Thus, we have multinomial distributions.

$$\{Y_{i1k\ell+}, Y_{i2k\ell+}, N_{ik\ell} - \sum_{j=1}^{2} Y_{ijk\ell+}\} | N_{ik\ell}, p_{i1\ell+}, p_{i2\ell+}$$

$$\sim Multinomial(N_{ik\ell}; p_{i1\ell+}, p_{i2\ell+}, 1 - \sum_{j=1}^{2} p_{ij\ell+}), \qquad (1)$$

$$\{Y_{i3k\ell-}, Y_{i4k\ell-}, Y_{i5k\ell-}, N_{ik\ell} - \sum_{j=3}^{5} Y_{ijk\ell-}\} | N_{ik\ell}, p_{i3\ell-}, p_{i4\ell-}, p_{i5\ell-}$$

$$\sim Multinomial(N_{ik\ell}; p_{i3\ell-}, p_{i4\ell-}, p_{i5\ell-}, 1 - \sum_{j=3}^{5} p_{ij\ell-}). \qquad (2)$$

Next, let $H_{i(\ell)}$ denote the latent health of the $i(\ell)th$ site; and θ_s and $\beta_{j(s)}$ respectively denote the metric group effect and individual metric effect (both unknown) in the regression model. Then, the linear predictor in the LHFI framework is.

$$v_{ij\ell+} = \log \frac{p_{ij\ell+}}{1 - p_{i1\ell+} - p_{i2\ell+}}, \quad j = 1, 2, \qquad (3)$$

$$v_{ij\ell-} = \log \frac{1 - p_{i3\ell-} - p_{i4\ell-} - p_{i5\ell-}}{p_{ij\ell-}}, \quad j = 3, 4, 5, \qquad (4)$$

$$v_{ij\ell s} = H_{i(\ell)} + \theta_s + \beta_{j(s)}, \quad s = +, -. \qquad (5)$$

For Equation 5, we model θ_s as a fixed effect and take $\theta_+ = 0$ (as is customary when considering one of the categories as baseline) to ensure model identifiability, and we model site health and metric effects as random. Note that there is overlap and thus dependency between the two multinomials of Equations 1 and 2. This dependency is crudely accounted for by θ_s; similarly, metric effects $\beta_{j(s)}$ crudely account for the dependency among vs within

---

Table 2. Metrics based on the definition of ITI, used to construct preliminary LHFIs for the Richibucto estuary.

Metric Number	ITI Abundance Metric*	Preconceived Association with Health
1	suspension feeders: feed on detritus from the water column and usually lack sediment grains in their stomach contents	+
2	interface/surface detrital feeders: obtain the same types of food as suspension feeders but usually from the upper 0.5 cm of the sediment	+
3	deposit feeders: invertebrates (including carnivores); generally feed from the top few cm of the sediment and feed on encrusted mineral aggregates, deposit particles or biological remains	±
4	specialized environment feeders: mobile burrowers that feed on deposited organic material; all adapted to live in highly anaerobic sediment	−

*As described in [35].

group s. Thus, for these preliminary models, we assumed independent mean-zero Gaussian $\beta_{j(s)}s$, while allowing for unequal variances across j.

Finally, the latent regression of $H_{i(\ell)}$ is.

$$H_{i(\ell)} = \alpha_0 + \boldsymbol{\alpha}' \mathbf{x}_{i(\ell)} + \varepsilon_{i(\ell)}, \qquad (6)$$

$$\varepsilon_{i(\ell)} | \sigma_{H\ell}^2 \overset{\text{ind}}{\sim} Normal(0, \sigma_{H\ell}^2). \qquad (7)$$

where $\mathbf{x}_{i(\ell)}$ is the vector of a given combination of the aforementioned covariates, α_0 and $\boldsymbol{\alpha}$ are the unknown coefficients of the corresponding latent regression, and $\varepsilon_{i(\ell)}$ is the normally distributed regression error with unknown variance $\sigma_{H\ell}^2$ that may vary over months. In practice, covariate transformation might be necessary to satisfy the linearity of Equation 6. Covariates (possibly transformed) are then centered to reduce dependence among the αs. For a given covariate that is not an interaction, centered data are produced by subtracting from the raw covariate data a constant that is (approximately) equal to the observed covariate mean (averaged over i,ℓ). The "centered interaction" between two covariates is taken to be the product of two centered covariates. For example, the centered $x_{i(\ell)}$ terms in Equation 6 corresponding to log-SC, log-depth, and their interaction would be computed, respectively, as $\log(SC_{i(\ell)}) - \overline{\log(SC)}$, $\log(depth_{i(\ell)}) - \overline{\log(depth)}$, and $[\log(SC_{i(\ell)}) - \overline{\log(SC)}] \times [\log(depth_{i(\ell)}) - \overline{\log(depth)}]$.

The formulation of Equations 6 and 7 does not depend on the biotic metrics. Thus, for LHFI-A-I, we include additional ITI elements that correspond to Equations 1–5. Briefly, the combined framework is as follows. The ITI counterparts of Equations 1–5 are based on partitioning ITI metrics into "+" and "−" groups according to Table 2. For combining AMBI and ITI metric effects, we replace $\beta_{j(s)}$ with $\beta_{j(m \times s)}$ where $m = \{AMBI, ITI\}$. This reformulated β is decomposed as the sum of three components: a fixed effect due to m, a fixed effect due to the interaction between m and s, and a random effect due to j nested inside $(m \times s)$. For our preliminary models, these latter random effects took the role of $\beta_{j(s)}s$ from the AMBI-only case. Appendix S2 presents more details on the LHFI-A-I framework.

We implemented the above modeling framework using Markov chain Monte Carlo (MCMC) techniques. Several covariates and interactions exhibited a statistically significant relationship with health (Bayesian credible intervals that excluded the regression coefficient value of 0 had a credible level that was reasonably high, e.g., ≥ 0.8). For LHFI-A, two best-fitting models were identified among those investigated: one with abiotic covariates log-SC, log-depth, their interaction, and salinity; and another with the single covariate DD. The remaining covariates, including month, were found to be insignificant or confounded with others. We observed little evidence that β variances were unequal, and thus assumed constant variance $\sigma_{\beta A}^2$ for AMBI metric effects when formulating the integrated LHFI-A-I model (see Appendix S2). Under this formulation, there were three best LHFI-A-I models: two corresponded to the same sets of significant covariates as those for LHFI-A, and another model with covariates log-depth, log-SC, and their interaction. For both LHFI-A and LHFI-A-I, we observed some evidence that $\sigma_{H,9} < \sigma_{H,10}$.

However, our attempts to include covariates from various best-fitting models together in a single LHFI-A or LHFI-A-I model were unsatisfactory. In such combined models, DD remained highly significant, while all other covariates and their interactions were no longer significant at a reasonable credible level. Indeed, salinity and DD are highly correlated (Table 4), and the two cannot be simultaneously significant due to collinearity. However, no strong correlation exists among log-depth, log-SC, and DD (Table 4), and so why did DD eclipse all others in a combined model, despite nonDD covariates being significant when DD was absent? As well, while relationships between health and covariates were quite strong for the LHFI-A models, they were less clear for LHFI-A-I models (significance at credible levels $\approx$ 60–85% in the best-fitting LHFI-A-I models, as opposed to > 90% for LHFI-A). This indicated that the extra data from ITI metrics weakened the overall relationship between health and covariates. One possible explanation for this phenomenon is that the LHFI construct was appropriate for describing health using AMBI metrics and the available covariates, but ITI metrics have weak ecological relevance to Richibucto. This is plausible from a qualitative perspective, in light of our prior beliefs about health drivers as stated above. One remedy is to determine additional covariates that can be more appropriately paired with ITI metrics, then model these alongside the original covariates. However, this would require further field activities, and is beyond the scope of our current paper. Thus, for the remainder of this paper, we focus on addressing the domination of DD for LHFI-A only.

Indeed, the above preliminary models might be improved upon. Specifically, distance likely contained much less measurement error than the other covariates, it being easier to measure with precision than the environmental covariates which are intrinsically more variable in nature. With a simplistic LHFI model, the effect from distance on health could therefore manifest itself more clearly than effects from other covariates even if all of them were equally important in a qualitative sense. Given our prior beliefs about AMBI metrics being more relevant to Richibucto, and the fact that environmental covariates contained ecological information that distance did not, a more sophisticated LHFI modeling framework may be helpful in providing a common thread through health, DD and the other ecologically relevant covariates.

To this end, we proceed to determine if either of the following helps to clarify the nature of the relationship among latent health and the available covariates: (1) Introduce a covariance structure for the metric effects, instead of independence which was assumed for the preliminary models to reduce computational burden; (2) introduce additional level(s) to the regression hierarchy based upon the known associations between the available covariates. These steps pertain to different parts of the LHFI model, and thus we treat each as a stand-alone investigation. Note that even if ITI metrics had shown to be highly relevant to estuary health in Richibucto, introducing extra model complexity to LHFI-A-I models can be impractical for proper inference via MCMC. This is because AMBI and ITI metrics are dependent according to their definitions, so that extra model parameters are required to account for this. Even when such dependence is only informally accounted for by the fixed-effects terms in the LHFI-A-I formulation in Appendix S2, one can see the substantial extra complexity that is required.

Extending LHFI-A via a Nontrivial Covariance Structure for Metric Effects

Recall that our preliminary LHFI-A models provided little evidence that β variances were unequal; subsequently we took $\boldsymbol{\Sigma} = \sigma_\beta^2 \mathbb{I}$, where $\mathbb{I}$ is the identity matrix. To generalize $\boldsymbol{\Sigma}$, we now replace independence of metric effects by.

Table 3. Distance downstream (km) for Richibucto sites.

Site	1	2	3	4	5	6	7	8	9
Distance	0	1.164	1.298	1.731	2.179	1.686	2.463	2.970	3.433
Site	10	11	12	13	14	15	16	17	18
Distance	3.790	3.358	3.880	4.119	3.642	4.179	4.701	5.060	5.448

$$\beta|\Sigma \sim MVN(\mathbf{0},\Sigma), \quad \beta=\begin{bmatrix}\beta_+\\\beta_-\end{bmatrix}, \quad \mathbf{S}=\begin{bmatrix}\Sigma_+ & \Sigma_\pm\\\Sigma'_\pm & \Sigma_-\end{bmatrix} \quad (8)$$

where $\boldsymbol{\beta}_+ \equiv [\beta_{1(+)},\beta_{2(+)}]'$, $\boldsymbol{\beta}_- \equiv [\beta_{3(-)},\beta_{4(-)},\beta_{5(-)}]'$, $\sum$ is the unknown covariance matrix for $\boldsymbol{\beta}$, and "MVN" denotes the multivariate normal distribution. Thus, $\sum_+ (2\times2)$, $\sum_- (3\times3)$, and $\sum_\pm (2\times3)$ denote the covariance matrices for metric groups positively and negatively related to health, and their cross-covariance matrix, respectively. (Note that any $\sum$ structure in Equation 8 necessarily differs from the posterior covariance structure for $\boldsymbol{\beta}$.) With a small dataset from 18 Richibucto sites each with only 2 to 3 replicate grab samples, a practical concern is that a general $\sum$ may be only weakly identifiable depending on the complexity of the covariance structure (see [15] for a discussion on lack of identifiability in Bayesian inference). This issue was encountered in [9] when a fully unstructured $\sum$ was assumed for a freshwater benthic dataset that also involved 18 sites with 3 replicates per site, but with nine metrics altogether. To avoid weak identifiability, one could consider a structured $\sum$, as discussed in Appendix S3, to reduce the number of unknown parameters; a special case is the block diagonal $\sum$ with $\sum_\pm = \mathbb{O}$ in Equation 8. To further reduce inferential burden, we additionally assume $\sigma_{H\ell} \equiv \sigma_H$ to be constant over months. This assumption effectively removes the index ℓ from the entire model, and may be justified by the possibility that the preliminary evidence for $\sigma_{H,9} < \sigma_{H,10}$ was related to the lopsided abundance of data from September (13 sites) compared to October (5 sites).

Overall, statistical inference is focused on ecologically pertinent parameters, namely, $\boldsymbol{H}$, $\boldsymbol{\alpha}$, and σ_H; $\sum$ and other parameters are regarded as nuisance. For Bayesian inference, we use relatively diffuse distributions as priors for α_0, elements of $\boldsymbol{\alpha}$, θ_- (univariate Gaussians, each with mean 0 and variance 100), and σ_H^2 (inverse-Gamma with unit shape and scale). Diffuseness of priors reflects the fact that in the absence of data, we have no clear perception of

Table 4. Sample correlation coefficients among covariates for latent health of Richibucto sites.

	DD	salinity	log-depth	log-SC
DD	1	0.88	0.16	−0.47
salinity		1	0.23	−0.33
log-depth			1	−0.41
log-SC				1

the properties of the corresponding unknown quantities. In general, diffuseness reduces the need for justification of prior distributional assumptions. To complete the Bayesian modeling hierarchy, we must specify the priors for $\sum_+$ and $\sum_-$. The most general form is for each to be unstructured, and thus we take.

$$\sum_+ \sim IW_2, \quad \sum_- \sim IW_3 \quad (9)$$

where IW_d is the inverse of a $d \times d$ random Wishart matrix with d degrees of freedom and scale matrix equal to the identity, which is a relatively diffuse prior for a $d \times d$ unstructured covariance matrix. Then, one can take advantage of existing MCMC software such as OpenBUGS [21] for straightforward implementation of the LHFI model, although in our experience, nontrivial hierarchical centering is essential to improve MCMC mixing [9,34]. Altogether, the extended model comprises Equations 1–9 with $\sigma_{H,9} = \sigma_{H,10}$.

Extending LHFI-A via an Extra Level in the Latent Regression

Extra model complexity can be introduced also through an additional level in the regression of latent health on covariates. Specifically, although the strong correlation between salinity and DD reflects ecological reasoning for coastal sea waters entering an estuary, it is the only clear empirical relationship detected among the available covariates. Therefore, instead of considering salinity and DD to be complementary covariates, we now take salinity as a response of DD, and in turn, latent health as a response of salinity and the remaining covariates identified in the preliminary analyses as statistically significant (Figure 2). Then, the LHFI model comprises Equations 1–8 with $\sigma_{H\ell} \equiv \sigma_H$ (see subsection above), plus.

$$x_{sal,i} = \alpha_{DD}x_{DD,i} + \delta_i, \quad (10)$$

$$\delta_i|\sigma_\delta^2 \overset{iid}{\sim} Normal(0,\sigma_\delta^2) \quad (11)$$

where $\mathbf{x}_i$ in Equation 6 denotes the vector of centered covariates for site i including salinity $x_{sal,i}$ (and possibly other covariates) but excluding DD $x_{DD,i}$. Hence, Equations 6 and 10 can be collapsed into

$$H_i = \alpha_0 + \alpha'_{-sal}\chi_{-sal,i} + \alpha_{sal}\alpha_{DD}x_{DD,i} + \alpha_{sal}\delta_i + \varepsilon_i \quad (12)$$

where $\boldsymbol{\alpha}_{-sal}$ is $\boldsymbol{\alpha}$ with α_{sal} (and α_{DD}) removed, and similarly for $\mathbf{x}_{-sal,i}$. Thus, Equation 12 regards salinity as an implicit covariate, so that when latent health is explicitly regressed on $\mathbf{x}_{-sal,i}$ and DD, the implicit covariate decomposes the total error variation into

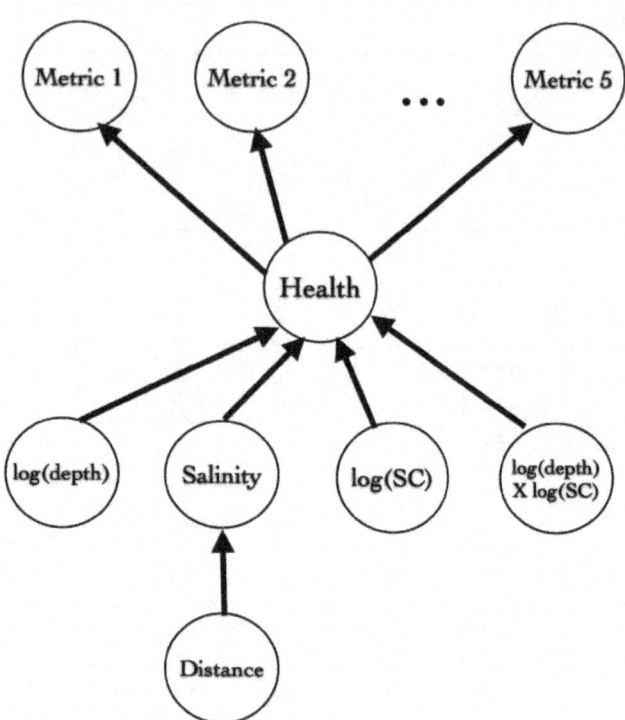

Figure 2. Regressing salinity on distance downstream as an additional level in the hierarchical latent health model. Distance downstream (DD) is the driver of salinity. Salinity and other covariates are complementary drivers of latent health, which is a driver of AMBI metrics.

$Var(\alpha_{sal}\delta_i + \varepsilon_i | \alpha_{sal}, \sigma_\delta^2, \sigma_H^2) = \alpha_{sal}^2 \sigma_\delta^2 + \sigma_H^2$. Hence, a smaller ratio $\sigma_H^2 / (\alpha_{sal}^2 \sigma_\delta^2 + \sigma_H^2)$ reflects a higher contribution from the implicit covariate towards explaining the total error variation of the latent health regression.

We again employ an inverse-Gamma prior with unit shape and scale for σ_H^2, and also for σ_δ^2. Univariate Normal(0, 100) priors are employed for θ_-, α_0, and elements of $\boldsymbol{\alpha}$, with one exception: $Cor(\alpha_{sal}, \alpha_{DD}) = \rho \neq 0$ is additionally considered, where $\rho \sim Unif(-1, 1)$ a priori.

Results

Results of our preliminary LHFI-A and LHFI-A-I models already appear under **Identifying Drivers of Estuary Health**. Below, we report the results of our extended LHFI-A models.

Health Inference as a Whole is Robust to Metric Covariance Structure

We considered the LHFI-A model with covariates log-depth, log-SC, their interaction, and DD, which were identified from our preliminary models as the most statistically relevant covariates while assuming $\sum = \sigma_\beta^2 \mathbb{I}$. Our Bayesian estimates for parameters of main interest and corresponding credible intervals then were compared to their preliminary counterparts. The result was that increasing complexity of the LHFI-A model through a nontrivial $\sum$ did not lead to a noticeable difference in the significance of the covariates or the posterior mean of H_i. In general, extra model complexity could lead to overfitting, which in turn leads to weaker model inference. In light of the concern over weak identifiability as

explained above, we would expect weaker model inference to manifest itself in the form of MCMC mixing difficulties for $\sum$ despite having employed hierarchical centering. However, this was not the case for our analysis, as two independently generated MCMC chains mixed readily after a manageable burn-in. In particular, although parameters of $\sum$ could require a burn-in of up to approximately 20,000 iterations (Figure 3), all other model parameters each required a burn-in of only 1,000 or less (Figure 4). (For a given model, inference for model parameters as a whole was always based on the longest burn-in required.) Instead, weaker inference was apparent only in the form of slightly larger posterior dispersions for H_i and certain nuisance parameters when compared to the case of $\sum = \sigma_\beta^2 \mathbb{I}$. Therefore, neither the relative health rankings among sites (even accounting for wider credible intervals) nor the identification of significant health drivers was affected by assuming a more complex structure for $\sum$. Our investigation here suggests that the inference for latent health $\boldsymbol{H}$ associated with the Richibucto system is reasonably robust to the prior covariance structure in Equation 8 for metric effects $\boldsymbol{\beta}$. Consequently, for model parsimony, we regard $\sum = \sigma_\beta^2 \mathbb{I}$ (as we had originally assumed) to be adequate for these Richibucto data.

Simultaneously Significant Covariates in Two-Level Health Regression

Table 5 and Figures 5 and 6 present inference summaries assuming $\sum = \sigma_\beta^2 \mathbb{II}$ and $\sigma_{H\ell} \equiv \sigma_H$ for various LHFI-A models. Models (3)–(5) each comprises two levels of covariates (Equations 1–8 and 10–11). Models (1) and (2), provided for comparison, each comprises a single level of covariates (Equations 1–8 only). Posterior means for latent health along with their 95% posterior credible intervals (CIs) appear in Figure 5; those for α_0, $\beta_{j(s)}$, θ_s, σ_H, and σ_β appear in Figure 6. Note that in addition to $\boldsymbol{H}$, $\boldsymbol{\alpha}$, and σ_H, here σ_δ from the extra level in Models (3)–(5) is also a parameter of ecological interest.

It is evident from the 95% CIs in Table 5 that, when DD is considered a driver of salinity, the two are simultaneously relevant to explaining latent health. In fact, 0 is excluded from the 99% CI (not shown) for both α_{sal} and α_{DD} in each of Models (3)–(5), suggesting very high credibility for the covariates in a two-level structure. The CIs from Model (5) suggest that the interaction $(log - depth) \times (log - SC)$ is an additional credible driver of latent health, complementing the explanatory capacity of salinity-on-DD. This is the first time that a scientifically sensible model has been successfully constructed to rigorously identify the collective explanatory capacity of salinity, DD, depth, and SC–all regarded a priori as qualitatively important–towards site health in the Richibucto ecosystem.

For Models (3)–(5), the posterior mean for the ratio $\sigma_H^2 / (\alpha_{sal}^2 \sigma_\delta^2 + \sigma_H^2)$ ranges respectively from around 0.55 to 0.65; corresponding 95% CIs span from 0.3+ to 0.8+. These moderately sized figures suggest that the salinity-on-DD structure is desirable for the model hierarchy, decomposing the total latent regression error variance into nontrivial components. Among Models (3)–(5), the former exhibits a smaller variance ratio, but not by much. Thus, despite the high credibility of the correlation between α_{sal} and α_{DD} in Model (4) and of the influence on health from (the interaction between) depth and SC in Model (5), the least complex Model (3) provides slightly clearer evidence for the explanatory capacity of the two-level structure. In terms of the model's predictive power, the least and most complex among the three models share the same deviance information criterion (DIC) [30]

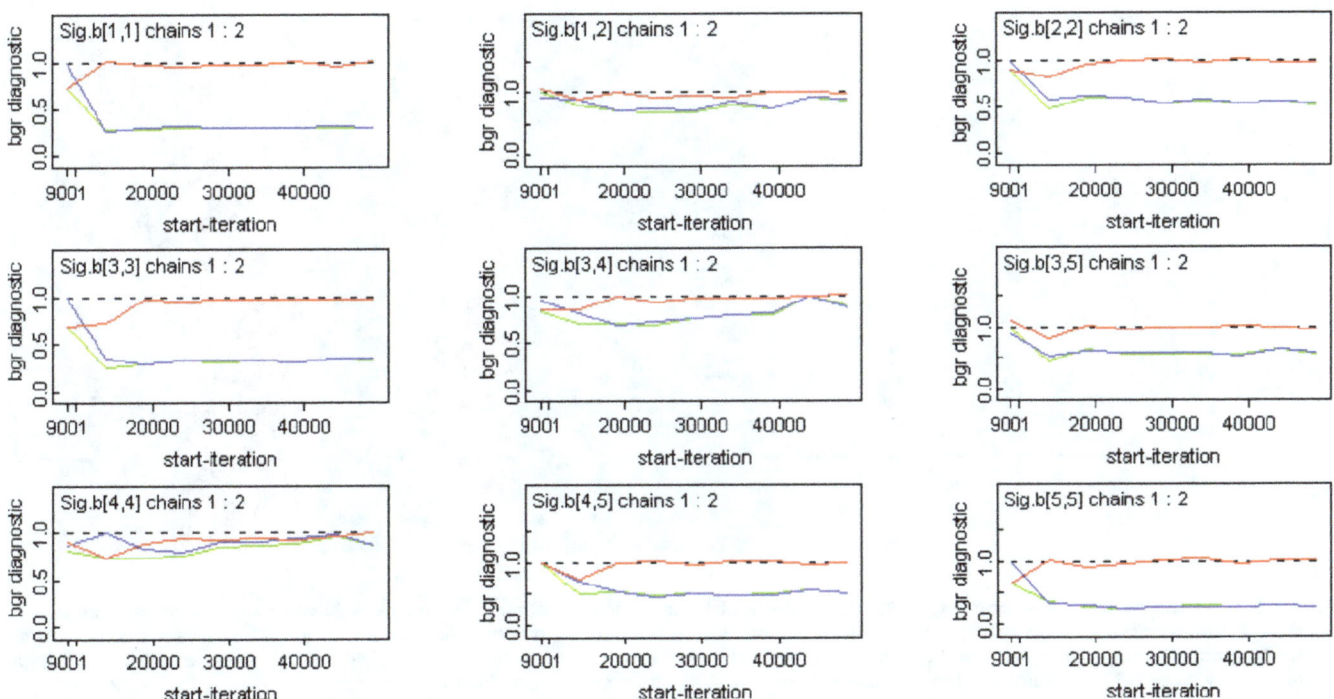

Figure 3. Brooks-Gelman-Rubin (BGR) diagnostics for the block diagonal Σ. BGR plots are presented for the MCMC samples of each nonzero matrix element Σ_{ij} (axis label Sig.b[i,j] in plot), i.e., the $(i,j)th$ element of $\sum$. Convergence is suggested by a red curve approaching 1, together with green and blue curves approaching the same constant [34].

which is slightly smaller (better) than that of Model (4). This predictive power corresponds to observed AMBI metrics (not latent health) being predicted by the model. To assess the model's predictive ability for latent health, one could conduct a simulation study in which unobservable H_i values are generated then

estimated, although such an approach for hierarchical models has its shortcomings [22] or requires intensive computations [13] that may be impractical.

Instead, we compare CIs for H_i among models; in Figure 5, they appear nearly identical across all Models (1)–(5), i.e., the

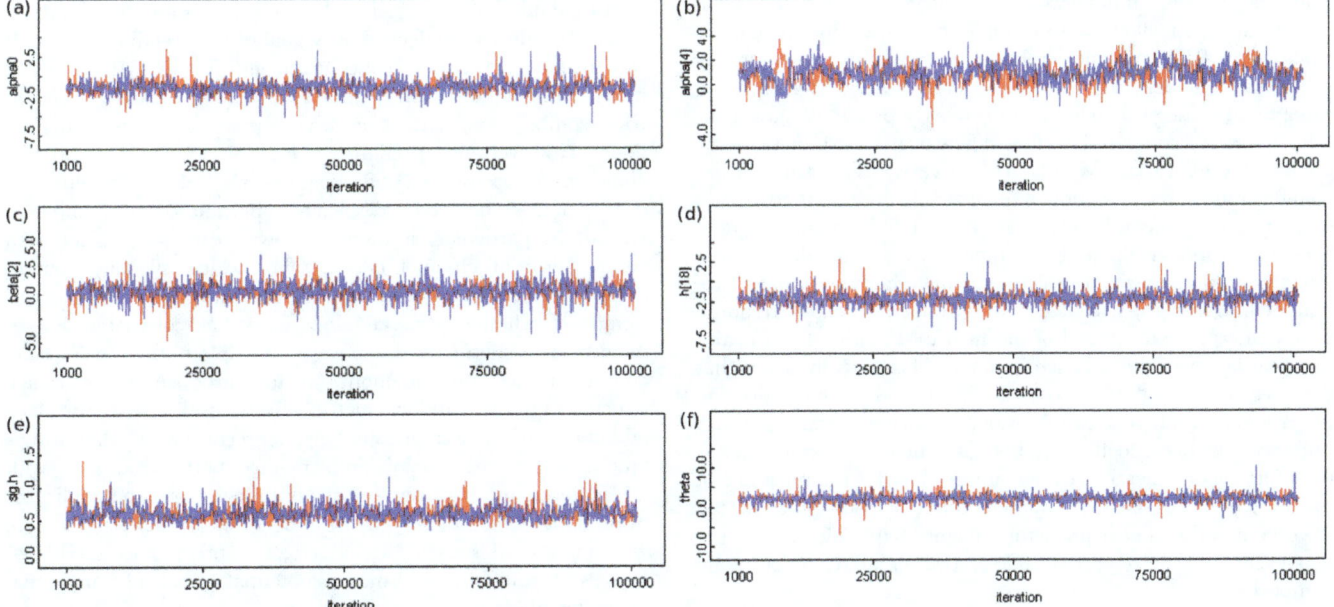

Figure 4. Trace plots of two independent MCMC chains. The two chains for selected parameters from fitting Equations 3–9 (assuming $\sigma_{H\ell} \equiv \sigma_H$) are shown in red and blue, both thinned by 100 iterations. (a) Intercept α_0. (b) Regression coefficient of the (centered) interaction $(log-depth) \times (log-SC)$. (c) Random effect β_2. (d) Latent health H_{18}. (e) Standard deviation σ_H. (f) Fixed effect θ_-. Trace plots for all other $non \sum$ model parameters show similar patterns that suggest convergence after a burn-in of merely 1,000.

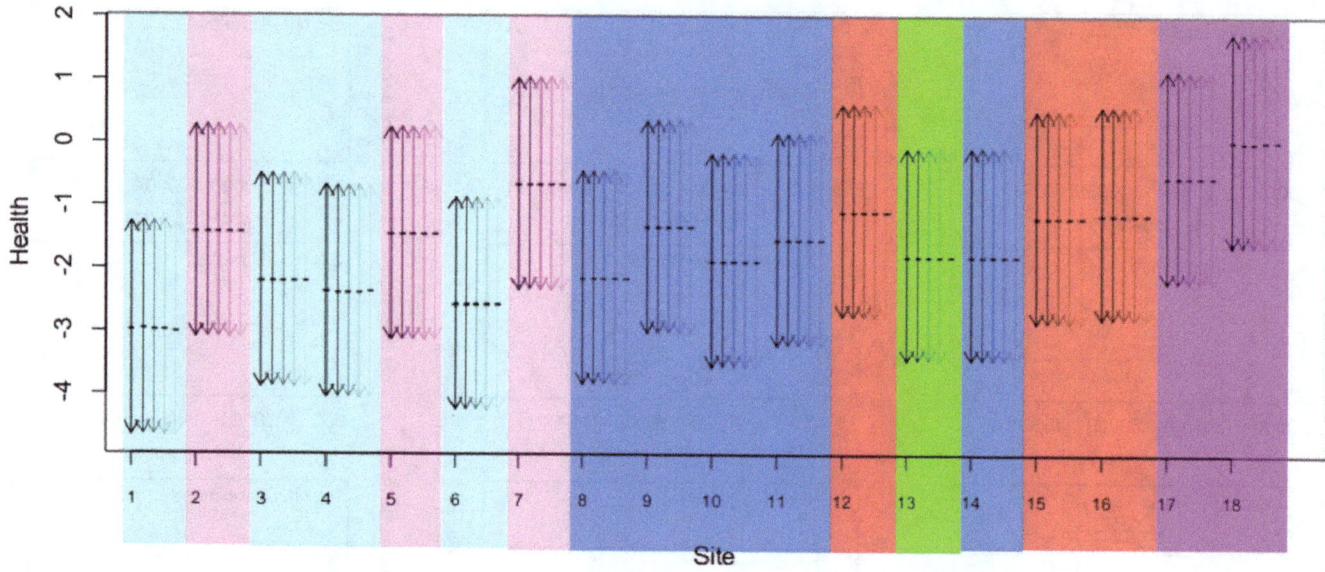

Figure 5. LHFI scores and corresponding 95% CIs of site health. Health scores (posterior means marked by "–") and CIs (arrows from dark to light) are based respectively on Models (1), (2), (3), (4), and (5) of Table 5. Lu et al. [20] partition Richibucto sites into six groups according to their benthic community composition: red (lower channel: Sites 12, 15, 16), pink (upper channel: Sites 2, 5, 7), violet (estuarine mouth: Sites 17, 18), blue (lower shallow: Sites 8–11, 14), turquoise (upper shallow: Sites 1, 3, 4, 6), and green (other: Site 13).

inference for health is essentially equally credible across various models. Within models, the relative ranking of sites according to their LHFI scores and associated CIs do not coincide with the clustering by Lu et. al [20], which was based on similarity in benthic community composition, and subsequently identified to be highly correlated with site location. Our results indicate that the LHFI approach does not merely represent community composition or site location; instead, it rigorously and comprehensively models biotic indicators, abiotic drivers, the abstract notion of health, and the relationship among them. Note that the health CIs from the 18 sites mutually overlap, suggesting that the small dataset does not allow us to distinguish sites according to their health at a 95% credible level; this was also the case for our preliminary models, all with single-level covariates. Despite (i) suboptimal distinguishability and (ii) weaker predictive power for AMBI metrics compared to the single-covariate Models (1)–(2), our two-level-covariate Models (3)–(5) clearly resolved the earlier counterintuitive phenomenon of qualitatively important covariates not being simultaneously significant. Indeed, (i) is an improvement over conventional methods in quantitative rigor due to the integrated manner from which our uncertainty estimates are obtained. Moreover, (ii) is of secondary concern when the response of key interest is H instead of the metrics Y. Aside from nearly identical latent health CIs across models, Figure 6 indicates that the five models perform equally well with respect to the uncertainty (width of CIs) of various dispersion parameters and nuisance regression coefficients, but with one exception: two-level-covariate models clearly yield less uncertainty for the inference of σ_β (Figure 6 (f)). As this parameter directly contributes to the uncertainty of the linear predictor ν, Figure 6 (f) indicates that the two levels can lead to more reliable prediction inference for faunal composition.

Discussion

Unlike conventional multimetric health indices, the integrated LHFI approach employs hierarchical generalized linear mixed modeling to yield health scores, assess the influence of health

drivers, and provide their associated uncertainty, all in a single, unified analysis for a given model. LHFI models can be tailored to different types of aquatic ecosystems through health metrics and environmental covariates that are specific to these systems. For example, while salinity can appear as a driver in an estuarine LHFI model, it would not be meaningful in a freshwater LHFI model (e.g., [17]) because of the lack of a salinity gradient in freshwater ecosystems.

For the Richibucto estuary, we constructed preliminary LHFI-A (with AMBI metrics only) and LHFI-A-I (with both AMBI and ITI metrics) models involving single-level covariates and independently distributed metric effects. A key goal of the preliminary models was to understand how biotic health indicators (AMBI and ITI metrics) might be driven by observed abiotic covariates, and in what combination of these covariates (main effects and interactions). However, our preliminary models lacked the important ability to rigorously identify relationships between health and drivers that are deemed ecologically important for the Richibucto system. In particular, if distance downstream were ignored, a combination of channel depth, salinity, and silt–clay content demonstrated high significance; distance downstream alone was significant when considered alongside other covariates. Subsequently, we considered two ways to explore this ecologically counterintuitive phenomenon: (a) to introduce a covariance structure on the random metric effects, and (b) to introduce additional levels of regression given preconceived relationships among the covariates. We implemented both (a) and (b) with AMBI biotic metrics only, but the approach would be applicable in principle to combining AMBI and ITI biotic metrics. Though, with merely 18 sites in Richibucto, our preliminary LHFI-A-I models suggested that ITI metrics potentially weakened any signal in the health-covariate relationship.

Based on our extended LHFI-A models, we have found (a) to be inconsequential to either the inference of latent health among Richibucto sites, or the lack of simultaneous statistical relevance of qualitatively important abiotic drivers of Richibucto health. On the other hand, (b) helped to rigorously express biological insight

Table 5. Selected summary statistics of posterior draws.

Model	Description	DIC	Parameter	Mean	Median	95% CI 2.5%	95% CI 97.5%
(1)	*DD* only	4380					
			αDD	0.37	0.37	**0.16**	**0.58**
			σ_β	1.48	1.03	0.35	5.32
			σ_H	0.63	0.61	0.45	0.89
(2)	*sal* only	4379					
			α_{sal}	0.39	0.39	**0.13**	**0.64**
			σ_β	1.48	1.02	0.35	5.32
			σ_H	0.67	0.66	0.48	0.95
(3)	*sal*-on-*DD* only; $Cor(\alpha_{sal},\alpha_{DD})=0$	4417					
			α_{sal}	0.39	0.39	**0.13**	**0.64**
			α_{DD}	0.77	0.77	**0.54**	**1.00**
			σ_β	1.12	1.01	0.59	2.31
			σ_δ	0.70	0.68	0.51	0.98
			σ_H	0.67	0.65	0.48	0.95
			$\frac{\sigma_H^2}{\alpha_{sal}^2\sigma_\delta^2+\sigma_H^2}$	0.54	0.55	0.31	0.76
(4)	*sal*-on-*DD* only; $Cor(\alpha_{sal},\alpha_{DD})=\rho$	4419					
			α_{sal}	0.49	0.49	**0.28**	**0.71**
			α_{DD}	0.59	0.59	**0.35**	**0.79**
			ρ	−0.94	−0.96	**−1.00**	**−0.78**
			σ_β	1.12	1.01	0.59	2.30
			σ_δ	0.74	0.72	0.53	1.07
			σ_H	0.68	0.66	0.49	0.96
			$\frac{\sigma_H^2}{\alpha_{sal}^2\sigma_\delta^2+\sigma_H^2}$	0.59	0.59	0.37	0.79
(5)	log-*depth*, log-*SC*, $(\log-depth)\times(\log-SC)$, and *sal*-on-*DD*; $Cor(\mathbf{a})=\mathbb{I}$	4417					
			α_{dep}	0.16	0.16	−0.63	0.94
			α_{sal}	0.42	0.42	**0.17**	**0.67**
			α_{SC}	−0.83	−0.83	−1.74	0.08
			α_{DD}	0.77	0.77	**0.54**	**1.00**
			$\alpha_{dep}\times SC$	1.88	1.88	**0.28**	**3.49**
			σ_β	1.12	1.01	0.59	2.29
			σ_δ	0.70	0.68	0.51	0.98
			σ_H	0.59	0.57	0.41	0.87
			$\frac{\sigma_H^2}{\alpha_{sal}^2\sigma_\delta^2+\sigma_H^2}$	0.63	0.63	0.37	0.87

Boldfaced CI limits indicate that the slope or correlation parameter differs from 0 with at least 95% credibility.

about these drivers: an additional level of covariates based upon the preconceived relationship between salinity and distance downstream has allowed the model to identify the simultaneous significance of distance and those abiotic covariates that our preliminary models had shown to be significant only when distance was excluded. Moreover, we have shown that model inference is more reliable overall when compared to single-level-covariate models. Thus, our two-level-covariate modeling frame-work more comprehensively exploits the ecological relationship among health, biotic metrics, and abiotic covariates, and it yields less uncertainty in model inference.

We implemented three variants of the two-level structure: (i) salinity-on-distance alone, with a priori independent regression coefficients and metric effects, (ii) same as (i) but assuming bivariate regression coefficients, and (iii) same as (i) but including channel depth and silt–clay content (both on the log scale), as well

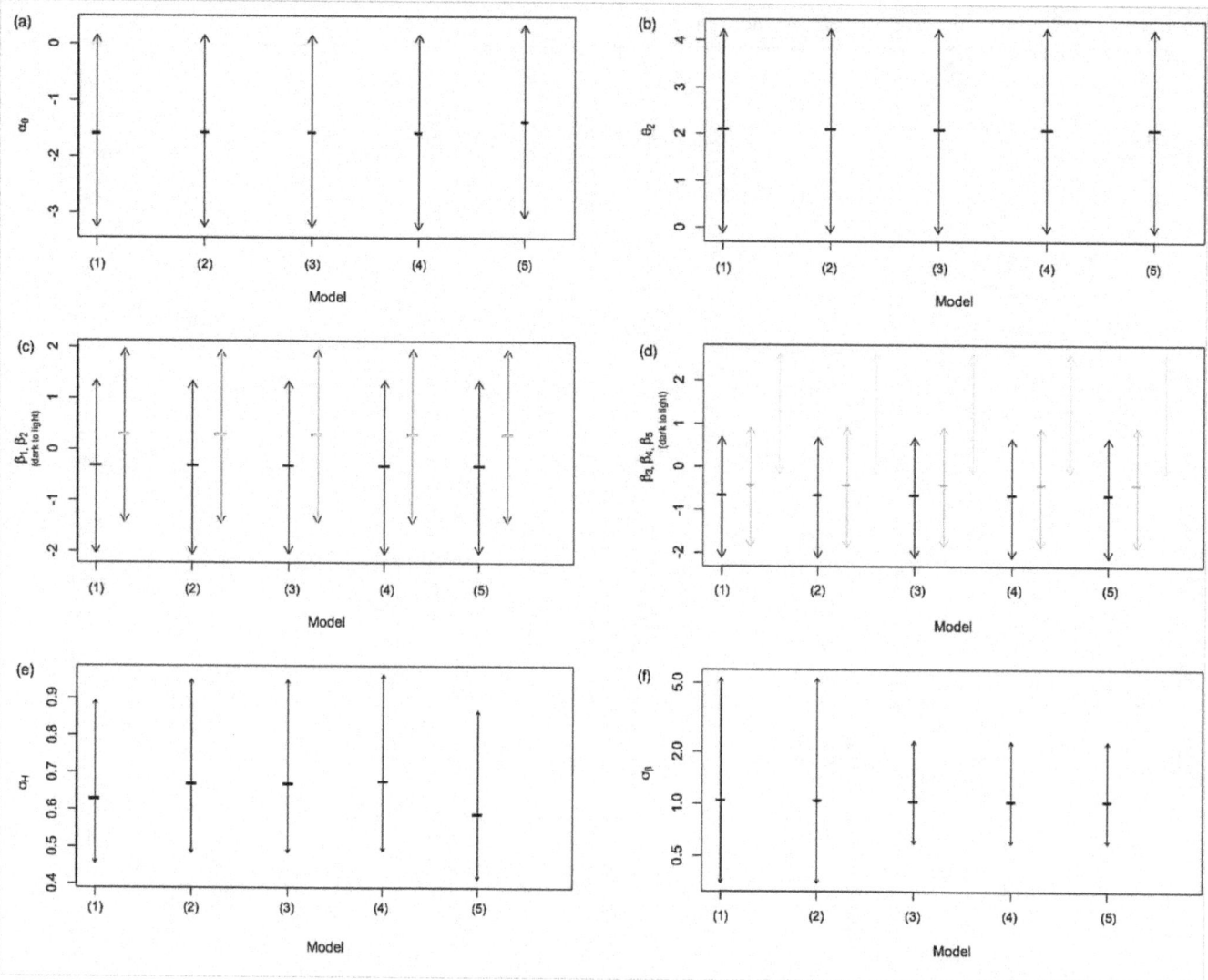

Figure 6. Posterior means and 95% CIs for dispersion parameters and nuisance regression coefficients. Posterior means are marked by "–" inside the CIs. (a) Intercept α_0. (b) Fixed effect θ_- (axis label θ_2). (c) Elements of β_+. (d) Elements of β_-. (e) Standard deviation σ_H. (f) Standard deviation σ_β (plotted on log-scale).

as their interaction. By decomposing the total latent regression error variance, the two-level structure successfully teased apart the explanatory contribution of salinity and distance, two highly collinear abiotic covariates. Overall, (i)–(iii) were almost equal in statistical performance, with slightly better predictive power of biotic metrics by (i) and (iii). Finally, while (i) corresponds to marginally stronger evidence for the two-level structure between salinity and distance to influence site health, (iii) confirms the simultaneous predictive power of all four ecologically relevant abiotic attributes. Our research demonstrates that LHFI modeling is flexible and can be an effective tool for assessing estuarine ecosystem health.

A technical note is that the LHFI framework is built on the fundamental principles of analysis-of-covariance, so that one can only interpret H_i values in a relative sense. However, Chiu et al. [17] explain that including in the study any site that is qualitatively preidentified as very healthy or very unhealthy would facilitate the interpretation of the magnitude of H_i for an individual i. This is slightly different from the approach in [33] which includes sites that span the gradient spectrum of individual covariates. In

general, any biologically relevant covariate that exhibits a substantial gradient across the domain of study should be a candidate for incorporation into the LHFI framework. On the other hand, a balance between model complexity and model parsimony is important to achieve reliable inference. For this reason, formal spatial models (point-reference or spatial random-effects models) were not considered for the small Richibucto dataset. Work is in progress at the Commonwealth Scientific and Industrial Research Organisation (CSIRO) to integrate spatial modeling into the LHFI framework for large spatial stream network datasets from eastern Australia. Like this paper, the work at CSIRO directly models the quantitative biotic metrics instead of their ordinal counterparts such as in [24].

Collecting and processing biotic data can be much more costly than abiotic data for the use in quantitative assessment of ecosystem health. Our research suggests that for the LHFI approach to rigorously distinguish sites according to AMBI and/ or ITI metrics as indicators of estuary health, (1) more than 18 sites and/or (2) measurements made on abiotic covariates with higher precision may be needed. Though, implementation of (1) alone may suffice if abiotic data are recorded at additional sites even without collecting and processing faunal samples. This is because in the case of Richibucto, our approach led to an LHFI-A model (case (iii) above) which identified various abiotic predictors that are ecologically sensible and statistically relevant. Thus, one can (a) design health restoration experiments that focus on perturbing the identified abiotic predictors only, and (b) expect posterior interpolation of biotic conditions to be reliable, given these predictors. In general, the unified LHFI approach facilitates formal statistical interpolation of biotic conditions at sites for which abiotic information is observed. Thus, this methodology addresses key operational considerations in ecosystem management: guidelines for intervention of abiotic attributes and cost effectiveness of health inference without substantial tradeoff for statistical power. In light of its practical and statistical advantages, the LHFI methodology could be an invaluable asset if adopted for biomonitoring protocols by conservation biologists and environmental resource managers.

Supporting Information

Appendix S1　Richibucto data collected and studied by Lu et al. [19].

Appendix S2　LHFI-A-I: integrating AMBI and ITI metrics.

Appendix S3　Some covariance structures for the random metric effects.

Acknowledgments

Part of this paper was written under a CSIRO Water for a Health Country Flagship Appropriation Research Project in the Ecosystems and Contaminants Theme. We thank Profs. J. Grant and M. Dowd (Dalhousie University) for their advice and support throughout the duration of this research. G.C. thanks the following people for their valuable comments: the Academic Editor and Reviewers of this paper, and Drs. M. Dobbie and A. Zwart (CSIRO Mathematics, Informatics and Statistics).

Author Contributions

Conceived and designed the experiments: GSC. Performed the experiments: GSC MAW. Analyzed the data: GSC MAW LL. Contributed reagents/materials/analysis tools: GSC MAW. Wrote the paper: GSC.

References

1. Bilyard GR (1987) The value of benthic infauna in marine pollution monitoring studies. Marine Pollution Bulletin 18: 581–585.
2. Dauer DM (1993) Biological criteria, environmental health and estuarine macrobenthic community structure. Marine Pollution Bulletin 16: 249–257.
3. Borja A, Franco J, Pérez V (2000) A marine biotic index to establish the ecological quality of soft-bottom benthos within European estuarine and coastal environments. Marine Pollution Bulletin 40: 1100–1114.
4. Bazairi H, Bayed A, Hily C (2005) Structure and bioassessment of benthic communities of a lagoonal ecosystem of the Atlantic Moroccan coast. Comptes Rendus Biologies 328: 977–990.
5. Cai W, Liu L, Meng W, Zheng B (2012) The suitability of AMBI to benthic quality assessment on the intertidal zones of Bohai Sea. Acta Scientiae Circumstantiae 32: 992–1000.
6. Medeiros JP, Chaves ML, Silva G, Azeda C, Costa JL, et al. (2012) Benthic condition in low salinity areas of the Mira estuary (Portugal): lessons learnt from freshwater and marine assessment tools. Ecological Indicators 19: 79–88.
7. Teixeira H, Weisberg SB, Borja A, Ranasinghe JA, Cadien DB, et al. (2012) Calibration and validation of the AZTI's marine biotic index (AMBI) for Southern California marine bays. Ecological Indicators 12: 84–95.
8. Muniz P, Hutton M, Kandratavicius N, Lanfranconi A, Brugnoli E, et al. (2012) Performance of biotic indices in naturally stressed estuarine environments on the Southwestern Atlantic coast (Uruguay): a multiple scale approach. Ecological Indicators 19: 89–97.
9. Word JQ (1980) Classification of benthic invertebrates into infaunal trophic index feeding groups. In: Bascom W, editor, Biennial Report for the Years 1979{1980. Long Beach CA: Southern Cali-fornia Coastal Water Research Project, 103–121.
10. Deegan LA, Finn JT, Ayvazian SG, Ryder-Kieffer CA, Buonaccorsi J (1997) Development and validation of an estuarine biotic integrity index. Estuaries 20: 601–617.
11. Smith RW, Bergen M, Weisberg SB, Cadien D, Dalkey A, et al. (2001) Benthic response index for assessing infaunal communities on the southern California mainland shelf. Ecological Applications 11: 1073–1087.
12. Rosenberg R, Blomqvist M, Nilsson HC, Cederwall H, Dimming A (2004) Marine quality assessment by use of benthic species-abundance distributions: a proposed new protocol within the European Union Water Framework Directive. Marine Pollution Bulletin 49: 728–739.
13. Kennedy R, Arthur W, Keegan BF (2011) Long-term trends in benthic habitat quality as deter-mined by multivariate AMBI and infaunal quality index in relation to natural variability: a case study in Kinsale Harbour, south coast of Ireland. Marine Pollution Bulletin 62: 1427–1436.
14. Johnston CA, Zedler JB, Tulbure MG, Frieswyk CB, Bedford BL, et al. (2009) A unifying ap-proach for evaluating the condition of wetland plant communities and identifying related stressors. Ecological Applications 19: 1739–1757.
15. Chiu G, Guttorp P (2006) Stream health index for the Puget Sound Lowland. Environmetrics 17: 285–307.
16. Dobbie M, Dail D (2013) Robustness and sensitivity of weighting and aggregation in constructing composite indices. Ecological Indicators 29: 270–277.
17. Chiu GS, Guttorp P, Westveld AH, Khan SA, Liang J (2011) Latent health factor index: a statis-tical modeling approach for ecological health assessment. Environmetrics 22: 243–255.
18. Wu M (2009) A latent health factor model for estimating estuarine ecosystem health. Unpublished Master's Thesis. Waterloo: University of Waterloo.
19. Lu L, Grant J, Barrell J (2008) Macrofaunal spatial patterns in relationship to environmental variables in the Richibucto estuary, New Brunswick, Canada. Estuaries and Coasts 31: 994–1005.
20. Hoff PD (2009) A first course in Bayesian statistical methods. Dordrecht: Springer.
21. Bhatt JP, Manish K, Pandit MK (2012) Elevational gradients in fish diversity in the Himalaya: water discharge is the key driver of distribution patterns. PLoS ONE 7: e46237.
22. Trebilco R, Halpern BS, Flemming JM, Field C, Blanchard W, et al. (2011) Mapping species richness and human impact drivers to inform global pelagic conservation prioritisation. Biological Conservation 144: 1758–1766.
23. Borja A, Rodríguez JG, Black K, Bodoy A, Emblow C, et al. (2009) Assessing the suitability of a range of benthic indices in the evaluation of environmental impact of fin and shellfish aquaculture located in sites across Europe. Aquaculture 293: 231–240.
24. Schliep EM, Hoeting JA (2013) Multilevel latent Gaussian process model for mixed discrete and continuous multivariate response data. arXiv: 1205.4163.
25. Banerjee S, Carlin BP, Gelfand AE (2004) Hierarchical modeling and analysis for spatial data. Boca Raton FL: Chapman and Hall/CRC.
26. Peet RK (1974) The measurement of species diversity. Annual Review of Ecology, Evolution, and Systematics 5: 285–307.
27. Pla L (2004) Bootstrap confidence intervals for the Shannon biodiversity index: A simulation study. Journal of Agricultural, Biological, and Environmental Statistics 9: 42–56.
28. Hurlbert SH (1971) The nonconcept of species diversity: A critique and alternative parameters. Ecology 52: 577–586.
29. Lunn D, Spiegelhalter D, Thomas A, Best N (2009) The BUGS project: evolution, critique, and future directions. Statistics in Medicine 28: 3049–3067.
30. Spiegelhalter DJ, Best NG, Carlin BR, van der Linde A (2002) Bayesian measures of model com-plexity and fit (with discussion). Journal of the Royal Statistical Society: Series B 64: 583–639.
31. Marshall EC, Spiegelhalter DJ (2007) Identifying outliers in Bayesian hierarchical models: a simulation-based approach. Bayesian Analysis 2: 409–444.
32. Dey DK, Gelfand AE, Swartz TB, Vlachos PK (1998) A simulation-intensive approach for checking hierarchical models. Test 7: 325–346.
33. Lopez RD, Fennessy MS (2002) Testing the oristic quality assessment index as an indicator of wetland condition. Ecological Applications 12: 487–497.
34. Brooks SP, Gelman A (1998) General methods for monitoring convergence of iterative simulations. Journal of Computational and Graphical Statistics 7: 434–455.
35. Cromey CJ, Nickell TD, Black KD (2002) DEPOMOD-modelling the deposition and biological effects of waste solids from marine cage farms. Aquaculture 214: 211–239.

Fish Oil Supplementation Alters the Plasma Lipidomic Profile and Increases Long-Chain PUFAs of Phospholipids and Triglycerides in Healthy Subjects

Inger Ottestad[1,2], Sahar Hassani[3,4], Grethe I. Borge[3], Achim Kohler[4,3], Gjermund Vogt[3], Tuulia Hyötyläinen[5], Matej Orešič[5], Kirsti W. Brønner[6], Kirsten B. Holven[2], Stine M. Ulven[1], Mari C. W. Myhrstad[1]*

1 Department of Health, Nutrition and Management, Faculty of Health Sciences, Oslo and Akershus University College of Applied Sciences, Oslo, Norway, 2 Department of Nutrition, Institute for Basic Medical Sciences, University of Oslo, Oslo, Norway, 3 Nofima, Norwegian Institute of Food, Fisheries and Aquaculture Research, Ås, Norway, 4 Centre for Integrative Genetics (CIGENE), Department of Mathematical Sciences and Technology, Norwegian University of Life Science, Ås, Norway, 5 VTT Technical Research Centre of Finland, Espoo, Finland, 6 TINE SA, Centre for Research and Development, Kalbakken, Oslo, Norway

Abstract

Background: While beneficial health effects of fish and fish oil consumption are well documented, the incorporation of n-3 polyunsaturated fatty acids in plasma lipid classes is not completely understood. The aim of this study was to investigate the effect of fish oil supplementation on the plasma lipidomic profile in healthy subjects.

Methodology/Principal Findings: In a double-blinded randomized controlled parallel-group study, healthy subjects received capsules containing either 8 g/d of fish oil (FO) (1.6 g/d EPA+DHA) (n = 16) or 8 g/d of high oleic sunflower oil (HOSO) (n = 17) for seven weeks. During the first three weeks of intervention, the subjects completed a fully controlled diet period. BMI and total serum triglycerides, total-, LDL- and HDL-cholesterol were unchanged during the intervention period. Lipidomic analyses were performed using Ultra Performance Liquid Chromatography (UPLC) coupled to electrospray ionization quadrupole time-of-flight mass spectrometry (QTOFMS), where 568 lipids were detected and 260 identified. Both t-tests and Multi-Block Partial Least Square Regression (MBPLSR) analysis were performed for analysing differences between the intervention groups. The intervention groups were well separated by the lipidomic data after three weeks of intervention. Several lipid classes such as phosphatidylcholine, phosphatidylethanolamine, lysophosphatidylcholine, sphingomyelin, phosphatidylserine, phosphatidylglycerol, and triglycerides contributed strongly to this separation. Twenty-three lipids were significantly decreased (FDR<0.05) in the FO group after three weeks compared with the HOSO group, whereas fifty-one were increased including selected phospholipids and triglycerides of long-chain polyunsaturated fatty acids. After seven weeks of intervention the two intervention groups showed similar grouping.

Conclusions/Significance: In healthy subjects, fish oil supplementation alters lipid metabolism and increases the proportion of phospholipids and triglycerides containing long-chain polyunsaturated fatty acids. Whether the beneficial effects of fish oil supplementation may be explained by a remodeling of the plasma lipids into phospholipids and triglycerides of long-chain polyunsaturated fatty acids needs to be further investigated.

Trial Registration: ClinicalTrials.gov NCT01034423

Editor: Stephane Blanc, Institut Pluridisciplinaire Hubert Curien, France

Funding: This study was supported by the Norwegian Research Council (184813/110) http://www.forskningsradet.no and the Nordic Centre of Excellence on Food, Nutrition and Health "Systems biology in controlled dietary interventions and cohort studies" (SYSDIET nr. 070014) http://www.sysdiet.fi. The funders had no role in study design, data collection and analysis, decision to publish, or preparation of the manuscript.

Competing Interests: Kirsti Wettre Brønner is a clinical nutritionist/Project manager in TINE SA R&D Center, Norway. The cod liver oil used in this study, TINE EPADHA OIL Oil 1200, was produced by Martitex AS, (Havnegata 17, 8400) Sortland Norway and provided by TINE SA. (At the time of the intervention) Maritex AS was a fully owned subsidiary of TINE SA.

* E-mail: mari.myhrstad@hioa.no

Introduction

Intake of fish and fish oil, containing n-3 fatty acids; eicosapentaenoic acid (EPA; 20:5) and docosahexaenoic acid (DHA; 22:6), is associated with beneficial health effects such as reduced risk of cardiovascular disease and sudden cardiac death [1–4]. The beneficial effects of marine n-3 fatty acids have been explained by decreased plasma triglycerides (TGs) [5,6], moderate reduction in blood pressure [7], reduced platelet aggregation [8,9], and protection against cardiac arrhythmias [10,11]. It has been suggested that bioactive lipid components may be important in mediating these effects, but the molecular mechanisms are still to a large extent unknown.

Cells, tissues and biological fluids contain tens of thousands of structurally different lipids, that fulfil multiple roles in cellular signalling, in membrane structure, and as fuel sources for many cell types [12]. The entire spectrum of lipids in a biological system, can be defined as the lipidome [13], which combines mass spectrometry technology and bioinformatics methods with traditional methods such as sample preparation, lipid extraction and separation. Lipidome analyses have revealed a diversity of lipid compounds in human plasma, which can be classified into six main lipid categories including fatty acyls, glycerolipids, glycerophospholipids, sphingolipids, sterol lipids and prenol lipids [14]. The major plasma lipids are the glycerolipids (TGs), glycerophospholipids (phospholipids) and sterol lipids which are transported in the lipoprotein particles [14,15].

In n-3 FA intervention studies fatty acids have been measured in different blood compartments such as in platelets and red blood cells, and in plasma cholesteryl esters, triglycerides and phospholipids. Lipidomic analysis now offers the opportunity to detect exact fatty acid composition of these individual lipids. [16]. Recently it was shown that the plasma lipidomic profile was altered in subjects with coronary heart disease after intake of fatty fish, and in subjects with metabolic syndrome after consumption of a healthy diet containing fatty fish, wholegrain products and bilberries [17,18]. Furthermore, profiling of the plasma lipids suggests a relationship between the composition of plasma lipids and diet [19,20], with diet-induced weight loss [21] and diet-related diseases such as diabetes mellitus [22]. This opens up the opportunity to identify new functional lipid biomarkers to detect and prevent diet-related diseases. We have however not been able to find studies showing the plasma lipidomic profile in healthy subjects after intake of fish oil.

We have previously reported that in the present study a daily intake of fish oil (1.6 g EPA+DHA/d) did not change the level of serum lipids, markers of oxidative stress, lipid oxidation or inflammation, whereas an increase in plasma EPA, DPA and DHA was observed after three and seven weeks of intervention in a randomized controlled study in healthy subjects [23]. The aim of this study was to apply a lipidomic strategy to further describe the effect of fish oil supplementation in healthy subjects.

Materials and Methods

Subjects

Healthy men and women between 18–50 years were recruited into this study. Detailed description of the protocol, participant recruitment and enrolment, inclusion and exclusion criteria, and compliance are described in details elsewhere [23]. In brief, exclusion criteria were total cholesterol >7.5 mmol/l, triglycerides >4 mmol/l, glucose >6.0 mmol/l, C-reactive protein (CRP) >10 mg/l, body mass index (BMI) $\geq$30 kg/m^2 and blood pressure ($\geq$160/100). The study was performed at the Akershus University College, Norway between September and December 2009.

Ethics Statement

Written informed consent was obtained from all participants and the protocol was approved by the Regional Committee of Medical Ethics (approval no. 6.2008.2215) and by the Norwegian Social Science Data Services (approval no. 21924), and was conducted in accordance with the Declaration of Helsinki.

Study Design

This study was a part of a randomized controlled double-blinded three-arm parallel group study, designed to investigate health effects from intake of fish oil [23]. In the present study, data from two of the intervention groups are included, as shown in Figure 1. Subjects in the present study were given 8 g/d of either fish oil (FO) or high oleic sunflower oil (HOSO), and each subject was taking 16 capsules/d minimum twice each day for seven weeks. Subjects in the fish oil group received capsules containing 0.7 g/d EPA+0.9 g/d DHA from cod liver oil (Gadidae sp., TINE EPADHA Oil 1200) provided by TINE SA (Oslo, Norway) and subjects in the control group received high oleic sunflower oil purchased from AarhusKarlshamn AB (Malmø, Sweden). The subjects were instructed to take the capsules with food (minimum two meals). The fatty acid composition in the oils has been described elsewhere [23].

The subjects met for visits and blood samples for the lipidome analyses were collected at 0, 3 and 7 weeks. Between the screening and baseline visit (week 0), the subjects conducted a four-week washout period, where foods containing marine n-3 fatty acids were avoided. During the first three weeks of the intervention the subjects conducted a fully-controlled isocaloric diet, provided with all food and beverages at Akershus University College, Norway. The mean daily intake during the fully controlled diet period (exclusive capsules, n = 33) was 9.1±2.3 MJ and the composition of the diet consumed was similar in both groups containing 24.4±1.1 percentage of energy (E%) from fat, of which 7.8±0.3 E% from SFA, 5.8±0.4 E% from MUFA and 4.8±0.9% PUFA. The carbohydrate content was 54.4±1.1 E%, including 4.6±0.9 E% from added sugar and 19.4±0.5 E% from protein. The food items provided in this study have previously been described [23]. The last four weeks of the intervention period the subjects returned to their habitual diet. Intake of fish, fish products, marine n-3 enriched food or dietary supplements was not allowed during the entire study period of 11 weeks. The study was registered at www.clinicaltrial.gov (IDno. NCT01034423).

Blood sampling

Subjects were told to refrain from alcohol consumption and vigorous physical activity the day prior to blood sampling. Venous blood samples were drawn after an overnight fast ($\geq$12 hours) at the same time (±2 h) and serum were kept at room temperature at 30 min before centrifuged (1500 g 12 min). EDTA-plasma was immediately placed on ice and centrifuged within 10 min (1500 g, 4°C, 10 min). N_2 flushed plasma samples were snap frozen and stored at −80°C until further analysis.

Routine laboratory analysis

Fasting serum hsCRP, total- cholesterol, LDL-cholesterol, HDL-cholesterol, triglycerides and glucose were measured by standard methods at a routine laboratory (Fürst Medical Laboratory, Oslo, Norway).

Lipidomic analyses

An aliquot (10 µL) of plasma sample was diluted with 10 µL of 0.15 M (0.9%) sodium chloride and 10 µL of internal standard mixture containing PC(17:0/0:0), PC(17:0/17:0), PE(17:0/17:0), PG(17:0/17:0)[rac], Cer(d18:1/17:0), PS(17:0/17:0) and PA(17:0/17:0) (Avanti Polar Lipids, Inc., Alabaster, AL, USA) and TG(17:0/17:0/17:0) and MG(17:0/0:0/0:0)[rac], DG(17:0/17:0/0:0)[rac] (Larodan Fine Chemicals) was added. The lipids were extracted using a mixture of HPLC-grade chloroform and methanol (2:1; 100 µL). The lower phase was collected (60 µL) and 10 µL internal standard mixture containing labeled PC(16:1/0:0-D$_3$), PC(16:1/16:1-D$_6$) and TG(16:0/16:0/16:0-^{13}C3) was added.

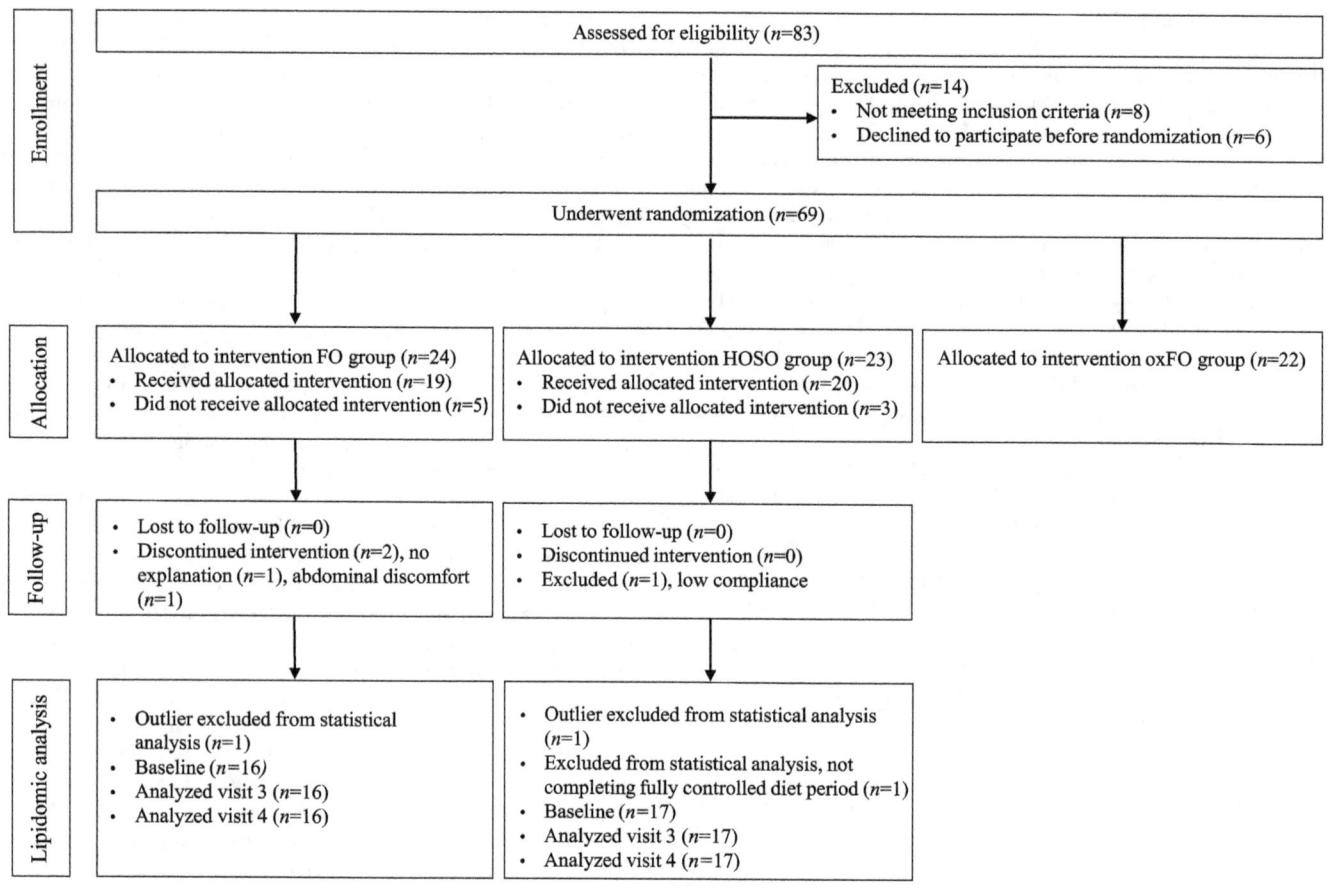

Figure 1. Flow chart of the study showing subjects enrolled, lost during follow-up and number of subjects included in the statistical analysis at baseline and after three and seven weeks of fish oil supplementation. FO group, fish oil group; HOSO, high oleic sunflower oil group; oxFO, oxidized fish oil group (not included in the present study).

The extracts were analyzed on a Waters Q-Tof Premier mass spectrometer combined with an Acquity Ultra Performance LC^{TM} (UPLC) in randomized order. The column (at 50°C) was an Acquity UPLC^{TM} BEH C18 2.1×100 mm with 1.7 μm particles. The solvent system included A: ultrapure water with 1% 1 M NH$_4$Ac and 0.1% HCOOH, and B: LC/MS grade acetonitrile/isopropanol (1:1) with 1% 1M NH$_4$Ac and 0.1% HCOOH. The gradient started from 65% A/35% B, reached 80% B in 2 min, 100% B in 7 min and remained there for 7 min. The flow rate was 0.400 ml/min and the injected amount was 2.0 μl (Acquity Sample Organizer, at 10°C). Reserpine was used as the lock spray reference compound. The lipid profiling was carried out using ESI+ mode and the data was collected at mass range of m/z 300–1200 with scan duration of 0.2 sec. The data was processed by using MZmine2 software [24] and the lipid identification was based on an internal spectral library.

The data processing included peak detection, integration, peak alignment, normalization and identification. Lipids were identified

Table 1. BMI and serum lipids at baseline and after three weeks of intervention with fish oil (n = 16) and high oleic sunflower oil (n = 17).

| Parameter | Fish oil | | Sunflower oil | | P-value* | P-value** |
	Baseline	3 wk	Baseline	3 wk		
BMI (kg/m²)	22±3	22±3	23±3	23±3	0.25	0.53
Triglycerides (mmol/l)	0.9±0.4	0.9±0.3	1.1±0.7	1.1±0.4	0.46	0.68
Total-cholesterol (mmol/l)	4.6±0.8	4.4±0.6	4.9±0.9	4.6±1.0	0.27	0.36
LDL-cholesterol (mmol/l)	2.5±0.8	2.4±0.8	2.7±0.6	2.6±0.6	0.35	0.64
HDL-cholesterol (mmol/l)	1.5±0.3	1.4±0.4	1.5±0.4	1.4±0.4	1	0.86

*Independent t- test for between groups at baseline.
**Independent t- test for changes between groups after three weeks.

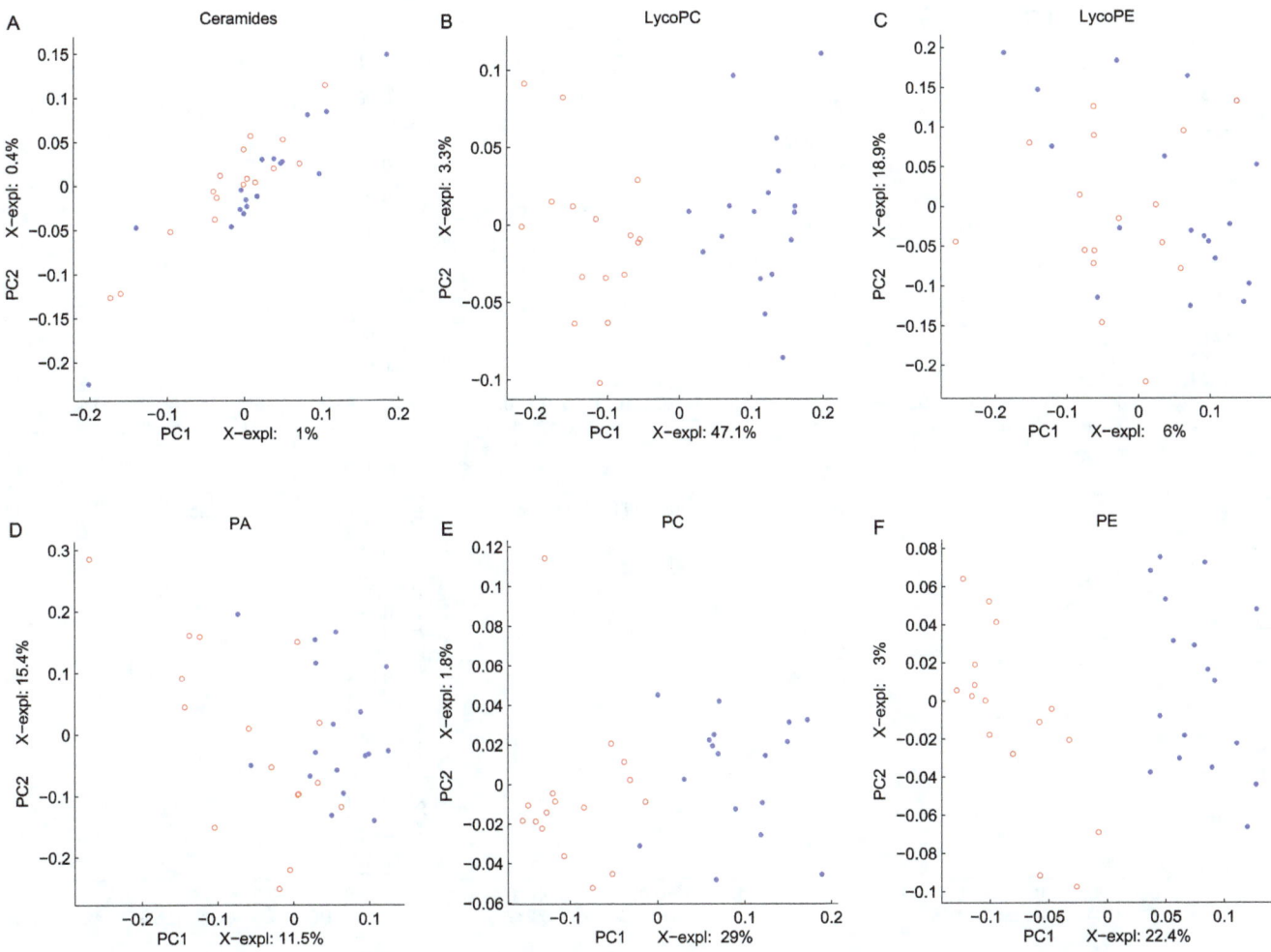

Figure 2. Multi-Block Partial Least Squares Regression (MBPLSR) analysis of the data after three weeks of intervention. First and second PLSR components of block scores of ceramides, lysoPC, lysoPE, PA, PC and PE are shown (A–F). The samples of each intervention group are presented as blue (HOSO group) or red (FO group) circles. The (un-validated) explained variances are shown on the axes.

using an internal spectral library. The data was normalized using internal standards representative of each class of lipid present in the samples: the intensity of each identified lipid was normalized by dividing it with the intensity of its corresponding standard and multiplying it by the concentration of the standard. All monoacyl lipids except cholesterol esters, such as monoacylglycerols and monoacylglycerophospholipids, are normalized with PC(17:0/0:0), all diacyl lipids except ethanolamine phospholipids are normalized with PC(17:0/17:0), all ceramides with Cer(d18:1/17:0), all diacyl ethanolamine phospholipids with PE(17:0/17:0), and TG and cholesterol esters with TG(17:0/17:0/17:0).

Statistical analyses

Sample size was calculated using expected change in plasma n-3 fatty acids as described as previously described [23]. Multi-Block Partial Least Squares Regression (MBPLSR) analysis was used for exploring the sample and variable variation patterns in the data [25] where each lipid class was defined as one individual block [26] resulting in 11 blocks of descriptor variables in total (i.e. $[\mathbf{X}^1,\mathbf{X}^2,...,\mathbf{X}^{11}]$). The multi-block set of descriptor variables were organized in the following order: Ceramides as block one ($\mathbf{X}^1$), lysophosphatidylcholines (lysoPC) as block two ($\mathbf{X}^2$), lysopho-

sphatidylethanolamines (lysoPE) as block three ($\mathbf{X}^3$), phosphatidic acid (PA) as block four ($\mathbf{X}^4$), phosphatidylcholines (PC) as block five ($\mathbf{X}^5$), phosphatidylethanolamines (PE) as block six ($\mathbf{X}^6$), phosphatidylglycerols (PG) as block seven ($\mathbf{X}^7$), phosphatidylser-ines (PS) as block eight ($\mathbf{X}^8$), sphingomyelins (SM) as block nine ($\mathbf{X}^9$), triglycerides (TG) as block ten ($\mathbf{X}^{10}$) and sums of lipid classes together with phosphatidylinositol (PI) as block eleven ($\mathbf{X}^{11}$) (a separate block was not assigned to PI class since it contained only one lipid). An intervention group indicator variable was used as response variable (*y*-variable). In order to estimate both the influence of the total amount of lipids in each lipid class and simultaneously the influence of the relative variation within each lipid class, each block was normalized by a division by the total amount of lipids in the corresponding class. The total amounts of lipids of each lipid classes were then used as an additional block and named "the sums of lipids". Subsequently, plasma lipids were transformed by taking the log2 ratio (baseline adjusted log2 values) after three and seven weeks in both the FO and the HOSO group. After this, each block (lipid class) was set on the same footing prior to MBPLSR analysis by block scaling as described in [26]. Model-validation and testing of the influence of each lipid class to the global model was done by cross-validation as described in

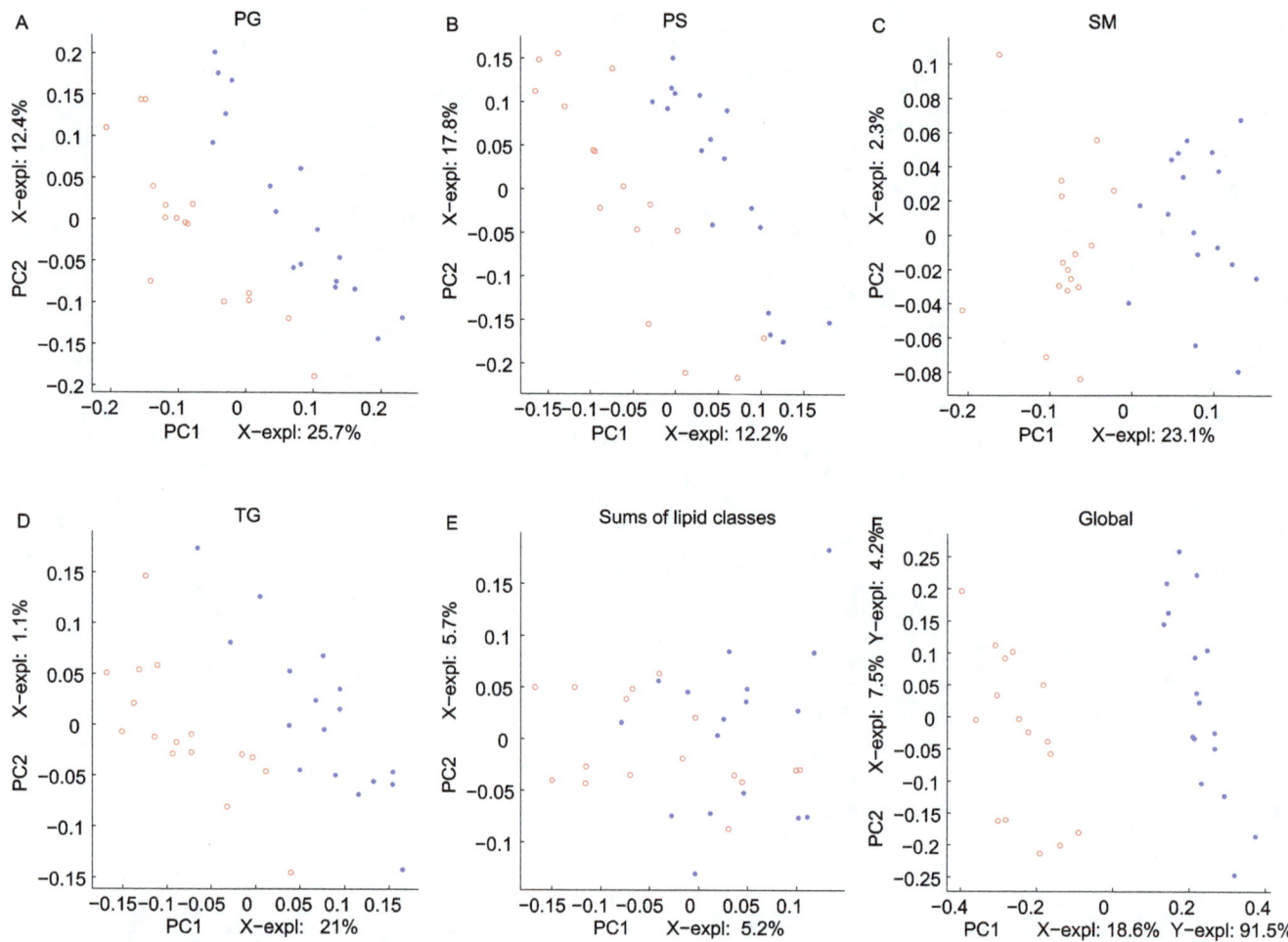

Figure 3. Multi-Block Partial Least Squares Regression (MBPLSR) analysis of the data after three weeks of intervention. First and second PLSR components of block scores of PG, PS, SM, TG, the sums of lipid classes and global scores are shown (A–F). The samples of each intervention group are presented as blue (HOSO group) or red (FO group) circles. The (un-validated) explained variances are shown on the axes.

[26].Variable significance testing for the difference between the groups at baseline and after the intervention (baseline adjusted values) was done by cross-validation [25,26] of the multivariate MBPLSR model and by univariate testing using Student's- t test. For the univariate testing the log2 ratios were used and False Discovery Rate (FDR) corrected q-values were computed using the R package 'qvalue'. Two subjects were detected as outliers by the MBPLSR models and were therefore excluded from the further analysis. Baseline characteristics were analyzed using (baseline adjusted values) Student's- t test and Mann Whitney U test (serum triglycerides) when data was normally and not normally distributed, respectively. The significance level was set to 5% (two-sided) and the power of the test was chosen to be 80%. Data in Table 1 are presented as mean ± SD. All univariate analyses were performed using SPSS for windows (SPSS, version 19.0) and multivariate analyses were performed using in-house-written and standard MATLAB routines (MATLAB version 7.8).

Results

Characteristics of the subjects

A total of 33 normal weight healthy subjects (n = 8 men and n = 25 women) completed this study. The subjects were young (28±8 years), and with serum lipids within the normal range as

shown in Table 1. No differences in age, BMI or serum lipids were observed between the FO group (n = 16) and the HOSO group (n = 17) at baseline (Table 1). Serum lipids and BMI were not significantly changed between the groups after three (Table 1) or seven weeks of intervention (data not shown).

Plasma lipidomic profile

A total of 568 lipids were detected and quantified in the plasma samples of the two intervention groups. Of these lipids, 260 were identified, including the following lipid classes; ceramides, sphingomyelins (SM), lysophosphatidylcholines (lysoPC), lysophosphatidylethanolamines (lysoPE), phosphatidic acids (PA), phosphatidylcholines (PC), phosphatidyethanolamines (PE), phosphatidylglycerols (PG), phosphatidylinositols (PI), phosphatidyserines (PS) and triglycerides (TG). In the present study, the identified lipids were included in the statistical analysis.

MBPLSR was performed with lipid class blocks as a multi-block **X** and intervention group indicator variable as y-variable. The lipid class block variation patterns after three weeks of intervention are shown in Figure 2 and Figure 3A–D. The lipidomic profiles of the two intervention groups were well separated in the global sample variation pattern (global score plot) (Figure 3F). The first principal component accounted for most of the separation of the two groups and explained 91.5% of the total variance in the y-

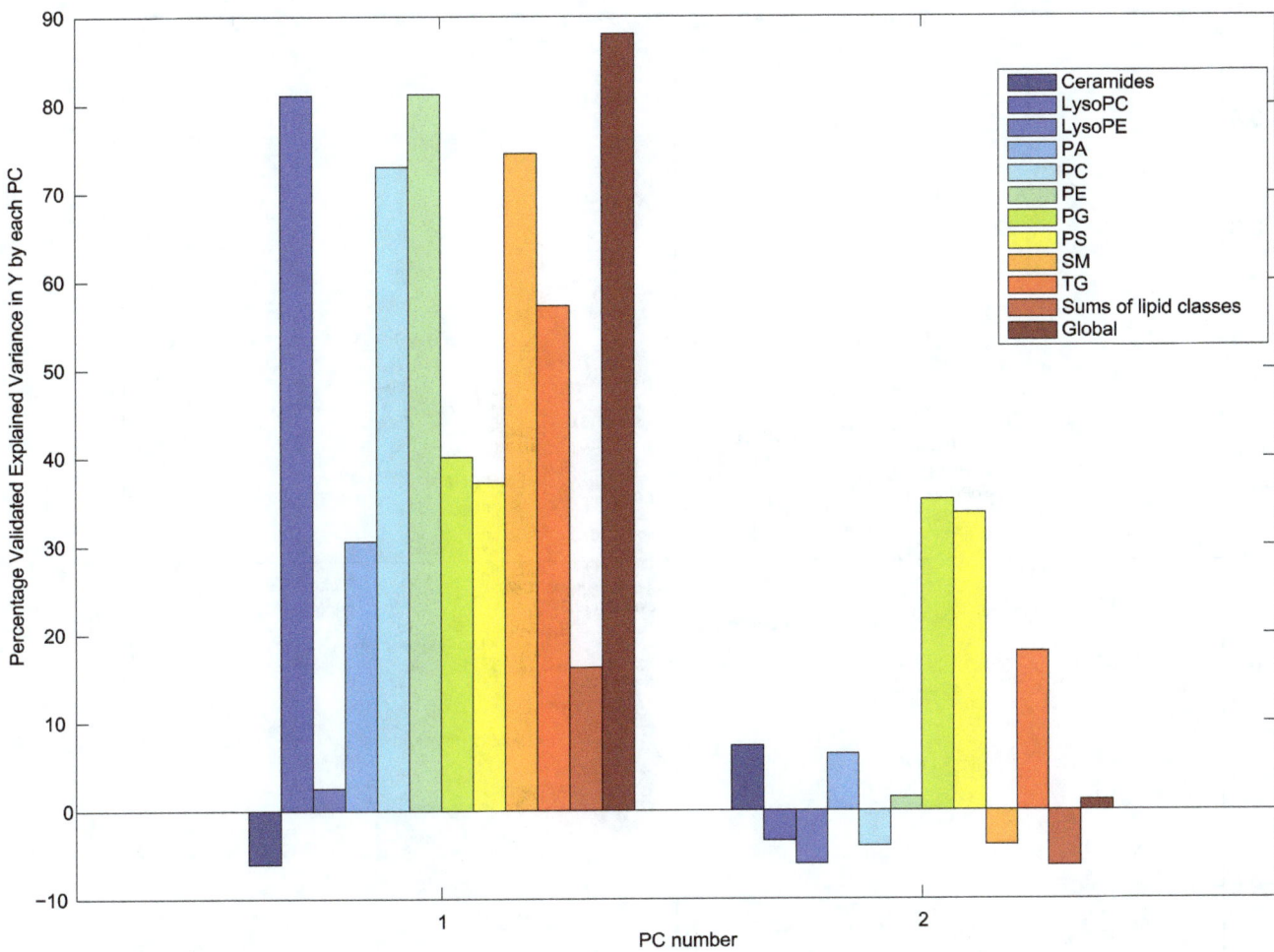

Figure 4. Cross-validated explained variance in Y. Bar plots of the validated explained variances in **Y** for each block and for the global model using data obtained after three weeks of intervention are presented.

variable. The explained block variances are shown on the respective axes. Several lipid class blocks (lysoPC, PC, PE, PG, PS, SM and TG) showed a clear separation of the FO and the HOSO group, whereas ceramides, lysoPE and PA lipids did not separate the FO and the HOSO group (Figure 2 and Figure 3A–D). In addition, the sums of lipid classes did not separate the FO and the HOSO group (Figure 3E) showing that the differences between FO and HOSO group can be explained by remodelling within lipid classes rather than by changes in the total amount of lipids in each class. A similar separation of the groups and patterns in block score plots were observed also after seven weeks of intervention (data not shown). To further analyze the plasma lipid profile and to identify specific lipids contributing to the separation, we decided to use the data obtained after completing a three weeks fully controlled diet period. After three weeks of intervention, the contribution of each lipid class block to the prediction of the group indicator variable was validated by calculating root mean squared errors of cross-validation per block. The validated explained variances for the first two components are shown in Figure 4. The first principal component of the lysoPC, PC, PE and SM lipid classes described most of the group separation and explained more than 70% of the y-variance. The second component of the PG, PS and TG lipids explained the separation of the two intervention groups further accounting for 19–37% of the y-variance. By

significance testing using cross-validation and Jack-knifing [26] a total of 75 lipids were identified as significant for the separation of the two intervention groups after three weeks using a two principal component model (Table S1). By investigating the validated root mean squared error as a function of components a two component model was selected (Figure S1).

To further identify the specific lipids that contributed to the distinction of the FO and the HOSO group correlation loading plots were studied [26]. The correlation loading plots in Figures 5 and 6 show that several phospholipids and TGs containing long-chain PUFAs including lysoPC(20:5), lysoPC(22:6), PC(36:5), PC(40:6), PE(38:5), TG(50:4), TG(52:6), TG(52:7), TG(54:8), TG(56:7), TG(56:8), TG(56:9), TG(58:6), TG(58:8), TG(58:9), and TG(58:10) were strongly positively correlated to intake of FO supplementation.

A strong positive correlation was also observed between intake of FO supplementation and lipids of long-chain and lower double bond content such as SM(18:0/24:0), TG(59:2) and TG(52:0). Only PC(40:4e), PC(37:4)/PE(40:4), PC(38:5) and PC(34:2) were found negatively correlated to the FO group.

In order to describe altered lipids in the FO group compared to the HOSO group unpaired t-test was performed. In the FO group, 74 lipids were significantly altered (FDR<0.05) compared with the HOSO group after three weeks of intervention, and 51 out of

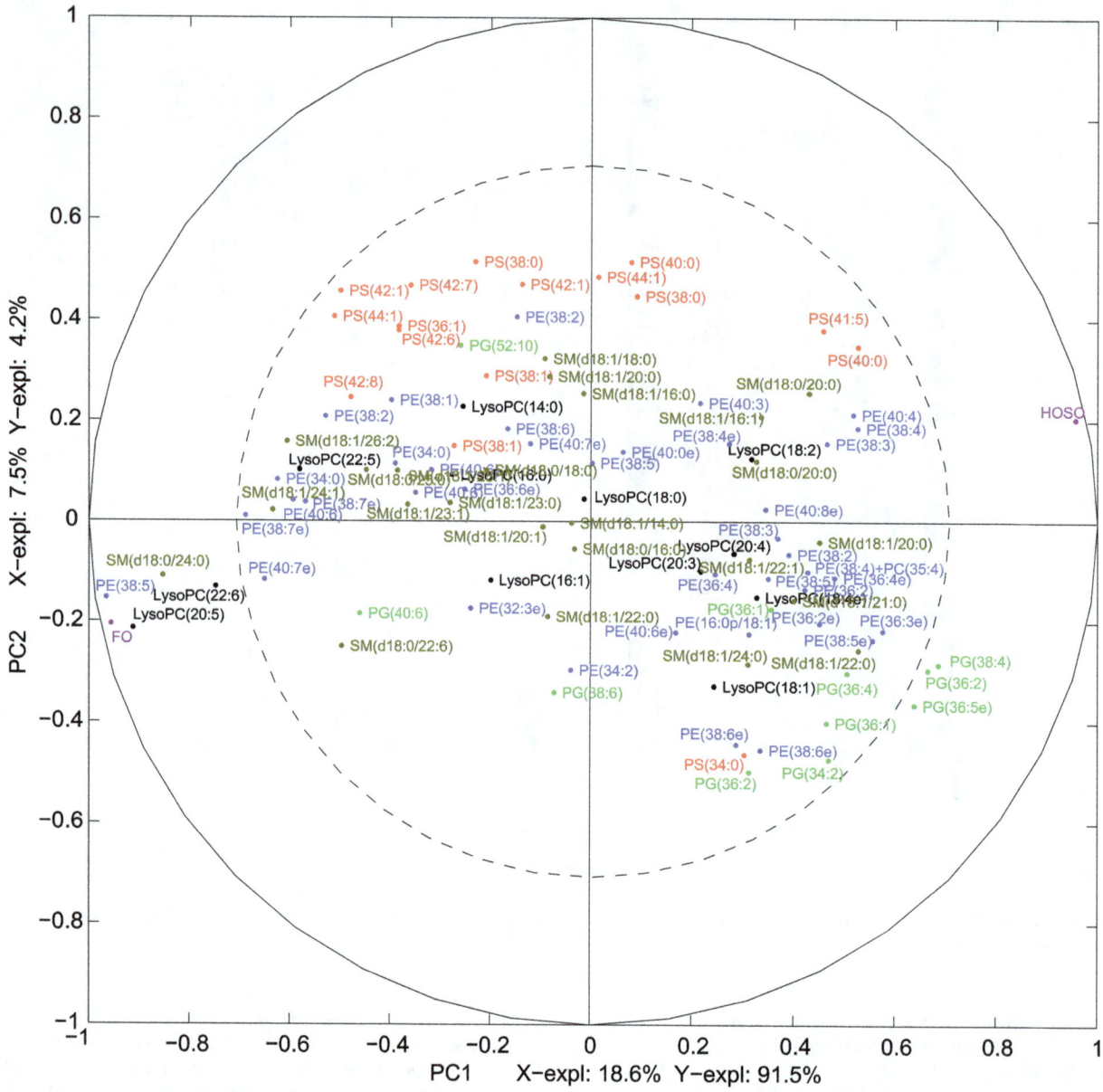

Figure 5. Multi-Block Partial Least Squares Regression (MBPLSR) analysis of the data after three weeks of intervention. Correlation loading plot for the variables contributing to the separation of the FO and the HOSO group after three weeks are shown for LycoPC, PE, PG, PS and SM. The (un-validated) explained variances in **X** and **Y** are shown on the axes.

these 74 lipids were significantly increased. Several phospholipids and TGs containing long-chain PUFAs were increased in the FO group, compared to the HOSO group. Significantly altered lysoPC, PC, PE, PA, PG, PS, PI, SM and TG lipids are shown in Tables 2 and 3. Furthermore, 49 lipids were identified as significantly altered in the FO group compared to the HOSO group by both unpaired t-test and MBPLSR (Table S1). After seven weeks of intervention 58 significant altered lipids were identified in the FO group compared to the HOSO group, and 33 out of these 58 lipids were significantly altered after both three and seven weeks (data not shown).

Discussion

We have investigated the effect of fish oil supplementation on the plasma lipidomic profile in healthy subjects. A clear distinction

of the lipidomic profile was obtained between the FO and the HOSO group after three and seven weeks of intervention. The lipid classes that contributed to the separation of the intervention groups were LysoPC, PC, PE, PG, PS, SM and TG. FO supplementation especially increased phospholipids and TGs of long-chain PUFAs, but the total concentration of the lipids within each lipid class remained unchanged and did not differ in the FO compared to the HOSO group. The clear distinction between the FO and the HOSO group was observed after a fully-controlled isocaloric diet period for three weeks and it was also evident after the subjects had continued on their habitual diet for additionally four weeks. HOSO was used as control oil due to a similar lipid composition as the background diet, and the expected change in plasma lipidome after intake of HOSO were thought to be small.

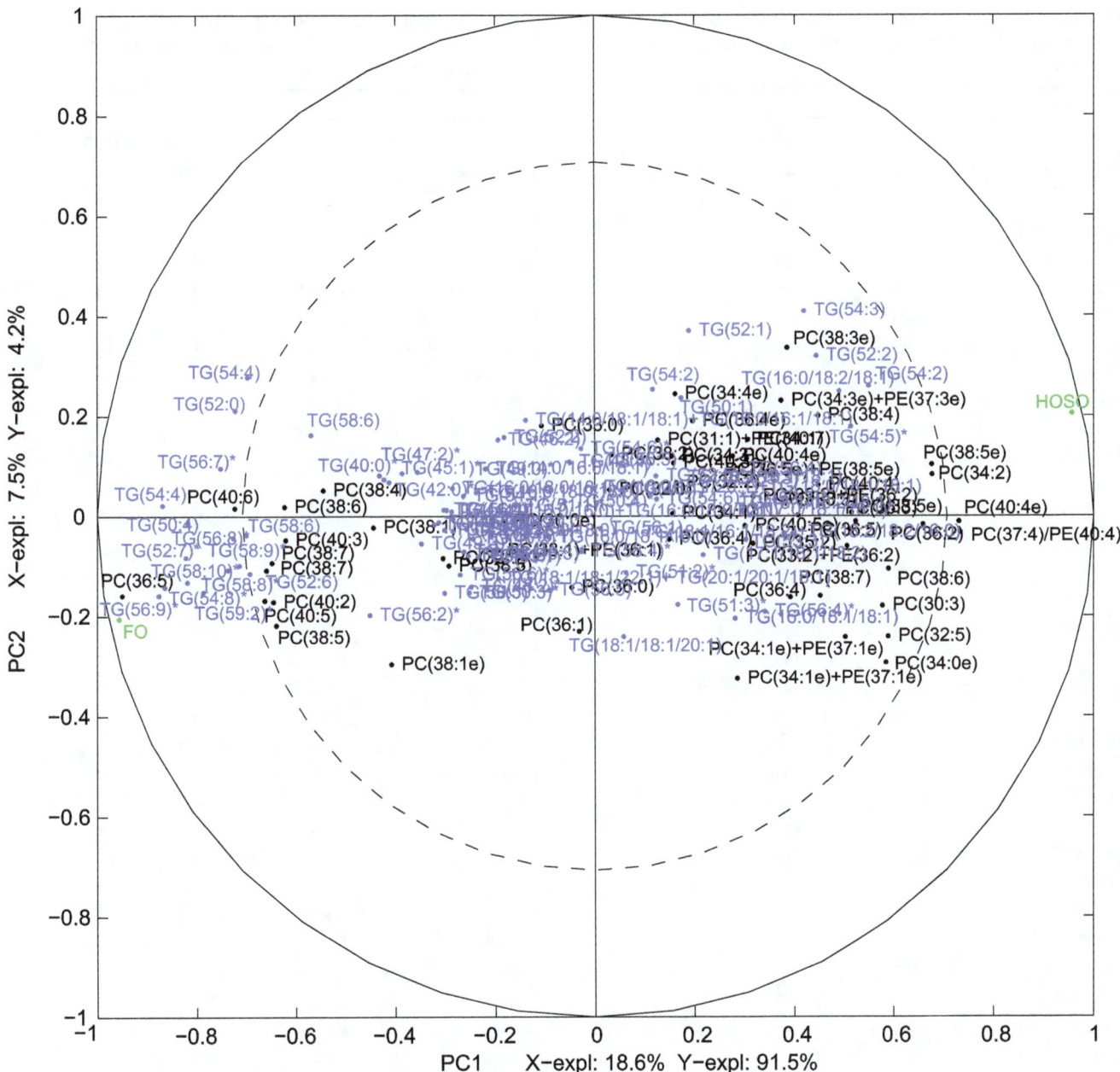

Figure 6. Multi-Block Partial Least Squares Regression (MBPLSR) analysis of the data after three weeks of intervention. Correlation loading plot for the variables contributing to the separation of the FO and the HOSO group after three weeks are shown for TG and PC. The (un-validated) explained variances in **X** and **Y** are shown on the axes.

By using MBPLSR, co-variation patterns in sample and variable space for the different lipid classes was studied. MBPLSR is a method based on latent variables, where by using only few latent variables the problem of over-fitting and false discovery is minimized. For MBPLSR analysis data blocks were organized and normalized, such that remodelling effects in each lipid class and changes in total amounts of lipids per class could be studied separately. A possible consequence of organizing the data into lipid classes is that patterns in the dataset related to other features maybe overlooked. We also included univariate analyses were the data was not organized into lipid classes. The univariate analyses will produce many false significant lipids due to the high number of variables and therefore false discovery adjusted q-values were

computed. The majority of the significant altered lipids identified with univariate analyses were also identified as changed with multivariate analyses, indicating that the organization into lipid classes did not influenced the identification of significant altered lipids.

Recent results from a dietary intervention study have shown that fish intake increased TGs of long-chain PUFA similar to our results, and that fish consumption for eight weeks increased plasma long-chain TGs in subjects with coronary heart disease [18]. Interestingly, this effect was significant after intake of lean fish and not fatty fish [18]. A healthy diet rich in whole grain products, fish and bilberries significantly changed multiple TGs incorporating long-chain PUFAs after 12 weeks intervention in subjects with

Table 2. Significantly altered lipids (FDR<0.05) in the fish oil (FO) group compared to the sunflower oil group (HOSO) after three weeks of intervention.

Lipid Name	q-value	Fold change from baseline	
		HOSO	FO
LysoPC(20:5)	<0.001	0.80	4.35
LysoPC(22:5)'	0.025	0.91	1.67
LysoPC(22:6)	0.003	0.94	1.89
PA(38:5e)	0.006	1.05	0.75
PE(38:4)	0.026	1.11	0.88
PE(38:4)+PC(35:4)	0.029	1.27	0.90
PE(38:5)	<0.001	0.92	3.29
PE(38:5e)	0.042	1.09	0.75
PE(38:7e)	<0.001	1.17	2.58
PE(38:7e)	0.013	1.11	1.40
PE(40:4)	0.027	1.10	0.78
PE(40:6)	0.013	1.10	1.75
PE(40:7e)	0.010	1.08	1.51
PG(36:2)	0.015	0.96	0.68
PG(36:5e)	0.028	1.02	0.80
PG(38:4)	0.013	1.24	0.89
PG(40:6)	0.003	1.11	1.91
PI(40:7)	0.024	1.25	0.96
PS(36:1)	0.001	1.14	1.84
PS(38:0)	0.031	1.18	1.67
PS(38:1)	0.010	1.11	1.58
PS(38:1)	0.019	1.26	1.83
PS(41:5)	0.031	1.17	0.96
PS(42:1)	0.003	1.09	1.86
PS(42:6)	0.001	0.96	1.54
PS(42:7)	0.001	0.92	1.39
PS(42:8)	0.001	1.36	2.51
PS(44:1)	0.001	1.08	1.86
SM(d18:0/20:0)	0.029	1.08	0.81
SM(d18:0/22:6)	0.015	1.01	1.44
SM(d18:0/24:0)	<0.001	0.93	2.62
SM(d18:1/26:2)	0.015	1.10	1.43

Table 3. Significantly altered lipids (FDR<0.05) in the fish oil (FO) group compared to the sunflower oil (HOSO) group after three weeks of intervention.

Lipid Name	q-value	Fold change from baseline	
		HOSO	FO
PC(30:3)	0.015	0.85	0.6
PC(32:5)	0.024	1.03	0.77
PC(36:3)	0.031	0.97	0.73
PC(36:5)	<0.001	0.80	4.00
PC(37:4)/PE(40:4)	0.021	1.07	0.83
PC(38:1)	0.025	1.05	1.60
PC(38:1e)	0.026	0.91	1.39
PC(38:4)	0.007	2.35	9.08
PC(38:5)	<0.001	1.12	3.97
PC(38:5e)	0.006	1.2	0.96
PC(38:6)	0.006	1.15	1.52
PC(38:6)	0.037	0.89	0.69
PC(38:7)	0.001	1.09	2.27
PC(38:7)	0.001	1.09	2.25
PC(40:2)	<0.001	0.99	3.54
PC(40:3)	0.001	0.93	1.74
PC(40:4)	0.029	1.09	0.79
PC(40:4e)	0.015	1.01	0.78
PC(40:5)	<0.001	0.89	1.57
PC(40:6)	<0.001	1.04	1.71
TG(50:4)	<0.001	1.04	3.57
TG(52:0)	0.001	1.03	2.52
TG(52:2)	0.029	1.09	0.79
TG(52:6)	0.004	0.96	2.86
TG(52:7)	<0.001	0.80	4.55
TG(54:2)	0.001	1.27	0.59
TG(54:3)	0.029	1.26	0.83
TG(54:4)	<0.001	0.96	2.34
TG(54:4)	0.003	0.97	1.73
TG(54:5)	0.022	1.51	0.82
TG(54:8)	<0.001	0.98	5.02
TG(56:2)	0.026	1.02	1.93
TG(56:4)	0.038	4.50	1.47
TG(56:7)	<0.001	1.09	3.02
TG(56:8)	<0.001	1.27	2.71
TG(56:9)	<0.001	1.01	4.76
TG(58:10)	<0.001	1.17	4.14
TG(58:6)	0.003	1.17	2.10
TG(58:6)	0.023	0.99	1.70
TG(58:8)	<0.001	1.36	4.07
TG(58:9)	0.001	1.48	4.05
TG(59:2)	<0.001	1.08	2.73

impaired glucose metabolism [17]. Fish oil supplementation was previously found to reduce the total plasma TG concentration by selectively reducing short chain fatty acids and to increase various phospholipids [27]. Thus, it is reasonable to assume that intake of fish and fish oil causes a remodeling of plasma TG species towards more long-chained fatty acids. Our results demonstrate that this remodeling occurs in healthy subjects where the total serum TG level and the BMI are unchanged.

We observed an increase in several phospholipids incorporating n-3 PUFAs, such as lysoPC(20:5) and lysoPC(22:6) in the FO group compared to the HOSO group. An association between n-3 FA intake and changes in lysoPC has previously been described [18,28], and in accordance with our results, lysoPC(20:5) was significantly increased in subjects with impaired glucose metabolism after a healthy diet containing fatty fish [17]. In contrast, fatty fish consumption for eight weeks in subjects with CVD decreased the total concentration of lysoPC [18]. In addition, Block and colleagues found that FO supplementation increased the EPA and

DHA species of lysoPC in healthy individuals [28]. The potential health effect of altering the blood plasma concentration of EPA and DHA lysoPC compounds is uncertain. However, the biological functions of lysoPC compounds are assumed to vary with the degree of saturation and acyl length [28] and LysoPC has been suggested as the major carrier of DHA to brain tissues [29].

Three out of four significantly altered SM lipids were increased in the FO group compared to the HOSO group. SM lipids are the most dominant circulating sphingolipid representing 88% of the total concentration in blood, whereasceramides account for approximately 3% [30]. SM in blood is key components and exists predominantly in the hydrophobic outer layer of lipoprotein particles. Of the lipoprotein particles, the VLDL particle contains the highest amount of SM lipids [31]. However, the localization, distribution and role SM lipid species among the lipoprotein particles is still obscure.

In the present study, FO supplementation was not associated with changes in plasma PA, lysoPE and ceramides, indicating that n-3 PUFA is selectively incorporated into other lipid classes. Ceramides have been associated with inflammation and cardiovascular disease [32,33]. However, high content of specific C_{24} ceramides have been linked to less atherogenic lipoprotein particles in healthy subjects [31]. Lankinen et al. observed that the total concentration of ceramides decreased after fatty fish consumption for eight weeks [18]. The discrepancies observed between these studies may be due to differences in the study population and design, or due to lack of specific bioactive components in fish oil which are normally present in fish.

Lipid profiling has identified a relation between lipid acyl chain structure and risk of disease [22]. The present study shows that fish oil supplement increases the level of lipids such as TG(56:9), TG(58:10), LysoPC(22:6) and PC(38:6). These lipid species were recently associated with decreased risk of diabetes, when lipidome analyses were applied to plasma obtained from participants in the Framingham heart cohort study [22]. In that study a higher carbon number and higher double bond content were associated with decreased risk of diabetes. Thus, long-chain highly unsaturated TGs that have been associated with diabetes risk reduction were increased after intake of fish oil in the present study.

Whether the beneficial effects of fish oil supplementation may be explained by a remodeling of the plasma lipids into TGs and PLs of long-chain PUFAs, needs to be further investigated. However, PUFAs incorporated into TGs and PLs may reach

tissues, cells and lipoproteins by a selective lipid exchange [34]. In the tissues, EPA and DHA can be incorporated into membranes and cause alterations in signaling pathways and the formation of lipid mediators that are important in inflammation [35,36]. In addition, EPA and DHA or their oxidation products have the ability to activate transcription factors both in the liver and in other metabolic active tissues and increase the expression of target genes involved in lipid metabolism and inflammation [37–40]. Altering the lipid composition of lipoprotein particles can also contribute to modulation of the lipoprotein particles [15], including altered spatial distribution of lipids and therefore also alternation of the function [41,42].

In conclusion, fish oil supplementation for three and seven weeks alter the plasma lipidomic profile markedly compared to intake of high-oleic sunflower oil. The selective elevation of TGs and phospholipids of high carbon number and double bond content may represent beneficial effects of fish oil supplementation in healthy subjects. Future studies are needed in order to elucidate the health benefits of incorporation of long-chain PUFAs into selective phospholipids classes and TGs.

Supporting Information

Figure S1   Global Root Mean Square Error plot of Y (RMSE$_Y$). RMSE of Y for the global model is plotted as a function of the number of components. Detailed explanation for the plot is given in [26].

Table S1   Significantly altered lipids (Multivariate analyses, p<0.05) in the fish oil (FO) group compared to the sunflower oil (HOSO) group after three weeks intervention. The corresponding q-values from univariate analyses are also given.

Author Contributions

Conceived and designed the experiments: IO GIB GV KWB SMU KBH MCWM. Performed the experiments: IO GB GV KWB SMU KBH MCWM. Analyzed the data: IO SH GIB AK MCWM. Contributed reagents/materials/analysis tools: IO SH AK GIB TH MO MCWM. Wrote the paper: IO SH AK MCWM. Revised the manuscript critically and final approval: IO SH GIB AK GV TH MO KWB SMU KBH MCWM.

References

1. Skeaff CM, Miller J (2009) Dietary fat and coronary heart disease: summary of evidence from prospective cohort and randomised controlled trials. Ann Nutr Metab 55: 173–201.

2. Kris-Etherton PM, Harris WS, Appel LJ (2003) Fish consumption, fish oil, omega-3 fatty acids, and cardiovascular disease. Arterioscler Thromb Vasc Biol 23: e20–30.

3. Gruppo Italiano per lo Studio della Sopravvivenza nell'Infarto miocardico (1999) Dietary supplementation with n-3 polyunsaturated fatty acids and vitamin E after myocardial infarction: results of the GISSI-Prevenzione trial. Lancet 354: 447–455.

4. Yokoyama M, Origasa H, Matsuzaki M, Matsuzawa Y, Saito Y, et al. (2007) Effects of eicosapentaenoic acid on major coronary events in hypercholesterolaemic patients (JELIS): a randomised open-label, blinded endpoint analysis. Lancet 369: 1090–1098.

5. Hartweg J, Perera R, Montori V, Dinneen S, Neil HA, et al. (2008) Omega-3 polyunsaturated fatty acids (PUFA) for type 2 diabetes mellitus. Cochrane Database Syst Rev: CD003205.

6. Harris WS (1997) n-3 fatty acids and serum lipoproteins: human studies. Am J Clin Nutr 65: 1645S–1654S.

7. Geleijnse JM, Giltay EJ, Grobbee DE, Donders AR, Kok FJ (2002) Blood pressure response to fish oil supplementation: metaregression analysis of randomized trials. J Hypertens 20: 1493–1499.

8. Knapp HR (1997) Dietary fatty acids in human thrombosis and hemostasis. Am J Clin Nutr 65: 1687S–1698S.

9. Hornstra G (2001) Influence of dietary fat type on arterial thrombosis tendency. J Nutr Health Aging 5: 160–166.

10. Christensen JH, Gustenhoff P, Korup E, Aaroe J, Toft E, et al. (1997) n-3 polyunsaturated fatty acids, heart rate variability and ventricular arrhythmias in post-AMI-patients. A clinical controlled trial. Ugeskr Laeger 159: 5525–5529.

11. Nodari S, Metra M, Milesi G, Manerba A, Cesana BM, et al. (2009) The role of n-3 PUFAs in preventing the arrhythmic risk in patients with idiopathic dilated cardiomyopathy. Cardiovasc Drugs Ther 23: 5–15.

12. Gross RW, Han X (2011) Lipidomics at the interface of structure and function in systems biology. Chem Biol 18: 284–291.

13. Seppanen-Laakso T, Oresic M (2009) How to study lipidomes. J Mol Endocrinol 42: 185–190.

14. Quehenberger O, Armando AM, Brown AH, Milne SB, Myers DS, et al. (2010) Lipidomics reveals a remarkable diversity of lipids in human plasma. J Lipid Res 51: 3299–3305.

15. Kontush A, Chapman MJ (2010) Lipidomics as a tool for the study of lipoprotein metabolism. Curr Atheroscler Rep 12: 194–201.

16. Fekete K, Marosvolgyi T, Jakobik V, Decsi T (2009) Methods of assessment of n-3 long-chain polyunsaturated fatty acid status in humans: a systematic review. Am J Clin Nutr 89: 2070S–2084S.

17. Lankinen M, Schwab U, Kolehmainen M, Paananen J, Poutanen K, et al. (2011) Whole grain products, fish and bilberries alter glucose and lipid metabolism in a randomized, controlled trial: the sysdimet study. PLoS ONE 6: e22646.

18. Lankinen M, Schwab U, Erkkila A, Seppanen-Laakso T, Hannila ML, et al. (2009) Fatty fish intake decreases lipids related to inflammation and insulin signaling–a lipidomics approach. PLoS ONE 4: e5258.

19. Harris WS (1989) Fish oils and plasma lipid and lipoprotein metabolism in humans: a critical review. J Lipid Res 30: 785–807.

20. Hodge AM, Simpson JA, Gibson RA, Sinclair AJ, Makrides M, et al. (2007) Plasma phospholipid fatty acid composition as a biomarker of habitual dietary fat intake in an ethnically diverse cohort. Nutr Metab Cardiovasc Dis 17: 415–426.

21. Schwab U, Seppanen-Laakso T, Yetukuri L, Agren J, Kolehmainen M, et al. (2008) Triacylglycerol fatty acid composition in diet-induced weight loss in subjects with abnormal glucose metabolism–the GENOBIN study. PLoS ONE 3: e2630.

22. Rhee EP, Cheng S, Larson MG, Walford GA, Lewis GD, et al. (2011) Lipid profiling identifies a triacylglycerol signature of insulin resistance and improves diabetes prediction in humans. J Clin Invest 121: 1402–1411.

23. Ottestad I, Vogt G, Retterstol K, Myhrstad MC, Haugen JE, et al. (2011) Oxidised fish oil does not influence established markers of oxidative stress in healthy human subjects: a randomised controlled trial. Br J Nutr: doi:10.1017/S0007114511005484

24. Pluskal T, Castillo S, Villar-Briones A, Oresic M (2010) MZmine 2: modular framework for processing, visualizing, and analyzing mass spectrometry-based molecular profile data. BMC Bioinformatics 11: 395.

25. Wangen LE, Kowalski BR (1989) A multiblock partial least squares algorithm for investigating complex chemical systems. Journal of Chemometrics 3: 3–20.

26. Hassani S, Martens H, Qannari M, Hanafi M, Kohler A (2011) Model validation and error estimation in multi-block partial least squares regression. Chemometrics and Intelligent Laboratory Systems 10.1016/j.chemolab.2011.06.001.

27. McCombie G, Browning LM, Titman CM, Song M, Shockcor J, et al. (2009) Omega-3 oil intake during weight loss in obese women results in remodelling of plasma triglyceride and fatty acids. Metabolomics 5: 363–374.

28. Block RC, Duff R, Lawrence P, Kakinami L, Brenna JT, et al. (2010) The effects of EPA, DHA, and aspirin ingestion on plasma lysophospholipids and autotaxin. Prostaglandins Leukot Essent Fatty Acids 82: 87–95.

29. Lagarde M, Bernoud N, Brossard N, Lemaitre-Delaunay D, Thies F, et al. (2001) Lysophosphatidylcholine as a preferred carrier form of docosahexaenoic acid to the brain. J Mol Neurosci 16: 201–204; discussion 215–221.

30. Hammad SM (2011) Blood sphingolipids in homeostasis and pathobiology. Adv Exp Med Biol 721: 57–66.

31. Hammad SM, Pierce JS, Soodavar F, Smith KJ, Al Gadban MM, et al. (2010) Blood sphingolipidomics in healthy humans: impact of sample collection methodology. J Lipid Res 51: 3074–3087.

32. Pfeiffer A, Bottcher A, Orso E, Kapinsky M, Nagy P, et al. (2001) Lipopolysaccharide and ceramide docking to CD14 provokes ligand-specific receptor clustering in rafts. Eur J Immunol 31: 3153–3164.

33. de Mello VD, Lankinen M, Schwab U, Kolehmainen M, Lehto S, et al. (2009) Link between plasma ceramides, inflammation and insulin resistance: association with serum IL-6 concentration in patients with coronary heart disease. Diabetologia 52: 2612–2615.

34. Shearer GC, Savinova OV, Harris WS (2011) Fish oil - How does it reduce plasma triglycerides? Biochim Biophys Acta. doi:10.1016/j.bbalip.2011.10.011

35. Calder PC (2006) n-3 polyunsaturated fatty acids, inflammation, and inflammatory diseases. Am J Clin Nutr 83: 1505S–1519S.

36. Sijben JW, Calder PC (2007) Differential immunomodulation with long-chain n-3 PUFA in health and chronic disease. Proc Nutr Soc 66: 237–259.

37. Jump DB (2008) n-3 polyunsaturated fatty acid regulation of hepatic gene transcription. Curr Opin Lipidol 19: 242–247.

38. Tontonoz P, Spiegelman BM (2008) Fat and beyond: the diverse biology of PPARgamma. Annu Rev Biochem 77: 289–312.

39. Kliewer SA, Sundseth SS, Jones SA, Brown PJ, Wisely GB, et al. (1997) Fatty acids and eicosanoids regulate gene expression through direct interactions with peroxisome proliferator-activated receptors alpha and gamma. Proc Natl Acad Sci USA 94: 4318–4323.

40. Hong C, Tontonoz P (2008) Coordination of inflammation and metabolism by PPAR and LXR nuclear receptors. Curr Opin Genet Dev 18: 461–467.

41. Yetukuri L, Soderlund S, Koivuniemi A, Seppanen-Laakso T, Niemela PS, et al. (2010) Composition and lipid spatial distribution of HDL particles in subjects with low and high HDL-cholesterol. J Lipid Res 51: 2341–2351.

42. Yetukuri L, Huopaniemi I, Koivuniemi A, Maranghi M, Hiukka A, et al. (2011) High density lipoprotein structural changes and drug response in lipidomic profiles following the long-term fenofibrate therapy in the FIELD substudy. PLoS ONE 6: e23589.

Preference Alters Consumptive Effects of Predators: Top-Down Effects of a Native Crab on a System of Native and Introduced Prey

Emily W. Grason[1,2]*, **Benjamin G. Miner**[1,2]

1 Western Washington University, Biology Department, Bellingham, Washington, United States of America, **2** Shannon Point Marine Center, Anacortes, Washington, United States of America

Abstract

Top-down effects of predators in systems depend on the rate at which predators consume prey, and on predator preferences among available prey. In invaded communities, these parameters might be difficult to predict because ecological relationships are typically evolutionarily novel. We examined feeding rates and preferences of a crab native to the Pacific Northwest, *Cancer productus*, among four prey items: two invasive species of oyster drill (the marine whelks *Urosalpinx cinerea* and *Ocenebra inornata*) and two species of oyster (*Crassostrea gigas* and *Ostrea lurida*) that are also consumed by *U. cinerea* and *O. inornata*. This system is also characterized by intraguild predation because crabs are predators of drills and compete with them for prey (oysters). When only the oysters were offered, crabs did not express a preference and consumed approximately 9 juvenile oysters crab^{-1} day^{-1}. We then tested whether crabs preferred adult drills of either *U. cinerea* or *O. inornata*, or juvenile oysters (*C. gigas*). While crabs consumed drills and oysters at approximately the same rate when only one type of prey was offered, they expressed a strong preference for juvenile oysters over drills when they were allowed to choose among the three prey items. This preference for oysters might negate the positive indirect effects that crabs have on oysters by crabs consuming drills (trophic cascade) because crabs have a large negative direct effect on oysters when crabs, oysters, and drills co-occur.

Editor: Louis-Felix Bersier, University of Fribourg, Switzerland

Funding: This work was also funded in part by NSF award OCE-0741372, and a grant from the Washington Sea Grant Program, University of Washington, pursuant to National Oceanic and Atmospheric Administration Award No. NA07OAR4170007. The views expressed in the manuscript are those of the authors and do not necessarily reflect the views of NOAA or any of its sub-agencies. Additional financial support was received from the Western Washington University Office of Sponsored Programs and Biology Departments. The funders provided financials support only, and had no role in study design, data collection and analysis, decision to publish, or preparation of the manuscript.

Competing Interests: The authors have declared that no competing interests exist.

* E-mail: egrason@u.washington.edu

Introduction

Predation is a major force structuring un-invaded [1,2] and invaded [3,4] communities. Directly, predators can limit [5,6], regulate [7], or extirpate [8] prey populations through consumption. These effects can, in turn, be propagated indirectly through the community in many ways, including trophic cascades, indirect facilitation, and apparent competition [9]. Direct and indirect effects of predation can engender quantitative, as well as qualitative, shifts in community composition. For instance, predators can enhance community diversity by consuming competitively dominant prey, thereby reducing interspecific competition and facilitating the persistence of competitively inferior prey [10]. The population and community effects of predators depend on the number of prey they consume, and their preference.

The trophic dynamics of an invaded food web are not always predictable because predator-prey interactions are typically evolutionarily novel. Predators might not be able to recognize an invasive prey as potential food, or overcome defenses of invasive prey [11]. Even where predators are generalists and invasive prey are ecologically similar to natives, an invaded trophic chain might not function in the same way as a comparable chain comprised only of co-evolved natives [12]. Prey preference in the European green crab, *Carcinus maenas*, destabilized competitive interactions between native and non-native clams, allowing the established non-native gem clam, *Gemma gemma*, to become invasive [13]. Appropriate management of invasions therefore requires investigation of the pathways by which predators impact the community.

We investigated a commercially and ecologically important food web with a native predator, mediated by invasive intermediate prey (Figure 1). In Washington State, two invasive species of oyster drill, the marine whelks *Ocenebra inornata* and *Urosalpinx cinerea*, are common predators of oysters. Because drills threaten the recovery efforts of a rare native oyster (*Ostrea lurida*) and yield of commercial oyster harvest, the effects of drills as predators of oysters have been well studied [14–16]. Drills of both species generally prefer to prey on small oysters, and can consume juveniles at a rate of 0.3 oysters per day [16] potentially limiting oyster populations at this life history stage. However, relatively little is known about predatory control that native crabs, which are common predators of marine snails, exert on this system (but see [12]). As predators of both drills and oysters, crabs could impact oysters directly by preying on them, and indirectly by consuming drills and releasing oysters from drill predation (Figure 1). Where all three species coincide, the relative importance of direct and indirect effects of crabs on oysters will be determined by which prey, drills or oysters, the crabs prefer. The drills' habitat overlaps with that of several native decapod crabs, including the red rock crab, *Cancer productus* [17]. A large-clawed generalist predator, *C. productus* preys on native whelks and can strongly influence the community structure of intertidal habitats [18].

This system is also characterized by asymmetric intraguild predation (IGP): crabs are both predators of, and competitors with, drills, but drills do not prey on crabs [19]. Similar systems are common in native intertidal communities where generalist crabs and seastars feed on both predatory snails and bivalve prey of snails (e.g., [20,21]). Additionally, models predict unintuitive dynamics caused by IGP in systems where the intraguild predator is a non-native species [22], and demonstrate that IGP by non-natives can accelerate the speed of invasion [23]. Less is known about IGP in systems where the intraguild prey is a non-native. Therefore, to estimate the potential for both direct and indirect effects of crabs on oysters, and the dynamics of IGP, we investigated crab feeding rates and preferences for invasive drills, and two species of oyster. This information will be critical to predicting the trajectory of this system, and will help enhance our understanding of the role of IGP in invaded communities.

Materials and Methods

In two experiments conducted at Shannon Point Marine Center, in Anacortes, WA, we estimated feeding rates and preferences of the native red rock crab, *Cancer productus*, for two species of juvenile oysters (*Crassostrea gigas*, and *Ostrea lurida*) and their invasive drill predators (*Urosalpinx cinerea*, and *Ocenebra inornata*). We initially tested whether the rates of consumption

and preference of crabs differed for the two oyster species. We then investigated crab feeding on both species of invasive drill and non-native *C. gigas*.

System

Both Atlantic (*U. cinerea*) and Japanese oyster drills (*O. inornata*, synonyms include *Ceratostoma inornatum*, and *Ocinebrellus inornatus*) were unintentionally introduced to the Pacific Northwest by the early 1920's [24]. Both drills arrived as hitchhikers on cultch (newly settled larvae on shell material) of non-native oysters imported to buoy the oyster-growing industry – *U. cinerea* from the east coast of the United States on *Crassostrea virginica* (Eastern oyster), and *O. inornata* from the Asian Pacific Ocean on *Crassostrea gigas* (Pacific oyster). Because they lack a larval planktonic stage, drills are dispersed primarily through human-mediated transport. Notably, native whelks are not considered a pest in oyster beds [24], so we did not include them in our study.

While at least five species of oyster are currently grown and harvested in Washington, we were particularly interested in the effects of crabs on non-native *C. gigas* and native *O. lurida*. The Pacific oyster, *C. gigas*, is the most widely introduced, and commercially valuable oyster species worldwide, and has also established naturally recruiting, wild populations in bays and inland waters of Puget Sound and coastal Washington [25]. In contrast, populations of *O. lurida*, the only native oyster species in Washington, collapsed in the late 1800's due to overharvesting and pollution, and have not sustained a harvestable wild population since [26]. Drills are pests to the oyster industry, costing shellfish growers money in control and eradication efforts, as well as in decreased revenues. Additionally, drills could be inhibiting re-covery efforts aimed at restoring wild populations of the rare and ecologically important native *O. lurida* [16,27].

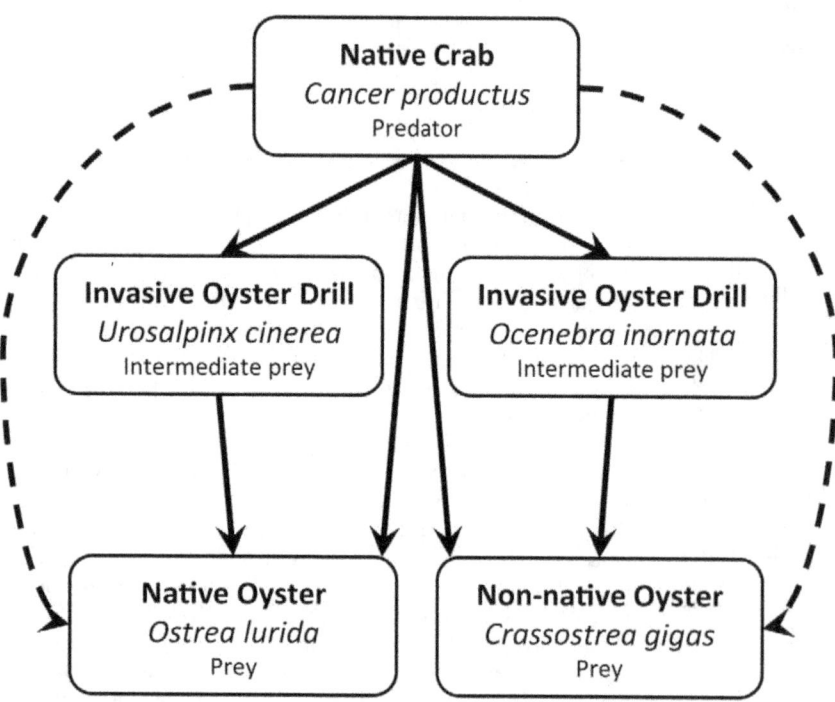

Figure 1. Diagram of trophic interactions in our study system. Diagram of trophic interactions among native red rock crabs, *Cancer productus*, two species of invasive oyster drill, *Urosalpinx cinerea* and *Ocenebra inornata*, and two oyster species, native *Ostrea lurida* and non-native *Crassostrea gigas*. Solid lines represent direct interactions, dashed lines indirect interactions with all arrows pointing in the direction of the trophic effect.

Due to their relatively thin shells, juveniles of both oyster species are most vulnerable to drill predation [16]. Both drills consume small individuals of *C. gigas* and *O. lurida* at similar rates of about 0.25–0.30 oysters per drill per day [16]. Preference for the two species of drills between these two species of oyster has already been established and so was not tested here. Despite divergent evolutionary histories with these oyster prey, both drills prefer *C. gigas* to *O. lurida* oysters of similar size [16]. Population growth of oysters depends on at least some recruits reaching reproductive size without being consumed. Survival of oyster populations, therefore, could be strongly affected by predation on juvenile oysters by adult drills (E. Grason, unpublished data). Buhle and Ruesink [16] hypothesized that crab predation might limit distribution of drills, and therefore their effects on oysters, but there is, as yet, no experimental support for this hypothesis.

Animal Collection and Husbandry

Both male and female red rock crabs, *C. productus*, were collected by hand and trap from beaches and docks around Anacortes, WA. We assumed crabs had no experience with either species of drill or oyster because those organisms or their remains were not found in the areas where crabs were collected. Neither oysters nor invasive drills were present at these sites, likely because oysters have not historically been cultured at these sites. This reduces the probability that crabs had prior experience with any of the species of prey used in our experiments and allows us to infer that preferences are innate. While in the lab, crabs were maintained in flow-through aquaria, on a diet of mussels (*Mytilus* sp.) and frozen fish fillet (*Tilapia* sp.). Crabs were not starved prior to the feeding experiments.

Atlantic drills, *U. cinerea*, were collected from naturally recruiting *C. gigas* reefs in the southeastern corner of Willapa Bay, WA. Japanese drills, *O. inornata*, were collected from commercial *C. gigas* beds owned by Taylor Shellfish Farms in West Samish Bay, WA. Drill species were maintained in separate, closed 140 L aquaria, and allowed to feed *ad libitum* on mussels and *C. gigas* juveniles.

Juvenile oysters of both species were obtained as seed (singles) from Taylor Shellfish Farm hatcheries. At the time of experiments, oysters were of a size at which they could typically be out-planted by commercial or recreational growers (*C. gigas* = 2.7±0.4 cm, *O. lurida* = 2.5±0.2 cm). Oysters were held in sea-tables and had access to a limited amount of plankton that came in with the natural seawater. We supplemented this diet with commercial shellfish diet (Shellfish Diet 1800-Reed Mariculture) at least once a week.

Preference Experiments

For the purpose of this study, preference was defined as a deviation of feeding behavior (proportion of prey consumed of one type) in the presence of choice compared to feeding behavior without choice [28]. Therefore each experiment included treatments in which crabs were offered one prey type only, and one treatment where all prey types were offered simultaneously in equal abundance. This design has the advantage of providing researchers with several relevant estimates of feeding rate and clearly differentiates between preference and electivity (for a discussion of "preference" versus "electivity" see [29]). Electivity can cause diet changes in a predator via "prey switching", which occurs when predators attack different prey depending on relative prey abundance [30]. Such "switching" behavior can be ecologically relevant, but was not tested in this experiment, as we chose to focus on preference as a trait of crabs, rather than how contexts affect the interaction between crabs and the two types of prey.

Two experiments were conducted to estimate preference and feeding rates among the prey items. We first estimated *C. productus* feeding rates on and preferences among juvenile oysters, Pacific (*C. gigas*) and Olympia (*O. lurida*) oysters (September, 2009). Then, we estimated *C. productus* feeding rates on and preferences for juvenile oysters, (*C. gigas*) and both species of drill (*U. cinerea* and *O. inornata*) (October, 2009). Conducting these experiments separately, rather than as a single experiment with all four prey types, allowed for greater replication. In this way we improved resolution in determining crab preference between the two species of oyster because we thought that a general preference for oysters over drills might obscure any difference between the two oyster species. Comparing preference among adult drills and juvenile oysters was ecologically relevant to oyster restoration efforts as well as aquaculture scenarios. Oysters in both experiments were marked with enamel to facilitate correct species identification. Different individual crabs were used in each of the experiments. In the first experiment, crabs (mean carapace width ± SE: 106.7±2.1 mm) were randomly assigned to one of three treatments (n = 12): (1) 25 *C. gigas* only, (2) 25 *O. lurida* only, or (3) 25 *C. gigas* and 25 *O. lurida*. In the second experiment, crabs (mean carapace width ± SE: 107.2±2.2 mm) were randomly assigned to one of four treatments (n = 14): (1) 25 *C. gigas* only, (2) 25 *O. inornata* only, (3) 25 *U. cinerea* only, or (4) 25 *C. gigas*, 25 *O. inornata*, and 25 *U. cinerea*. We used only *C. gigas* in this experiment because there was no preference between oyster species in our first experiment and this species is more readily available and not of conservation concern. Observation confirmed that there was no predation by drills on oysters during this experiment, and all oyster mortality was due to crab predation.

One day prior to each experiment, individual crabs were placed in separate flow-through bins with 10 individuals of their assigned prey types to allow feeding behavior to stabilize [31]. At the start of the experiment, we removed all waste, uneaten food, and shell material from the bins, and added individuals of each prey type appropriate for the treatment of each bin. The number of surviving prey of each type was recorded at 24 hours without replacement. In one replicate, a crab consumed all available prey of one type (all individuals of *C. gigas* consumed by one crab in choice treatment of oyster preference experiment).

Analysis

To determine whether crabs preferred one type of prey, we developed a new method, which is described below. Currently there is debate about how best to statistically test for preference, and the proposed methods all have benefits and drawbacks [28,32–35]. With all available methods, it is difficult to test for preference with more than two species. Because we wanted to test for preference for more than two prey species, we developed an alternative method carefully considering the concerns raised in previous studies [26,30–33].

To test for preference, we used the interaction term of a two-factor ANOVA with prey type and choice both modeled as fixed factors. Prey type had two levels in the first experiment (two species of oyster) and three levels in the second experiment (two species of drill and one species of oyster), and choice had two levels in both experiments (whether crabs had a choice of prey or not). The response variable, the proportion of prey consumed, was calculated as follows. In the no-choice bins, we randomly grouped one replicate from each prey type together and calculated the proportion of prey consumed for each prey type out of the total prey consumed for the group. In the choice treatment, we calculated the proportion of prey consumed for each prey type out of the total prey consumed in a single replicate bin. If the proportion of each prey type consumed was different when crabs were offered a choice, versus when crabs were only offered a single type, the interaction term of a two-factor ANOVA with prey type and choice as fixed factors would be significant. However, because of our calculations of the response variable, the data were not independent – the proportions of each prey type consumed in each replicate where crabs were allowed a choice were, by definition, constrained to total to 1.0. Therefore, use of the F-statistic distribution would calculate an incorrect probability of type 1 error. We therefore generated a null distribution of F-ratios for the interaction between choice and prey type by randomly assigning prey types to bins in both the no choice and choice treatments for each experiment (10,000 iterations). We used the generated null distribution to calculate the probability of a type I error for the observed F-ratio for the interaction between choice and prey type.

To ensure that the results were not particular to a random pairing of bins in the no choice treatment, we randomly re-paired bins in the no choice treatment and re-ran the analysis 1,000 times. Distributions of P values of the interaction between choice and prey type were generated and compared to an alpha value of 0.05 to determine whether crabs preferentially fed on the prey species. All analyses were conducted in R [36].

Results

Oyster Preference Experiment

There was no evidence that crabs preferred one species of juvenile oyster to the other (Figure 2). Crabs consumed a similar number of *C. gigas* as juvenile *O. lurida* when they were offered only one species and denied a choice, but consumed slightly more *C. gigas* than *O. lurida* when offered a choice. However, all of the P values of the interaction between prey type and choice generated in the random pairing of bins were greater than 0.05, suggesting that the interaction term is not significant. Crabs, therefore, do not consume a different proportion of juvenile *C. gigas* and *O. lurida* in the presence and absence of choice.

Pacific Oysters and Drills Preference Experiment

There was evidence that crabs preferred non-native oysters to drills (Figure 3). When offered only one type of prey, crabs did not consume prey types at different rates. An average of 7 juvenile *C. gigas*, 7 *O. inornata*, or 9 *U. cinerea* were consumed per crab per day in the absence of choice. However, in treatments in which crabs were allowed to choose from among the three prey types, oysters were disproportionately preyed on compared to drills–crabs consumed nearly 6 times as many oysters as either species of drill. Approximately 99.9% of the P values of the interaction between factors prey species and choice generated by randomly pairing bins were less than 0.05, indicating that crabs consumed a significantly different proportion of oysters in the presence and absence of choice.

Discussion

Whether or not crabs demonstrated a preference among prey choices depended on the types of prey offered. In the first experiment, when both prey were oysters, the predator expressed no preference for either the native or the invasive oyster. In the second experiment, when both species of drill were offered along with one species of oyster, crabs strongly preferred *C. gigas*, oysters, to either species of drill. This difference is likely not a product of different handling times for each prey species, as all prey types were consumed at relatively similar rates when offered in single species treatments. It is probable that the preference for *C. gigas*

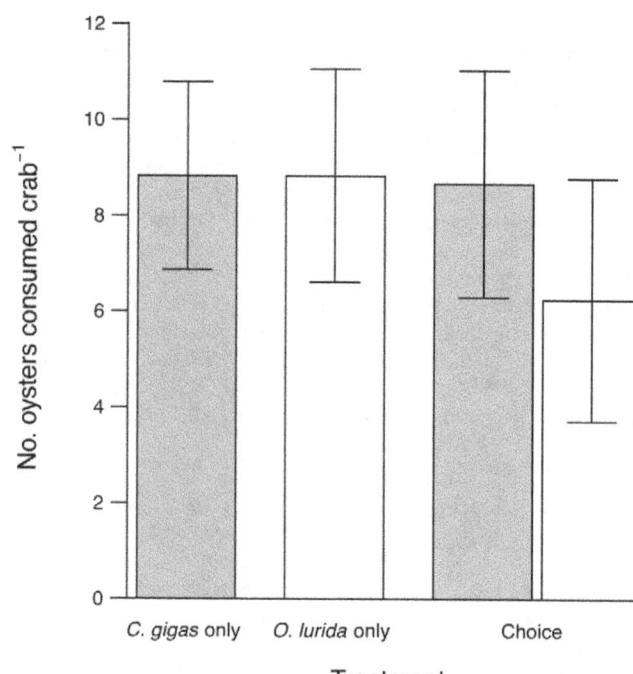

Figure 2. Crab feeding rates and preference on native and non-native oysters. Number (mean ± SE) of oysters consumed by crabs over 24 hours in the oyster preference experiment. Gray bars are *Crassostrea gigas*, Pacific oysters, and white bars are *Ostrea lurida*, Olympia oysters. Crabs were randomly assigned to one of three treatments (n = 12): (1) 25 *C. gigas* only, (2) 25 *O. lurida* only, or (3) the treatment labeled "Choice", 25 *C. gigas* and 25 *O. lurida*.

results from a relatively greater energy yield per unit effort required to obtain the food from oysters as opposed to well-armored drills.

At sites where crabs, drills, and oysters co-occur, crab preference could cause a direct negative effect on juvenile oysters that negates the positive indirect effect of crabs eating drills. In our treatment that allowed crabs to choose, crabs consumed an average of about 6 oysters and 2 drills (one of each species) per day. Individual drills consume at most approximately 0.3 juvenile oysters per day [16]. In the presence of choice, therefore, the direct negative effect of an individual crab on oysters is approximately -6 oysters per day, while the positive indirect effect is +0.6 oysters per day. The net effect of crabs in the system is still strongly negative: 5.4 oysters removed per crab per day. Therefore, despite the fact that they can be highly efficient and motivated predators on drills, *C. productus* is likely to exert stronger direct consumptive effects on oysters. Long-term dynamics in this system will depend on population responses of oysters to predation by both crabs and drills. For instance, predation on oysters could facilitate a population increase in crabs that could then impact drills negatively via apparent competition, or deplete oysters to the extent that crabs switch their search image to prey primarily on drills.

As an intraguild predator, therefore, *C. productus* interacts more strongly with drills as a competitor for oysters than as a predator. It is notable that this crab-drill-oyster system does not have the characteristics of systems in which IGP is believed to be stabilizing. In asymmetric IGP systems, a condition for stability is that the intraguild predator is the stronger competitor for the extraguild resource [37]. In this study, not only is crab-drill predation unidirectional, but it also seems likely, based on differences in per

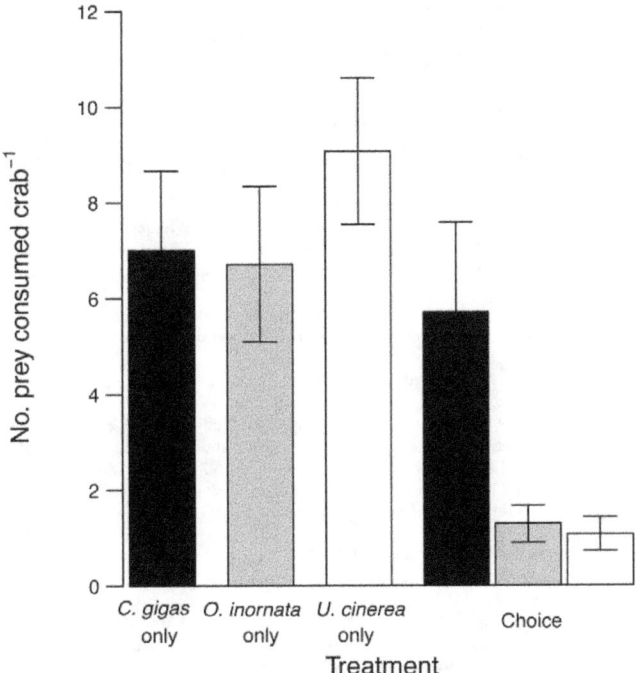

Figure 3. Crab feeding rates and preference on oyster drills and *Crassostrea gigas*. Number (mean ± SE) of prey consumed by crabs over 24 hours in the drill/oyster preference experiment. Black bars are *Crassostrea gigas*, Pacific oysters, gray bars are *Ocenebra inornata*, Japanese drills, and white bars are *Urosalpinx cinerea*, Atlantic drills. Crabs were randomly assigned to one of four treatments (n = 14): (1) 25 *C. gigas* only, (2) 25 *Ocenebra inornata* only, (3) 25 *Urosalpinx cinerea* only, or (4) the treatment labeled "Choice", 25 *C. gigas*, 25 *O. inornata*, and 25 *U. cinerea*.

capita feeding rates, that crabs are better at exploiting oysters than drills. Other researchers have pointed out that models of IGP stability have required systems of closed populations, a condition which is clearly not met in many marine habitats where species often have widely-dispersing pelagic larvae [21]. Thus the population growth rate of oysters in our system might be decoupled from local community and population dynamics. Alternatively, theory predicts that the three species might be able to co-exist exist in habitats where alternative food sources not important in the crab's diet (e.g., barnacles) could sustain drill populations [38].

Our laboratory experiments did not account for spatial and temporal heterogeneity that will affect the strength of these interactions in the field. We purposely eliminated structure in our

mesocosms to prevent drills from using refugia to avoid predation, because both drill species reduce feeding and increase use of refugia when they detect chemicals released by *C. productus* consuming conspecific drills [39]. This design enabled us to better isolate consumptive (predation) from non-consumptive (intimidation) effects of crabs in this system. However, in the field, we would expect that structural complexity could further reduce the rate at which crabs are able to prey on drills, and would also reduce the rate at which feeding drills consume oysters. Therefore, our estimates are likely conservative, as such non-consumptive effects would only increase the relative importance of the direct effects of crab predation on oysters.

Notwithstanding the evidence presented here that crabs are unlikely to reduce drill densities in oyster beds, other researchers have noticed that in Willapa Bay, WA, there is an overall negative correlation between abundance of crabs and oyster drills [17]. This might be explained by the fact that crabs do still prey on drills at low levels, even when oysters are present. Additionally, along with emigration and predation, inducible defenses of drills in response to crabs likely carry fitness costs that explain lower drill densities where *C. productus* is present.

Our study suggests the interesting possibility that oysters facilitated the invasion of both species of drill, not only as a vector (non-native oysters) and food source (both native and non-native oysters), but also by reducing the potential for biotic resistance by native crabs. Both drills were originally introduced simultaneously with oysters, and, notably, are almost entirely restricted to oyster beds in Washington State. This granted drills a degree of enemy release, at least in the short term, because crabs preferentially prey on oysters when they have a choice. The corollary to this idea is that while *C. productus* might not strongly affect drill populations in oyster beds, crabs could help limit the range of invasive drills to oyster beds. Where oysters are rare, it is possible that crabs will switch to consuming relatively more drills, and crabs could thereby provide greater biotic resistance against drill incursion into these habitats. This provides one way that context-dependent species interactions, such as those mediated by preference, could be particularly important in invaded systems.

Acknowledgments

We are very grateful for logistical support for this research contributed by Taylor Shellfish Farms and the Pacific Northwest Shell Club. Helpful feedback on manuscripts was provided by B. Bingham and M. Peterson.

Author Contributions

Conceived and designed the experiments: EG BM. Performed the experiments: EG. Analyzed the data: BM. Contributed reagents/materials/analysis tools: EG. Wrote the paper: EG.

References

1. Paine RT (1974) Intertidal community structure - experimental studies on relationship between a dominant competitor and its principal predator. Oecologia 15: 93–120.

2. Hixon MA, Beets JP (1993) Predation, prey refuges, and the structure of coral-reef fish assemblages. Ecological Monographs 63: 77–101.

3. Salo P, Korpimaki E, Banks PB, Nordstrom M, Dickman CR (2007) Alien predators are more dangerous than native predators to prey populations. Proceedings of the Royal Society B-Biologial Sciences 274: 1237–1243.

4. Gherardi F, Acquistapace P (2007) Invasive crayfish in Europe: the impact of *Procambarus clarkii* on the littoral community of a Mediterranean lake. Freshwater Biology 52: 1249–1259.

5. Fitzner RE, Eberhardt LE, Rickard WH, Gray RH (1994) Great Basin Canada goose nesting on the mid-Columbia River, Washington: An historical perspective and update, 1981–1990. Northwest Science 68: 37–42.

6. Gurevitch J, Morrison JA, Hedges LV (2000) The interaction between competition and predation: A meta-analysis of field experiments. American Naturalist 155: 435–453.

7. Korpimaki E, Krebs CJ (1996) Predation and population cycles of small mammals - A reassessment of the predation hypothesis. Bioscience 46: 754–764.

8. Kavanagh RP (1988) The impact of predation by the Powerful owl, *Ninox-strenua*, on a population of the Greater glider, *Petauroides-volans*. Australian Journal of Ecology 13: 445–450.

9. Wootton JT (1994) The nature and consequences of indirect effects in ecological communities. Annual Review of Ecology and Systematics 25: 443–466.

10. Paine RT (1966) Food web complexity and species diversity. The American Naturalist 100: 65–75.

11. Carlsson NOL, Sarnelle O, Strayer DL (2009) Native predators and exotic prey - an acquired taste? Frontiers in Ecology and the Environment 7: 525–532.

12. Kimbro DL, Grosholz ED, Baukus AJ, Nesbitt NJ, Travis NM, et al. (2009) Invasive species cause large-scale loss of native California oyster habitat by disrupting trophic cascades. Oecologia 160: 563–575.

13. Grosholz ED (2005) Recent biological invasion may hasten invasional meltdown by accelerating historical introductions. PNAS 102: 1088–1091.

14. Hanks JE (1957) The rate of feeding of the common oyster drill, *Urosalpinx cinerea* (Say), at controlled water temperatures. Biological Bulletin 112: 330–335.

15. Chew KK (1960) Study of food preference and rate of feeding of Japanese Oyster Drill, *Ocinebra japonica* (Dunker). Washington, D.C.: United States Fish and Wildlife Service.

16. Buhle ER, Ruesink JL (2009) Impacts of invasive oyster drills on Olympia oyster (*Ostrea lurida* Carpenter 1864) recovery in Willapa Bay, Washington, United States. Journal of Shellfish Research 28: 87–96.

17. Holsman KK, McDonald PS, Armstrong DA (2006) Intertidal migration and habitat use by subadult Dungeness crab *Cancer magister* in a NE Pacific estuary. Marine Ecology Progress Series 308: 183–195.

18. Yamada SB, Boulding EG (1996) The role of highly mobile crab predators in the intertidal zonation of their gastropod prey. Journal of Experimental Marine Biology and Ecology 204: 59–83.

19. Polis G, Myers C, Holt R (1989) The Ecology and Evolution of Intraguild Predation - Potential Competitors That Eat Each Other. Annual Review of Ecoloty and Systematics 20: 297–330.

20. Menge B, Lubchenco J, Gaines S, Ashkenas L (1986) A Test of the Menge-Sutherland Model of Community Organization in a Tropical Rocky Intertidal Food Web. Oecologia 71: 75–89.

21. Navarrete SA, Menge BA, Daley BA (2000) Species interactions in intertidal food webs: Prey or predation regulation of intermediate predators? Ecology 81: 2264–2277.

22. Courchamp F, Langlais M, Sugihara G (1999) Cats protecting birds: modelling the mesopredator release effect. Journal of Animal Ecology 68: 282–292.

23. Hall RJ (2011) Eating the competition speeds up invasions. Biology Letters 7: 307–311.

24. Chapman WM, Banner AH (1949) Contributions to the life history of the Japanese oyster drill, *Tritonalia japonica* (= *Ceratostoma inornatum*), with notes on other enemies of the Olympia oyster, *Ostrea lurida*. Washington Department of Fisheries.

25. Ruesink JL, Lenihan HS, Trimble AC, Heiman KW, Micheli F, et al. (2005) Introduction of non-native oysters: Ecosystem effects and restoration implications. Annual Review of Ecology Evolution and Systematics 36: 643–689.

26. White J, Ruesink JL, Trimble AC (2009) The nearly forgotten oyster: *Ostrea lurida* Carpenter 1864 (Olympia Oyster) history and management in Washington State. Journal of Shellfish Research 28: 43–49.

27. Cook AE, Shaffer JA, Dumbauld BR, Kauffman BE (2000) A plan for rebuilding stocks of Olympia oysters (*Ostreola conchaphila*, Carpenter 1857) in Washington State. Journal of Shellfish Research 19: 409–412.

28. Underwood AJ, Chapman MG, Crowe TP (2004) Identifying and understanding ecological preferences for habitat or prey. Journal of Experimental Marine Biology and Ecology 300: 161–187.

29. Singer MC (2000) Reducing ambiguity in describing plant-insect interactions: "preference", "acceptability" and "electivity." Ecology Letters 3: 159–162.

30. Murdoch W (1969) Switching in General Predators. Experiments on Predator Specificity and Stability of Prey Populations. Ecological Monographs 39: 335–354.

31. Jackson AC, Underwood AJ (2007) Application of new techniques for the accurate analysis of choice of prey. Journal of Experimental Marine Biology and Ecology 341: 1–9.

32. Manly BFJ (1993) Comments on design and analysis of multiple-choice feeding-preference experiments. Oecologia 93: 149–152.

33. Underwood AJ, Clarke KR (2005) Solving some statistical problems in analyses of experiments on choices of food and on associations with habitat. Journal of Experimental Marine Biology and Ecology 318: 227–237.

34. Taplin RH (2007) Experimental design and analysis to investigate predator preferences for prey. Journal of Experimental Marine Biology and Ecology 344: 116–122.

35. Underwood AJ, Clarke KR (2007) More response on a proposed method for analysing experiments on food choice. Journal of Experimental Marine Biology and Ecology 344: 113–115.

36. R Development Core Team (2009) R: A Language and Environment for Statistical Computing. Vienna, Austria: R Foundation for Statistical Computing. Available:http://www.R-project.org.

37. Holt RD, Polis GA (1997) A theoretical framework for intraguild predation. Am Nat 149: 745–764.

38. Holt RD, Huxel GR (2007) Alternative prey and the dynamics of intraguild predation: Theoretical perspectives. Ecology 88: 2706–2712.

39. Grason EW, Miner BG (2012) Behavioral plasticity in an invaded system: non-native whelks recognize risk from native crabs. Oecologia 169: 105–115.

5

Optimisation of Mesh Enclosures for Nursery Rearing of Juvenile Sea Cucumbers

Steven W. Purcell[1,2]*, **Natacha S. Agudo**[2]

1 National Marine Science Centre, Southern Cross University, Coffs Harbour, New South Wales, Australia, **2** Natural Resources Management, The WorldFish Center, Penang, Malaysia

Abstract

Mariculture of tropical sea cucumbers is promising, but the nursery rearing of juveniles is a bottleneck for farming and sea ranching. We conducted four medium-scale experiments lasting 3–6 weeks, using thousands of cultured juvenile sandfish *Holothuria scabra,* to optimise nursery rearing in mesh enclosures in earthen seawater ponds and to test rearing in enclosures in the sea. In one experiment, survival in fine-mesh enclosures (1 m³; 660-μm mesh) related nonlinearly to juvenile size, revealing a threshold body length of 5–8 mm for initial transfer from hatchery tanks. Survival in enclosures within ponds in the other experiments ranged from 78–97%, and differences in growth rates among experiments were explained largely by seasonal differences in seawater temperatures in ponds. Stripped shadecloth units within fine-mesh enclosures increased feeding surfaces and improved growth rates by >15%. On the other hand, shading over the enclosures may lower growth rates. Following the rearing in fine-mesh enclosures, small juveniles (0.5 to 1 g) were grown to stocking size (3–10 g) in coarse-mesh enclosures of 1-mm mesh. Sand or mud added to coarse-mesh enclosures did not significantly improve growth compared to controls without sediment. Survival of sandfish juveniles in coarse-mesh enclosures set on the benthos within seagrass beds differed between two sheltered bays and growth was slow compared to groups within the same type of enclosures in an earthen pond. Our findings should lead to significant improvement in the cost-effectiveness of rearing sandfish juveniles to a stocking size compared to established methods and highlight the need for further research into nursery systems in the sea.

Editor: Senjie Lin, University of Connecticut, United States of America

Funding: This study was funded by the Australian Centre for International Agricultural Research (ACIAR), through project FIS/1999/025, and the three Provinces of New Caledonia. The funders had no role in study design, data collection and analysis, decision to publish, or preparation of the manuscript.

Competing Interests: The authors have declared that no competing interests exist.

* E-mail: steven.w.purcell@gmail.com

Introduction

Sea cucumbers (Echinodermata, Holothuroidea) have attracted global interest for mariculture and multi-trophic culture systems because of their high market value and ability to thrive on the waste products of fish and shellfish [1–3]. They are now cultured at commercial scales in China, Japan, Maldives, Madagascar, Australia, New Caledonia, Palau, Mexico and Vietnam [4]. One temperate species, *Apostichopus japonicus,* is cultured in billions while four tropical species have been produced in hundreds of thousands [4]. Sea cucumbers are firstly cultured in tanks and the newly settled juveniles grow on diatom-covered plates [5–7]. They must then be grown in nursery systems to a larger (i.e. fingerling or 'sluglet') size for stocking into earthen seawater ponds or into the sea for stock restoration, sea ranching, or farming in sea pens or cages. For tropical species like the sandfish (*Holothuria scabra*), this means to a body weight of 1–20 g for stocking into ponds [8–10], or to 3–15 g for culture at sea in pens or for sea ranching [11,12]. However, a lack of published technology on nursery systems has hindered cost-effective commercial production of juveniles [4].

Sandfish can attract up to US$6 per kg fresh weight (gutted) but their growth to a large commercial size, e.g. 0.7–1 kg, would take at least 2 years in earthen ponds [4,8,13] or 3+ years in coastal seagrass beds [11,12]. Survival of sandfish from released juveniles (3–15 g) to minimum market size (300–500 g) in sea pens or in a

sea ranch appears to be generally in the range of 10–30% [4,11,12]. To compensate for long grow-out cycles and potentially low survival, sea cucumber juveniles must be produced cheaply to a competent size for stocking [4,14]. Culture systems for sea cucumbers are relatively labour intensive, especially in the hatchery [7,15]. Mariculture profitability is therefore predicated on systems that allow juvenile sandfish to be transferred, at an early stage, out of hatchery tanks and into nursery units where they survive well and grow to a size for stocking in a short timeframe [4].

Battaglene [6] and Pitt [16] described the early technology of culturing sandfish larvae and rearing the newly-settled juveniles on stacked plates, or on sand, within tanks. However, competition for space becomes the most significant constraint for grow-out of juveniles to a stocking size, especially where surface area of hatchery tanks is limited. Battaglene et al. [17] found in Solomon Islands that growth of sandfish juveniles is impeded once densities reach 225 g m^{-2} of tank floor space; a threshold corroborated in Vietnam (R. Pitt, pers. comm.) and New Caledonia [7]. Earthen ponds offer expansive nursery area and a natural source of detritus as food but juvenile sandfish survive poorly when placed directly on pond substrates at a small size [18].

Mesh enclosures in the sea or in earthen ponds are one solution for scaling up production of juvenile sandfish because the costs of mesh and water exchange are relatively low [4,19]. Pitt and Duy

[18] conducted the first pioneering work of growing sandfish juveniles in mesh enclosures within earthen ponds. This nursery system involved two steps: small juveniles of a few mm body length from hatchery tanks are grown in fine-mesh enclosures (mesh size ~450 μm) to about 20 mm body length (almost 1 g body weight), then transferred to coarse-mesh enclosures of 1-mm mesh and grown further to a competent size for stocking into ponds or into the sea.

In the present study, we sought to develop optimal methods for scaling up the production of sandfish juveniles in mesh enclosures within earthen ponds. We used replicated experiments with cultured sandfish juveniles to (1) determine the effects of shading and increased surface area of mesh in fine-mesh enclosures on growth and survival, (2) identify the minimum size at which small juveniles could be transferred from hatchery tanks into fine-mesh enclosures in ponds, (3) compare growth among substrate types in coarse-mesh enclosures, and (4) assess the viability of sea-based mesh enclosures as a nursery system. Our findings provide new technology, which should help to improve cost effectiveness of rearing juvenile sandfish for land-based mariculture, sea farming, sea ranching and restocking. The findings and experimental approach should be valuable for optimising mariculture of other tropical sea cucumbers.

Materials and Methods

Production of Juveniles

Sandfish juveniles for all experiments were produced at a hatchery at Saint Vincent (21.926°S, 166.083°E) near Boulouparis, New Caledonia, using established methods [6,7,16]. Induced spawnings during summer months comprised >10 male and female broodstock, hence the juveniles were of mixed parentage. Sea cucumber larvae were reared in 1000-l and 2000-l cylindrical conical-bottom tanks where they settled on the tank walls and on coated fibreglass plates after about 10 rearing days. After 4 weeks, newly-settled juveniles were transferred to 7000-l indoor (90% shade greenhouse) raceway tanks where they were fed daily with dried feeds [7]. Different batches of juveniles from 1–20 mm in length were used for experiments in fine-mesh enclosures and coarse-mesh enclosures (also called 'hapas' and 'bag nets', respectively, *sensu* [18]). We used some newly settled animals directly from the larval tanks for experiments requiring juveniles <10 mm in length.

General Procedures

We used two types of supple mesh enclosures: $1 \times 1 \times 1$ m fine-mesh enclosures made of 660 μm transparent nylon mesh, and $1 \times 1 \times 1$ m coarse-mesh enclosures made of 1 mm black Tentex™ mesh (a heavy gauge, plastic-coated mosquito netting). Each enclosure had a mesh floor but no cover (i.e. no top mesh), and presented a sleeve on the upper edge of the side walls with tie-off cords at 1-m intervals. These were set up in earthen seawater ponds of 0.13–1.14 ha next to the hatchery, and held upright by tying to metal posts externally. The ponds we used were owned by the site owner, IFREMER, who gave permission for the facilities. Seawater in ponds was 0.7 to 1.0 m deep and renewed by constant input from a reservoir. Shrimp (*Litopenaeus stylirostris*) were regularly cultured in these ponds for separate studies and some of the present experiments were conducted in a pond with shrimp at low stocking densities of ~5 ind m^{-2}. The ponds therefore had background levels of organic waste from previous shrimp culture.

All mesh enclosures were held approximately 10 cm above the sandy-mud substratum in ponds to mitigate low-oxygen and high-temperature conditions at the sediment interface, which could occur after water stratification from heavy rains [7]. Mesh enclosures in all experiments were placed in haphazard orientation, approximated 1–2 m apart. The upper rim of mesh enclosures was ~30 cm above the water surface, excluding shrimp from interacting directly with sea cucumbers.

Prior to each experiment, the mesh enclosures were conditioned in the pond for 3–7 days to acquire a natural biofilm and detritus. Although the sea cucumbers fed on the detritus, excessive fouling was routinely brushed off the upper 50 cm of the inside and outside mesh of the enclosures to allow water exchange.

Sea cucumber juveniles from hatchery tanks and raceways were distributed evenly, according to body sizes, into replicate groups held temporarily in buckets unless stated otherwise. We believed that the smaller juveniles (i.e. <10 cm long) for experiments using fine-mesh enclosures would be injured by blotting and weighing out of seawater because these were known to be less robust than larger juveniles [18]. Therefore, we used body length as a measure of initial animal size for the first two experiments with fine-mesh enclosures. Body length was a satisfactory measure of body size for juvenile sandfish since these measurements explained 91% of the variation in weights among individuals (see *Results*). For the juveniles in the fine-mesh enclosures, average initial body lengths of a random sample ($n = 20$) of individuals were measured to ±1 mm using a ruler set under a Petri dish with seawater in which the animals were held. As the animals were disturbed, the measurements are contracted body length. For small juveniles (<1 g in weight), weights of individuals for examining the relationship between body weight and length were measured on a high-precision electronic balance to 0.001 g (Mettler Toledo, Model AG204).

The larger juveniles used for the last two experiments (*Effects of sediment type and sediment depth*; *Enclosures at sea vs pond*) were firstly held together in a bare tank overnight to defecate sediments and detritus. Those groups were then drained and blotted dry on a damp cloth for 30 sec, weighed as a group to ±0.1 g with an electronic balance (Masscal Model NJW-3000), and assigned randomly to coarse-mesh enclosures. For all experiments, we acclimatised juveniles to pond conditions by gradually adding pond water to the buckets over 10 min before placing them into the enclosures. We randomised replicates among experimental units using random number tables or pulling numbered pieces of paper from a hat.

Water parameters within mesh enclosures were measured every 2–7 days at 5 cm above the bottom mesh surface so that the measurements were close to the enclosure floor where sea cucumbers resided most but not corrupted by precipitated detritus or faeces. We measured dissolved oxygen (to ±0.1 mg L^{-1}) and temperature (to ±0.1°C) with a YSI™ handheld D.O. meter and salinity (to ±1 ppt) with an Eclipse refractometer.

At the end of experiments, mesh enclosures were removed from the ponds and the contents transferred through a sieve in seawater to collect the remaining juvenile sea cucumbers. The sea cucumbers were counted twice and weighed, as described, as groups to ±0.1 g and/or individual body lengths measured with a ruler to 1 mm; lengths in all cases are contracted body length. The number of juveniles counted in each group was used to calculate proportionate survival and final average body weight. Additional measurements are described later. In the last experiment, where the juveniles had been used in a previous experiment, they were combined into a bare tank, left overnight, then re-randomised into new groups, as described.

Effects of Increasing Mesh Surfaces and Shading

We examined the effects of shade and extra surface area of mesh within fine-mesh enclosures on growth and survival of small juveniles. Sandfish larvae apparently settle onto seagrass blades in the wild and the small juveniles feed on epiphytic matter on the surfaces of the blades until about 6–9 mm in length [20,21]. In addition, lower light levels have been shown to extend foraging periods of juveniles [22,23]. We constructed 'stripped shadecloth' units using a 3.5 m×0.7 m band of woven coated polyethylene fabric ('shadecloth'), weighted to the floor of enclosures with sand-filled polypropylene tubes (Fig. 1). Strips of ~10 cm were cut across ~80% of the width of the shadecloth and made buoyant at the upper edge with foam, such that the strips mimicked the erect structure of seagrass when set in the enclosures. The units were laid in a zig-zag fashion within the enclosures and acted to increase the surface area that the small juveniles could feed upon. Within fine-mesh enclosures, the units provided 60% more surface area on which detritus could accumulate and sea cucumber juveniles could climb.

During summer, 24 fine-mesh enclosures were set in one pond ('pond H') then assigned randomly to combinations of two orthogonal factors: shade (a single layer of 70% shadecloth on top of fine-mesh enclosures; two levels: with and without) and stripped shadecloth (two levels: with and without), giving six replicates per combination (Fig. 2). We conditioned one stripped shadecloth unit in each of 12 fine-mesh enclosures for 3 days prior to the experiment. Recently-settled juvenile sea cucumbers for this experiment had not been used in previous experiments. Initial body lengths of the juveniles ranged from 4 to 13 mm, and the average was 6 mm (±2 mm s.d.) based on measurement of a single random sample ($n=20$) of individuals, as described. This average body length converts (see *Results*) to approximately 0.05 g in weight. We then prepared 24 groups of 200 juveniles, i.e., 4,800 juveniles total. We distributed the larger and smaller individuals as evenly as possible among the 24 groups before randomly allocating and transferring the groups to the replicate fine-mesh enclosures. We did not weigh each group at the start of this experiment, for reasons explained earlier. During this 3-week experiment, juveniles fed only on the natural detritus and fouling in the fine-mesh enclosures. Daily measurements of daytime water temperature averaged 31.6°C, range: 30–33°C, and dissolved oxygen averaged 5.2 mg l^{-1}. At the end of the 3 weeks, the number of juveniles surviving in each group was counted and each group was weighed, as described.

Juvenile Size for Transfer to Fine-mesh Enclosures

We aimed to determine the minimum size at which small sea cucumber juveniles could be transferred from hatchery tanks to fine-mesh enclosures in ponds, based on their survival across groups of varying starting size classes. During summer, 20 fine-

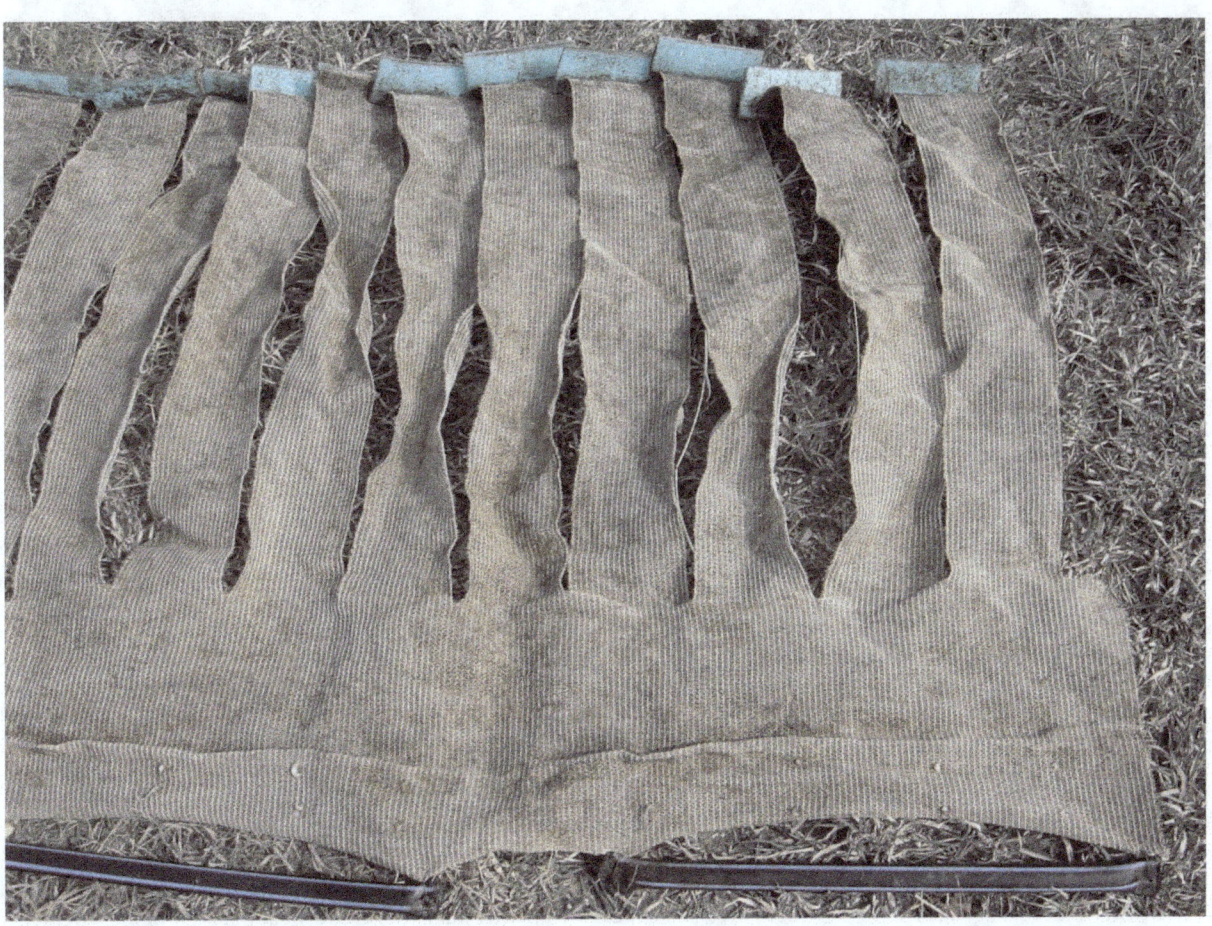

Figure 1. Stripped shadecloth unit. Photograph of the end of a stripped shadecloth unit used to increase the surface area of mesh inside fine-mesh enclosures. The sand-filled polypropylene tubes at the bottom of the photograph weigh down the woven polyethylene fabric while the foam pieces at the top kept the strips upright in the water.

Figure 2. Fine-mesh enclosures. Photograph of 1-m³ fine-mesh enclosures (660 μm mesh) in the earthen pond during the first experiment (Effects of increasing mesh surfaces and shading), showing some replicates shaded with dark shadecloth and some with stripped shadecloth inside.

mesh enclosures were set up in one pond ('pond H'). Based on the first experiment (*Effects of increasing mesh surfaces and shading*), each fine-mesh enclosure contained a 3.5 m×0.7 m stripped-shadecloth unit but no cover (top mesh) for shade.

These were different juveniles from the experiment on *Effects of increasing mesh surfaces and shading* and had not been used in other experiments. Recently settled and small juvenile sea cucumbers averaging 1.3–18.2 mm in length were placed into 20 groups of 200 individuals of similar length; i.e. each group represented a different length class. Random samples of 15 individuals from each replicate group were transferred to Petri dishes with seawater temporarily for measuring body lengths, as described. The average precision (SE/mean) of initial body lengths within groups was 6%. We did not weigh each group at the start of the experiment, as explained earlier.

The 20 groups of juveniles were allocated randomly to the 20 fine-mesh enclosures and transferred into them. Again, feed was not added to the fine-mesh enclosures and the juveniles fed solely on natural detritus fouling. After 3 weeks, juveniles were collected, taking care to check the stripped shadecloth units. Juveniles surviving in each group were counted and the groups were weighed, as described. We also measured the lengths of 15 randomly chosen sea cucumbers from each group. Daytime water

temperatures averaged 29.6°C and dissolved oxygen in the fine-mesh enclosures was 6 mg l⁻¹ throughout the experiment.

Effects of Sediment Type and Sediment Depth

This two-factor experiment during winter examined differences in growth and survival of larger juveniles in coarse-mesh enclosures (Fig. 3) among treatments with different quantities of added sediment, either sand or mud. Into one pond ('pond G1'), we set 24 coarse-mesh enclosures of 1 m² (floor area) directly onto the pond floor in haphazard orientations. Treatment combinations of sediment type (two levels: river *sand*, averaging 325 μm grain size and conditioned with Algamac 2000™; and *mud*, collected from the pond floor) and sediment depth (three levels: 1, 3, and 5 mm, distributed by volume) were then allocated randomly to coarse-mesh enclosures (*n* = 3), leaving six coarse-mesh enclosures without sediment as controls.

We used juveniles from the hatchery that had not been used in other experiments. One week after the coarse-mesh enclosures were set up with their respective treatments, juveniles averaging 0.56 g in weight (range: 0.3–1.1 g) were allocated into 24 groups of 50 individuals. We took care to homogenise size distributions among groups before the random allocation to experimental units and the range in group weights was 23–32 g. We weighed each

Figure 3. Coarse-mesh enclosures. Photograph of some 2 m×2 m×1-m-high coarse-mesh enclosures (1-mm mesh) in an earthen pond.

group (of 50 animals) separately to ±0.1 g, as described earlier, and used initial group weight as a covariate to test potential bias of differences in initial groups' weights on growth performance. The groups were weighed, allocated randomly to the coarse-mesh enclosures and juveniles transferred into each, as described. No food was added to the coarse-mesh enclosures. We allowed the experiment to run for 6 weeks before collecting, counting and weighing the juveniles in groups, as described.

Enclosures at Sea vs. Pond

In autumn, we aimed to compare growth and survival of juvenile sea cucumbers over 5 weeks in coarse-mesh enclosures in an earthen pond with those placed in the sea within sheltered bays. We firstly conditioned 4 replicate coarse-mesh enclosures of 1 m² (floor area) for two days in an earthen pond ('pond G2') and in each of two shallow subtidal (~0.3 m depth at zero tidal datum) seagrass beds at Ouano (21.845°S, 165.812°E) and at Baie de Saint-Vincent (21.925°S, 166.080°E), New Caledonia. The corners of the coarse-mesh enclosures were staked to the natural substratum. Due to a 1–1.5 m tidal flux, we gathered and bound the upper part of coarse-mesh enclosures in bays to exclude entry of predators and suspended that part in the water by an attached float.

Small individuals from the *Juvenile size for transfer to fine-mesh enclosures* experiment were used for this experiment after 8 days of being mixed together and fed in a raceway tank to recover and one night together in a bare tank to expel sediments. A total of 1,200 juveniles averaging 0.38 g body weight, ranging from 0.15 to 1 g (equivalent to 10 to 26 mm in length), were distributed evenly by body weight into groups of 100 individuals. We individually weighed 10 individuals sampled randomly from each replicate group at the start of the experiment. Each replicate group of 100 juveniles was weighed separately, prior to randomisation among replicates and treatments, as described, showing a range in group

weights from 36–39 g. We then randomly assigned groups to the replicates and placed them into the coarse-mesh enclosures on the same day. No sediment or food was added to the coarse-mesh enclosures, so juveniles only had the naturally accumulated detritus upon which to feed. The upper 50 cm of mesh on all coarse-mesh enclosures were brushed cleaned once, half-way through the experiment to remove excess detritus fouling on the mesh walls to promote water exchange in the enclosures. At the completion, the coarse-mesh enclosures were retrieved and the juveniles surviving in each group were counted and groups were weighed, as described.

Statistical Analyses

We used Levene's test to examine homogeneity of variances of data from the experiments, and transformed data as appropriate (specified below). For the experiment on *Effects of increasing mesh surfaces and shading*, we compared proportionate survival between levels in treatments (with or without shade; with or without stripped shadecloth; fixed factors) using a two-way ANOVA on arcsine $\sqrt{y}$ transformed data. Differences in final mean individual weight, water temperature and oxygen concentration between treatment levels were tested using two-way ANOVA tests. Mean final individual weight per group (i.e. total final group weight divided by total number surviving, for each group) and survival per group were used as the data for analyses. Heteroscedastic data ($P=0.012$) of final mean individual weight should not be problematic since the experiment was balanced with sufficient replication [24].

For the experiment on *Juvenile size for transfer to fine mesh enclosures*, we examined the relationship between survival and the juvenile size (body length) for transfer of sea cucumbers to fine-mesh enclosures using nonlinear regression modelling. Mean initial body length of individuals per group (i.e. averages of the measurements of 15 sampled animals from each of the 20 groups) and the survival per group were used as the data for analyses ($n = 20$). We fitted the data to 298 standard functions using DataFit 8.0 software (Oakdale Engineering) and used Akaike's Information Criterion (AIC) to determine the most appropriate function [25].

For the experiment on *Effects of sediment type and sediment depth*, two-way orthogonal analysis of covariance (ANCOVA) tests were used to compare growth (mean individual weight gain) and survival among substrate type and substrate depths using initial mean body weight within each group as the covariate. Mean weight gain of animals (i.e. total group weight divided by total number surviving, minus the initial mean individual starting weight for each group) and survival per experimental unit (enclosure) were used as the data for analyses ($n = 3$; $N = 18$). Based on non-significant difference in mean growth of juveniles among substrate depths, we then pooled data among substrate depths within both substrate types ($n = 9$) in order to compare mean individual weight gain of juveniles with that of control groups ($n = 6$) using a one-way ANCOVA, with the same covariate. ANCOVA is essentially an ANOVA on data adjusted by the regression slope of the response variable on the covariate (e.g., measurements of an initial condition of replicates) and provides a test of the covariate's effect on the response [26]. In this study, the ANCOVA tests accounted for the potential effect of variations in initial mean animal weight among groups on their subsequent survival and average growth.

For the last experiment on *Enclosures at sea vs pond*, we used one-way ANCOVA tests to compare growth and survival of juvenile sea cucumbers between locations at sea and in a pond, using initial mean body weight as a covariate. Mean weight gain of animals (i.e. total final group weight divided by total numbers surviving, minus the initial mean individual starting weight for each group) and survival per experimental unit (enclosure) were used as the data for analyses ($n = 4$; $N = 12$). For those tests, we used untransformed data since errors were unconstrained [24] and data were homoscedastic.

Results

Length and Weight Measurements

Precise measurements (lengths ± 1 mm and weights ± 0.01 g) of 303 juveniles, ranging from 6 to 70 mm in body length, from various experiments provided growth functions for converting measurements of length to weight and vice versa:

$$w = 0.000614 * l^{2.407}$$

$$l = 22.826 * w^{0.370}$$

where: l is length in mm and w is weight in g.

Body length provided a reasonable prediction ($F_{1, 301} = 2877$, $P < 0.001$, $r^2 = 0.91$) of the drained body weight of these hatchery-produced juveniles (Fig. 4). The scaling coefficient of 2.407 shows negative allometric growth in sandfish; i.e. the juveniles become more slender (weigh relatively less for their length) as they grow and increase in body length.

Effects of Increasing Mesh Surfaces and Shading

Survival averaged 97% and was similar between levels of both main effects and the interaction (F-values $= 0.51$–2.89, P-values $= 0.11$–0.48). Since this experiment was in mid-summer, growth of the juveniles was rapid for their body size. Final mean individual weights among treatments averaged 1.6 to 2.0 g, which

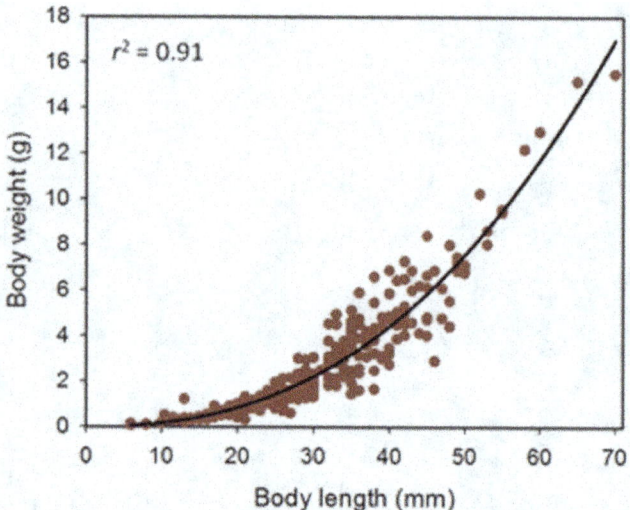

Figure 4. Body length vs body weight of juveniles sandfish. Scatterplot shows relationship (growth curve) of body length with body weight for sandfish juveniles of 6–70 mm in length. Juveniles were drained and blotted dry on a damp cloth for 30 sec before being measured and weighed.

are greater than needed for transfer to coarse-mesh enclosures. Considering the converted initial individual average body weight of 0.05 g, average growth rates equate to 0.07–0.10 g day^{-1}. Final body weights of sandfish juveniles were more variable in fine-mesh enclosures with stripped shadecloth (Fig. 5). Mean final body weight was significantly higher in the fine-mesh enclosures with stripped shadecloth than those without ($F_{1, 20} = 6.22$, $P = 0.02$). Juveniles in unshaded fine-mesh enclosures attained marginally non-significantly greater weights than those in shaded fine-mesh enclosures ($F_{1, 20} = 3.60$, $P = 0.07$). The stripped shadecloth improved growth of sandfish juveniles by more than 15% compared to fine-mesh enclosures without them.

Seawater temperatures in the pond were high (30–33°C) and water parameters were consistent within treatments but different between them (Table 1). Shade lowered the average water temperatures by 0.14–0.18°C in fine-mesh enclosures and this effect was statistically significant ($F_{1, 20} = 96.6$, $P < 0.001$) and consistent between fine-mesh enclosures with and without stripped shadecloth (interaction: $F_{1, 20} = 1.79$, $P = 0.20$). Dissolved oxygen in fine-mesh enclosures averaged 0.24–0.45 mg l^{-1} higher in unshaded fine-mesh enclosures than shaded fine-mesh enclosures ($F_{1, 20} = 28.0$, $P < 0.001$). The average dissolved oxygen concentrations were 0.84–1.06 mg l^{-1} higher without stripped shadecloth than with ($F_{1, 20} = 208.8$, $P < 0.001$), and the interaction of main effects was non-significant ($F_{1, 20} = 2.58$, $P = 0.12$).

Juvenile Size for Transfer to Fine-mesh Enclosures

The recovery rate (survival) of juvenile sea cucumbers from the fine-mesh enclosures was highly variable among groups and related significantly to initial mean body lengths of the animal groups ($F_{1, 18} = 168.3$, $P < 0.001$) (Fig. 6). The modelled relationship ($r^2 = 0.90$) suggests that more than 50% of juveniles longer than 5 mm at initial transfer could be expected to be recovered from similar fine-mesh enclosures after 3 weeks.

Growth rates, using initial lengths converted to weights, were highly variable among groups, due to the wide range in initial sizes. Growth averaged 0.022 g ind^{-1} day^{-1} for groups of juveniles with mean starting lengths >5 mm. The sandfish

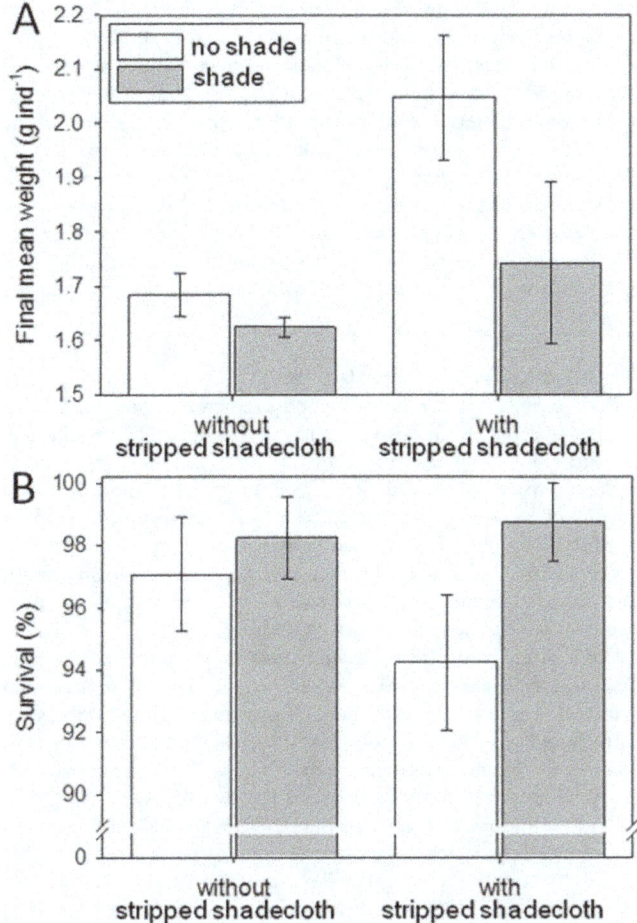

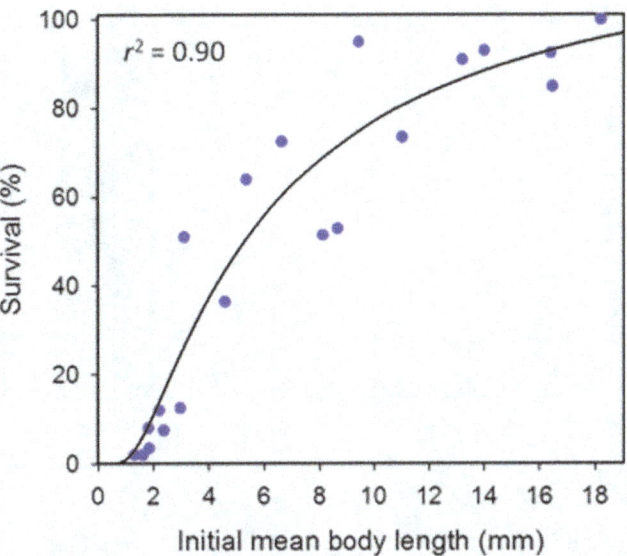

Figure 6. Effects of sediment type and sediment depth. Scatterplot of survival of juvenile sandfish in fine-mesh enclosures across various initial mean group body lengths, three weeks after transfer from hatchery tanks; $n = 20$ fine-mesh enclosures. The fitted curve is the most appropriate nonlinear function by AIC selection.

Figure 5. Effects of increasing mesh surfaces and shading. Bar graphs of (A) average individual body weights and (B) average survival of sandfish in fine-mesh enclosures with and without stripped shadecloth, and with (shaded) and without (open) shade; $n = 6$. Error bars are 1 standard error of the mean.

juveniles increased their body weights by 2 to 35 times their initial body weights (converted from body length) over the 21 days.

Effects of Sediment Type and Sediment Depth

Seawater temperatures were low in the pond (mean: 21.3°C) because this experiment was conducted in winter and, consequently, growth of the sandfish was slow relative to previous experiments. From a starting individual average weight of 0.56 g, juveniles averaged only 1.13 g after six weeks. Across all treatments, the average growth rate was 0.013 g ind^{-1} day^{-1}.

Survival averaged 78%. There were no significant differences in growth of juveniles between substrate types ($F_{1, 11} = 0.26$, $P = 0.62$), nor among substrate depths ($F_{2, 11} = 1.78$, $P = 0.21$), nor the interaction between the two factors ($F_{2, 11} = 0.78$, $P = 0.48$). The initial group weights (the covariate) did not significantly affect the differences in growth among replicate groups ($F_{1, 11} = 0.04$, $P = 0.85$). Likewise, neither substrate type nor depth affected survival of juveniles (F-values = 0.13–0.96, P-values = 0.41–0.73) and initial group weights had no significant effect on survival ($F_{1, 11} = 0.19$, $P = 0.68$). After pooling the non-significant term of substrate depths, to enable a one-way ANCOVA test to include control groups (no sediment), we found no significant differences in growth rates of juveniles with sand or mud compared to juveniles with no sediment ($F_{2, 20} = 1.88$, $P = 0.18$).

Enclosures at Sea vs. Pond

The average individual growth of sandfish juveniles differed significantly among the three locations ($F_{2, 8} = 176$, $P < 0.001$) (Fig. 7) and the effect of starting weight on growth was non-significant (covariate: $F_{1, 8} = 2.9$, $P = 0.13$). Growth varied little among the replicate enclosures within locations and differed between the two sites at sea and was markedly low in both compared to the other experiments (average growth rates: 0.006

Table 1. Average water parameters in fine-mesh enclosures within the pond in the first experiment (*Effects of increasing mesh surfaces and shading*).

	Without stripped shadecloth		With stripped shadecloth	
	Without shade	**With shade**	**Without shade**	**With shade**
Temperature (°C)	31.8±0.1	31.6±0.1	31.7±0.1	31.6±0.1
Dissolved oxygen (mg l^{-1})	5.9±0.1	5.4±0.1	4.8±0.1	4.6±0.1

Error estimates are standard errors of averages for the four treatment combinations.

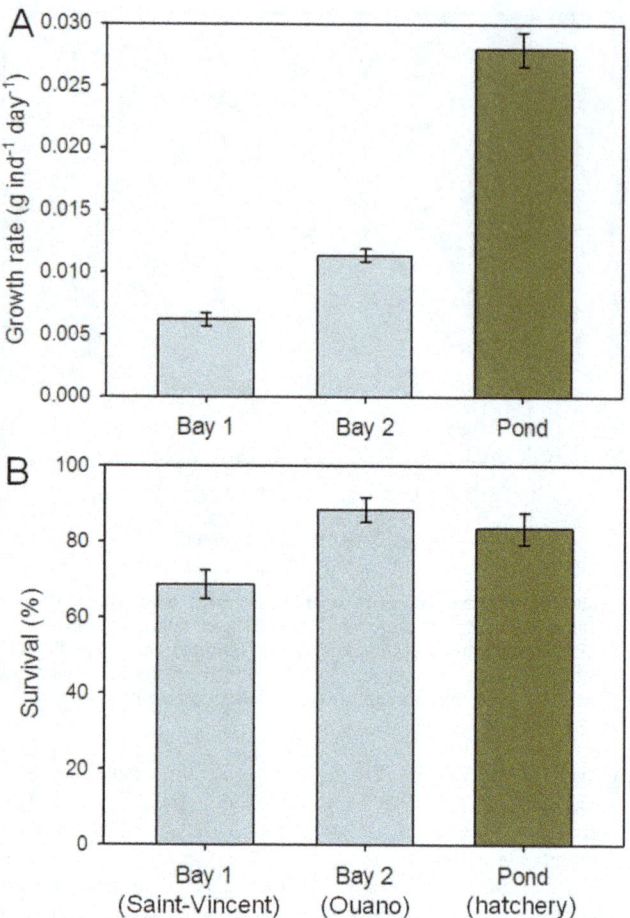

Figure 7. Enclosures at sea vs. pond. Bar graphs comparing (A) average growth rates and (B) average survival of sandfish in coarse-mesh enclosures at two sites at sea and those in an earthen pond; $n = 4$. Error bars are 1 standard error of the mean.

the availability of natural food and water quality inside fine-mesh enclosures [27]. Our two-stage nursery system should provide better water exchange by using coarse-mesh enclosures as a second stage, reducing the need for frequent cleaning, but at the expense of additional materials and transfer of juveniles. Regardless of the approach, our experimental findings should be invaluable to increasing the cost-effectiveness of production. Our experiment on *Effects of increasing mesh surfaces and shading* in enclosures furnishes an innovation that should provide 15% faster growth rates of newly-transferred juvenile sandfish, which should reduce the time, and cost, of producing larger juveniles for farming, sea ranching or restocking.

Length and Weight Measurements

Body length was a relatively good predictor of body weight of hatchery-produced sandfish juveniles in this study. The result may be surprising, considering the previous finding that length explained 61% [31] and 79% [32] of the variation in body weights of adult sandfish and 73% of the variation in weights of sandfish juveniles [17]. We used body length as an initial measure of the size of sandfish in the first two experiments (in fine-mesh enclosures) to avoid disproportionate stress on very small juveniles from the out-of-water weighing procedures.

While adult sandfish have deep transverse wrinkles in their body [33] and can contract like an accordion, juveniles lack such wrinkles. Despite the lack of concurrence with findings of Battaglene et al. [17], we posit that body weight may be predicted from body length more accurately in sandfish juveniles than adults. A close relationship between the two measures ($r^2 = 0.95$) was also found for hatchery-produced *Holothuria fuscogilva* juveniles [34]. Indeed, body length is often used as a convenient descriptor of size of small (<1 g) juvenile sea cucumbers in hatcheries [10,17,21,35–37].

Early Nursery of Small Juveniles

We devised the stripped shadecloth units as a means of increasing surface area of mesh in the fine-mesh enclosures on which the juveniles could feed and to increase surfaces on which microalgae and detritus could accumulate. Newly-settled sandfish can be found on seagrass blades in the wild and appear to begin living on the sediment surface at around 6–9 mm in length [20,21]. Our finding of significantly better growth of sandfish juveniles in fine-mesh enclosures with stripped shadecloth units indicates that the animals either received more detrital food matter with these units and/or foraged better when provided with more surface area of mesh. This innovation should markedly improve the rate of growing small juvenile sandfish of 5–10 mm in length (c.a. 0.03–0.16 g) to sizes of about 1.0 g (22 mm). For early nursery rearing of sandfish juveniles of 0.05–1 g, the growth rates of 0.1 g ind^{-1} day^{-1} that we found for juveniles in unshaded fine-mesh enclosures with stripped shadecloth are comparable to growth rates of similar-sized sandfish in the Philippines within mesh enclosures in a pond ([10]; average: 0.07–0.18 g ind^{-1} day^{-1}) and in floating enclosures at sea ([19]; range: 0.004–0.12 g ind^{-1} day^{-1}). We did not examine whether the stripped shade-cloth could continue to improve growth of sandfish juveniles larger than 1 g because they generally climb less on upright surfaces [17] but that could be a useful subject for further research.

Studies on the diurnal burying cycle and effect of light on the surface foraging behaviour of sandfish juveniles [20,22,23] suggested that they may feed over longer periods and, hence, grow faster in shaded tanks. On the other hand, growth rates and survival of sandfish juveniles have been shown to decline in shaded compared to unshaded tanks [17]. We found that shading did not

and 0.011 g ind^{-1} day^{-1}, respectively). Growth of juveniles in coarse-mesh enclosures within the pond was significantly higher than those at sea, but still relatively low (average growth rate: or 0.028 g ind^{-1} day^{-1}).

Survival differed significantly among locations ($F_{2,\ 8} = 5.72$, $P = 0.03$) (Fig. 7) and was not affected significantly by starting weight of the juveniles (covariate: $F_{1,\ 8} = 0.25$, $P = 0.63$). Average survival of juveniles in coarse-mesh enclosures at sea at Ouano (89%) was similar to that of juveniles in coarse-mesh enclosures within the earthen pond at the hatchery (84%), but survival was much lower in coarse-mesh enclosures at sea at Saint-Vincent (69%). Encouragingly, the coarse-mesh enclosures at sea were undamaged and prevented the entry of predators.

Discussion

Mesh enclosures in ponds are a cost-efficient nursery system for other marine animals [27–29]. They are cheaper than tanks, and aquaculture ponds offer natural productivity of organic detritus and cheap supply of water under relatively stable conditions for rearing sandfish juveniles.

Recently, fine-mesh enclosures have been used in Vietnam as the sole unit for growing sandfish juveniles to a competent size for stocking [30]. However, cleaning of fouled fine-mesh enclosures incurs high labour costs and fouling reduces water flow, decreasing improve the growth of sandfish juveniles in fine-mesh enclosures and slowed growth rates, although the effect was marginally non-significant.

When to Transfer of Juveniles from Hatchery Tanks to Mesh Enclosures?

Early transfer of small juveniles from hatchery tanks to less resource-intensive nursery systems reduces production costs but sandfish juveniles do not seem to handle being transferred to pond conditions at very small sizes [18]. Our modeling of survival across size classes of sandfish juveniles suggests that the animals should be 5–8 mm body length before being transferred from hatchery tanks to fine-mesh enclosures in ponds. Above this threshold, the survival rate is similar or better than in nursery tanks in the hatchery [7,15,18].

A large proportion of the smallest sea cucumbers (<5 mm in length) unrecovered from the fine-mesh enclosures may have conceivably escaped by squeezing through the mesh. Such escapement could have biased our growth rate estimations, since the starting weight of larger remaining individuals would change the calculated weight gain. However, this potential bias would be relatively small and was minimized in our experiment on *Juvenile size for transfer to fine-mesh enclosures* by precise (6% variation) size classes and we do not report growth rates of the smallest size classes. Finer mesh enclosures of 450 µm used in the nursery of sandfish juveniles in Vietnam [18] could reduce or prevent escapement of juveniles of just a few mm in length. However, such mesh was unavailable to us at the time of our study and we considered that water exchange in the fine-mesh enclosures would become impeded with finer mesh. In view that Gamboa et al. [10] found low escapement rates (as low as 4%) of small juvenile sandfish (some just 3 mm in length) from some enclosures of larger (1-mm) mesh, mortality of our small juveniles may be the more likely explanation of low recovery rates from our fine-mesh enclosures than escapement.

Nursery to Stocking Size

Once cultured sandfish reach about 1 g in weight, a behavioral change finds them preferring, and growing better on, sandy or muddy-sand substrates [17]. Juvenile sandfish in the wild appear to need muddy-sand in which to bury, presumably to avoid diurnal predators [22,23]. We found no benefit in growth or survival of sandfish juveniles between coarse-mesh enclosures with sand or pond mud and no sediment, nor among different sediment depths. It is possible that sand or pond mud could have improved growth significantly had the experiment been conducted in summer, when growth was faster. Indeed, Juinio-Meñez et al. [19] reported higher weight gain of sandfish juveniles in tanks with sand than in bare tanks, although methods and results were not presented. Our finding suggests that, at least for cultured sandfish in New Caledonia, sand may not be needed for, and does not

improve, the assimilation of organic detritus. However, from an ecological perspective, providing juvenile sandfish with sand in the couple weeks before stocking into the sea may still serve as 'behavioural conditioning' to encourage diurnal burying, which could conceivably improve predator avoidance and post-release survival rates [19,23].

Nursery Systems at Sea?

In cases where suitable earthen ponds are unavailable, the nursery rearing of juvenile sea cucumbers might need to be in the sea. Unfortunately, 1-mm mesh coarse-mesh enclosures gathered at the top and attached to the benthos do not appear to be a viable nursery system (Fig. 7). The slow growth rates of sandfish juveniles that we measured in replicate coarse-mesh enclosures at both protected bays would stifle production of juveniles for stocking programs. Fouling on the coarse-mesh enclosures in seagrass beds at sea may have prevented a sufficient supply of detritus to juveniles or the inside of the coarse-mesh enclosures was too shaded because they were gathered and tied at the top. The mediocre survival of sandfish juveniles in coarse-mesh enclosures at one of the bays further suggests that nursery systems in the sea may be riskier than in ponds. Recently, Juinio-Meñez et al. [19] tested different enclosure systems in the Philippines for rearing sandfish juveniles starting at 4–10 mm over 1–2 months and also found poor survival (18%) of juveniles in enclosures set on the sea bed compared to a floating enclosures (12–44%) or enclosures in an earthen pond (57–73%). They reported modest growth rates in the floating enclosures of up to 0.018 g ind^{-1} day^{-1} in one trial and a range in individual growth of 0.003–0.118 g ind^{-1} day^{-1} in a second experiment. Further research needs to examine the design and maintenance of such floating enclosures to improve growth rates. If mesh enclosures at sea is the only viable nursery strategy, sites should also be evaluated experimentally to find ones that give good growth and survival of the juveniles.

Acknowledgments

We thank the staff of WorldFish Center and the Secretariat of the Pacific Community for their administrative support. We thank IFREMER staff at Saint-Vincent for helping with water supply, ponds and facilities. Valuable assistance was given by Pascal Blazer, Bernard Blockmans, Eric Danty, Johann Le Dreau, Mateo Simutoga and Eric Vigne. We thank Matthew Slater and two anonymous reviewers for their comments, which improved the manuscript. This article was supported by the Marine Ecology Research Centre, Southern Cross University.

Author Contributions

Conceived and designed the experiments: SWP. Performed the experiments: NSA SWP. Analyzed the data: SWP NSA. Contributed reagents/materials/analysis tools: SWP NSA. Wrote the paper: SWP NSA. Preparation of figures: SWP NSA. Procurement of research funds and reporting: SWP.

References

1. Ahlgren MO (1998) Consumption and assimilation of salmon net pen fouling debris by the red sea cucumber *Parastichopus californicus*: implications for polyculture. J World Aqua Soc 27: 133–139.

2. Palzat DL, Pearce CM, Barnes PA, McKinley RS (2008) Growth and production of California sea cucumbers (*Parastichopus californicus* Stimpson) co-cultured with suspended Pacific oysters (*Crassostrea gigas* Thunberg). Aquaculture 275: 124–137.

3. Slater MJ, Jeffs AG, Carton AG (2009) The use of the waste from green-lipped mussels as a food source for juvenile sea cucumber, *Australostichopus mollis*. Aquaculture 292: 219–224.

4. Purcell SW, Hair C, Mills D (2012) Sea cucumber culture, farming and sea ranching in the tropics: progress, problems and opportunities. Aquaculture 368: 68–81.

5. Yanagisawa T (1998) Aspects of the biology and culture of the sea cucumber. In: De Silva SS, editor. Tropical Mariculture. London: Academic Press. 292–308.

6. Battaglene SC (1999) Culture of tropical sea cucumbers for stock restoration and enhancement. Naga – The ICLARM Quarterly 22: 4–11.

7. Agudo N (2006) Sandfish Hatchery Techniques. Noumea: Australian Centre for International Agricultural Research, Secretariat of the Pacific Community and the WorldFish Center. 44 p.

8. Bell J, Agudo N, Purcell S, Blazer P, Simutoga M, et al. (2007) Grow-out of sandfish *Holothuria scabra* in ponds shows that co-culture with shrimp *Litopenaeus stylirostris* is not viable. Aquaculture 273: 509–519.

9. Duy NDQ (2012) Large-scale sandfish production from pond culture in Vietnam. In: Hair C, Pickering T, Mills D, editors. Asia–Pacific Tropical Sea

Cucumber Aquaculture. ACIAR Proceedings No. 136. Canberra: ACIAR. 34–39.

10. Gamboa RU, Aurelio RA, Ganad DA, Concepcion LB, Abreo NAS (2012) Small-scale hatcheries and simple technologies for sandfish (*Holothuria scabra*) production. In: Hair C, Pickering T, Mills D, editors. Asia–Pacific Tropical Sea Cucumber Aquaculture. ACIAR Proceedings No. 136. Canberra: ACIAR. 63–74.

11. Purcell SW, Simutoga M (2008) Spatio-temporal and size-dependent variation in the success of releasing cultured sea cucumbers in the wild. Rev Fish Sci 16: 204–214.

12. Robinson G, Pascal B (2012) Sea cucumber farming experiences in south-west Madagascar. In: Hair C, Pickering T, Mills D, editors. Asia–Pacific Tropical Sea Cucumber Aquaculture. ACIAR Proceedings No. 136. Canberra: ACIAR. 142–155.

13. Purcell SW (2004) Rapid growth and bioturbation activity of the sea cucumber *Holothuria scabra* in earthen ponds. Proc Australasian Aquaculture 2004: 244.

14. Raison CM (2008) Advances in sea cucumber aquaculture and prospects for commercial culture of *Holothuria scabra*. CAB Reviews 3(82): 1–15.

15. Pitt R, Duy NDQ (2003) How to produce 100 tonnes of sandfish. SPC Beche-de-mer Inf Bull 18: 15–17.

16. Pitt R (2001) Review of sandfish breeding and rearing methods. SPC Beche-de-mer Inf Bull 14: 14–21.

17. Battaglene SC, Seymour JE, Ramofafia C (1999) Survival and growth of cultured juvenile sea cucumbers, *Holothuria scabra*. Aquaculture 178: 293–322.

18. Pitt R, Duy NDQ (2004) Breeding and rearing of the sea cucumber *Holothuria scabra* in Viet Nam. In: Lovatelli A, Conand C, Purcell S, Uthicke S, Hamel J-F, Mercier A, editors. Advances in Sea Cucumber Aquaculture and Management. FAO Fisheries Technical Paper 463. Rome: FAO. 333–346.

19. Juinio-Meñez MA, de Peralta GM, Dumalan RJP, Edullantes CM, Catbagan TO (2012) Ocean nursery systems for scaling up juvenile sandfish (*Holothuria scabra*) production: ensuring opportunities for small fishers. In: Hair C, Pickering T, Mills D, editors. Asia–Pacific Tropical Sea Cucumber Aquaculture. ACIAR Proceedings No. 136. Canberra: ACIAR. 57–62.

20. Mercier A, Battaglene SC, Hamel JF (2000) Periodic movement, recruitment and size-related distribution of the sea cucumbers *Holothuria scabra* in Solomon Islands. Hydrobiologia 440: 81–100.

21. Mercier A, Battaglene SC, Hamel JF (2000) Settlement preferences and early migration of the tropical sea cucumber *Holothuria scabra*. J Exp Mar Biol Ecol 249: 89–110.

22. Mercier A, Battaglene SC, Hamel JF (1999) Daily burrowing cycle and feeding activity of juvenile sea cucumbers *Holothuria scabra* in response to environmental factors. J Exp Mar Biol Ecol 239: 125–156.

23. Purcell SW (2010) Diel burying by the tropical sea cucumber *Holothuria scabra*: effects of environmental stimuli, handling and ontogeny. Mar Biol 157: 663–671.

24. Underwood AJ (1997) Experiments in Ecology: Their Logical Design and Interpretation Using Analysis of Variance. Cambridge: Cambridge University Press. 504 p.

25. Burnham KP, Anderson DR (1998) Model Selection and Inference: A Practical Information-Theoretic Approach. New York: Springer-Verlag. 488 p.

26. Quinn GP, Keough MJ (2002) Experimental design and data analysis for biologists. Cambridge: University Press. 556 p.

27. Bhujel RC (2000) A review of strategies for the management of Nile tilapia (*Oreochromis niloticus*) broodfish in seed production systems, especially hapa-based systems. Aquaculture 181: 37–59.

28. Barman BK, Little DC (2011) Use of hapas to produce Nile tilapia (*Oreochromis niloticus* L.) seed in household foodfish ponds: A participatory trial with small-scale farming households in Northwest Bangladesh. Aquaculture 317: 214–222.

29. Rodriguez EM, Bombeo-Tuburan I, Fukumoto S, Ticar RB (1993) Nursery rearing of *Penaeus monodon* (Fabricius) using suspended (hapa) net enclosures installed in a pond. Aquaculture 112: 107–111.

30. Mills DJ, Duy NDQ, Juinio-Menez MA, Raison CM, Zarate JM (2012) Overview of sea cucumber aquaculture and sea-ranching research in the South-East Asian region. In: Hair C, Pickering T, Mills D, editors. Asia–Pacific Tropical Sea Cucumber Aquaculture. ACIAR Proceedings No. 136. Canberra: ACIAR. 22–31.

31. Conand C (1989) Les holothuries aspidochirotes du lagon de Nouvelle-Calédonie: biologie, écologie et exploitation. PhD Thesis. Brest: Université de Bretagne Occidentale. 393 p.

32. Skewes T, Dennis D, Burridge C (2000) Survey of *Holothuria scabra* (sandfish) on Warrior Reef, Torres Strait, January 2000. CSIRO Division of Marine Research Final Report. Brisbane: CSIRO. 28 p.

33. Massin C, Uthicke S, Purcell SW, Rowe FWE, Samyn Y (2009) Taxonomy of the heavily exploited Indo-Pacific sandfish complex (Echinodermata: Holothuriidae). Zool J Linn Soc 155: 40–59.

34. Purcell SW, Tekanene M (2006) Ontogenetic changes in colouration and morphology of white teatfish, *Holothuria fuscogilva*, juveniles at Kiribati. SPC Beche-de-mer Inf Bull 23: 29–31.

35. Liu X, Zhu G, Zhao Q, Wang L, Gu B (2004) Studies on hatchery techniques of the sea cucumber, *Apostichopus japonicus*. In: Lovatelli A, Conand C, Purcell S, Uthicke S, Hamel J-F, Mercier A, editors. Advances in Sea Cucumber Aquaculture and Management. FAO Fisheries Technical Paper 463. Rome: FAO. 287–29.

36. So JJ, Hamel JF, Mercier A (2010) Habitat utilisation, growth and predation of *Cucumaria frondosa*: implications for an emerging sea cucumber fishery. Fish Manag Ecol 17: 473–484.

37. Mercier A, Ycaza RH, Espinoza R, Arriaga-Haro VM, Hamel JF (2012) Hatchery experience and useful lessons from *Isostichopus fuscus* in Ecuador and Mexico. In: Hair C, Pickering T, Mills D, editors. Asia–Pacific Tropical Sea Cucumber Aquaculture. ACIAR Proceedings No. 136. Canberra: ACIAR. 79–90.

6

Climate Change, Precipitation and Impacts on an Estuarine Refuge from Disease

Jeffrey Levinton[1]*, **Michael Doall**[1], **David Ralston**[2], **Adam Starke**[3], **Bassem Allam**[3]

1 Department of Ecology and Evolution, Stony Brook University, Stony Brook, New York, United States of America, **2** Woods Hole Oceanographic Institution, Woods Hole, Massachusetts, United States of America, **3** School of Marine and Atmospheric Sciences, Stony Brook University, Stony Brook, New York, United States of America

Abstract

Background: Oysters play important roles in estuarine ecosystems but have suffered recently due to overfishing, pollution, and habitat loss. A tradeoff between growth rate and disease prevalence as a function of salinity makes the estuarine salinity transition of special concern for oyster survival and restoration. Estuarine salinity varies with discharge, so increases or decreases in precipitation with climate change may shift regions of low salinity and disease refuge away from optimal oyster bottom habitat, negatively impacting reproduction and survival. Temperature is an additional factor for oyster survival, and recent temperature increases have increased vulnerability to disease in higher salinity regions.

Methodology/Principal Findings: We examined growth, reproduction, and survival of oysters in the New York Harbor-Hudson River region, focusing on a low-salinity refuge in the estuary. Observations were during two years when rainfall was above average and comparable to projected future increases in precipitation in the region and a past period of about 15 years with high precipitation. We found a clear tradeoff between oyster growth and vulnerability to disease. Oysters survived well when exposed to intermediate salinities during two summers (2008, 2010) with moderate discharge conditions. However, increased precipitation and discharge in 2009 reduced salinities in the region with suitable benthic habitat, greatly increasing oyster mortality. To evaluate the estuarine conditions over longer periods, we applied a numerical model of the Hudson to simulate salinities over the past century. Model results suggest that much of the region with suitable benthic habitat that historically had been a low salinity refuge region may be vulnerable to higher mortality under projected increases in precipitation and discharge.

Conclusions/Significance: Predicted increases in precipitation in the northeastern United States due to climate change may lower salinities past important thresholds for oyster survival in estuarine regions with appropriate substrate, potentially disrupting metapopulation dynamics and impeding oyster restoration efforts, especially in the Hudson estuary where a large basin constitutes an excellent refuge from disease.

Editor: John Murray Roberts, Heriot-Watt University, United Kingdom

Funding: Funding was provided by the Hudson River Foundation, grant number 00607A, and the New York State Department of Environmental Conservation (MOU 2008). The funders had no role in study design, data collection and analysis, decision to publish, or preparation of the manuscript.

Competing Interests: The authors have declared that no competing interests exist.

* E-mail: Levinton@life.bio.sunysb.edu

Introduction

Estuaries are biologically productive, supporting rich fisheries and diverse habitats, including oyster reefs, sea grass meadows, and vast expanses of fringing marshes. But their very richness coincides with human habitation, which has resulted in damage from pollution, overfishing and habitat destruction. Mid-Atlantic estuarine fisheries have severely declined from habitat alterations, pollution and overfishing [1,2,3]. For example, the loss of a key species, the eastern oyster *Crassostrea virginica*, has had significant effects on estuarine ecosystems of eastern and Gulf Coast North America [4,5]. Oysters and other estuarine bivalves affect estuarine water quality by removing particles [4,6] and influencing nitrogen cycling [7]. Oyster reefs also create three-dimensional benthic habitat that enhances diversity of other suspension feeders and offers important refuge from predators [5,8].

Oyster growth and disease rates vary substantially along the estuarine salinity transition between fresh and marine waters.

Oysters on the Atlantic and Gulf Coasts exposed to marine salinity are readily infected by two diseases. The parasite *Haplosporidium nelsoni*, or MSX, caused 90–95% mortality in the eastern oyster *Crassostrea virginica* in Delaware Bay in the 1950s [9]. MSX likely arrived in eastern North America from Japan, perhaps through an intermediate oyster host, and has spread to the eastern oyster [10] from Florida to Nova Scotia. The infection period is seasonal and disease can be reduced by moving oysters to lower salinities where survival of MSX is poor [9]. However, oyster growth and reproductive success decreases in lower salinities, and survival rates decrease below 5 psu [11]. Oysters have faster growth rates in higher salinities, but MSX infections decrease survival [9,12], with a few exceptions of evolved resistance to the disease [13]. MSX infections occur in the mesohaline and polyhaline zones of estuaries, but infection rates are much lower and often absent and oysters can grow in oligohaline zones of 6–12 psu [9,14].

A similar tradeoff between growth and disease exists for the other major oyster disease, the alveolate protistan *Perkinsus marinus* known as Dermo [15,16,17,18]. First discovered on the Gulf Coast of the United States [19], it has spread to the northeast and is a major source of mortality in marine waters. Increases of coastal sea surface temperature over the past few decades [20], especially in the form of winter warming, have facilitated the disease's northward spread [21,22]. Like MSX, Dermo does not thrive in oligohaline salinities [23,24].

In oligohaline waters, oysters grow slowly but have refuge from disease and from marine predators like whelks, oyster drills, flatworms, and starfish. In watersheds with controlled discharge, experiments have suggested that periods of increased river flow can temporarily reduce oyster disease, with enhanced growth during subsequent lower discharge periods [25,26]. In natural estuaries, seasonal and interannual variability in river discharge leads to continuous variation in salinity. High discharge during freshets will lower salinity at a location, but droughts will increase salinity and potentially increase disease susceptibility [27]. At the upper end of an estuary, increases in discharge may negatively impact oyster survival by reducing the frequency and duration of oligohaline conditions, making habitat that was formerly estuarine into a tidal freshwater river.

Oyster restoration is a priority in many estuaries of eastern North America, and in particular the Hudson River estuary [28]. The Hudson River estuary once supported among the richest oyster grounds in eastern North America [29], but signs of overfishing appeared early in the 19th century, and urban pollution hastened the decline in the early 20th century [29]. Jamaica Bay supported thousands of oyster fishers through the 19th century [30], but oyster populations are now negligible there due to pollution, habitat disturbance and the 1938 hurricane. The Tappan Zee-Haverstraw Bay (TZ-HB) region is a focus of restoration efforts in the Hudson due to historic oyster cultivation in that part of the estuary. In the 18th and 19th centuries Haverstraw Bay supported commercial oyster fisheries [29]. In the 1950s, a time of below-average rainfall over the past century, the Flower and Sons Oyster Company moved their operation to the TZ-HB Bay region and raised juvenile oysters with high growth rates and survival [31]. These results raised hopes that the broad shallow waters of TZ-HB with suitable bottom substrate and high benthic population densities would be well suited for oyster restoration [32].

Oyster restoration objectives include not only reestablishment of fisheries, but also revitalization of a critical element of the estuarine ecosystem for increasing biodiversity and improving water quality. The current poor state of eastern oyster populations has led to skepticism for restoration potential [33], despite some successful efforts [34]. Climate change is a one potential challenge for restoration. Increased sea surface temperature has facilitated the northward extension of Dermo [22] and MSX [35], threatening oyster habitat in polyhaline and oceanic salinities. Regional shifts in timing and magnitude of precipitation with climate change will alter river discharges and estuarine salinities. Current climate models predict an increase in precipitation in the northeast U.S. of 5 to 8% in the next few decades and up to 30 percent by the end of the century with increases most likely in the winter and spring [36,37,38]. Climate projections also suggest greater variability in streamflow with more frequent high and low discharge periods [36,37,38]. The shifts in magnitude and timing of precipitation and discharge will affect the salinity distributions in estuaries and therefore the habitat, growth, and vulnerability of oyster populations and associated species. While estuarine oysters can tolerate freshwater during the winter, very low salinities cause high degrees of physiological stress under spring and summer temperature conditions [11].

We are examining the Hudson River estuary, once a major oyster grounds and now a focus for restoration. We have combined regional studies of oyster performance (growth rate and survival) and estuarine modeling to predict physical conditions and potential impacts on oyster habitat under a regime of increased precipitation with climate change.

Results

We investigated oyster performance in coastal and estuarine regions to evaluate tradeoffs between performance and disease occurrence. We compared growth, survival, reproduction, and disease occurrence at coastal sites from eastern Long Island, New York USA to western Raritan Bay, New Jersey USA and at sites in the TZ-HB section of the Hudson River estuary (river km 42–58) (Figure 1). We quantified shell growth and disease prevalence (Dermo) of overwintered oysters that were transplanted from one hatchery (Fishers Island, New York USA) to replicate floating cages at 9 sites in 2008, and to 5 of these sites in 2009 (Figure 2, see "Materials and Methods"). Shell growth showed a strong positive correlation with salinity (Figure 2). In contrast, Dermo was far less prevalent in the lower salinity sites of the estuary than in coastal sites. These results are consistent with the expected tradeoff between growth and disease with salinity, documenting the tradeoff more completely and at higher latitudes than previous work.

The prevalence of MSX was low at most sites during 2008 and 2009. MSX was responsible for substantial mortality in 2008 at one site in the lower Hudson estuary (Pier 40, "P40" in Figure 1, with a mean cumulative mortality of 43%). This elevated mortality due to MSX occurred near the mouth of the estuary, a location with higher salinities and greater salinity variability than the upper estuary sites in TZ-HB.

Both 2008 and 2009 had higher than average precipitation (measured at Albany) and discharge in the Hudson River, but the timing of the high discharge period appears to be critical. At the TZ-HB sites salinities were in the range of 5–10 psu through July and August of 2008. We found generally high survival rates (Figure 3a), albeit with low growth rates (Figure 2b). During a discharge event in August 2008 salinity decreased below 2 psu at the site farthest up-estuary (Ossining, "OS" in Figure 1), corresponding with a mortality increase of ca. 30 percent.

Precipitation and river discharge during the summer months of 2009 were greater than in 2008, with lower salinities at the estuarine stations and much greater mortality in TZ-HB (Figure 3b). In contrast, mortality was minimal at the coastal sites in the study. In 2009, salinities in TZ-HB dropped to nearly 0 and remained around 3 psu for most of the summer. The populations farthest up-estuary ("OS") died off completely and two other TZ-HB sites ("I" and "WI") had significant mortality. Mortality at "I" and "WI" decreased in October as river discharge declined and salinity increased.

A limited extension of the observations into 2010 offers additional evidence of the sensitivity of oyster growth and survival in TZ-HB to summer river discharge. Soft tissue growth, shell height growth, and survival were measured at the Washington Irving Boat Club in Tarrytown ("WI") during the summers of 2008, 2009 and 2010 (Table 1). In 2009, summer precipitation and discharge were high: precipitation at Albany (averaged May 1 to September 1) was the highest in the 132 year record and average discharge in the Hudson ranked 13th in the 93 year record. Average summer precipitation and discharge in 2008 and 2010 were significantly lower. Correspondingly, oyster growth at the Tarrytown site was much less and mortality was greater during the wet summer of 2009 than 2008 or 2010.

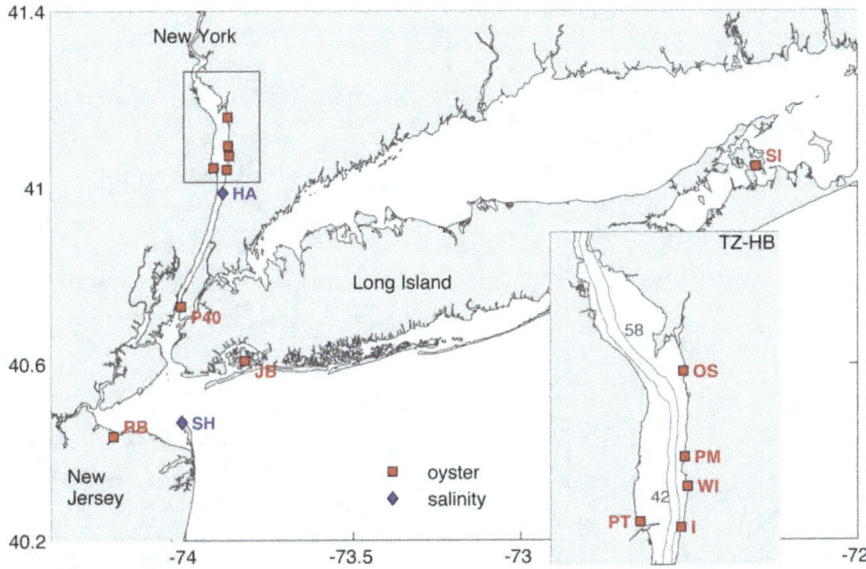

Figure 1. Map showing observation locations. Red squares are oyster test stations: OS = Ossining, PM = Philips Manor, PT = Piermont, WI = Tarrytown (Washington Irving Boat Club), I = Irvington, P40 = Pier 40, SI = Shelter Island, JB = Jamaica Bay, RB = Raritan Bay, New Jersey. Blue diamonds show salinity measurement locations (data in Figure 3b): HA = Hastings (USGS), SH = Sandy Hook (NOAA). The inset focuses on the Tappan Zee-Haverstraw Bay (TZ-HB) region; also noted for reference are the along estuary distances of 42 km and 58 km (from the Battery at the southern end of Manhattan) and 10-m isobath at the transition between the channel and shoals.

To relate conditions during the observations to longer-term estuarine variability, we use a numerical model of the circulation and salinity in the Hudson River estuary that has previously been validated against observations [39,40]. A hindcast of the salinity variability over the past ca. 90 years was made using the available discharge and tidal records. The model calculates the vertical salinity structure as well as the along-estuary distribution, and here we focus on salinities in the relatively shallow regions (depths less than about 3 m) on the east side of TZ-HB where leases for oyster culture were maintained in the 1950s and where restoration is most likely [40]. Estuarine salinity depends inversely on discharge – as discharge increases, salt is pushed toward the mouth and salinity decreases.

The model suggests that during high discharge periods, salinities in TZ-HB are frequently low enough to limit oyster growth and even survival (Figure 4). For example, averaging over the 90-year record, salinities in summer months in TZ-HB were in the range of 3 to 7 psu. In contrast, average salinities during the 5 years with the highest annual precipitation were on average 2 to 3 psu lower during the summer months. Model results in TZ-HB are also shown for 2008 and 2009. The increased precipitation during the late spring and summer of 2009 lead to decreased salinities during the summer, with salinities similar to the average conditions during historically high discharge years. While the annual precipitation and discharge were greater in 2008 than in 2009, the high discharge period in 2008 was during the typical spring freshet rather than during the summer months of oyster recruitment.

Salinities from upper (river km 58) and lower (river km 42) TZ-HB over the full historical simulation demonstrate the inverse dependence between summer salinity (average July salinity shown

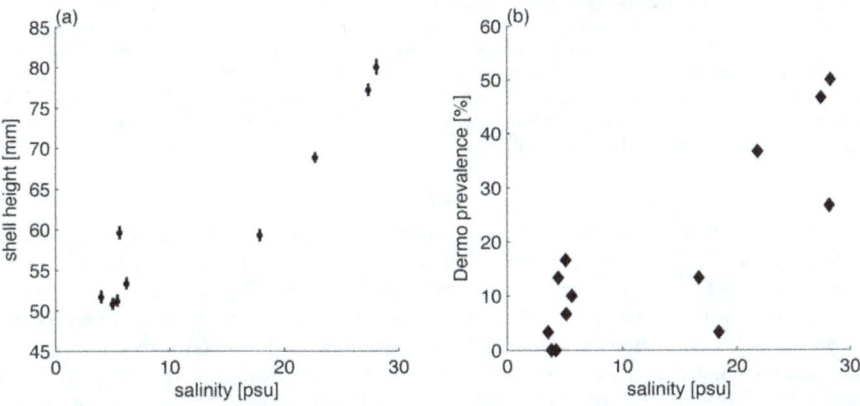

Figure 2. Growth and disease prevalence as a function of salinity. (a) Relationship of mean oyster shell height to salinity ($r^2 = 0.89$, in samples collected in October 2008, after 3 months of growth from a mean starting height of 51.7 mm);vertical bars show standard error. (b) Prevalence of Dermo in oysters (30 per site) from 9 sites taken from coastal and TZ-HB sites in September, 2008, and 4 sites from coastal and TZ-HB sites in August, 2009 ($r^2 = 0.67$).

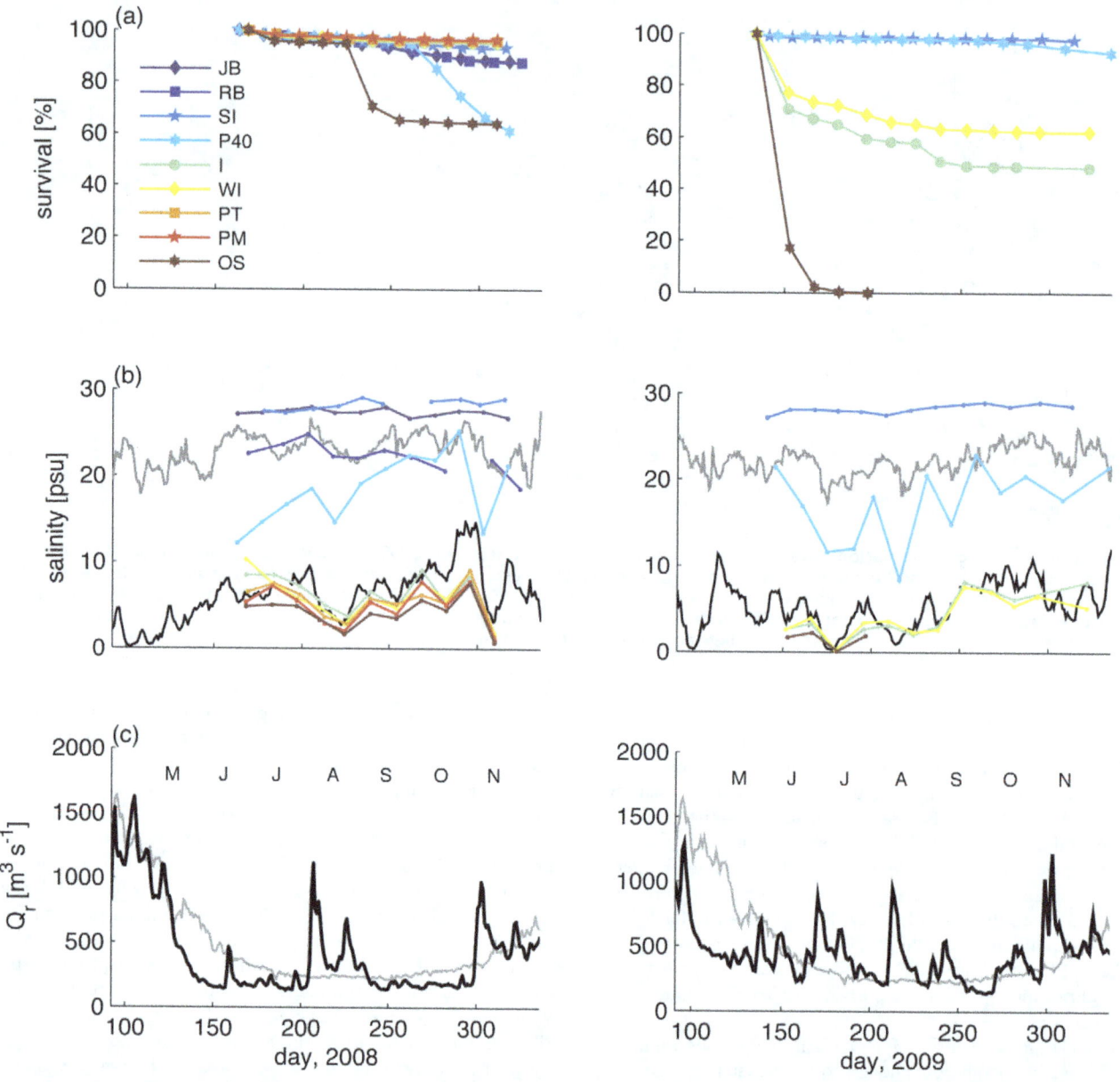

Figure 3. Survival patterns, salinity variation, and river discharge. (a) Left: Survivorship of oysters grown in summer 2008 at a series of coastal and oligohaline sites in Tappan Zee-Haverstraw Bay. The decline at Ossining, the lowest-salinity TZ-HB site, was associated with a drop of salinity while the decline at Pier 40 was associated with a major infection of MSX. Right: Survivorship of oysters grown in the summer of 2009 (only 5 of the 2008 sites were investigated), comparing TZ-HB with two of the coastal sites studied in 2008. (b) Salinities in 2008 (left) and 2009 (right). Continuous, tidally filtered surface salinities are shown for Sandy Hook NJ (NOAA station # 8531680, SH in Figure 1) and Hastings NY (USGS station # 01376304, HA in Figure 1) (grey and black lines, respectively); oyster sites were sampled biweekly. (c) River discharge in 2008 (left, dark line) and 2009 (right), as compared to average seasonal discharge pattern for the period 1918–2009 (light grey line).

here) and mean annual discharge (Figure 5a). Conditions in the upper bay range from about 6 psu to essentially fresh, while the lower bay ranges from about 10 to less than 3 psu. The mean annual discharge and the mean discharge of the 5 years with greatest precipitation are indicated with markers on the abscissa for reference. The model results indicate that during high discharge years, only a very limited region of TZ-HB would retain sufficiently high salinities in summer to provide suitable oyster habitat.

Mean annual discharge in the Hudson River depends primarily on regional precipitation (Figure 5b). Most current climate models project increases in precipitation in the U.S.

Northeast in the coming decades, with the greatest increases during winter and spring [36]. Climate models predict a range of outcomes for summer precipitation [36,37]. Oyster restoration prospects are sensitive to these projections, as summer is the season of oyster larval recruitment. The total projected increase in precipitation in the Northeast over the coming century is about 25 percent, similar to the difference between the average annual precipitation over the past 90 years and the average of the 5 highest precipitation years. If predictions of increased precipitation hold, particularly during the summer months, then decreases in salinity in TZ-HB may be detrimental to oyster survival and therefore restoration.

Table 1. Soft tissue growth (g), shell height growth (cm), and survival (percent), relative to salinity during the growing season (numerical model estimates for June 1–September 22 of 2008, 2009, and 2010) at the Washington Irving Boat Club in Tarrytown ("WI" in Figure 1).

Year	Tissue±S.E. (N)	Height±S.E. (N)	Survival (%)	Days over 5 psu
2008	0.40±0.03(40)	8.62±0.81(80)	95.9	61
2009	0.14±0.01(40)	4.71±0.56(80)	62.3	6
2010	0.49±0.04(20)	24.02±0.79(60)	87.9	83

In addition to the local salinity, the rate of salinity change can be a source of physiological stress [11] and is a factor in oyster restoration. During high discharge, the estuarine salinity distribution compresses, and variability in salinity from tidal and meteorological forcing increases at any particular point due to the sharper salinity gradient. The temporal variability in salinity would be exacerbated by projected increases in the intensity of extreme precipitation events with climate change [36,37]. Overall, the projected increase in precipitation and discharge would be expected to shift the location of suitable habitat for oyster growth. The bathymetry of the Hudson is such that appropriate water depths and appropriate substrate for oyster growth are sparse down-estuary from TZ-HB, so the total area with suitable water column and benthic conditions could be expected to decrease with a shift in the salinity distribution toward the mouth. Within the TZ-HB region, extensive suitable bottom areas exist that would support oyster growth and widespread larval recruitment has been observed there [41].An additional consideration is that increases in water temperature may exacerbate negative impacts of disease in oysters, particularly at coastal sites.

Discussion

At present, the Tappan Zee – Haverstraw Bay region of the Hudson estuary provides suitable benthic habitat for oysters and a likely refuge from Dermo and MSX diseases. However, increased mortality in TZ-HB during the high discharge summer months of 2009 suggest that projected increases in precipitation with climate change may reduce salinities in this region below thresholds for oyster survival. Our modeling results suggest that discharges

consistent with precipitation in future climate scenarios could decrease salinities in the region to levels below the threshold for oyster survival. The seasonal timing of precipitation and discharge remains a critical uncertainty in this assessment. While climate models generally agree that precipitation is likely to increase during winter and spring in the Northeast [36,37], uncertainty remains for the summer months that are important for oyster growth, spawning, and larval dispersal. Historically, high annual average precipitation correlates with lower salinities in July due to longer, higher volume freshets (Figure 5a). Whether the trend continues depends on the future partitioning of precipitation between snow and rain and its effect on the timing of river discharge. Independent of the seasonal distribution, projections of increased variability in streamflow [36,37] are likely to be a stressor to oyster communities at the upstream margins of estuaries.

Restoration of oyster populations in TZ-HB could have important implications for oysters throughout the Hudson-Raritan region. If populations could be restored, larvae from TZ-HB Bay might be exported to coastal sites in years when coastal populations with higher vulnerability to disease and predators fail to reproduce or survive. We found oysters recruiting to our cages in TZ-HB in the late summer of 2008, but could not determine if the larvae came from within the bay or from down estuary. Our observations in 2009 showing no recruitment in Jamaica Bay or the New York Harbor region suggests that the recruitment within TZ-H may have been indigenous. Thus the possibility for a metapopulation of interacting disease-prone, but high growth rate oysters on the coasts and low growth rate but disease-free oysters in the TZ-HB region could provide temporal reinforcement and promote overall survival of the regional oyster metapopulation [42]. A model of connectivity has not yet been developed for this region, but restoration efforts would depend on maintaining a metapopulation of rapidly growing and disease-resistant local populations. In Chesapeake Bay, connections of similar distances have been shown to be feasible according to modeling studies [43].

A broader assessment of effects of regional precipitation shifts on oyster populations in estuaries in eastern North America and the Gulf Coast could relate results to metapopulation design to maximize oyster recruitment and survival [44]. Salinity structure in Chesapeake Bay, for example, is driven by variation of discharge in the major tributaries, particularly the Susquehanna [36]. Anticipated increases in precipitation from climate change may cause major losses of oysters and estuarine habitat as salinity decreases, particularly in tributaries in the middle of the bay where

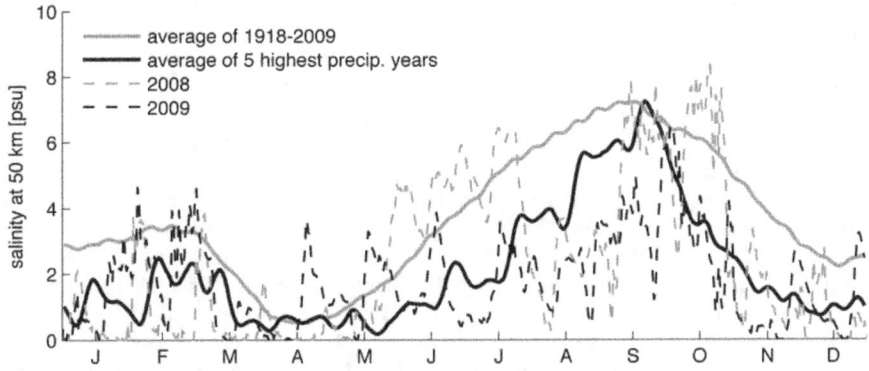

Figure 4. Model simulations of salinity at TZ-HB site (river km 50). Shown are average conditions over the entire period 1918–2009, and average conditions during the 5 years of that period with the greatest annual precipitation. Model output is averaged by year-day and filtered with a 5-day running average, Daily average salinities from the model at the same location are shown for 2008 and 2009.

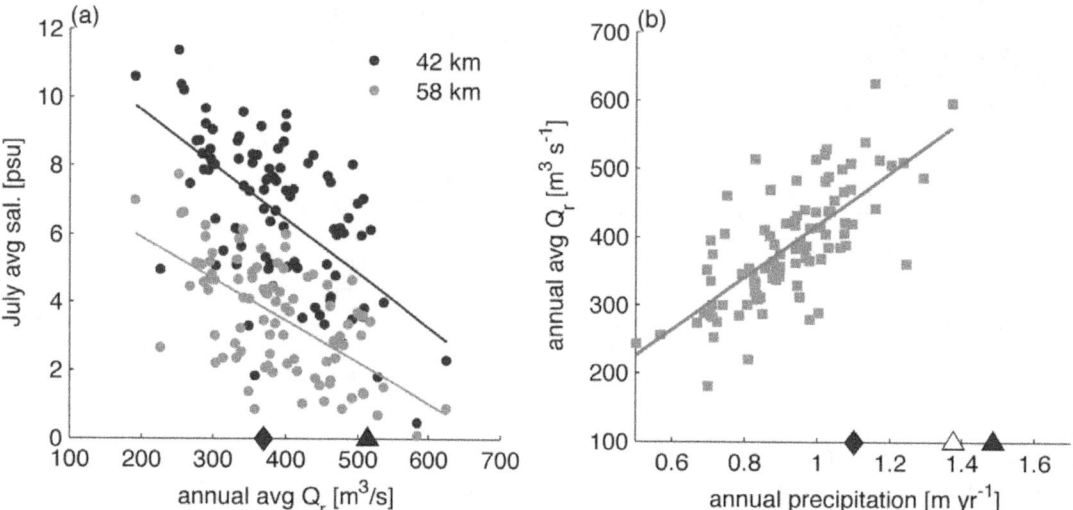

Figure 5. (a) Average salinity in July in lower (42 km) and upper (58 km) TZ-HB from model results against average discharge for the water year. Analysis of covariance shows slopes to be not distinguishable in value over data from 90 years (F = 1.89, p~0.17) but trend lines are significantly displaced (F = 175.23, p<0.001). The diamond marker indicates the median discharge over the period, the closed triangle corresponds with discharge averaged for five wettest years. (b) Relationship between annual rainfall at Albany NY and annual river discharge ($r^2 = 0.57$). The diamond and closed triangle markers are as in (a). The open triangle is a crude projection of the precipitation with climate change (25% increase).

isohalines may shift seaward by as much as 55 km [36]. Delaware Bay has a small watershed and increased rainfall might have a salutary effect, driving low salinity waters and disease refuge into the shallow bay. Previous droughts were associated with expanded mortality from MSX as saline water moved into the upper reaches of Delaware and Chesapeake Bays [9,45]. In general, the impacts of climate change on estuarine oyster populations will depend on how the modified salinity distribution corresponds to the location of suitable benthic habitat. The uncertainty of seasonal effects on changes in rainfall [36] will strongly affect our predictions of potential for oyster restoration.

The summer of 2009 was notable for increased precipitation and discharge during the late spring and summer, but climate predictions suggest increased precipitation may become more common in the future. In the Hudson, the shoals of Tappan Zee and Haverstraw Bay may evolve from a refuge from disease to an inhospitable habitat for oysters, eliminating a crucial component of a larger metapopulation. Even a decade of rainy years, such as the past decade in the Hudson, could hinder restoration efforts. Oyster restoration planning should take into consideration the response of the oligohaline transition between estuarine and fresh waters to potential shifts in forcing with climate change, in particular the magnitude and seasonal timing of discharge. The resilience of restored estuarine oysters may depend on the availability and proximity of suitable benthic substrate for colonization with shifts in the salinity regime. Significant uncertainty remains among predictions of climate change impacts on precipitation, as well as for other potential factors in oyster survival such as water temperature and sea level rise. Restoration efforts could address this uncertainty by focusing on estuarine regions that would allow for translation of the oysters in response to shifts in forcing and by continuously monitoring environmental conditions and oyster population response to better inform subsequent restoration efforts.

Similar effects of climate change on the spread of disease have been widely noted [46] and may portend major reorganization of natural communities in future decades. In the Hudson, the transitional zone of Tappan Zee-Haverstraw Bay and its

vulnerability may provide lessons for estuaries throughout the world. The simultaneous effects of climate change on disease and physiological adaptations may give insight to the effect of regional climate change in other transitional environments.

Materials and Methods

Eastern oysters, *Crassostrea virginica*, were placed in plastic mesh grow-out bags (14 mm mesh size) supported in wire cages suspended 1–2 meters below the surface at nine sites throughout the coastal New York, New Jersey, and Tappan Zee-Haverstraw Bay region (Figure 1, Table 2). Two semi-rigid, rectangular shaped (dimensions of 94×43×7.6 cm) grow-out bags were placed in each wire cage. 300 oysters were placed in each grow-out bag, resulting in a starting density of 742 oysters m^{-2}. Oysters were purchased from the Fishers Island Oyster Farm and were spawned and settled in the summer of 2007 (data for Figure 2a) and overwintered before being transferred to the cages in June 2008. Oysters used in cages in 2009 were spawned and settled in the summer of 2008, overwintered and placed in cages in late May 2009 (data for Figure 2b, 1b). In coastal sites, three replicate cages (6 grow-out bags) were used, located about one meter apart. At Tappan Zee-Haverstraw Bay sites, two cages (4 grow-out bags) were each maintained one meter or more from the other. In both years, oyster height was measured with a random sample of 20 oysters from each sample bag without replacement every two weeks from June-November. We report oyster shell height for the October sampling. Since shell size was the same for all starting samples, the final mean shell height for a locality is a measure of shell growth. All live and dead oysters were counted to calculate survivorship. Cages and bags were cleaned of fouling organisms once every 2 weeks when measurements were taken.

Temperature was monitored with in situ temperature loggers (TidbiT v2 temp loggers from Onset Corporation) attached to one cage at each of the 9 localities. Temperature was registered every 15 minutes. Salinity, temperature and dissolved oxygen were measured biweekly at cage depth using a YSI model 85 environmental TSO meter.

Table 2. Localities, keys to localities, and GPS coordinates.

Locality Key	Site Description	GPS Location
OS	Ossining, Westerly Marina, Bulkhead	N 41°09.521' W 073°52.321'
PM	Philips Manor, Beach Club, floating dock	N 41°05.648' W 073°52.240'
PT	Piermont, Cornetta Marina	N 41°02.708' W 073°54.884'
WI	Tarrytown, Washington Irving Boat Club, permanent dock	N 41°04.320' W 073°52.077'
I	Irvington, Irvington Boat and Kayak Club, floating dock	N 41°02.463' W 073°52.450'
P40	Pier 40, cages suspended over the south side of the vessel *Lilac*	N 40°43.835' W 074°00.776
SI	Shelter Island, Log Cabin Creek in Mashomack Preserve, floating dock	N 41°02.832' W 072°18.020'
JB	Jamaica Bay, residence in Broad Channel, NY, floating dock	N 40°36.414' W 073°49.302
RB	Raritan Bay, New Jersey, Brown's Point Marina, floating dock	N 40°26.147' W 074°12.786'

Disease was assessed for occurrence and intensity of occurrence of MSX and Dermo in the laboratory. A sample of 30 oysters was tested once a year at each site in September. Oysters were dissected and biopsies of mantle and rectum tissues were incubated in Ray's fluid thioglycollate medium (RFTM) for the detection of *P. marinus* [19]. Following incubation (1 week), biopsies were stained with Lugol's iodine and examined using a light microscope for the presence of enlarged, black stained parasite cells. Infection intensity was ranked (0–5) following a scale assessing the relative abundance of parasite cells in tissues (0: no infection, 5:heavy infection) [47]. MSX detection was performed using standard histopathology procedures. Briefly, a transverse slice of tissue roughly between 3 and 5 mm in thickness was made through the central region of the visceral mass to include digestive organs, gonads, as well as gill and mantle tissues. Tissue sections were

placed in histo-cassettes and fixed in 10% buffered formalin. Following fixation, tissue samples were dehydrated and embedded in paraffin, sectioned (5 to 6 μm in thickness), and mounted on histology slides. MSX infection intensity was ranked as light, moderate or heavy based on the abundance of parasite cells in tissue sections and following general guidelines [48].

The numerical model is an unsteady, quasi-2d solution for the along-estuary velocity and salinity distributions. The model has been previously applied to and validated for the Hudson River estuary based on comparisons with high resolution observations in a single year [45] and against observations over several decades, corresponding with simulations presented here [44]. The model was forced with river discharge upstream (USGS station #01358000 from 1946 to present, #01357500 from 1917, and #01335754, from 1887) and with tidal water level downstream (NOAA stations #8518750, #8531680, and #8534720). The model calculates the vertical structure of velocity and salinity at discrete points along the thalweg of the estuary (dx = 1 km). We extract model salinities at depths corresponding to the bed elevation on the shoals where the oyster sites were located. Precipitation observations were taken from Albany, NY (NCDC WBANID #14735 and #14796).

Work on oysters was done with permission under permits to the New York-New Jersey Baykeeper Oyster Gardener Program (Raritan Bay and Jamaica Bay) and, for the other sites, under New York State Fish and Wildlife Scientific Collecting License number 1257 to Jeffrey Levinton.

Acknowledgments

We thank Shauna Kuhn, Abigail Cahill, Pat Lyons, Megan Flennikan, and Sarah Winnicki for assistance in field work. We thank David Wethey for critical input.

Author Contributions

Conceived and designed the experiments: JL MD DR AS BA. Performed the experiments: JL MD DR AS BA. Analyzed the data: JL MD DR AS BA. Contributed reagents/materials/analysis tools: JL MD DR BA. Wrote the paper: JL MD DR. Performed disease analysis: BA. Performed physical modeling: DR.

References

1. Jackson JBC, Kirby MX, Berger WH, Bjorndal KA, Botsford LW, et al. (2001) Historical overfishing and the recent collapse of coastal ecosystems. Science 293: 629–637.

2. USEPA (1998) Condition of the Mid-Atlantic Estuaries. Washington, D.C.: Office of Research and Development, USEPA.

3. Kemp WM, Boynton WR, Adolf JE, Boesch DF, Boicourt WC, et al. (2005) Eutrophication of Chesapeake Bay: historical trends and ecological interactions. Marine Ecology Progress Series 303: 1–29.

4. Newell RIE (1988) Ecological changes in Chesapeake Bay: Are they the result of overharvesting the American oyster, *Crassostrea virginica*? Understanding the estuary: Advances in Chesapeake Bay Research Proceedings of a Conference. Solomons MD: Chesapeake Bay Research Consortium. pp 536–546.

5. Coen LD, Brumbaugh RD, Bushek D, Grizzle R, Luckenbach MW, et al. (2007) Ecosystem services related to oyster restoration. Marine Ecology Progress Series 341: 303–307.

6. Officer CB, Smayda TJ, Mann R (1982) Benthic filter feeding: A natural eutrophication control. Mar Ecol Prog Ser 9: 203–210.

7. Newell RIE, Fisher TR, Holyoke RR, Cornwell JC (2005) Influence of eastern oysters on nitrogen and phosphorus regeneration in Chesapeake Bay, USA. In: Dame R, Olenin S, eds. The Comparative Roles of Suspension Feeders in Ecosystems. Netherlands: Springer.

8. Coleman FC, Williams SL (2002) Overexploiting marine ecosystem engineers: Potential consequences for biodiversity. Trends in Ecology and Evolution 17: 40–43.

9. Haskin HH, Ford SE (1982) *Haplosporidium nelsoni* (MSX) on Delaware Bay seed oyster beds: a host-parasite relationship along a salinity gradient. Journal of Invertebrate Pathology 40: 388–405.

10. Burreson EM, Stokes NA, Friedman CS (2000) Increased Virulence in an Introduced Pathogen: *Haplosporidium nelsoni* (MSX) in the Eastern Oyster *Crassostrea virginica*. Journal of Aquatic Animal Health 12: 1–8.

11. Shumway SE (1996) Natural environmental factors. In: Kennedy VS, Newell RIE, Eble AE, eds. The Eastern Oyster *Crassostrea virginica*. College ParkMD: Maryland Sea Grant. pp 467–513.

12. Kraeuter JN, Ford S, Canzonier W (2003) Increased biomass yield from Delaware Bay oysters (*Crassostrea virginica*) by alternation of planting season. Journal of Shellfish Research 22: 39–49.

13. Haskin HH, Ford SE (1979) Development of resistance to *Minchinia nelsoni* (MSX) mortality in laboratory-reared and native oyster stocks in Delaware Bay. Marine Fisheries Review Jan–Feb: 54–63.

14. Mann R, Evans DA (2004) Site selection for oyster habitat rehabilitation in the Virginia portion of the Chesapeake Bay: A commentary. Journal of Shellfish Research 23: 41–49.

15. Fisher WS, Newell RIE (1986) Salinity effects on the activity of granular hemocytes of American oysters, *Crassostrea virginica*. Biological Bulletin 170: 122–134.

16. Mackin JG, Owen HM, Collier A (1950) Preliminary note on the occurrence of a new protistan parasite, *Dermocystidium marinum*, n. sp. in *Crassostrea virginica* (Gmelin). Science 111: 328–329.

17. Dungan CF, Hamilton RM (1995) Use of a tetrazolium-based cell proliferation assay to measure effects of in vitro conditions on *Perkinsus marinus* (Apicomplexa) proliferation. Journal of Eukaryotic Microbiology 42: 375–398.

18. McCollough CB, Albright BW, Abbe GR, Barker LS, Dungan CR (2007) Aquisition and progression of *Perkinsus marinus* infections by species-pathogen-

free juvenile oysters (*Crassostrea virginica* Gmelin) in a mesohaline Chesapeake Bay tributary. Journal of Shellfish Research 26: 465–477.

19. Ray SM (1952) A culture technique for the diagnosis of infection with *Dermocystidium marinum* (Mackin, Owen, and Collier) in oysters. Science 114: 360–361.

20. Nixon SW, Granger S, Buckley BA, Lamont M, Rowell B (2004) A one hundred and seventeen year coastal water temperature record from Woods Hole, Massachusetts. Estuaries and Coasts 27: 397–404.

21. Cook T, Folli M, Klinck J, Ford S, Miller J (1998) The relationship between increasing sea-surface temperature and the northward spread of *Perkinsus marinus* (Dermo) disease epizootics in oysters. Estuarine, Coastal and Shelf Science 46: 587–597.

22. Ford SE, Smolowitz R (2007) Infection dynamics of an oyster parasite in its newly explanded range. Marine Biology 151: 119–133.

23. Ray S, Mackin JG, Boswell JL (1953) Quantitative measurement of the effect on oysters of disease caused by *Dermocystidium marinum*. Bulletin of Marine Science Gulf and Caribbean 5: 6–33.

24. Craig A, Powell EN, Fay RR, Brooks JM (1989) Distribution of Perkinsus marinus in Gulf coast oyster populations. Estuaries 12: 82–91.

25. La Peyre MK, Gossman B, La Peyre JL (2009) Defining optimal freshwater flow for oyster production: Effects of freshet rate and magnitude of change and duration on eastern oysters and Perkinsus marinus infection. Estuaries and Coasts 35: 522–534.

26. La Peyre MK, Nickens AD, Volety AK, Tolley GS, La Peyre JF (2003) Environmental significance of freshets in reducing *Perkinsus marinus* infection in eastern oysters *Crassostrea virginica*: potential management applications. Marine Ecology Progress Series 248: 165–176.

27. Albright BW, Abbe GR, McCollough CB, Barker LS, Dungan CF (2007) Growth and mortality of dermo-disease-free juvenile oysters (*Crassostrea virginica*) at three salinity regimes in an enzootic area of Chesapeake Bay. Journal of Shellfish Research 26: 451–463.

28. Bain J, Lodge J, Suszkowski DJ, Botkin D, Brash A, et al. (2007) Target Ecosystem Characteristics for the Hudson Raritan Estuary: Technical Guidance for Developing a Comprehensive Ecosystem Restoration Plan. New York: Hudson River Foundation.

29. Franz DR (1982) An historical perspective on mollusks in Lower New York Harbor, with emphasis on oysters. In: Meyer GF, ed. Ecological Stress and the New York Bight: Science and Management. Columbia SC: Estuarine Research Federation. pp 181–197.

30. Black FR (1981) Jamaica Bay: A History. Washington, D.C.: U.S. National Parks Service. pp 1–104.

31. Bromley A (1953–1954) The oyster and the brothers Flower; the Hudson River and private enterprise combine to write a new story. The New York State Conservationist. pp 4–9.

32. Ristich SS, Crandall M, Fortier J (1972) Benthic and epibenthic macroinvertebrates of the Hudson River : I. Distribution, natural history and community structure. Estuarine and Coastal Marine Science 5: 255–266.

33. Mann R, Powell EN (2007) Why oyster restoration goals in the Chesapeake Bay are not and probably cannot be achieved. Journal of Shellfish Research 26: 905–917.

34. Schulte DM, Burke RP, Lipcius RN (2009) Unprecedented restoration of a native oyster metapopulation. Science 325: 1124–1128.

35. Hofmann E, Ford S, Powell E, Klinck J (2001) Modeling studies of the effect of climate variability on MSX disease in eastern oyster (*Crassostrea virginica*) populations. Hydrobiologia 460: 195–212.

36. Najjar RG, Pyke CR, Adams MB, Breitberg D, Hershner C, et al. (2010) Potential climate-change impacts on the Chesapeake Bay. Estuarine, Coastal and Shelf Science 86: 1–20.

37. Hayhoe K, Wake CP, Huntington TG, Luo L, Schwartz MD, et al. (2007) Past and future changes in climate and hydrological indicators in the US Northeast. Cimate Dynamics 28: 381–407.

38. Najjar RG, Walker HA, Anderson PJ, Barron EJ, Bord RJ, et al. (2000) The potential impacts of climate change on the mid-Atlantic coastal region. Climate Research 14: 219–233.

39. Ralston DK, Geyer WR (2009) Episodic and long-term sediment transport capacity in the Hudson River estuary. Estuaries and Coasts 32: 1130–1151.

40. Ralston DK, Geyer WR, Lerczak JA (2008) Subtidal salinity and velocity in the Hudson River Estuary: Observations and modeling. Journal of Physical Oceanography 38: 753–770.

41. Starke AF (2010) Restoration of the Hudson River oyster: A physiological and spatial assessment of *Crassostrea virginica*'s restoration in the Hudson River, NY. Stony Brook: Stony Brook University. 66 p.

42. Lipcius RN, Eggleston DB, Schreiber SJ, Seitz RD, Shen J, et al. (2008) Importance of metapopulation connectivity to restocking and restoration of marine species. Reviews in Fisheries Science 16: 101–110.

43. North EW, Shlag Z, Hood RR, Li M, Zhong L, et al. (2008) Veritcla swimming behavior influences the dispersal of simulated oyster larvae in a coupled particle-tracking and hydrodynamic model in Chesapeake Bay. Marine Ecology Progress Series 359: 90–115.

44. Hofmann E, Bushek D, Ford S, Guo X, Haidvogen D, et al. (2009) Understanding how disease and environment combine to structure resistance in estuarine bivalve populations. Oceanography. pp 213–231.

45. Burreson EM, Ford SE (2004) A review of recent information on the Haplosporidia, with special reference to *Haplosporidium nelsoni* (MSX disease). Aquatic Living Resources 17: 499–517.

46. Harvell CD, Mitchell CE, Ward JR, Altizer S, Dobson A, et al. (2002) Climate warming and disease risks for terrestrial and marine biota. Science 296: 2158–2162.

47. Mackin JG (1962) Oyster disease caused by *Dermocystidium marinum* and other microorganisms in Louisiana. Publications of the Institute of Marine Science, University of Texas 7: 132–229.

48. Ford SE, Figueras AJ (1988) Effects of sublethal infection by the parasite *Haplosporidium nelsoni* (MSX) on gametogenesis, spawning, and sex ratios of oysters in Delaware Bay, USA. Diseases of Aquatic Organisms 4: 121–133.

A New Cryptic Species of South American Freshwater Pufferfish of the Genus *Colomesus* (Tetraodontidae), Based on Both Morphology and DNA Data

Cesar R. L. Amaral[1,2]*, Paulo M. Brito[1], Dayse A. Silva[2], Elizeu F. Carvalho[2]

1 Department of Zoology, Universidade Estadual do Rio de Janeiro, Rio de Janeiro, Brazil, 2 Department of Ecology, Universidade Estadual do Rio de Janeiro, Rio de Janeiro, Brazil

Abstract

The Tetraodontidae are an Acantomorpha fish family with circumglobal distribution composed of 189 species grouped in 19 genera, occurring in seas, estuaries, and rivers between the tropical and temperate regions. Of these, the genus *Colomesus* is confined to South America, with what have been up to now considered only two species. *C. asellus* is spread over the entire Amazon, Tocantins-Araguaia drainages, and coastal environments from the Amazon mouth to Venezuela, and is the only freshwater puffers on that continent. *C. psittacus* is found in coastal marine and brackish water environments from Cuba to the northern coast of South America as far south as to Sergipe in Brazil. In the present contribution we used morphological data along with molecular systematics techniques to investigate the phylogeny and phylogeography of the freshwater pufferfishes of the genus *Colomesus*. The molecular part is based on a cytochrome C oxidase subunit I dataset constructed from both previously published and newly determined sequences, obtained from specimens collected from three distinct localities in South America. Our results from both molecular and morphological approaches enable us to identify and describe a new *Colomesus* species from the Tocantins River. We also discuss aspects of the historical biogeography and phylogeography of the South American freshwater pufferfishes, suggesting that it could be more recent than previously expected.

Editor: Dorothee Huchon, Tel-Aviv University, Israel

Funding: This study was supported by the Brazilian National Counsel of Technological and Scientific Development and Fundação Carlos Chagas Filho de Amparo à Pesquisa do Estado do Rio de Janeiro. The funders had no role in study design, data collection and analysis, decision to publish, or preparation of the manuscript.

Competing Interests: The authors have declared that no competing interests exist.

* E-mail: crlamaral@yahoo.com.br

Introduction

The Tetraodontidae is an Acantomorpha fish family with circumglobal distribution composed of 189 species in 19 genera, occurring in seas, estuaries, and rivers between the tropical and temperate regions [1]. They are mainly characterized by their typical four large dental plates; the ability to inflate their body in stressful situations; the presence of the neurotoxin Tetrodotoxin/Saxitoxin in its tissues, being responsible for numerous cases of fatal poisoning in many countries, including Brazil; and by having the smallest genome among vertebrates, therefore being considered as a model for the genome evolution of the group.

Among the Amazonian taxa exploited by the ornamental fish industry in South America are those of *Colomesus* [2], a genus confined to South America, with what is presently considered two species, *C. asellus* and *C. psittacus*. *C. asellus* [3] is spread in the entire Amazon, Tocantins-Araguaia drainages, and coastal environments from the Amazon mouth to Venezuela, being the only freshwater puffers on that continent. *C. psittacus* [4] is found in coastal marine and brackish water environments from Cuba and the northern coast of South America to Sergipe in Brazil.

Mainly located in tropical and subtropical regions all around the world, including the Amazon region, the ornamental fish industry is one of the largest transporters of live animals and plants with an annual trade volume estimated at U\$15–25 billion [5–7], in a scenario where species identification problems, mainly related to border biosecurity are not rare.

The DNA barcode is a widely accepted tool for species determination mainly due to its enhanced attention on standardization and data validation [8], being a rapid and low cost method of identification [9]. The use of DNA barcoding techniques has been utilized in many taxa, including bacteria, birds, bivalves, butterflies, fishes, flies, macroalgae, mammals, spiders, sprigtails, and also for plants [10–27].

The DNA barcode technique for Metazoans uses a short (~650 bp) and standardized gene region from the mitochondrial 5′ region of the cytochrome C oxidase subunit I (COI) for a rapid and cost-effective animal identification. This has been demonstrated to be an effective fish identification tool in numerous situations, including consumer protection [28–30], fisheries management/conservation [31], border biosecurity in the ornamental fish trade [5], and in the identification of overlooked or cryptic species [32].

Here we used both morphological and molecular methodologies in an integrative taxonomical approach to investigate the diversity of the Amazonian freshwater pufferfishes of the genus *Colomesus* based on specimens collected from three distinct populations from both Brazil and Peru. Additionally, we describe a new *Colomesus*

Figure 1. Map of South America showing the northern hydrology and the localities where the specimens were collected (grey and green marks).

species from the Upper Tocantins drainage based on both morphological and molecular data.

Methods

Specimens of *Colomesus asellus* were collected from three distinct populations with about 2200 km of mean distance separating them. The collection localities were Ilha do Mosqueiro, Belém, Brazil; Upper Tocantins River - Porto Nacional, Tocantins, Brazil; and Nanay River - Iquitos, Peru (Figure 1).

Ethics Statement

No statement from an ethics committee was necessary, and the manuscript did not involve any endangered or protect species. All samples were extracted from dead specimens collected with

appropriate permissions under authorization number 22512 issued by SISBIO/Instituto Chico Mendes de Conservação da Biodiversidade. We used the ice-slurry method for killing following [33] as they are tropical warm water species and the collected specimens are all smaller than 5 cm SL. All specimens were preserved in alcohol. The reported localities do not include protected areas.

Nomenclatural Acts

The electronic version of this document does not represent a published work according to the International Code of Zoological Nomenclature (ICZN), and hence the nomenclatural acts contained in the electronic version are not available under that Code from the electronic edition. Therefore, a separate edition of

Table 1. Taxonomic sampling and accession numbers.

Taxon	Accession No.	Taxon	Accession No.
Família Triodontidae			JQ841396
Triodon macropterus	AP009170		JQ840304
Família Diodontidae			GU225449
Diodon holocanthus	AP009177	Sphoeroides annulatus	GU440524
Chilomycterus reticulatus	AP009188	Sphoeroides testudineus	KC959927*
Família Tetraodontidae			KC959928*
Lagocephalus laevigatus	AP011934		GU225665
	JQ365394		GU225453
	JQ365392		GU225450
	JQ365395		JQ842706
	JQ365393		JQ843064
	KC959926*		JQ840306
Lagocephalus inermis	FJ434549		GU225664
	GU804920		JQ365576
Lagocephalus lunaris	DSFSG9111		JQ365575
Lagocephalus lagocephalus	AP011933		JQ365574
Lagocephalus guentheri	HQ149858		GU440524
	JF493722	Colomesus psittacus	KC959923*
	JF493724		KC959924*
Lagocephalus wheeleri	JF952772		KC959925*
	AP009538	Colomesus asellus	KC959904*
	FJ434551		KC959907*
Lagocephalus spadiceus	EU595163		KC959908*
	EU595161		KC959909*
Takifugu ocellatus	AP009536		KC959910*
Takifugu poecilonotus	AP009539		KC959911*
Takifugu snyderi	AP009531		KC959913*
Takifugu oblongus	AP009535		KC959914*
Takifugu pardalis	AP009528		KC959915*
Takifugu niphobles	AP009526		KC959916*
Takifugu porphyreus	AP009529	Colomesus tocantinensis	KC959905*
Tetraodon biocellatus	KC959929*		KC959906*
Tetraodon nigroviridis	KC959930*		KC959912*
Sphoeroides pachygaster	EU074597		KC959917*
	EU074598		KC959918*
	EU869843		KC959919*
	CSFOM07310		KC959920*
	JF494544		KC959921*
	JF494541		KC959922*
	EU869842		
	EU869839		
	AP006745		
Sphoeroides greeleyi	JQ365572		
Sphoeroides nephelus	JQ842698		
	JQ842695		
	JQ842699		
Sphoeroides spengleri	JQ842704		
	JQ842695		
	JQ842701		
	JQ841395		

(*)Sequences newly determined in this study.

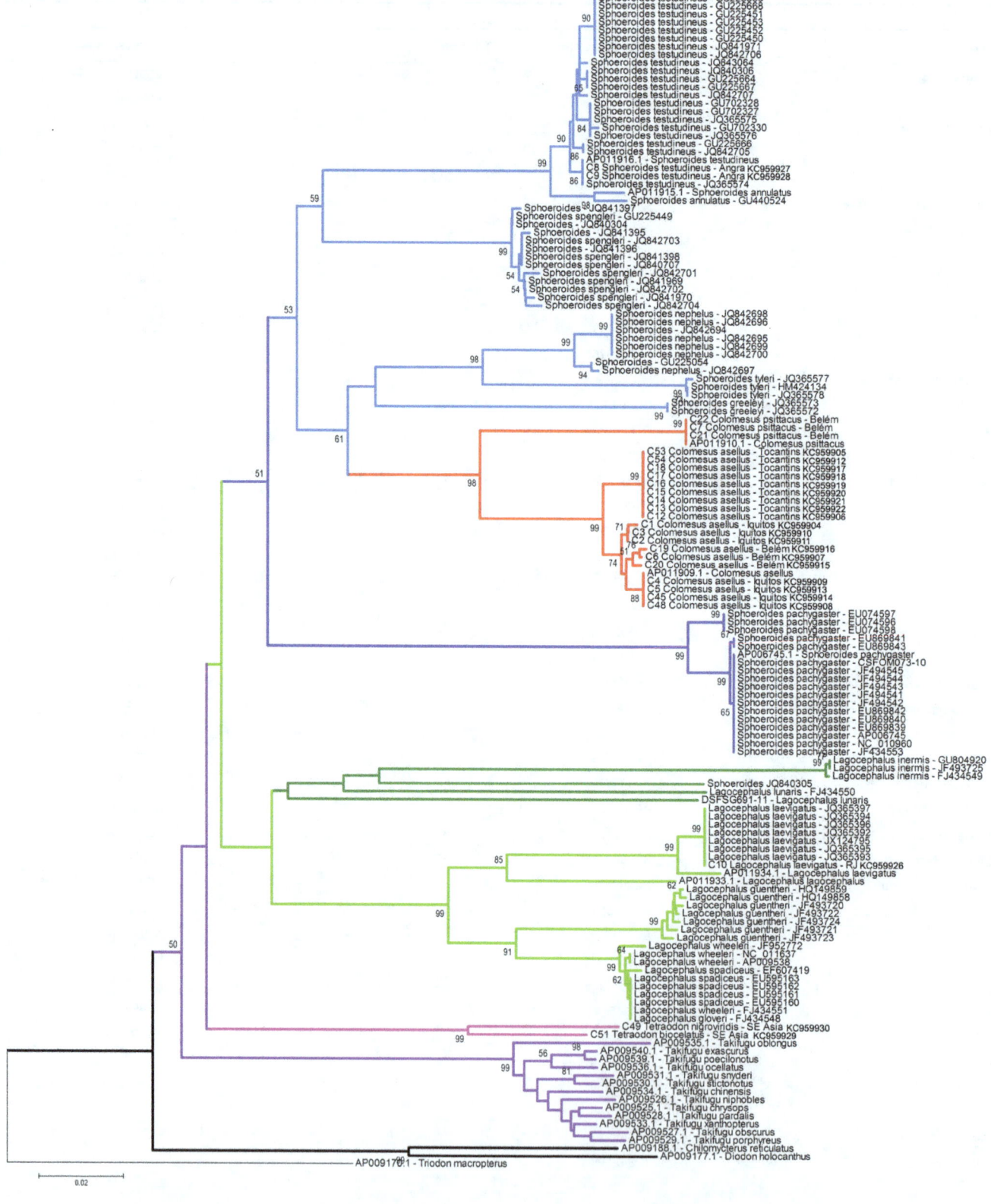

Figure 2. Neighbor-Joining tree based on the barcode region of the COI. The numbers near the branches represent bootstrap probabilities higher than 50%.

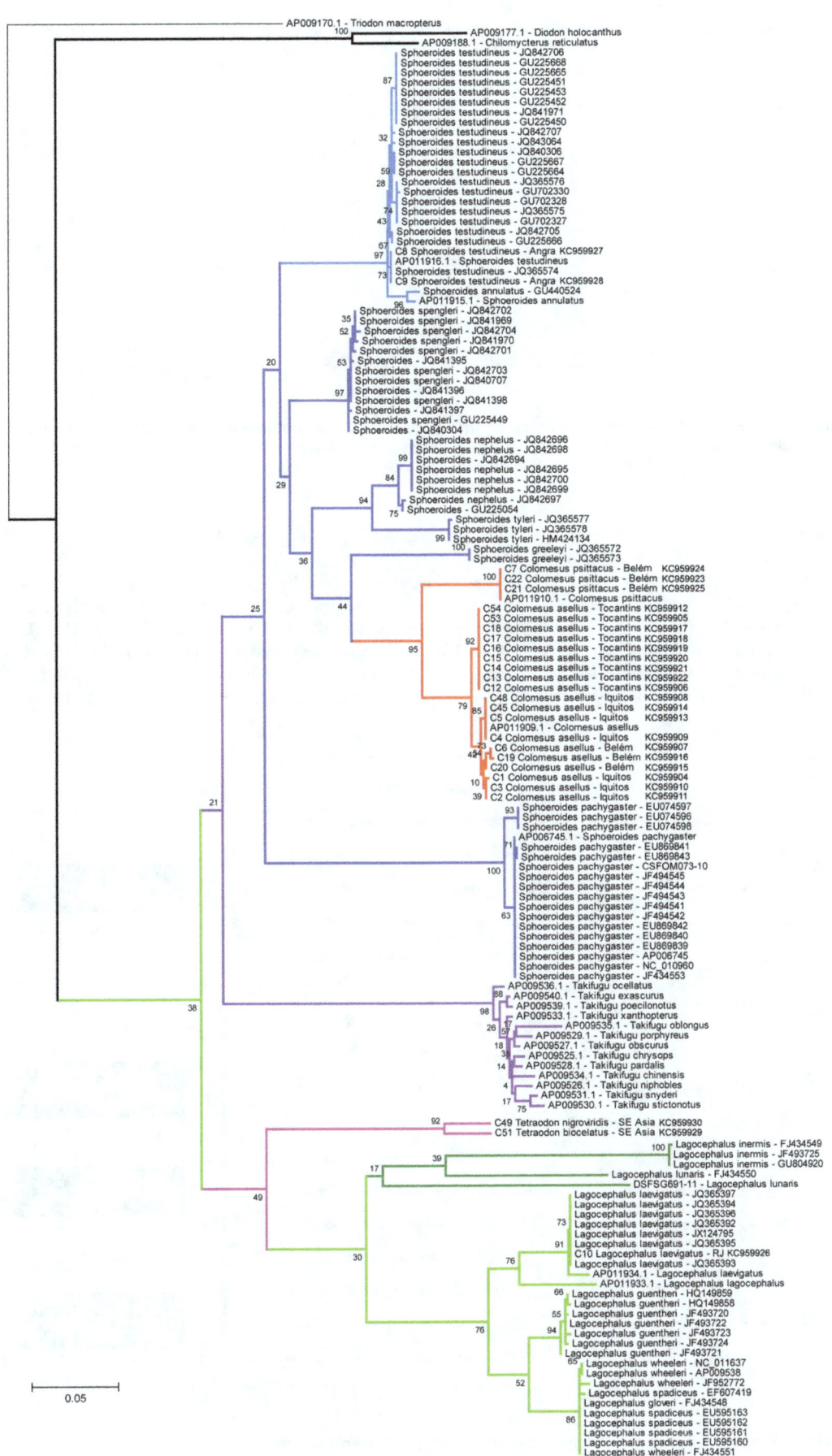

Figure 3. Maximum-likelihood phylogeny based on the barcode region of the COI marker. The numbers near the branches represent the bootstrap probabilities.

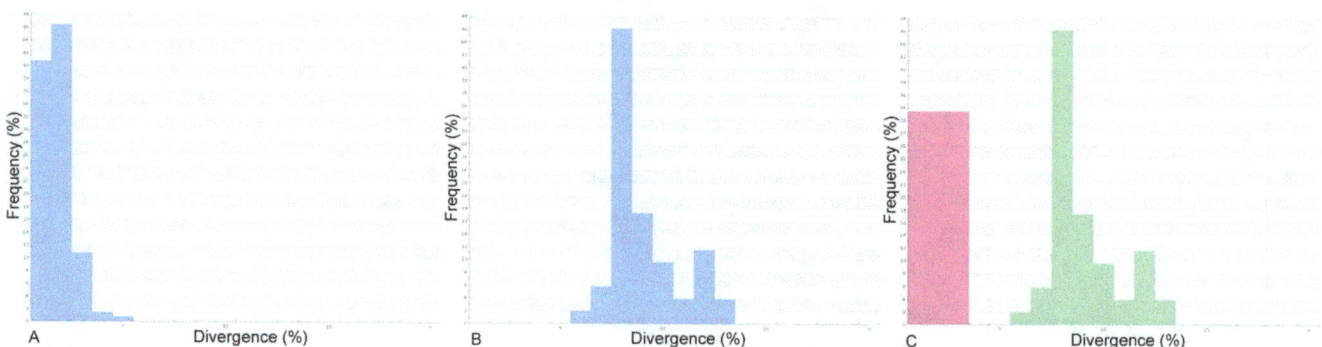

Figure 4. Distribution of K2P distances (%) for COI: A) within species; B) within genera; C, normalized distribution of K2P distance (%) within species. The analyses included the following taxa: *Tetraodon nigroviridis*, *Tetraodon biocellatus*, *Sphoeroides testudineus*, *Lagocephalus laevigatus*, *Colomesus asellus*, *Colomesus psittacus*, and the freshwater *Colomesus* from the Tocantins drainage.

this document was produced by a method that assures numerous identical and durable copies, and those copies were simultaneously obtainable (from the publication date noted on the first page of this article) for the purpose of providing a public and permanent scientific record, in accordance with Article 8.1 of the Code. The separate print-only edition is available on request from PLoS by sending a request to PLoS ONE, 1160 Battery Street, Suite 100, San Francisco, CA 94111, USA along with a check for $10 (to cover printing and postage) payable to "Public Library of Science".

In addition, this published work and the nomenclatural acts it contains have been registered in ZooBank, the proposed online registration system for the ICZN. The ZooBank LSIDs (Life Science Identifiers) can be resolved and the associated information viewed through any standard web browser by appending the LSID to the prefix "http://zoobank.org/". The LSID for this

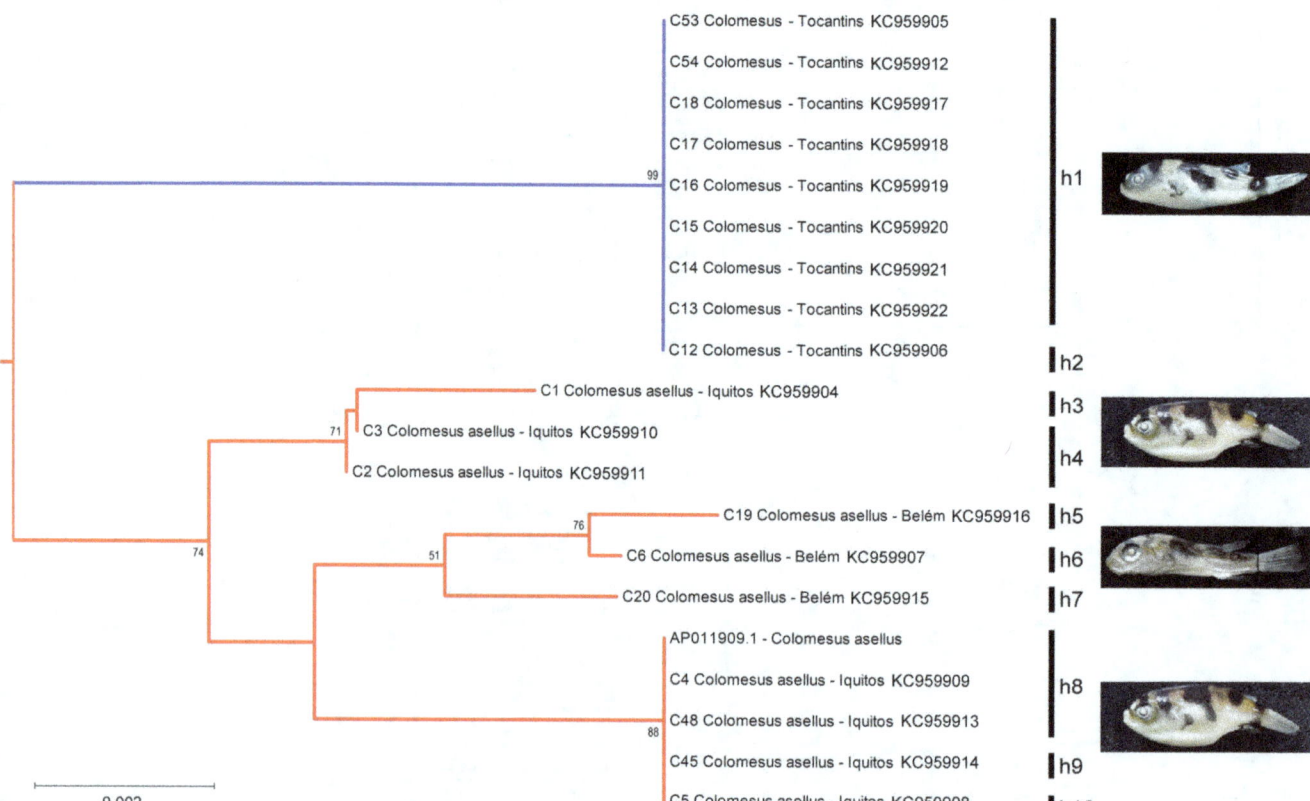

Figure 5. Neighbor-Joining phylogeny of the freshwater *Colomesus* and haplotype determination.

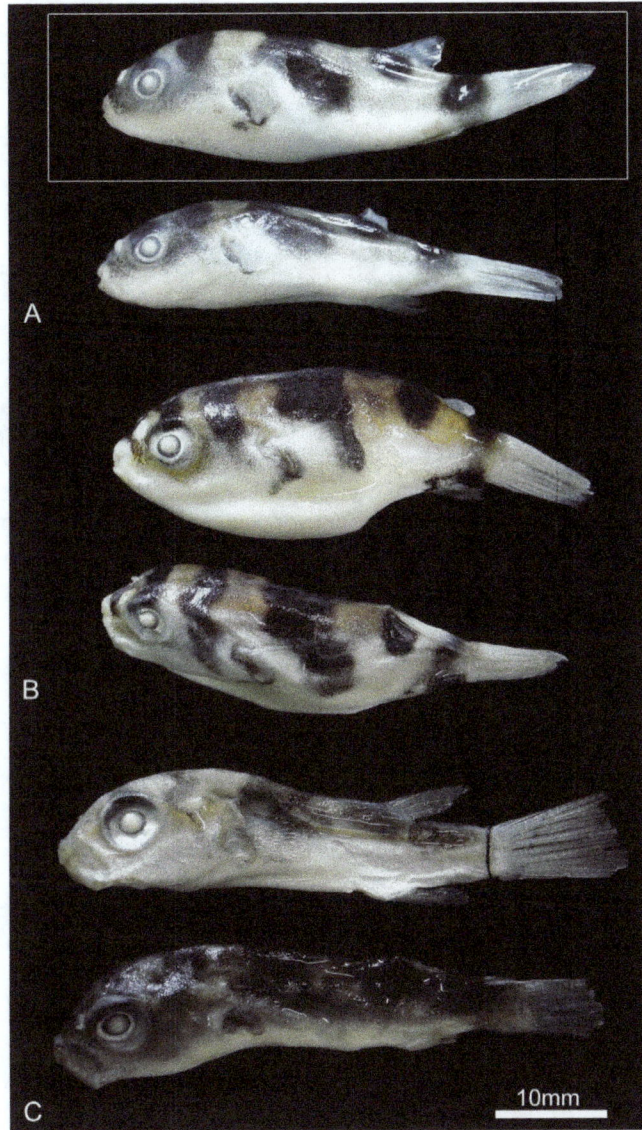

Figure 6. External morphology of the genus *Colomesus*. A) *Colomesus tocantinensis* **nov. sp.** – Tocantins (holotype PNT.UERJ.405 highlighted in white); B) *Colomesus asellus* – Iquitos; C) *Colomesus asellus* – Belém.

publication is: urn:lsid:zoobank.org:pub:033A323A-18F7-4788-8405-32D78BF65B13.

Morphological Analyses

Specimens from all three localities were cleared and stained following the methodology of [34].

Molecular Analyses

The molecular systematic analyses used newly determined sequences obtained from the mitochondrial barcode marker COI as well as previously published sequences obtained from the NCBI and BOLD databases.

For the newly determined sequences, a fragment of epaxial musculature was submitted to the standard protocol for DNA extraction and purification from the Qiagen QIAamp DNA FFPE Tissue kit. The fragments were amplified and sequenced using the primers VF2_t1 and FishR2_t1 [35–37]. All primers were

Table 2. Morphometric and meristic data of the type series of *Colomesus tocantinensis* nov. sp.

Register	SL	HL	PR	DR	AR	CR	IOL
PNT.403	30.84	10.75	15	10	9	11	5.8
PNT.404	34.9	11.83	15	10	9	11	5.9
PNT.405*	29.62	10.37	16	10	9	11	5.47
PNT.395	30.35	11.25	15	10	9	11	6.35
PNT.396	29.28	11.1	16	9	9	11	6.78
PNT.397	29.59	10.79	16	10	9	11	5.48
PNT.398	29.46	10.6	15	10	9	11	5.77
PNT.399	30.66	11.99	16	10	9	11	5.55
PNT.400	32.92	11.75	15	10	9	11	6.03
PNT.401	27.02	9.61	16	10	9	11	5.12

SL, standard length; HL, head length; PR, pectoral fin rays; DR, dorsal fin rays; AR, anal fin rays; CR, caudal fin rays; IOL, interorbital length. (*)Holotype.

appended with M13 tails on sequencing reactions. The PCR profile consisted of 2 min at 95°C, 35 cycles of 30 sec at 94°, 40 sec at 52°C, and 1 min at 72°C, with a final extension step for 10 min at 72°C. Sequencing reactions were performed with the use of the BigDye® Terminator v.3.1 Cycle Sequencing kit (Applied Biosystems, Inc.), with 25 cycles of 10 sec at 95°C, 5 sec at 50°C and 4 min at 60°C. Sequencing products were processed in an ABI 3500 capillary system (Applied Biosystems, Inc.).

The chromatograms were checked and aligned using the BioEdit 7.053 [38] software with its built-in ClustalW routine [39]. The alignment was visually inspected for accuracy and to minimize missing data. All the newly determined sequences are available at the BOLD database (http://www.barcodinglife.com) under the project acronym PUFER. The GenBank accession numbers for all newly determined and previously published sequences used in the present manuscript are summarized in Table 1. The dataset consisted of a 651 bp COI matrix, and we used the MEGA 5.06 software [40] to determinate the TN93+G+I as the most appropriate model of sequence evolution based on the Akaike criterion (AIC) [41].

The neighbor-joining (NJ) and maximum-likelihood (ML) trees that encompass the genera *Triodon*, *Diodon*, *Chilomycterus*, *Lagocephalus*, *Tetraodon*, *Takifugu*, *Sphoeroides*, and *Colomesus*, were constructed using the MEGA 5.06 software [40].

The neighbor-joining sequence divergences were calculated based on the Kimura Two Parameter (K2P) distance model [42] on BOLD workbench and MEGA 5.06 software [40]. The haplotype determination was carried with the use of the server FaBox (http://birc.au.dk/software/fabox/).

Results and Discussion

The neighbor-joining (NJ) and maximum-likelihood (ML) result trees are presented in Figures 2 and 3, respectively. The genus *Colomesus* was recovered as monophyletic inside the group formed by the sampled *Sphoeroides* species, in except for *Sphoeroides pachygaster*. *Lagocephalus* was recovered in a basal phylogenetic position in relation to *Sphoeroides* and *Colomesus*, therefore corroborating recent results such as those presented by [43–45].

Colomesus was recovered deep inside the group formed by the remaining *Sphoeroides* species, therefore suggesting *Sphoeroides* as paraphyletic, with *S. pachygaster* being recovered as basal in relation

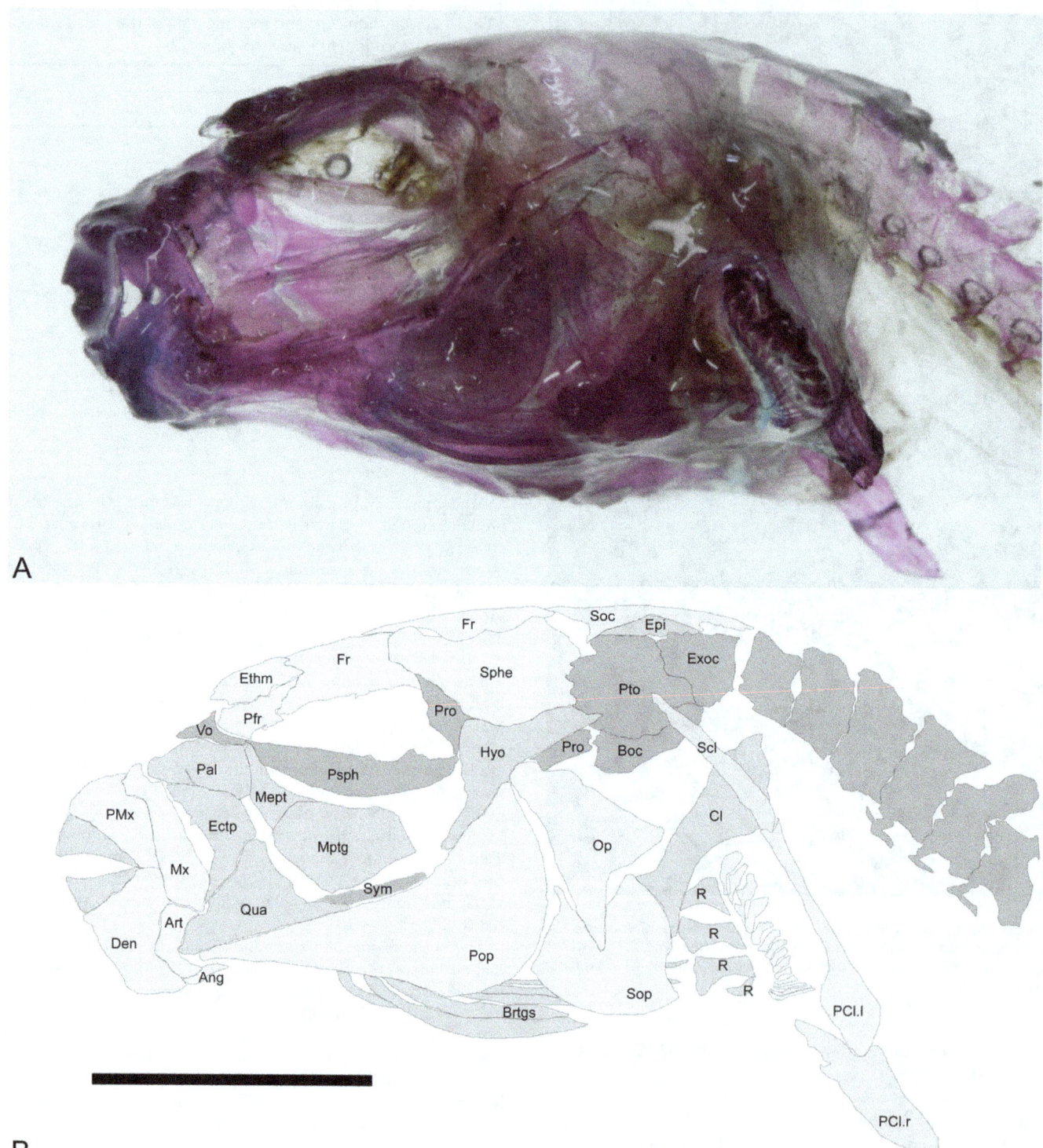

Figure 7. *Colomesus tocantinensis* **nov. sp. (PNT.UERJ.398).** A) left photograph of the head; B) anatomical interpretations. Abbreviations: Ang, angular; Art, articular; Boc, basioccipital; Brstgs, branchiostegals; Cl, cleithrum; Den, dentary; Epi, epiotic; Ethm, ethmoid; Exo, exoccipital; Fr, frontal; Hyo, hyomandibula; Ecptg, ectopterygoid; Mept, mesopterygoid; Mtptg, metapterygoid; Mx, Maxilla; Op, opercle; Pal, palatine; PCl.l/r, ventral post-cleithrum left and right; Pfr, prefrontal; PMx, premaxilla; Pop, preopercle; Pro, prootic; Psph, parasphenoid; Pto, pterotic; Qua, quadrate; R, radials; Scl, supracleithrum; Soc, supraocciptal; Sop, subopercle; Sym, sympletic; Vo, vomer. Scale bar equals 5 mm.

to all the remaining *Sphoeroides* species in all the analyses. Additionally, *Colomesus* was also recovered as the sister-taxa of the group formed by the species *Sphoeroides nephelus*, *S. tyleri*, and *S.*

greeleyi in the NJ result, although it was recovered as the sister-taxa of *S. greeleyi* in the ML results.

In the same way, *Colomesus* is clearly distinguishable from the group formed by all the *Sphoeroides* species mainly by the banded

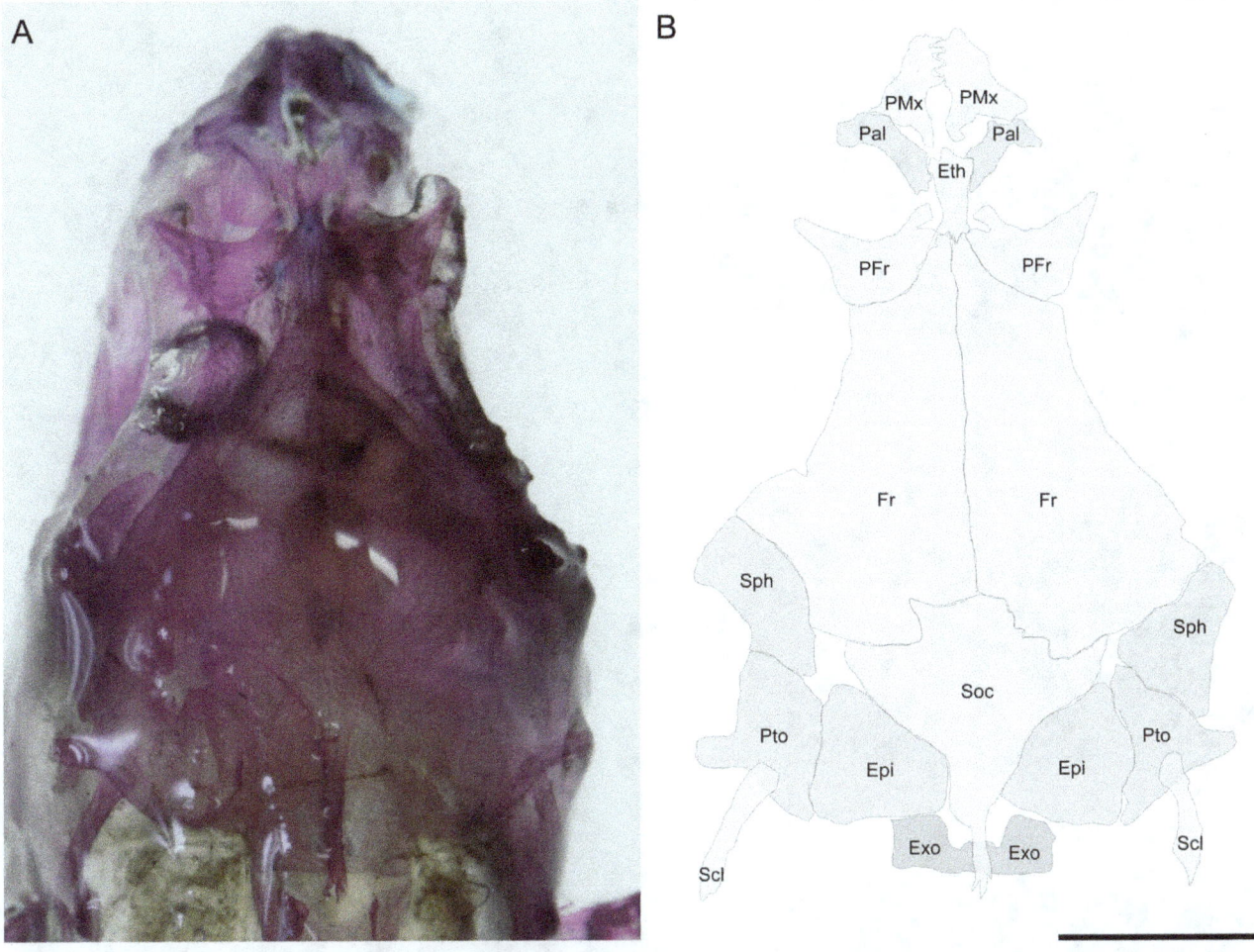

Figure 8. *Colomesus tocantinensis* **nov. sp. (PNT.UERJ.398).** A) top photograph of the head; B) anatomical interpretations. Abbreviations: Epi, epiotic; Ethm, ethmoid; Exo, exoccipital; Fr, frontal; Pal, palatine; Pfr, prefrontal; PMx, premaxilla; Pto, pterotic; Scl, supracleithrum; Soc, supraocciptal; Sphe, sphenotic. Scale bar equals 5 mm.

pigmentation pattern present in all the *Colomesus* species; the presence of two lateral lines, with the ventral line running the full length of the caudal peduncle; and the absence of an upraised horizontal ridge of skin ventrolaterally along the caudal peduncle. The color pattern was used by [46], along with pectoral fin ray counts, the presence of a dark bar underside of caudal peduncle, and the presence of dermal flaps across the chin, to distinguish between what at that time were considered to be the only two species of the genus, the marine/estuarine *C. psittacus*, and the freshwater *C. asellus*. The presence of a dark bar on the underside of the caudal peduncle is a prominent feature for specimens of *C. asellus* from Iquitos, but this bar is present or not in specimens from both Belém and Tocantins. Dermal flaps were observed in all specimens from both Iquitos and Belém, but such flaps were not observed in any of the examined specimens from the Tocantins drainage.

DNA Barcode and Deep Sequence Divergence

COI amplicons were obtained from all the specimens included in the analyses. The obtained sequences clearly identified both previous accepted *Colomesus* species (*C. asellus* and *C. psittacus*), therefore being in accordance with the previous morphological diagnose presented by [46].

The K2P divergence distances between congeneric species ranged from 5.557% to 12.394% with a mean distance of 8.546%, while the uncorrected K2P distance ranged from 0 to 4.472% within species. The mean K2P distance within the analyzed populations was 0.657% and the mean normalized distance within species is 1.079% (Figure 4).

Deep sequence divergence was observed regarding the freshwater *Colomesus* from the Tocantins drainage (Figure 5). The mean sequence divergence of the specimens from both Belém and Iquitos was estimated at 1.079%, while the Tocantins distances ranged from 1.955% to 3.063%, with a mean distance of 2.166%. The observed sequence divergence values together with the congruence observed from both molecular and morphological phylogenetic approaches used here suggest the existence of an overlooked species within the genus *Colomesus*.

A new *Colomesus* species from the Tocantins River, Brazil

Systematics. **Tetraodontiformes** *sensu* Tyler, 1980 [47]

Tetraodontidae *sensu* Santini & Tyler, 2003 [45]

Colomesus Gill, 1885 [2]

Colomesus tocantinensis **nov. sp.** urn:lsid:zoobank.org:act:9B8ACCB5-FF55-4514-901B-6366FB6EA307

Derivation of name. The specific epithet *tocantinensis* refers to the type locality, Porto Nacional, State of Tocantins, Brazil.

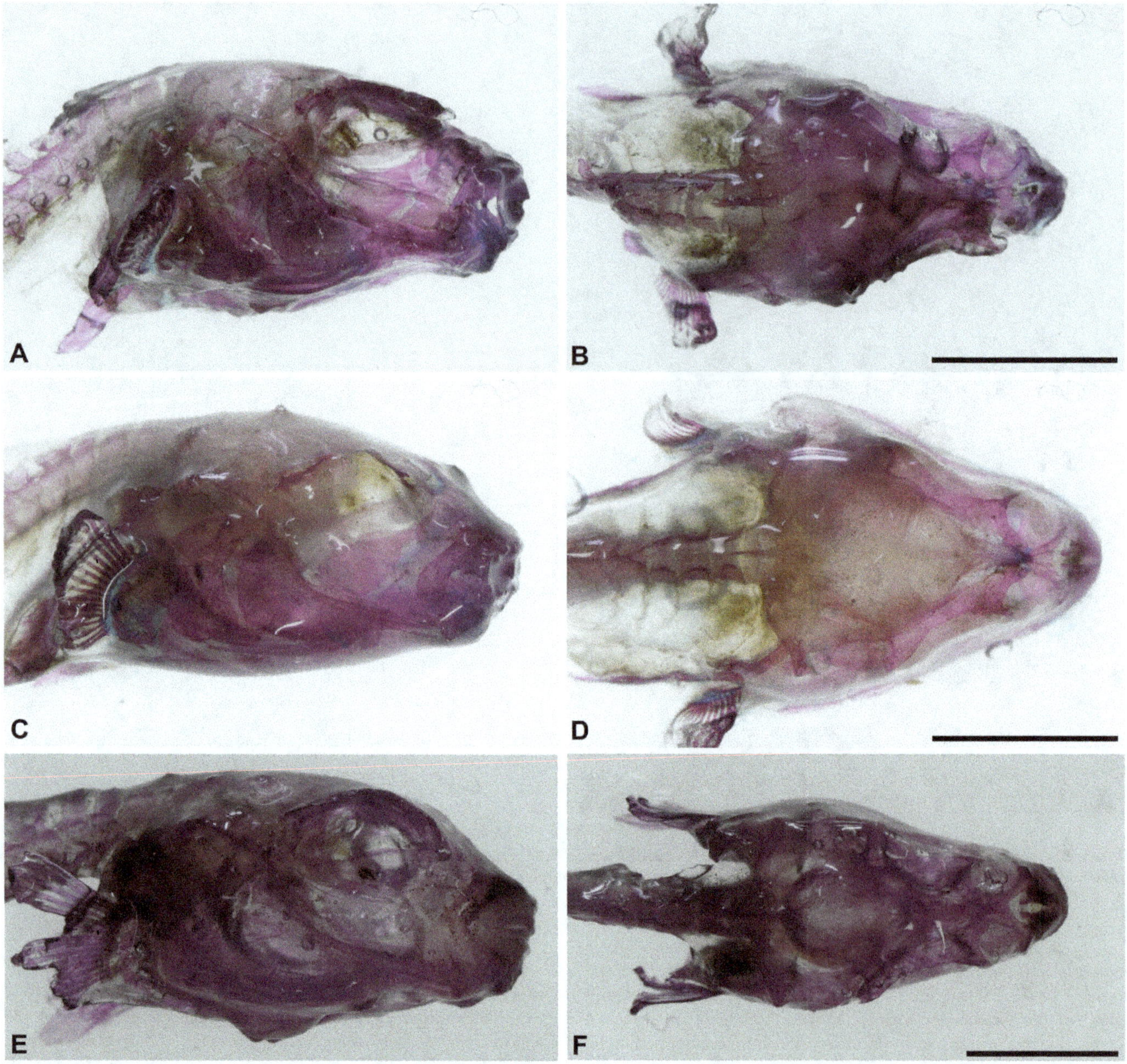

Figure 9. Right and top photographs from cleared-and-stained specimens of: A–B) *Colomesus tocantinensis* **nov. sp. – Tocantins (PNT.UERJ.398); C–D)** *Colomesus asellus* **– Iquitos (PNT.UERJ.470); E–F)** *Colomesus asellus* **– Belém (PNT.UERJ.386).**

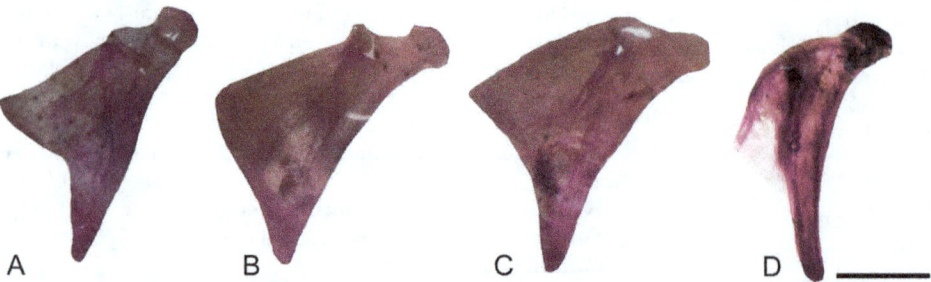

Figure 10. Isolated opercles from: A) *Colomesus tocantinensis* **nov. sp. – Tocantins (PNT.UERJ.398); B)** *Colomesus asellus* **– Iquitos (PNT.UERJ.470); C)** *Colomesus asellus* **– Belém (PNT.UERJ.386); D)** *Colomesus psittacus* **– Belém (PNT.UERJ.387).** Scale bar equals 1 mm.

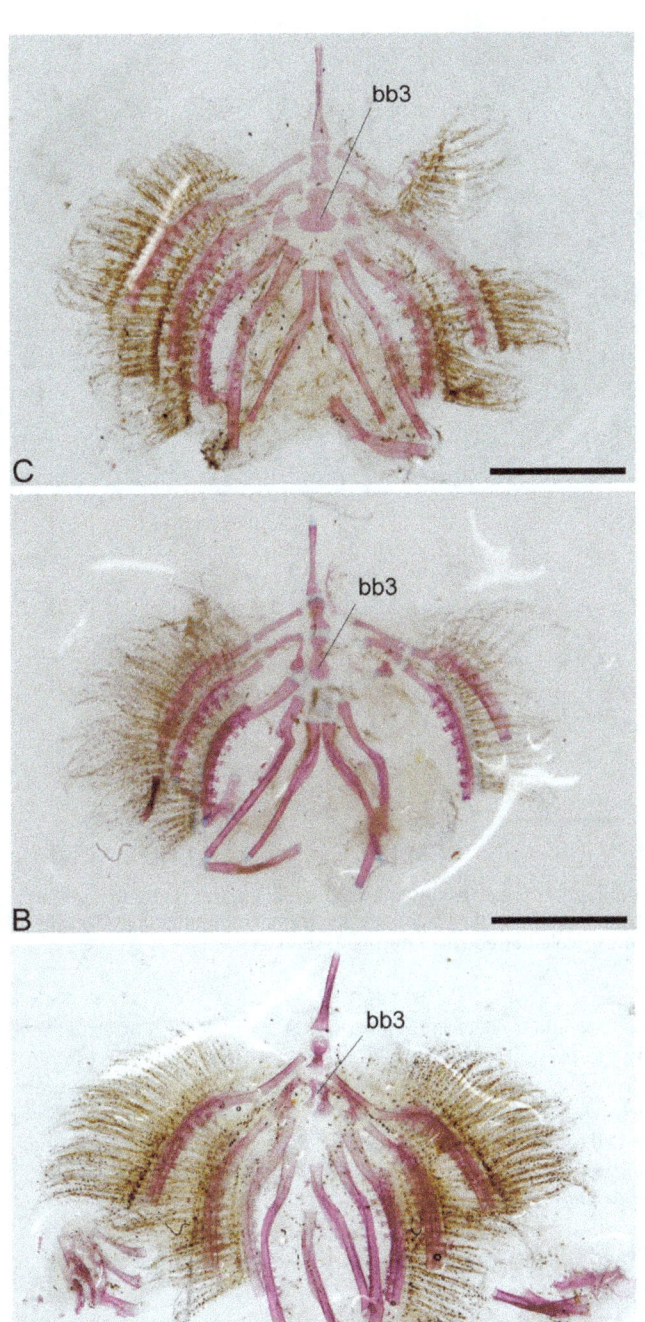

Figure 11. Isolated branchial apparatus from: A) *Colomesus tocantinensis* **nov. sp. – Tocantins (PNT.UERJ.404); B)** *Colomesus asellus* **– Iquitos (PNT.UERJ.470); C)** *Colomesus asellus* **– Belém (PNT.UERJ.386).** Scale bar equals 2.5 mm.

Holotype.   PNT.UERJ.405 (Figure 6).

Paratypes.   PNT.UERJ.396, PNT.UERJ.397, PNT.UERJ.398, PNT.UERJ.399, PNT.UERJ.400, PNT.UERJ.401, PNT.UERJ.402, PNT.UERJ.403, PNT.UERJ.404.

Type-locality.   The specimens are from the Tocantins River near Porto Nacional, State of Tocantins, Brazil.

Diagnosis.   *Colomesus* species diagnosed by six to seven basal pterygiophores and nine rays in the anal fin (*contra* ten to eleven in both *C. asellus* and *C. psittacus*); ten basal pterygiophores and rays in the dorsal fin (*contra* eleven for both *C. asellus* and *C. psittacus*); the

absence of dermal flaps across the chin (*contra* its presence uniquely in *C. asellus*); a caudal peduncle with eight vertebrae; and an opercle with a posterior ventral border subdivided in a ventral and a posterior region, the herein called "inverted V" shape (*contra* the triangular opercle exhibited by both *C. asellus* and *C. psittacus*).

Description.   The holotype (PNT.UERJ.405) is 29,62 mm SL (Figure 6), with 10,37 mm HL; the entire type-series ranges from 27.02 mm to 34.9 mm SL. The meristic and morphometric data of the type series is presented in Table 2. The extent of the dorsal and ventral lateral lines is similar to those found in *C. asellus*. The prickles extend along the dorsal, lateral, and ventral surfaces of the body, from the level of the eye to the origin of the dorsal fin.

The color pattern of *Colomesus tocantinensis* **nov. sp.** is essentially the same as that of *Colomesus asellus*, with five transverse dark bars across the dorsal region of the body. A dark blotch on the underside of the caudal peduncle, which is a state used by [46] to diagnose *Colomesus asellus*, *is* present or absent, being vestigial to unobservable or absent in several specimens. The interspaces between the dark bars are light yellow, with gradually decreasing pigmentation and becoming white in the ventral region (Figure 6). However, the light yellow to pale pattern presented by *C. tocantinensis* **nov. sp.** clearly contrasts with the gold-yellow pattern present in specimens from Iquitos and Belém.

The nasal sac is higher than that presented in the specimens of *C. asellus*. Two large lateral and anteromedial nostrils are present. They are similar to those found on *C. psittacus*, rather than the two small nostrils exhibited by *C. asellus*. The anterior surface of the nasal sac is smooth while the posterior surface of it is folded as in *C. psittacus*, exhibiting a "T-shaped" ridge with a relatively small dorsal flap. This flap seems much smaller than the one found on *C. asellus*, although more flexible when compared to *C. psittacus*.

The presence of dermal flaps across the chin is another character used by [46] to distinguish *C. asellus* from *C. psittacus*. No dermal flaps could be seen in the examined specimens from the Tocantins River, although they are always present in examined specimens from Iquitos and Belém.

The skull is partially similar to those found in *Colomesus asellus* described and figured by [46], although the frontals exhibit a wide posterior border and prominently participate in the orbital margin (Figures 7–9). The prefrontals are triangular and articulate medially with the ethmoid, which posteriorly articulates with the frontals and anteriorly with the palatines (Figure 8). The supraoccipital is roughly triangular and well developed, with an elongate posterior process which covers the first vertebrae (Figure 8). The sphenotics articulate postero-laterally with the frontals and, in the examined specimens, they neither contact nor closely approach the prefrontals. The lateral wing of the sphenotics is only partially developed (Figure 8). Posterior to the sphenotics, the pterotics (Figures 7 and 8) articulates posteriorly with the slender supracleithrum and medially with the epiotics, which articulates medially with the supraoccipital (Figure 9).

In lateral view, the skull is characterized by the wide preopercle with about 110 degrees between both horizontal and vertical rami (Figure 7), with the preopercular canal running along its anterior border, and by the opercle which is divided in two distinct regions, having ventral and posterior wings, the herein called "inverted V" shape, distinct from the condition found in all other examined specimens of *Colomesus* (Figure 10). The subopercle is sturdy, with two small dorsal processes.

The parasphenoid is elongate and does not exhibit any developed dorsal flange (Figures 7 and 9). The hyomandibula is roughly triangular and has a slender ventral region; its wide head articulates dorsally with the sphenotics, and its upper posterior edge with the anterior end of the opercle (Figure 7).

Figure 12. *Colomesus tocantinensis* **nov. sp. (PNT.UERJ.403).** A) left photograph of the unpaired fins and caudal endoskeleton; B) anatomical interpretations. Abbreviations: E, epural; H, dorsal hypural plate; H-H, ventral hypural plate fused with the ural centrum; Phy, parhypural; PU, pre-ural vertebrae; PbD, dorsal-fin basal pterigiophores; PbA, anal-fin basal pterigiophores. Scale bar equals 5 mm.

The palatine is wide and somewhat triangular, with a robust anterior process for the maxilla (Figure 7). The maxilla is robust, with an anterodorsal region articulating with the premaxilla and a posterior expanded region, medially concave for muscle insertion. The ectopterygoid articulates dorsally with the palatine and ventrally with the anterodorsal border of the quadrate. The metapterygoid is wide and composes almost the entire ventral orbital region (Figure 7). It articulates anteriorly with the mesopterygoid (Figure 7), and with the posterior end of the large and triangular quadrate (Figure 7). The quadrate exhibits a well-developed posteroventral spine articulating posteriorly with the slender symplectic (Figure 7), and anteriorly with the articular. The articular is "L" shaped and articulates anteriorly with the robust dentary and ventrally with the small angular (Figure 7).

Five branchiostegal rays (Figure 7) are present and the branchial apparatus is strikingly similar in all the examined specimens (Figure 11).

The pectoral girdle is robust and formed by a wide cleithrum, somewhat triangular and posteriorly expanded, articulating dorsally with the slender supracleithrum. The supracleithrum articulates ventrally with the two postcleithra; a slender dorsal postcleithrum, followed by the posteriorly expanded ventral postcleithrum (Figure 7). There are four radials and sixteen pectoral fin rays (Figure 7).

The axial skeleton has 19 vertebrae. The dorsal fin originates between vertebrae 7–8 and has ten basal pterygiophores and ten fin rays (Figures 12–14). The anal fin is located beneath the 9th vertebra and has six basal pterygiophores and nine fin rays.

The caudal skeleton (Figure 12) has a wide ural centrum formed by the preural centrum 1, the ural centrum, the ventral hypural plate, and the postero-dorsal expansion which articulates anteriorly with the almost triangular epural, and posteriorly with the dorsal hypural plate (Figure 12). Eleven caudal fin rays, five dorsal and six ventral, are present in all of the specimens, both the uppermost and the two lowermost rays are unbranched.

Phylogeography of the South American Freshwater Pufferfishes

Although the influence of marine incursions after the Miocene is still under debate, the Caribbean (or Miocene) marine incursion, via the Llanos Basin (Colombia-Venezuela), is well accepted based on both geological and paleontological evidence, suggesting that these incursions may have isolated marine taxa within the western South America freshwater environments [48–52]. This might be the case for the freshwater tetraodontids. As pointed by [53], this scenario predicts that the distribution of the marine sister groups of marine lineages should be related with the Caribbean or western Atlantic, the age of freshwater taxa should be coincident with marine incursions, and the biogeographic congruence should be observed among multiple unrelated taxa, conditions only partially filled by the genus *Colomesus*.

The timing of divergence between the brackish/marine *C. psittacus* and the freshwater *C. asellus* was recently discussed [45], based on a multiple loci approach including both nuclear and mitochondrial markers. The authors dated the split between 2,5-7My, therefore postdating the Miocene marine incursions usually

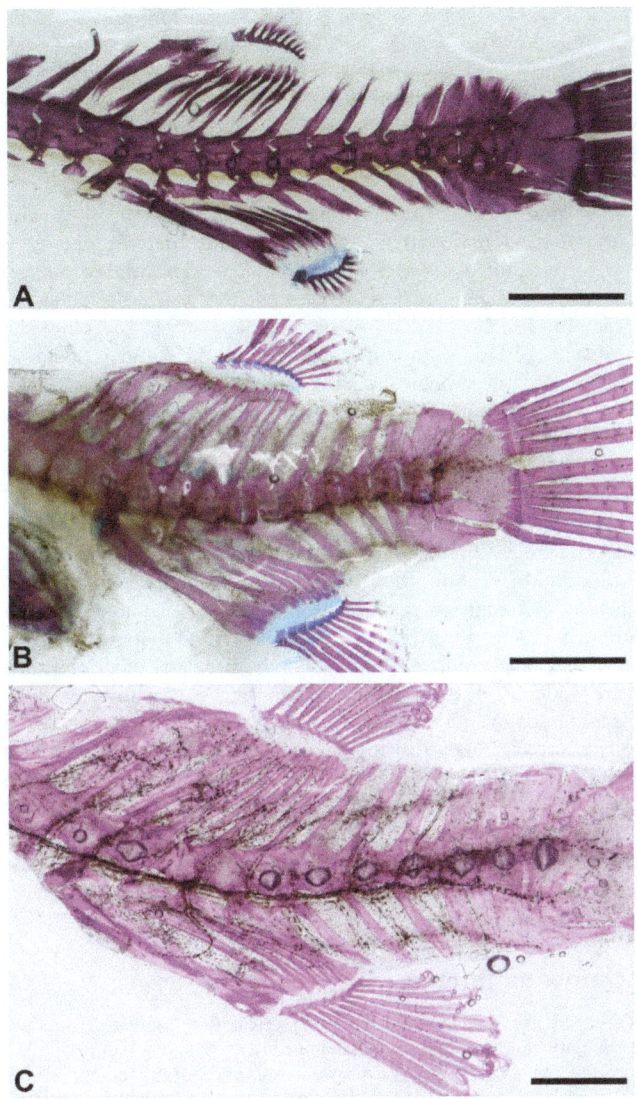

Figure 13. Left view photographs from cleared-and-stained specimens of: A) *Colomesus tocantinensis* **nov. sp. (PNT.UERJ.403); B)** *Colomesus asellus* **– Iquitos (PNT.UERJ.470); C)** *Colomesus asellus* **– Belém (PNT.UERJ.386).** Scale bar equals 5 mm.

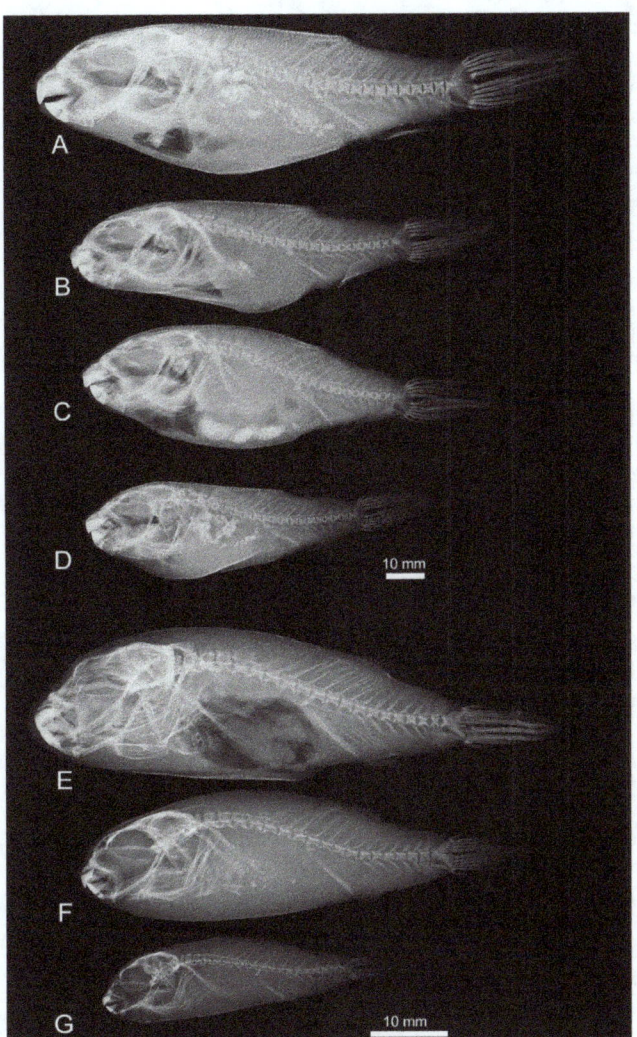

Figure 14. High-definition x-ray images of: A–D, *Colomesus psittacus* **USNM.393077; E–G,** *Colomesus asellus* **USNM.191569.** Scale bar equals 10 mm.

used to explain the presence of several marine groups within the western Amazon. In this sense, as observed by [45], the colonization carried by the tetraodontids in South America could be presumably related to the Pliocene global climate oscillations. Additionally, the basal split of the *Colomesus* from Tocantins and from Iquitos/Belém agrees with the general area cladogram of neotropical fishes presented by [54] in which the Xingu/Tocantins-Araguaia group was recovered in a basal position in relation to the groups from the lowlands of Western and Eastern Amazon.

It was recently proposed [55], based on the distribution of characiforms, that recent marine incursions would have isolated fish populations in upland terrains or refuges, where lineage divergence is maximized, followed by dispersal episodes back to the lowlands. The "museum hypothesis" predicts that lowlands exhibit a higher number of species, but lower levels of endemism,

than highlands, and that the upland refuges would represent areas of high endemism.

Looking on the molecular phylogeny of the serrasalmids *Pygocentrus* and *Serrasalmus*, [56] proposed a phylogenetic test which predicts that basal lineages in a phylogeny of widespread fishes would occur in highland areas, and lowland lineages would have originated only during the last 5 Ma. Additionally, [57] studying the genetics of *Symphysodon* cichlids, indicated the effects that marine incursions would have in population structure, stating that populations in upland terrains or refuges would exhibit reduced genetic variation, while populations in lowlands would represent multiple upland sources, therefore exhibiting a high level of genetic variation, and that populations in lowlands would show a demographic pattern of expansion.

Our results recovered the Upper Tocantins lineages in a basal position in relation to all the remaining specimens, with the sequences being collapsed in uniquely two haplotypes (Figure 5), the first one (h1), represented by eight sequences, and the second haplotype (h2), represented by a unique sequence. This suggests low genetic variation, at least among the studied sampling, and a history initially related with the eastern Amazon, followed by a subsequently expansion to the western South America.

The Tocantins-Araguaia Ichthyofauna

The Tocantins-Araguaia drainage is the fourth largest Brazilian drainage, draining part of the northern end of the Brazilian shield directly to the eastern end of the Amazon Basin. It exhibits a recent geomorphological history, within a still tectonically active sedimentary basin with recent subsidence episodes, which are related with the high load of sediments observed within the basin, leading the development of the Bananal Plain, in the lower part of the drainage, mainly during the Quaternary [58–59].

The Tocantins drainage, specially the Upper Tocantins River, is constantly regarded as an area of high endemism, with several fish taxa restricted to this area having been described, such as *Leporinus taeniofasciatus* (Anostomidae), *Sternarchorhynchus mesensis* (Apteronotidae), *Aspidoras albater*, *A. eurycephalus* (Callichthyidae), *Acestrocephalus maculosus*, *Astyanax unitaeniatus*, *Astyanacinus goyanensis*, *Creagrutus atrisignum*, *C. britskii*, *C. mucipu*, *C. saxatilis*, *Hyphessobrycon hamatus*, *Moenkhausia tergimaculata*, *Cetopsis caiapo*, *C. sarcodes* (Cetopsidae), *Characidium stigmosum* (Crenuchidae), *Pimelodella spelaea* (Heptapteridae), *Ancistrus aguaboensis*, *A. jataiensis*, *A. minutus*, *A. reisi*, *Corumbataia veadeiros*, *Hemiancistrus micrommatos*, *Hypostomus ericae*, *Gymnotocinclus anosteos*, *Lamontichthys avacanoeiro* (Loricariidae), *Apareiodon argenteus* and *A. cavalcante* (Parodontidae), *Cynolebias griseus*, *C. notatus*, *Rivulus planaltinus*, and *Simpsonichthys marginatus* (Rivulidae), the herein described *Colomesus tocantinensis* **nov. sp.** (Tetraodontidae), and *Ituglanis bambui* and *I. mambai* (Trichomycteridae).

In the same way, [60–62] pointed that the Tocantins River is the only river system related with the Amazon Basin, and draining shield areas which exhibit a considerable number of fish taxa known from the lowlands of the central and western Amazon. The faunal similarity between both basins is constantly related with the evolution of the Bananal Plain as a selective barrier for the lowland fauna and the upper part of the drainage, and the list of typical lowland ichthyofauna shared by the Tocantins-Araguaia and the Amazon drainages includes, as presented by [60], *Leporinus trifasciatus* (Anostomidae), *Arapaima gigas* (Arapaimatidae), *Auchenipterichthys coracoideus* (Auchenipteridae), *Mylossoma* spp., *Pygocentrus nattereri* (Characidae), *Cetopsis candiru*, *C. coecutiens* (Cetopsidae), *Cichla monoculus*, *C. pleiozona*, *C. kelberi* (Cichlidae), *Psectrogaster amazonica* (Curimatidae), *Thorachocharax stellatus* (Gasteropelecidae), *Osteoglossum bicirrhosum* (Osteoglossidae), *Pellona castelnaeana*, *Pristigaster cayana* (Pristigasteridae), and *Colomesus asellus* (Tetraodontidae). In this sense, even still under debate, it seems clear that the Tocantins-Araguaia drainage has a composite nature including both lowland and upland ichthyofauna, as argued by [62].

Conclusions

Based on a comprehensive analysis including both morphological and molecular methodologies using the cytochrome C oxidase I gene, we were able to discuss aspects of the phylogeny and phylogeography of the South American freshwater pufferfishes of the genus *Colomesus*.

Our molecular results based on the COI marker agrees with the recent results such as [43–45], and suggest that the genus *Sphoeroides* should be revised, mainly regarding the phylogenetic position recovered for the genus *Colomesus*, deeply nested within the *Sphoeroides* tree, and the basal position recovered for *S. pachygaster*. We plan further investigations along these lines to reconcile any conflicts between these molecular hypotheses presented herein and morphologically based interpretations [47] of the phylogeny of the taxa of *Colomesus*, *Sphoeroides*, and *Lagocephalus*.

The use of molecular systematic techniques together with morphological methodologies confirmed the identification of a new cryptic pufferfish species from the Upper Tocantins drainage, *Colomesus tocantinensis* **nov. sp.** Morphological features such as the color pattern, the absence of dermal flaps across the chin, the distinct 'inverted V' opercle shape, and the caudal peduncle morphology, all support the description of *Colomesus tocantinensis* **nov. sp.**, as a new pufferfish species from the South American freshwater drainages.

The timing of divergence between the marine/brackish species *Colomesus psittacus* and the freshwater group formed by *C. asellus* and *C. tocantinensis* **nov. sp.**, as recovered by [45], postdates the Miocene marine incursions usually used to explain the presence of tetraodontids within the Amazon freshwater environments. Therefore, it suggests that the freshwater colonization in South America, at least for the tetraodontids, could be more recent than previously expected. Additionally, together with the observed distribution of haplotypes, our results suggest that the history of tetraodontids into the Amazonian freshwater environments could be presumably related to the Pliocene global climate oscillations and its effects inside the eastern Amazon and subsequently to the western South America.

Finally, our results reinforce the Upper Tocantins drainage as an area of high endemism within the Tocantins-Araguaia drainage, although the composite nature of the entire drainage is unquestionable.

Acknowledgments

We would like to thank Dr. James C. Tyler (Smithsonian Institution, Washington) for his support and helpfull comments on the manuscript. We are also grateful to Dr. Francesco Santini (Universitá degli Studi di Torino), Dr. Dorothée Huchon (Tel Aviv University), and an anonymous reviewer for the valuable suggestions during the review of the manuscript, Dr. Leonor Gusmão and Dr. Antonio Amorim (Universidade do Porto) for the comments during the initial discussion of the results, Yuri Modesto (Universidade do Estado do Rio de Janeiro) for the specimens from the Tocantins drainage, Lúcio Paulo Machado and Diogo de Mayrink (Universidade do Estado do Rio de Janeiro) for the specimens from Iquitos; Ms. Sandra Raredon (Smithsonian Institution, Washington) for the x-rays of tetraodontids, Dr. Richard Pyle (Hawaii Biological Survey) for the LSID numbers, and Kleyton M. C. Severiano and Anna Carolina Chaves (Universidade do Estado do Rio de Janeiro) for the technical assistance.

Author Contributions

Conceived and designed the experiments: CRLA PMB DAS EFC. Performed the experiments: CRLA PMB DAS EFC. Analyzed the data: CRLA PMB DAS EFC. Contributed reagents/materials/analysis tools: CRLA PMB DAS EFC. Wrote the paper: CRLA.

References

1. Froese R, Pauly D (2012) *FishBase*. World Wide Web electronic publication. www.fishbase.org, 2012, version (02/2012).

2. Gill TN (1885) Synopsis of the Plectognath fishes. Proc United States Natl Mus 7: 411–427.

3. Müller J, Troschel FH (1849) Fische. In: Reisen in Britisch-Guiana in den Jahren, 1840–1844. Im Auftrag Sr. Mäjestat des Königs von Preussen ausgeführt von Richard Schomburgk. [Versuch einer Fauna und Flora von Britisch-Guiana.] 3. Berlin.

4. Bloch ME, Schneider JG (1801) Systema Ichthyologiae iconibus cx illustratum. Post obitum auctoris opus inchoatum absolvit, correxit, interpolavit Jo. Gottlob Schneider, Saxo. Berolini. Sumtibus Auctoris Impressum et Bibliopolio Sanderiano Commissum. M. E. Blochii, Systema Ichthyologiae.: i-lx +1–584, Pls. 1–110.

5. Collins RA, Armstrong KF, Meier R, Yi Y, Brown SDJ, et al. (2012) Barcoding and border Biosecurity: Identifying Cyprinid Fishes in the Aquarium Trade. PLoS ONE 7(1): e28381. doi:10.1371/journal.pone.0028381.

6. Padilla DK, Williams SL (2004) Beyond ballast water: aquarium and ornamental trades as sources of invasive species in aquatic ecosystems. Front Ecol Environ 2: 131–138.

7. Ploeg A, Bassleer G, Hensen R (2009) Biosecurity in the Ornamental Aquatic Industry. Maarssen: Orn Fish Intl. 148 p.

8. Mabragaña E, Díaz de Astarloa JM, Hanner R, Zhang J, González Castro M (2011) DNA Barcoding Identifies Argentine Fishes from Marine and Brackish Waters. PLoS ONE 6(12): e28655. doi:10.1371/journal.pone.0028655.

9. Golding GB, Hanner R, Hebert PDN (2009) Preface. Molecular Ecology Resources 9(Suppl. 1): iv–vi.

10. Hogg ID, Hebert PDN (2004) Biological identification of springtails (Collembola: Hexapoda) from the Canadian Arctic, using mitochondrial DNA barcodes. Can J Zoo 82: 749–754.

11. Barrett RDH, Hebert PDN (2005) Identifying spiders through DNA barcodes. Can J of Zoo 83: 481–491.

12. Hebert PDN, Cywinska A, Ball SL, deWaard JR (2003a) Biological identification through DNA barcodes. Proc Roy Soc London, Series B: Biol Sci 270: 313–321.

13. Hebert PDN, Ratnasingham S, deWaard JR (2003b) Barcoding animal life: cytochrome c oxidase subunit 1 divergences among closely related species. Proc Roy Soc London, Series B: Biol Sci (Suppl 1) 270: 96–99.

14. Janzen DH, Hajibabaei M, Burns JM (2005)Wedding biodiversity inventory of a large and complex Lepidoptera fauna with DNA barcoding. Phil Trans Roy Soc London, Series B, Biol Sci 1462: 1835–1846.

15. Hajibabaei M, Janzen DH, Burns JM, Hallwachs W, Hebert PDN (2006) DNA barcodes distinguish species of tropical Lepidoptera. Proc Natl Acad Sci U S A, 103: 968–971.

16. Lukhtanov VA, Sourakov A, Zakharov EV, Hebert PDN (2009) DNA barcoding Central Asian butterflies: increasing geographical dimension does not significantly reduce the success of species identification. Mol Ecol Res 9: 1302–1310.

17. Smith MA, Wood DM, Janzen DH, Hallwachs W, Hebert PDN (2007) DNA barcodes affirm that 16 species of apparently generalist tropical parasitoid flies (Diptera, Tachinidae) are not all generalists. Proc Natl Acad Sci U S A 104: 4967–4972.

18. Järnegren J, Schander C, Sneli JA, Ronningen V, Young CM (2007) Four genes, morphology and ecology: distinguishing a new species of Acesta (Mollusca; Bivalvia) from the Gulf of Mexico. Mar Biol 152: 43–55.

19. Ward RD, Zemlak TS, Innes BH, Last PR, Hebert PDN (2005) DNA barcoding Australia's fish species. Phil Trans Roy Soc London, Series B, Biol Sci 360: 1847–1857.

20. Hebert PDN, Stoeckle MY, Zemlak TS, Francis CM (2004) Identification of birds through DNA barcodes. PLoS Biol 2: 1657–1663.

21. Kerr KCR, Lijtmaer DA, Barreira AS, Hebert PDN, Tubaro PL (2009) Probing evolutionary patterns in neotropical birds through DNA barcodes. PloS ONE 4: e4379.

22. Clare EL, Lim BK, Engstrom MD, Eger JL, Hebert PDN (2007) DNA barcoding of Neotropical bats: species identification and discovery within Guyana. Mol Ecol Notes 7: 184–190.

23. Amaral AR, Sequeira M, Coelho MM (2007) A first approach to the usefulness of cytochrome c oxidase I barcodes in the identification of closely related delphinid cetacean species. Mar Fresh Res 58: 505–510.

24. Borisenko AB, Lim BK, Ivanova NV, Hanner RH, Hebert PDN (2008) DNA barcoding in surveys of small mammal communities: a field study in Suriname. Mol Ecol Res 8: 471–479.

25. Hollingsworth PM, Forrest LL, Spouge JL, Hajibabaei M, Ratnasingham S, et al. (2009) A DNA barcode for land plants. Proc Natl Acad Sci U S A 106: 12794–12797.

26. Saunders GW (2005) Applying DNA barcoding to red macroalgae: a preliminary appraisal holds promise for future applications. Phil Trans Roy Soc London, Series B, Biol Sci 360: 1879–1888.

27. Sogin ML, Morrison HG, Huber JA, Welch DM, Huse SM, et al. (2006) Microbial diversity in the deep sea and the underexplored 'rare biosphere'. Proc Natl Acad Sci U S A 103: 12115–12120.

28. Lowenstein JH, Amato G, Kolokotronis SO (2009) The real maccoyii: identifying tuna sushi with DNA barcodes - contrasting characteristic attributes and genetic distances. PLoS ONE 4: e7866.

29. Lowenstein JH, Burger J, Jeitner CW, Amato G, Kolokotronis SO, et al. (2010) DNA barcodes reveal species-specific mercury levels in tuna sushi that pose a health risk to consumers. Biol Lett 6: 692–695.

30. Cohen NJ, Deeds JR, Wong ES, Hanner RH, Yancy HF, et al. (2009) Public health response to puffer fish (tetrodotoxin) poisoning from mislabeled product. J Food Prot 72: 810–817.

31. Holmes BH, Steinke D, Ward RD (2009) Identification of shark and ray fins using DNA barcoding. Fish Res 95: 280–288.

32. Steinke D, Zemlak TS, Hebert PDN (2009) Barcoding Nemo: DNA-Based identifications for the Ornamental Fish Trade. PLoS ONE 4(7): e6300. doi:10.1371/journal.pone.0006300.

33. Blessing JJ, Marshall JC, Balcombe SR (2010) Humane killing of fishes for scientific research: a comparison of two methods. J Fish Biol. 76(10): 2571–2577. doi: 10.1111/j.1095–8649.2010.02633.x.

34. Song J, Parenti L (1995) Clearing and Staining Whole Fish Specimens for Simultaneous Demonstration of Bone, Cartilage, and Nerves. Copeia 1995(1): 114–118.

35. Palumbi SR (1996) Nucleic acids II: the polymerase chain reaction. In: Molecular Systematics, Hillis DM, Moritz C, Mable BK (Eds), Sinauer & Associates Inc., Sunderland, Massachusetts: 205–247.

36. Ward RD, Zemlak TS, Innes BH, Last PR, Hebert PDN (2005) DNA barcoding Australia's fish species. Phil Trans R Soc London. Series B, Biol Sci 360: 1847–1857.

37. Ivanova NV, Zemlak TS, Hanner RH, Hebert PDN (2007) Universal primer cocktails for fish DNA barcoding. Mol Ecol Notes 7: 544–548. doi: 10.1111/j.1471–8286.2007.01748.x.

38. Hall TA (1999) BioEdit: a user-friendly biological sequence alignment editor and analysis program for Windows 95/98/NT. Nucleic Acids Symposium Series 41: 95–98.

39. Thompson JD, Gibson TJ, Plewniak F, Jeanmougin F, Higgins DG (1994) The Clustal X windows interface: Flexible strategies for multiple sequence alignment aided by quality analysis tools. Nuc Acid Res 24: 4876–4882.

40. Tamura K, Peterson D, Peterson N, Stecher G, Nei M, et al. (2011) MEGA5: Molecular Evolutionary Genetics Analysis using Maximum Likelihood, Evolutionary Distance, and Maximum Parsimony Methods. Mol Biol Evol 28: 2731–2739.

41. Akaike H (1973). Information theory and an extension of the maximum likelihood principle. Proc. 2nd Inter. Symposium on Information Theory, Budapest: 267–281.

42. Kimura M (1980) A simple method for estimating evolutionary rate of base substitutions through comparative studies of nucleotide sequences. J Mol Evol 16: 111–120.

43. Yamanoue Y, Miya M, Doi H, Mabuchi K, Sakai H, et al. (2011) Multiple Invasions into Freshwater by Pufferfishes (Teleostei: Tetraodontidae): A Mitogenomic Perspective. PLoS ONE 6(2): e17410. doi:10.1371/journal.pone.0017410.

44. Santini F, Tyler JC (2003) A phylogeny of the families of fossil and extant tetraodontiform fishes (Acanthomorpha, Tetraodontiformes), Upper Cretaceous to recent. Zoo J Linn Soc 139: 565–617.

45. Santini F, Nguyen MTT, Sorenson L, Waltzek TB, Lynch Alfaro JW, et al. (2013) Do habitat shifts drive diversification in teleost fishes? An example from the pufferfishes (Tetraodontidae). J Evol Biol: 1–16.

46. Tyler JC (1964) A diagnosis of the two species of South American puffer fishes (Tetraodontidae, Plectognathi) of the genus Colomesus. Proc Acad Natl Sci Philad 116: 119–148.

47. Tyler JC (1980) Osteology, phylogeny, and higher classification of the fishes of the order Plectognathi (Tetraodontiformes). National Oceanic and Atmospheric Administration technical reports, Natl Mar Fish Serv 434: 1–422.

48. Nuttall CP (1990) A review of the Tertiary non-marine molluscan faunas of the Pebasian and other inland basins of north-western South America. Bull Brit Mus Natl Hist, Geol 45: 165–371.

49. Webb SD (1995) Biological implications of the Middle Miocene Amazon seaway. Science 269: 361–362.

50. Lovejoy NR (1997) Stingrays, parasites, and Neotropical biogeography: A closer look at Brooks, et al's hypotheses concerning the origins of Neotropical freshwater rays (Potamotrygonidae). Sys Biol 46: 218–230.

51. Lovejoy NR, Bermingham E, Martin AP (1998) Marine incursion into South America. Nature 396: 421–422.

52. Lovejoy NR, Albert JS, Crampton WGR (2006) Miocene marine incursions and marine/freshwater transitions: evidence from Neotropical fishes. J Sou Am Ear Sci 21: 5–13.

53. Bloom DD, Lovejoy NR (2011) The Biogeography of Marine Incursions in South America. In: Albert JS, Reis RE (eds). Historical biogeography of neotropical freshwater fishes. 406p.

54. Albert JS, Carvalho TP (2011) Neogene Assembly of Modern Faunas. In: Albert JS, Reis RE (eds). Historical biogeography of neotropical freshwater fishes. 406p.

55. Hubert N, Renno J-F (2006) Historical biogeography of South American freshwater fishes. J Biog 33: 1414–1436.

56. Hubert N, Duponchelle F, Nuñez J, Garcia-Davila C, Paugy D, et al. (2007) Phylogeography of the piranha genera Serrasalmus and Pygocentrus: Implications for the diversification of the Neotropical ichthyofauna. Mol Ecol 16: 2115–2136.

57. Farias IP, Hrbek T (2008) Patterns of diversification in the discus fishes (Symphysodon spp. Cichlidae) of the Amazon basin. Mol Phyl Evol 49: 32–43.

58. Saadi A (1993) Neotectônica da plataforma brasileira: Esboço e interpretação preliminares. Geonomos 1: 1–15.

59. Saadi A, Bezerra FHR, Costa RD, Igreja HLS, Franzinelli E (2005) Neotectônica da Plataforma Brasileira. In: Quaternário do Brasil, edited by Souza CRG, Suguio K, Oliveira AMS, Oliveira PE, 211–234. Ribeirão Preto: Holos Editora.

60. Jégu M, Keith P (1999) Lower Oyapock River as northern limit for the Western Amazon fish fauna or only a stage in its northward progression. Compt Rend Biol Acad Sci 322: 1133–1143.

61. Hrbek T, Seckinger J, Meyer A (2007) A phylogenetic and biogeographic perspective on the evolution of poeciliid fishes. Mol Phyl Evol 43: 986–998.

62. Lima FCT, Ribeiro AC (2011) Continental-Scale Tectonic Controls of Biogeography and Ecology. In: Albert JS, Reis RE (eds). Historical biogeography of neotropical freshwater fishes. 406p.

Spatial Variation in the Population Structure and Reproductive Biology of *Rimicaris hybisae* (Caridea: Alvinocarididae) at Hydrothermal Vents on the Mid-Cayman Spreading Centre

Verity Nye*, Jonathan T. Copley, Paul A. Tyler

Ocean and Earth Science, National Oceanography Centre, University of Southampton, Southampton, United Kingdom

Abstract

The dynamics and microdistribution of faunal assemblages at hydrothermal vents often reflect the fine-scale spatial and temporal heterogeneity of the vent environment. This study examined the reproductive development and population structure of the caridean shrimp *Rimicaris hybisae* at the Beebe and Von Damm Vent Fields (Mid-Cayman Spreading Centre, Caribbean) using spatially discrete samples collected in January 2012. *Rimicaris hybisae* is gonochoric and exhibits iteroparous reproduction. Oocyte size-frequency distributions (21–823 μm feret diameters) varied significantly among samples. Embryo development was asynchronous among females, which may result in asynchronous larval release for the populations. Specimens of *R. hybisae* from the Von Damm Vent Field (2294 m) were significantly larger than specimens from the Beebe Vent Field. Brooding females at Von Damm exhibited greater size-specific fecundity, possibly as a consequence of a non-linear relationship between fecundity and body size that was consistent across both vent fields. Samples collected from several locations at the Beebe Vent Field (4944–4972 m) revealed spatial variability in the sex ratios, population structure, size, and development of oocytes and embryos of this mobile species. Samples from the Von Damm Vent Field and sample J2-613-24 from Beebe Woods exhibited the highest frequencies of ovigerous females and significantly female-biased sex ratios. Environmental variables within shrimp aggregations may influence the distribution of ovigerous females, resulting in a spatially heterogeneous pattern of reproductive development in *R. hybisae*, as found in other vent taxa.

Editor: Athanassios C. Tsikliras, Aristotle University of Thessaloniki, Greece

Funding: VN was supported by a UK Natural Environmental Research Council (NERC) award (NE/F017774/1) to J. Copley and P. Tyler, which is gratefully acknowledged. The biological component of the 'Atlantis' Cayman cruise was supported by a National Science Foundation (NSF) award (OCE-1031050) and NASA ASTEP subcontract through WHOI (NNX09AB75G). The funders had no role in study design, data collection and analysis, decision to publish, or preparation of the manuscript.

Competing Interests: The authors have declared that no competing interests exist.

* E-mail: vn205@noc.soton.ac.uk

Introduction

Deep-sea chemosynthetic environments supporting chemosynthesis-based faunal assemblages are distributed widely but patchily throughout the global ocean.

Reproduction is therefore an essential process for the establishment and maintenance of isolated populations of specialist vent, seep and whale-fall fauna. Understanding the life histories of organisms inhabiting these insular environments is a prerequisite for understanding their ecology, population biology, dispersal, gene flow and biogeography [1,2,3,4].

More than 400 new faunal species have been described from deep-sea hydrothermal vents since the 1970s [5], and aspects of life-history biology have been elucidated in more than 90 species from vents, seeps, and whale falls (Copley, Nye et al., unpublished data). Studies have described a variety of reproductive traits and developmental modes in species from chemosynthetic environments, and revealed spatial and temporal patterns in the reproductive development of some species (see [2,6] for reviews).

The Mid-Cayman Spreading Centre (MCSC) is an ultraslow-spreading and geographically isolated ridge in the Caribbean that hosts two high-temperature hydrothermal vent fields [7]. The Beebe Vent Field (~4960 m) is situated on the axis of the MCSC and it consists of a sulfide mound (~80 m diameter, 50 m height) surmounted by several active black-smoker chimney complexes and areas of diffuse flow [7]. The Von Damm Vent Field (~2300 m) is a conical mound (~150 m diameter, 70 m height) venting clear, buoyant fluids, located off-axis (approximately 13 km away from the Beebe Vent Field) on the upper slopes of the Mount Dent oceanic core complex [7].

Research efforts on the fauna at MCSC vents have so far focused on the taxonomy, phylogenetics and assemblage compositions. The Beebe vent assemblage includes provannid gastropods, anemones and ophiuroids, whereas the faunal assemblage at the Von Damm Vent Field includes skeneid gastropods, hippolytid shrimp, lysianssid amphipods and tubeworms [7,8,9,10]. The alvinocaridid shrimp *Rimicaris hybisae* [8] is present and abundant at both known MCSC vent fields [8].

To date, more than 125 species representing 33 families of decapods have been reported from deep-sea chemosynthetic environments [11], yet the reproductive traits of only ten species have been described [12,13,14,15,16,17,18]. Reproductive patterns of decapods from chemosynthetic environments are thought to have strong phylogenetic constraints [2,12].

The family Alvinocarididae [19] is represented to date by 26 described species from eight genera and appears to be endemic to deep-sea chemosynthetic environments [8]. Alvinocaridid shrimp examined previously exhibit planktotophic development and gametogenesis characteristic of carideans [12,13,15,16,17]. Seasonal reproduction has been described in *Alvinocaris stactophila* [20] from the Brine Pool cold seep (650 m) in the Gulf of Mexico, where the seasonal peak in surface productivity and its export may be a cue for larvae to hatch [15].

Zonation in the population structure and reproductive biology of *Alvinocaris stactophila* has also been revealed at the Brine Pool ([15], Nye, unpublished data). Avoidance of sulfidic extremes by female crustaceans brooding embryos has been proposed for several taxa at vents and seeps (e.g., [14,15,18,21,22]). A similar explanation has been invoked to explain the apparent scarcity of ovigerous females of *Rimicaris exoculata* [23] in the immediate vicinity of black smokers at deep vents on the Mid-Atlantic Ridge (MAR) [13,24].

The aims of this study were therefore to: (1) examine variation in the population structure and reproductive features of *Rimicaris hybisae* between the Beebe and Von Damm vent fields; (2) assess spatial variation in the reproductive features of *R. hybisae* within the Beebe Vent Field; (3) discuss and compare the results with data available for other alvinocaridid species. This is the first study on the autecology of vent fauna from the Mid-Cayman Spreading Centre, and reveals a high degree of spatial variability in the population structure and reproductive features of this mobile species in the vent environment.

Materials and Methods

To assess spatial variation in population structure and reproductive features, samples of *Rimicaris hybisae* were collected from two vent fields at the Mid-Cayman Spreading Centre, Caribbean, during the 18th voyage (16th leg) of the RV 'Atlantis' (see Table 1). Samples were collected using a suction sampler attached to the remotely operated vehicle (ROV) 'Jason II'. Four samples were collected from different locations within the Beebe Vent Field (4944–4972 m; Table 1): J2-613-24 was collected from a large, high-density aggregation of shrimp at the base of a chimney; J2-619-15 was collected from a large, high-density aggregation of shrimp at the edge of a gulley; J2-613-19 was taken from a small, dense aggregation of shrimp, next to anemones, provannid gastropods and bacterial mats; J2-620-32 was taken in a peripheral area, dominated by anemones with sparse shrimp.

Two samples (J2-617-5 and J2-617-8) were collected from a large, high-density aggregation of shrimp at the Von Damm Vent Field. Although they were placed in two separate chambers of the multi-chamber suction sampler, these samples were collected within minutes of each other at the same location and depth (within a 1 m^2 area) and could not be discriminated spatially. Consequently these two samples were pooled (J2-617-5/8). Unfortunately constraints of expedition logistics precluded a replicate sample from this vent field. Specimens were fixed in 10% buffered seawater formalin for 48 h and stored in 70% isopropanol.

No specific permits were required for the described field studies. No specific permissions were required for these locations/ activities. The location is not privately-owned or protected in any way and the field studies did not involve endangered or protected species.

Carapace length (CL) of each shrimp was measured to the nearest 0.1 mm with Vernier callipers from the rear of the eye socket to the rear of the carapace in the mid-dorsal line. This is the standard measure of length for a shrimp and it is used herein as an indication of body size because it avoids errors associated with measuring a flexible abdomen [25].

The sex of each shrimp was determined under a Leica MZ8 dissecting microscope (*sensu* 8). For all males, the presence/absence of a spermatophoric mass was recorded. All females were categorised as either: brooding (brooding embryos on pleopods 1–4); hatched (with a matrix of empty embryo sacs attached to the pleopods); or female (neither brooding nor hatched).

The ovaries were dissected from the female shrimp and where oocyte size allowed, individual oocytes were removed from each ovary under a Leica EZ4 HD dissecting microscope. Images of oocytes were captured using a Leica EZ4 HD dissecting microscope. Packing of oocytes in ovaries often results in irregular oocyte shapes in *Rimicaris hybisae*. Consequently oocytes were laid flat and measured directly, rather than from histological sections, to ensure maximum cross-sectional areas were recorded (*sensu* [15]). Where female specimen numbers and condition allowed, the feret diameters and areas of 100 oocytes were measured in 30 females per sample (9 females for J2-620-32) using ImageJ. Feret diameter was used to standardise variations in oocyte shape. Images of oocytes were calibrated with measurements of a graticule slide at identical magnification.

Broods of embryos were removed from the pleopods of brooding females under a Leica MZ8 dissecting microscope. Within each brood, all embryos had developed synchronously and were at the same stage of development. The developmental stage of each brood was scored on the basis of morphological features (*sensu* [16]): early-stage embryos without features; mid-stage embryos with clear body differentiation; late-stage embryos with clear larval features (e.g. separation of the abdomen from the cephalothorax and developed eyes), including hatched larvae (Figure 1). Numbers of embryos per brood were counted to determine minimum realised fecundity (*sensu* [26]). Although it was not possible to guarantee that embryo batches were complete, embryos were attached firmly to each other and the mothers' pleopods (within which they were enclosed) and broods remained in tact post-sampling. Size-specific fecundity was calculated as number of embryos divided by carapace length (Table 2).

To determine mean embryo size, a sub-sample of ten broods at each developmental stage from both vent fields was selected at random. Embryos were laid flat and images of the embryos were captured using a Leica EZ4 HD dissecting microscope. The greater and lesser diameters of 100 embryos per brood were measured using ImageJ.

Frequencies of males and females in samples were tested for significant variation from a 1:1 sex ratio using χ^2 test with Yates' correction for one degree of freedom. In analyses of population structure and size-frequency distribution of females, brooding and hatched females were pooled as ovigerous females. To correct for variation in the sex ratio between samples, spatial variation in the population structure was assessed by comparing the ratio of ovigerous (brooding and hatched) females to non-ovigerous females, the ratio of brooding to hatched ovigerous females, and the ratio of males with spermatophores vs without (rather than the overall proportions). Frequencies of ovigerous females, brooding females and males with spermatophores were tested for significant variation from a 1:1 ratio between vent fields (Table 3) and between pairwise combinations of samples within the Beebe Vent Field (Table 4) using χ^2 test with Yates' correction for one degree

Table 1. *Rimicaris hybisae*: Sample and population data for 959 specimens used in this study.

Sample no.	Cruise	Sample method	Vent Field	Location	Depth (m)	Latitude (N)	Longitude (W)	Date (JD)	Total no. specimens	Males Total	M	SM	Females Total	F	BF	HF	Sex Ratio TM:TF	χ^2 (1 df)	Significance
J2-613-24	Atlantis 18_16	Jason II	Beebe	Beebe Woods	4971	18.546182	81.718086	12/01/2012	254	84	32	52	170	37	108	25	0.49:1	28.44	***
J2-619-15	Atlantis 18_16	Jason II	Beebe	Shrimp Gulley	4944	18.546563	81.717705	22/01/2012	118	48	36	12	70	42	28	0	0.69:1	3.74	NS
J2-613-19	Atlantis 18_16	Jason II	Beebe	Beebe Woods	4972	18.546974	81.718339	11/01/2012	96	44	22	22	52	47	5	0	0.85:1	0.51	NS
J2-620-32	Atlantis 18_16	Jason II	Beebe	Beebe Woods	4964	18.546929	81.718278	23/01/2012	94	76	48	28	18	16	0	2	4.2:1	34.56	***
			Beebe	All					562	252	138	114	310	142	141	27	0.81:1	5.78	*
J2-617-5/8	Atlantis 18_16	Jason II	Von Damm	Spire	2294	18.376630	81.798143	19/01/2012	397	141	103	38	256	147	77	32	0.55:1	32.74	***
				Total					959	393	241	152	566	289	218	59	0.69:1	30.85	***

BF, brooding female; F; female without brooding or recently hatched; H, female recently hatched (with a matrix of empty embryo sacs attached to the pleopods); M, male without spermatophore; NS = not significant; SM, male with spermatophore; TM, total males; TF, total females.

* = P value <0.05; ** = P value <0.01; *** = P value <0.001.

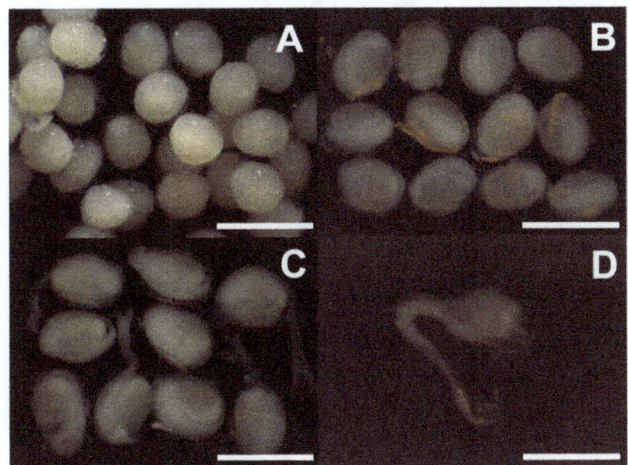

Figure 1. *Rimicaris hybisae*: **Embryo stages and larvae.** (A) Early-stage embryos; (B) mid-stage embryos; (C) late-stage embryos with larvae ready to hatch; (D) hatched larva. Scale bars = 1 mm.

of freedom. Population structure was examined using the size-frequency distribution of 959 individuals (Table 5).

Results

Population structure in January 2012

The gonads of *Rimicaris hybisae* are paired organs laying over the digestive gland of the cephalothorax. Of the 959 specimens of *Rimicaris hybisae* examined, 393 were identified as male (41%), resulting in an overall sex ratio that deviated significantly from 1:1 (393 males; 566 females) (Table 1). All males examined had only testes and all females studied have only ovaries. Of the 393 males, 152 (39%) were carrying spermatophores. Of the 566 females, nearly half (277, 49%) were either brooding embryos (218, 39%) or had hatched larvae recently (59, 10%).

The specimens ranged in body size (carapace length) from 5.2 mm (male, Von Damm) to 19.4 mm (male, Von Damm). The carapace length of the smallest male identified (5.2 mm, Von Damm) was less than that of the smallest females (6.2 mm, Beebe Woods); this indicated that any bias in sex ratio was not the result of immature males being misidentified as females. The carapace lengths of the smallest brooding female and smallest male with spermatophores were 8.5 and 6.9 mm respectively (both Beebe Woods).

The size-frequency distribution of the specimens displayed two modal peaks and a short tail of large specimen sizes (Figure 2A). The carapace lengths of the largest females identified were 17.2 and 18.1 mm at the Beebe and Von Damm Vent Fields respectively. The carapace lengths of the largest males identified were 17.4 and 19.4 mm at the Beebe and Von Damm Vent Fields respectively. The size-frequency distributions of all males and females in January 2012 were significantly different (Mann-Whitney U-test, $T = 168261.5$, $p<0.001$). Males were represented throughout the size-frequency distribution of the samples, but there were proportionally fewer large males resulting in a lower median body size in males at both vent fields (Table 5; Figure 2B, C). However, the size-frequency distributions of non-ovigerous (neither brooding nor hatched) females were not significantly different from males (Mann-Whitney U-test, $T = 99797.5$, $p>0.05$) and the ratio of males to non-ovigerous females did not deviate significantly from 1:1 (393 males: 289 non-ovigerous females,

$\chi^2 = 0$, 1 df, p>0.05). Ovigerous (brooding and hatched) females were confined to the peak and tail of large sizes and were significantly larger than non-ovigerous females at both vent fields (Table 5; Figure 3A, B; Mann-Whitney U-test, Beebe: $T = 15731.0$, $p<0.001$; Von Damm: $T = 19167.0$, $p<0.001$). Males with spermatophores were significantly larger than males without spermatophores at both vent fields (Table 5; Figure 3C, D; Mann-Whitney U-test, Beebe: $T = 168261.5$, $p<0.001$; Von Damm: $T = 3541.0$, $p<0.001$).

The brooded embryos of *Rimicaris hybisae* had a mean size of 0.64×0.48 mm in January 2012 (Table 2). There was no significant difference in embryo sizes between samples from the Beebe and Von Damm vent fields (Table 2; Mann-Whitney U-test, $T = 2250750.0$, $p>0.05$). Embryos in the early developmental stage were 0.58 ± 0.04 mm mean greater diameter and 0.46 ± 0.04 mm mean lesser diameter. The mean embryo size in the medium developmental stage was 0.64 ± 0.07 mm greater diameter and 0.48 ± 0.04 mm lesser diameter. Embryos in the most advanced stage were 0.69 ± 0.07 mm mean greater diameter and 0.51 ± 0.05 mm lesser diameter.

Spatial variation in reproductive features

Sex ratio. In January 2012 at the Von Damm Vent Field, 36% of specimens were male, resulting in a sex ratio that deviated significantly from 1:1 (Table 1). Males represented 45% of individuals sampled from the Beebe Vent Field, resulting in an overall sex ratio that deviated significantly from 1:1 (Table 1). However, there was significant variation in the sex ratio exhibited in different samples from the Beebe Vent Field (Table 1). Specimens from samples J2-613-19 (Beebe Woods) and J2-619-15 (Shrimp Gulley) did not deviate significantly from a 1:1 sex ratio. Sample J2-613-24 (Beebe Woods) showed significant female bias (67% of specimens were females), but the ratio of males to non-ovigerous females did not differ significantly from 1:1. A significant male bias was present in sample J2-620-32 (81% of specimens were males) (Table 1).

Population structure. To correct for variation in the sex ratio between samples, spatial variation in the occurrence of ovigerous females was assessed by comparing the ratio of ovigerous (brooding and hatched) females to non-ovigerous females (rather than the overall proportions). A significantly greater proportion of females were ovigerous at the Beebe Vent Field (54%) than at the Von Damm Vent Field (43%; Table 3). The majority of ovigerous females sampled from both vent fields were brooding, as opposed to females that showed evidence of having just hatched their brood (Table 3). However, brooding females represented a significantly greater proportion of ovigerous females at the Beebe Vent Field (84%) compared to the Von Damm Vent Field (71%; Table 3). The majority of males sampled from both vent fields were without spermatophores (Table 3). However, males with a spermatophoric mass accounted for a significantly greater proportion of the sampled male population at the Beebe Vent Field (45%) than at the Von Damm Vent Field (30%; Table 3). As a result of too few data points, it was not possible to determine the correlation between frequencies of ovigerous females, brooding females and males with spermatophores.

Within the Beebe Vent Field, the highest proportion of ovigerous females was 78% in sample J2-613-24 (Figure 4A). All samples were significantly different from each other in the proportion of ovigerous females with the exception of samples J2-613-19 and J2-620-32 (Table 4). All ovigerous females were brooding in samples J2-613-19 and J2-619-15, whereas 81% were brooding in J2-613-24 and all had hatched in J2-620-32 (Figure 4B). Brooding females accounted for a significantly greater

Table 2. Average minimum realised fecundity and embryo sizes of caridean shrimp from hydrothermal vents and cold seeps (updated from Ramirez-Llodra & Segonzac, 2006).

Species	Alvinocaris muricola (n = 9)	Alvinocaris stactophila (n = 55)	Alvinocaris stactophila (n = 65)	Alvinocaris lusca (n = 1)	Alvinocaris markensis (n = 1)	Chorocaris chacei (n = 1)	Mirocaris fortunata (n = 30)	Rimicaris exoculata (n = 2)	Rimicaris hybisae (n = 562)	Rimicaris hybisae (n = 397)
Site	Congo Basin (seep)	GoM: Brine Pool IMB (seep)	GoM: Brine Pool MMB (seep)	Galapagos Rift (vent)	MAR: Lucky Strike (vent)	MAR: Lucky Strike (vent)	MAR: Lucky Strike (vent)	MAR: Snake Pit & TAG (vent)	MCSC: Beebe (vent)	MCSC: Von Damm (vent)
Depth (m)	3113–3150	500		2500	1690	1690	1690	3480–3650	4944–4972	2294
CL (mm) Mean ± SD	20.7±2.3	3.77	3.91	11.45	13	16.8	7.16±0.18	17.05	10.2±1.5	13.9±2.3
Minimum realised fecundity (embryos) Mean ± SD	3130±1180.9	147	98	407	2007	2510	174.7±22.8	912	341.7±146.2	1054.2±229.8
Size-specific fecundity Mean ± SD	149.1±48.0	39	25	35	154	149	24.3	53	30.6±11.8	68.0±13.2
Mean embryo size (mm)	0.66×0.55	0.80	0.79×0.57	0.50×0.34	0.66×0.52	-	0.70×0.49	0.72×0.62	0.64×0.48	0.63×0.48
Reference	Ramirez-Llodra & Segonzac 2006	Copley & Young 2006	Copley & Young 2006	Van Dover et al. 1985	M. Segonzac, unpublished data	Ramirez-Llodra et al. 2000	Ramirez-Llodra et al. 2000	Williams & Rona 1986; Ramirez-Llodra et al. 2000	This paper	This paper

CL, carapace length; GoM, Gulf of Mexico; IMB, inner mussel bed site; MMB, middle mussel bed site; MAR, Mid-Atlantic Ridge; MCSC, Mid-Cayman Spreading Centre; N, number of females analysed; Size-specific fecundity, embryos/mm CL; SD, standard deviation.

Table 3. *Rimicaris hybisae*: Spatial variation in population structure between the Beebe and Von Damm vent fields, January 2012.

	Beebe	Von Damm	Ratio Beebe: Von Damm	χ^2 (1 df)	Significance
Proportion of females ovigerous in samples	54.2%	42.6%	1.27:1	13.47	***
Proportion of ovigerous females brooding	83.9%	70.6%	1.19:1	13.22	***
Proportion of males with spermatophore	45.2%	30.0%	1.51:1	18.23	***

Ovigerous was defined as brooding or hatched; proportion of females ovigerous refers to the ratio of ovigerous females to all females in samples.
*** = P value$<$0.001

proportion of ovigerous females at J2-619-15 than J2-613-24 (Figure 4B; $\chi^2 = 5.31$, 1 df, $p<0.05$), and at J2-613-24 than J2-620-32 (Figure 4B; $\chi^2 = 4.14$, 1 df, $p<0.05$). All other pairwise combinations of samples could not be tested for significant variation from a 1:1 ratio as a result of expected frequencies of zero.

The highest proportion of males with a spermatophoric mass was 62% in sample J2-613-24 (Figure 4C). All samples from the Beebe Vent Field were significantly different from each other in the proportion of males with spermatophores (Table 4, Figure 4C). As a result of too few data points, it was not possible to determine the correlation between frequencies of ovigerous females, brooding females and males with spermatophores.

Size-frequency distribution of shrimp

Overall, shrimp sampled from the Von Damm Vent Field were significantly larger than those from the Beebe Vent Field (Table 5; Figure 2; Mann-Whitney *U*-test, $T = 281949.0$, $p<0.001$). Although the January 2012 samples as a whole displayed two modal peaks and a short tail of large sizes, the peak of larger sizes was absent among samples from the Beebe Vent Field, where smaller shrimp were sampled with proportionally greater frequency (Figure 2). The peak of larger sizes was prominent among the samples from the Von Damm Vent Field, where smaller shrimp were sampled with proportionally lower frequency (Figure 2).

Both males and females were significantly larger at the Von Damm Vent Field than the Beebe Vent Field (Table 5; Figure 2; Mann-Whitney *U*-test, males: $T = 42656.0$, $p<0.001$; females: $T = 104412.5$, $p<0.001$). Females were represented throughout most of the size-frequency distribution at both vent fields, but there were proportionally greater large females at the Von Damm Vent Field (Figure 2; Mann-Whitney *U*-test, Beebe: $T = 61750.5$, $p<0.001$; Von Damm: $T = 25602.5$, $p<0.025$). Non-ovigerous females were significantly smaller at the Beebe Vent Field than the

Table 4. *Rimicaris hybisae*: Spatial variation in population structure within the Beebe Vent Field, January 2012.

Sample	J2-613-19	J2-613-24	J2-620-32	J2-619-15
J2-613-19	-	**914.90*****	**0.03 NS**	**71.08*****
J2-613-24	*4.30**	-	**43.67*****	**57.70*****
J2-620-32	*4.75**	*19.19****	-	**5.11****
J2-619-15	*11.02****	*19.19****	*5.07**	-

Results of χ^2 (1 df) analyses on proportions of females ovigerous (brooding or hatched; bold text) and proportions of males with spermatophores (italic text) in samples.
NS = not significant; * = P value$<$0.05; ** = P value$<$0.01; *** = P value$<$0.001

Von Damm Vent Field (Table 5; Figure 3; Mann-Whitney *U*-test, $T = 12744.5$, $p<0.001$), as were ovigerous females (Table 5; Figure 3; Mann-Whitney *U*-test, $T = 24202.0$, $p<0.001$). The smallest carapace length exhibited by an ovigerous female at the Von Damm Vent Field was 13.3 mm, compared with 8.5 mm at the Beebe Vent Field). Males with and without spermatophores were both significantly larger at the Von Damm Vent Field than the Beebe Vent Field (Table 5; Figure 3; Mann-Whitney *U*-test, males with spermatophores: $T = 5026.5$, $p<0.001$; males without spermatophores: $T = 18127.0$, $p<0.001$). The smallest carapace length shown by a male with a spermatophore was 12. 9 mm at the Von Damm Vent Field (compared with 6.9 mm at the Beebe Vent Field).

The size-frequency distributions of non-ovigerous (neither brooding nor hatched) females were not significantly different to males at the Beebe Vent Field (Table 5; Figure 2B, 3A; Mann-Whitney *U*-test, $T = 26586.0$, $p>0.05$) and the ratio of males to non-ovigerous females did not deviate significantly from 1:1 (138 males: 142 non-ovigerous females, $\chi^2 = 0$, 1 df, $p>0.05$). There was a significantly lower proportion of large non-ovigerous females compared with males at the Von Damm Vent Field (Table 5; Figure 2C, 3B; Mann-Whitney *U*-test, $T = 22364.5$, $p<0.001$). However, the ratio of males to non-ovigerous females did not deviate significantly from 1:1 (141 males: 147 non-ovigerous females; $\chi^2 = 0$, 1 df, $p>0.05$).

Overall, there was significant variation in the size-frequency distributions of shrimp collected from different locations within the Beebe Vent Field (Figure 5; Kruskal-Wallace multisample test, $H = 150.857$, 3 df, $p<0.001$). There was also significant variation in the size-frequency distributions of each sex between samples from the Beebe Vent Field (Kruskal-Wallace multisample test, males: $H = 42.8$, 3 df, $p<0.001$; females $H = 107.0$, 3 df, $p<0.001$).

Shrimp in sample J2-613-19 displayed the smallest median size for each sex (males: CL 9.1 mm, IQR 8.2–10.2; females: CL 9.5 mm, IQR 8.3–9.9). However, the median sizes of males and females in sample J2-613-19 were not significantly different from those in sample J2-620-32 (Figure 5; Dunn's Multiple Comparison Test, $p>0.05$). The median sizes and size-frequency distributions of males and females were not significantly different within sample J2-613-19 (Figure 5; Mann-Whitney *U*-test, $T = 2388.0$, $p>0.05$; Kolmogorov-Smirnof two-sample test, $D = 0.05$, $p>0.05$).

The median sizes and size-frequency distributions of males and females were not significantly different within sample J2-620-32 (Figure 5; Mann-Whitney *U*-test, $T = 723.0$, $p>0.05$; Kolmogorov-Smirnof two-sample test, $D = 0.12$, $p>0.05$). However, the number of females was low. The median sizes were CL 9.2 (IQR 8.4–10.5 mm) for males and 9.3 mm (IQR 7.4–10.3) for females.

Sample J2-613-24 exhibited a peak of large females and significantly greater median sizes for females than males (Figure 5; males: CL 10.3 mm, IQR 9.7–10.8; females: CL 10.8 mm, IQR 10.2–11.4; Mann-Whitney *U*-test, $T = 8779.5$,

Table 5. *Rimicaris hybisae*: Variation in body size (carapace length), January 2012.

Vent Field	Carapace length (mm)					
	Males			**Females**		
	All	Without spermatophore	With spermatophore	All	Non-ovigerous	Ovigerous
Beebe	10.1 (9.0–10.7)	9.7 (8.7–10.6)	10.2 (9.5–10.9)	10.6 (9.8–11.3)	9.9 (8.3–10.8)	10.9 (10.4–11.6)
Von Damm	14.9 (13.2–15.0)	13.8 (12.6–14.5)	14.8 (14.2–15.1)	14.7 (12.9–15.6)	13.3 (11.5–14.9)	15.3 (14.8–15.8)

Carapace length is shown as median (inter-quartile range).

$p < 0.001$). The distribution of sizes between males and females was significantly different (Kolmogorov-Smirnof two-sample test, $D = 0.25$, $p < 0.01$).

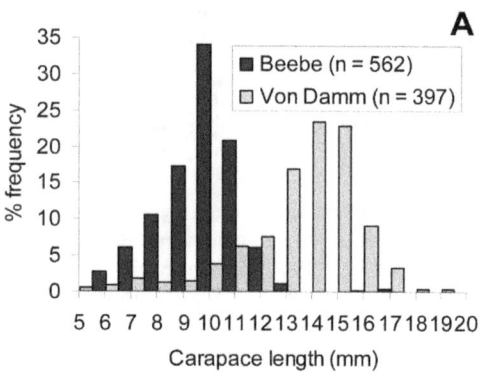

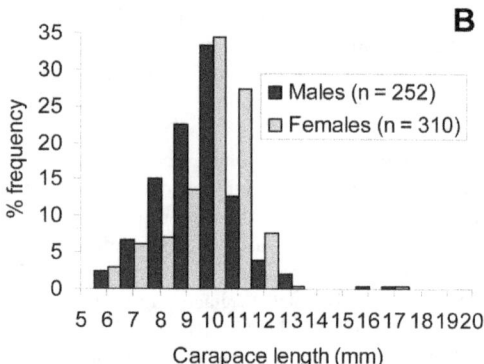

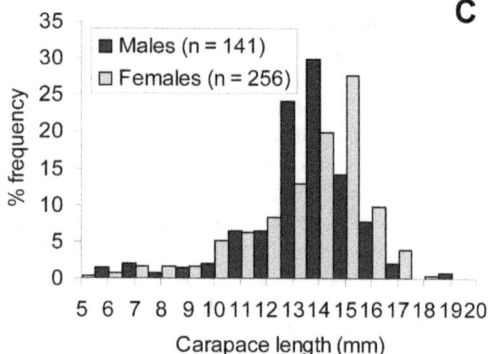

Figure 2. *Rimicaris hybisae*: **Size-frequency distribution, January 2012.** (A) All specimens; (B) Beebe Vent Field; (C) Von Damm Vent Field. n: no. of individuals measured.

Shrimp in sample J2-619-15 exhibited the largest median size for both males (CL 10.6 mm, IQR 10.1–11.2) and females (CL 11.2 mm, IQR 10.1–11.2) but the median size of males in J2-613-24 (CL 10.3 mm, IQR 9.7–10.8) was not significantly different (Figure 5; Dunn's Multiple Comparison Test, $p < 0.05$). Sample J2-613-15 exhibited a peak of large females and significantly greater median sizes for females than males (Figure 5; Mann-Whitney U-test, $T = 2267.5$, $p < 0.01$). The distribution of sizes between males and females was significantly different (Kolmogorov-Smirnof two-sample test, $D = 0.30$, $p < 0.05$).

Fecundity

The embryos in broods of *Rimicaris hybisae* formed a dense mass attached to pleopods 1–4 underneath the female abdomen. The broods were orange in colour and visible in video footage from the Beebe and Von Damm vent fields in January 2012. The greatest minimum realised fecundity determined among 218 brooding females examined from January 2012 was 1707 in an individual from the Von Damm Vent Field with a carapace length of 16.6 mm.

Specimens from the Von Damm Vent Field exhibited significantly greater fecundity than those from the Beebe Vent Field (Table 2; Von Damm vs J2-613-19, vs J2-613-24, vs J2-619-15, Kruskal-Wallace multisample test, $H = 139.8$, 3 df, $p < 0.001$; Dunn's Multiple Comparison Test, $p < 0.005$). The median number of embryos brooded by females from the Von Damm Vent Field was 1062 (IQR 931–1191). The median number of embryos brooded by females from the Beebe Vent Field ranged from 122 (IQR 20–304, J2-613-19, Beebe Woods) to 385 (IQR 357–439, J2-619-15, Shrimp Gulley). Differences in the fecundities of females among the samples from the Beebe Vent Field were not significant (Dunn's Multiple Comparison Test, $p > 0.05$).

Minimum realised fecundity correlated positively with carapace length (Figure 6A; Spearman correlation, $r = 0.77$, $p < 0.0001$). The slope of the relationship between $\log_e$-transformed fecundity and $\log_e$-transformed body size (carapace length) did not differ significantly among specimens from the Von Damm Vent Field and the Beebe Vent Field (Figure 6A; $F = 0.06$, $p < 0.05$). Overall, size-specific fecundity (related to carapace length) ranged from 1.2 to 102.8 embryos mm^{-1}. Although brooding females at the Von Damm Vent Field carried more embryos per mm carapace length than those at the Beebe Vent Field (Figure 6B), this difference could therefore be attributable to the larger body sizes of brooding females at Von Damm and a non-linear relationship between fecundity and body size, which appears to be consistent across the two vent fields.

Embryo developmental stages

Within each brood of the 218 brooding females analysed, the embryos had developed synchronously and were all at the same

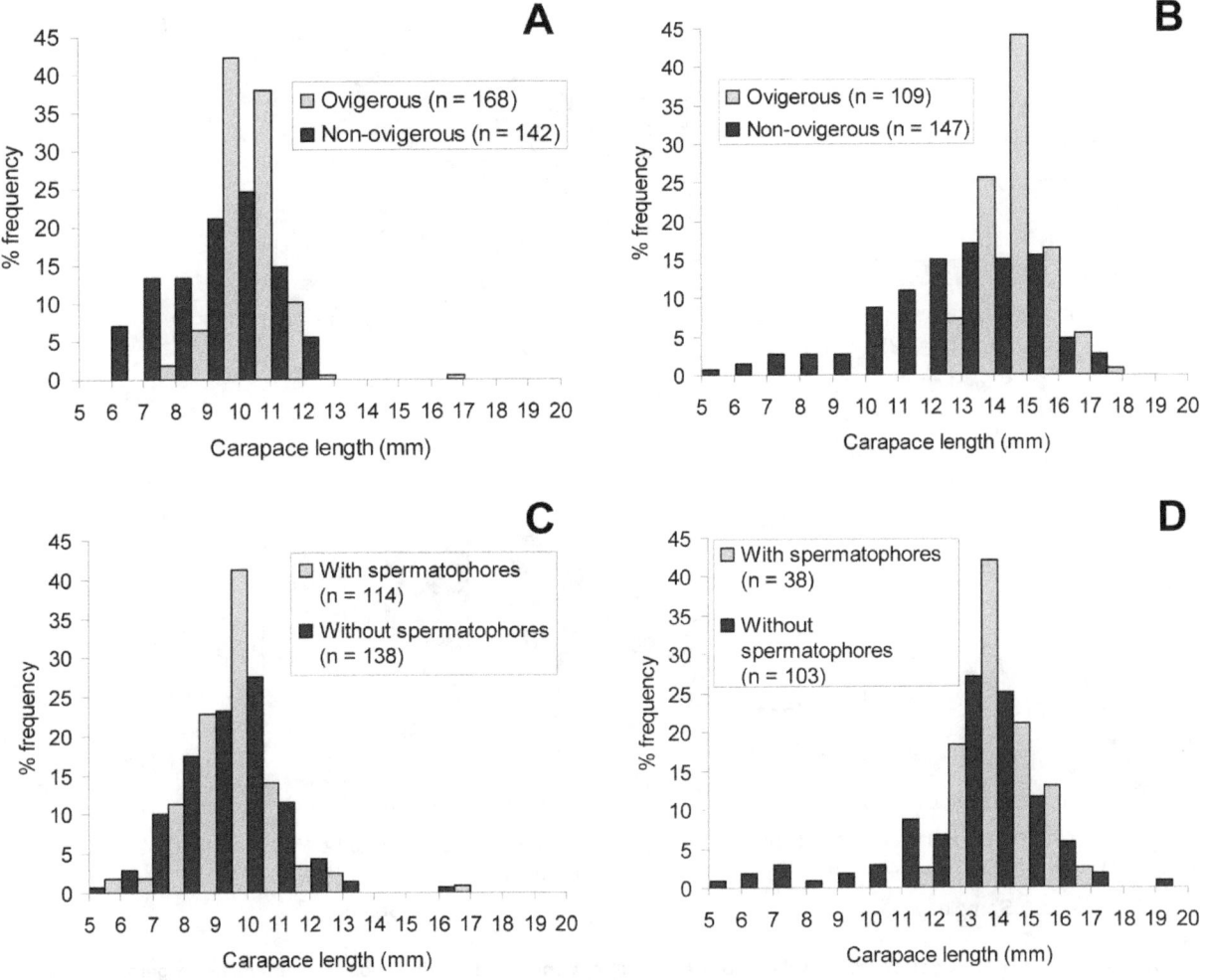

Figure 3. *Rimicaris hybisae:* **Size-frequency distribution of females and males, January 2012.** (A) Females, Beebe Vent Field; (B) females, Von Damm Vent Field; (C) males, Beebe Vent Field; (D) females, Von Damm Vent Field. n: no. of individuals measured.

stage of development (early, mid or late; Figure 1). However, there was no evidence of synchrony between the broods of different females. Overall, the majority of broods (48%) were in a medium developmental stage. One third (34%) exhibited an early stage of development, whereas advanced-stage broods were the minority (19%). The majority of broods at both the Von Damm and Beebe vent fields were in mid-stage development (60% and 41% respectively), and there were a greater proportion of broods at the early-developmental stage compared to the late stage (30% vs 10% Von Damm; 36% vs 24% Beebe).

In sample J2-613-19, 80% of broods were at the early stage and the remaining 20% were at the late stage (Figure 4D). In contrast, 47% of broods were at the mid-stage in sample J2-613-24; 34% were early-stage and 19% were late-stage (Figure 4D). Late-stage broods accounted for 43% of broods in sample J2-619-15, with 32% and 25% at the early and mid-stage broods respectively (Figure 4D). There were no brooding females in sample J2-620-32.

Oocyte size-frequency distribution

Females examined from January 2012 contained oocytes with feret diameters ranging from 21 μm (non-ovigerous female, CL 12.3, Von Damm) to 823 μm (non-ovigerous female, CL 16.7, Von Damm) (Figure 7). Specimens examined from both vent fields

included both ovigerous (brooding and hatched) and non-ovigerous females.

The median oocyte size in individuals from the Von Damm Vent Field was 82 μm (IQR 59–109). Three distinct oocyte sizes were apparent among females collected from the Von Damm Vent Field in January 2012 (Figure 7A). The larger oocytes belonged exclusively to large, non-ovigerous females (CL 14.4–17.5 mm). These specimens exhibited median oocyte sizes ranging from 581 μm (IQR 545–602) to 626.5 μm (IQR 589–665). The small oocytes belonged to ovigerous and non-ovigerous females (CL 11.6–17.2 mm); these specimens displayed a range of median oocyte sizes from 37 μm (IQR 33–41) to 120.5 μm (IQR 105.5–137). One individual (non-ovigerous female, CL 14.96 mm) showed a median oocyte size outside both these ranges (196.5 μm, IQR 169–246).

The median oocyte size in 99 individuals from the Beebe Vent Field was 86 μm (IQR 71–109). Three distinct oocyte sizes were apparent among females collected from the Beebe Vent Field in January 2012 (Figure 7B). Females examined in samples J2-619-15, J2-613-19 and J2-620-32 contained relatively small oocytes with feret diameters between 26 and 262 μm (Figure 7). Median oocyte sizes in individuals from sample J2-619-15 ranged from 66.5 μm (IQR 61–74.5) to 178 μm (IQR 162–191.5). Median oocyte sizes in the nine females from sample J2-613-19 ranged

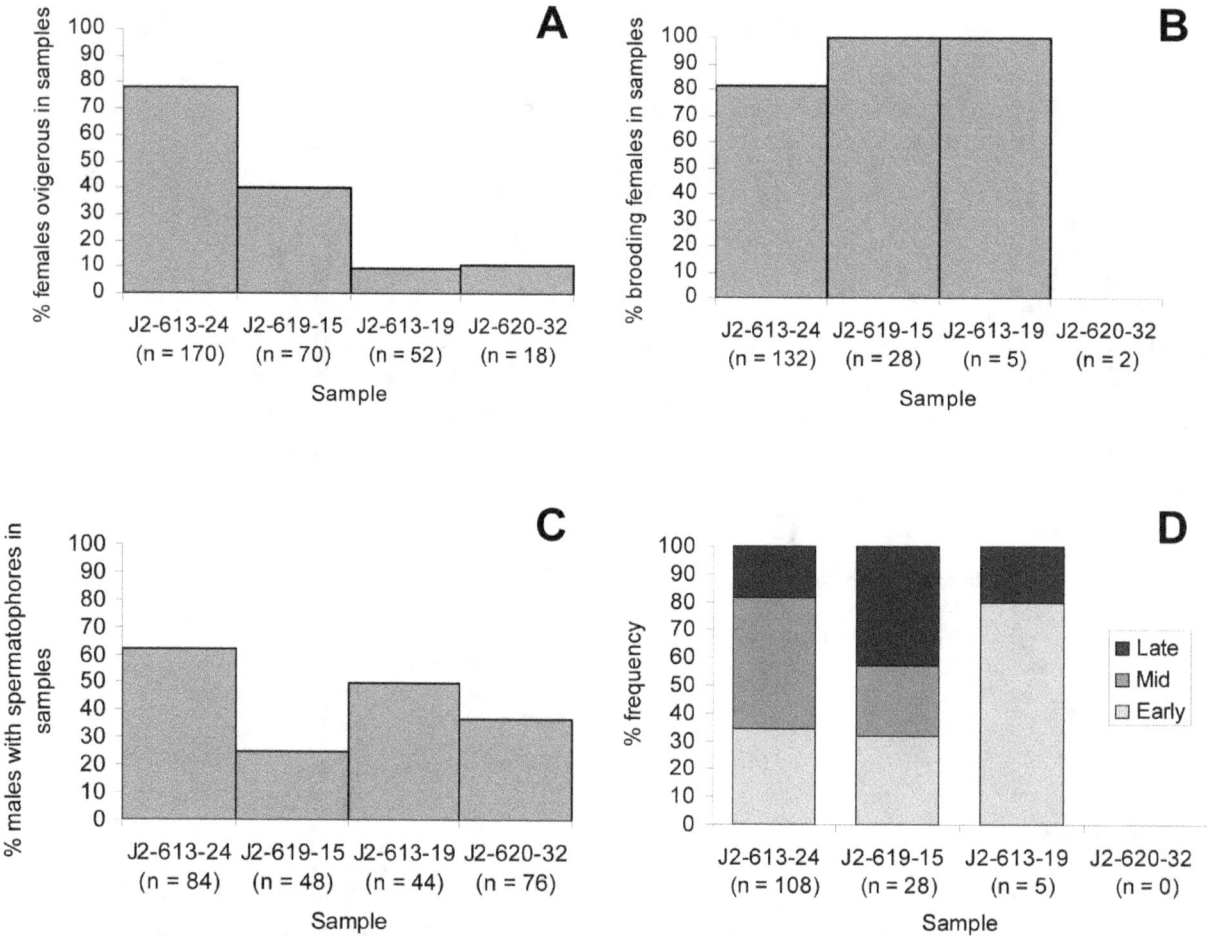

Figure 4. *Rimicaris hybisae*: **Spatial variation in samples from the Beebe Vent Field.** (A) in proportion of females ovigerous (defined as brooding embryos or hatched) in samples of females; (B) proportion of ovigerous females brooding in samples of ovigerous females; (C) proportion of males with spermatophores in samples of males; (D) proportion of embryos in samples of brooding females at each developmental stage. n: sample size (100%) in each case.

from 48 μm (IQR 42–57) to 186.5 μm (IQR 162–215.5). The range of median oocyte sizes in females from sample J2-620-32 was 38 μm (IQR 33–41.5) to 101 μm (IQR 90.5–117). Specimens examined from these samples included both ovigerous and non-ovigerous females. Three distinct oocyte sizes were evident among females from sample J2-613-24 (Figure 7C). Here the larger oocyte sizes belonged to non-ovigerous females (CL 9.2–10.3 mm). These individuals exhibited median oocyte sizes ranging from 527 μm (IQR 496–569) to 636 μm (IQR 586.5–683.5). The mid-size oocytes also belonged to non-ovigerous females (CL 9.6–10.7 mm); these specimens displayed median oocyte sizes ranging from 333 μm (IQR 315.5–358.5) to 388 μm (IQR 353–419.5). The smaller oocytes were from ovigerous and non-ovigerous females (CL 9–12.4 mm). These females revealed median oocyte sizes ranging from 40 μm (IQR 34.5–44) to 111 μm (IQR 84.5–136.5).

Overall, there was significant variation in oocyte sizes between samples (Kruskal-Wallis multi-sample test, $H = 371.8$, 4 df, $p < 0.01$) and no evidence of synchrony between samples. There were significant differences in oocyte sizes between every pairwise comparison of all five samples (Dunn's Multiple Comparison Test, $p < 0.05$).

Discussion

General features of reproduction in *Rimicaris hybisae*

The gonads of *Rimicaris hybisae* are similar to those of other caridean shrimp [16,13,27]. *Rimicaris hybisae* exhibits sexual dimorphism and is a gonochoric species, consistent with all other alvinocaridid species studied to date.

The maximum size of males analysed was larger than the maximum size of females of *Rimicaris hybisae*. Nevertheless, in January 2012 females exhibited a significantly larger median body size and greater size range than males with proportionally greater large females at both the Beebe and Von Damm vent fields. A larger size in females has been inferred for *Alvinocaris muricola* based on the maximum size of males vs females [16] and a larger size of females is a common feature of caridean shrimp [28]. However, the variance in the size of sexes in this study appears to be the result of spatial variation in the proportion of males and females in samples, rather than sexual dimorphism.

Males with spermatophores were significantly larger in carapace length than males without spermatophores in January 2012. Ovigerous (brooding and hatched) females were significantly larger than non-ovigerous females at both vent fields, whereas the size-frequency distributions of non-ovigerous females were not significantly different from males and the ratio of males to non-

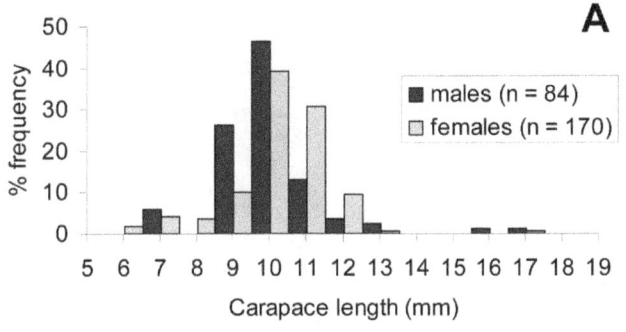

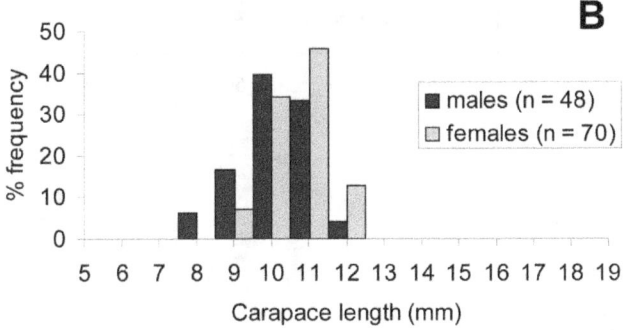

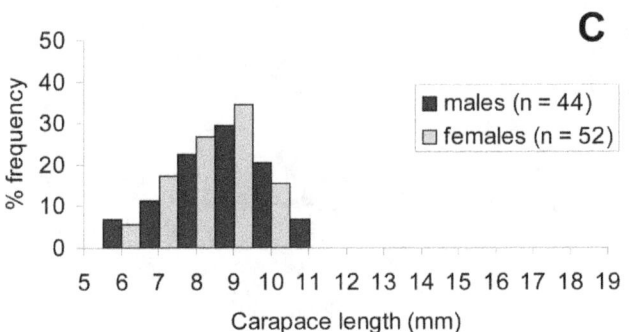

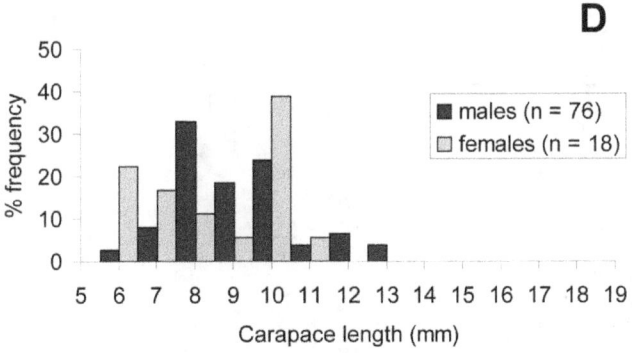

Figure 5. *Rimicaris hybisae*: **Size-frequency distribution within the Beebe Vent Field, January 2012.** (A) Sample J2-613-24; (B) sample J2-619-15; (C) sample J2-613-19; (D) sample J2-620-32. n: no. of individuals measured.

ovigerous females did not deviate significantly from 1:1. The larger size of ovigerous females has been reported previously for *Alvinocaris stactophila* from the Brine Pool cold seep [15]. A greater size of ovigerous females may be advantageous for embryo production where fecundity correlates positively with body size, as

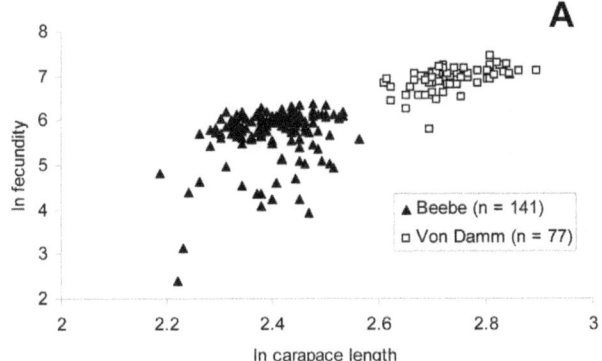

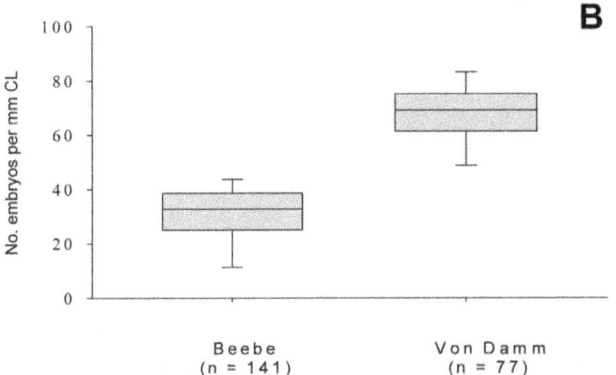

Figure 6. *Rimicaris hybisae*: **Fecundity (no. of embryos on pleopods), January 2012.** (A) Variation of $\log_e$-transformed minimum realised fecundity with $\log_e$-transformed carapace length; (B) corrected for body size (carapace length). n: no. of individuals measured.

in *Rimicaris hybisae*. A positive correlation between body size and fecundity has been determined for every alvinocaridid species in which these variables have been examined (*A. muricola* and *A. stactophila* [13,15]; *Mirocaris fortunata* [16]; *R. hybisae*, this study). This feature of carideans is a result of space availability between the pleopods for attachment of embryos [29].

The maximum oocyte size measured in *Rimicaris hybisae* females in January 2012 was 823 μm, which is greater than maximum oocyte sizes recorded for other alvinocaridid shrimp (*R. exoculata*: 500 μm [13], 601 μm [17]; *Alvinocaris muricola*: 515 μm [16]). However, the smaller mature oocytes in the range exhibited by *R. hybisae* (Figure 7) fall within the range recorded previously in the literature. Although it is possible that mature oocytes attain a greater size in *R. hybisae* than in other alvinocaridids for which information is available, an alternative explanation is that the true size range of mature oocytes has not been captured yet in other species, given smaller sample sizes from which data have been generated in other species.

Embryos had developed synchronously within (but not between) broods in January 2012. The developmental stages of embryos observed in *Rimicaris hybisae* were similar to those described for *Alvinocaris muricola* [16]. One experimental study found that the embryos of two brooding specimens of *Alvinocaris* sp. hatched after 73 and 79 days of rearing, with complete hatching of the brood over 7 and 14 days [30]. Nevertheless, it remains to be determined how long alvinocaridid females incubate their young for *in situ*, how long embryos take to develop from one stage to the next and whether these features are variable between different environ-

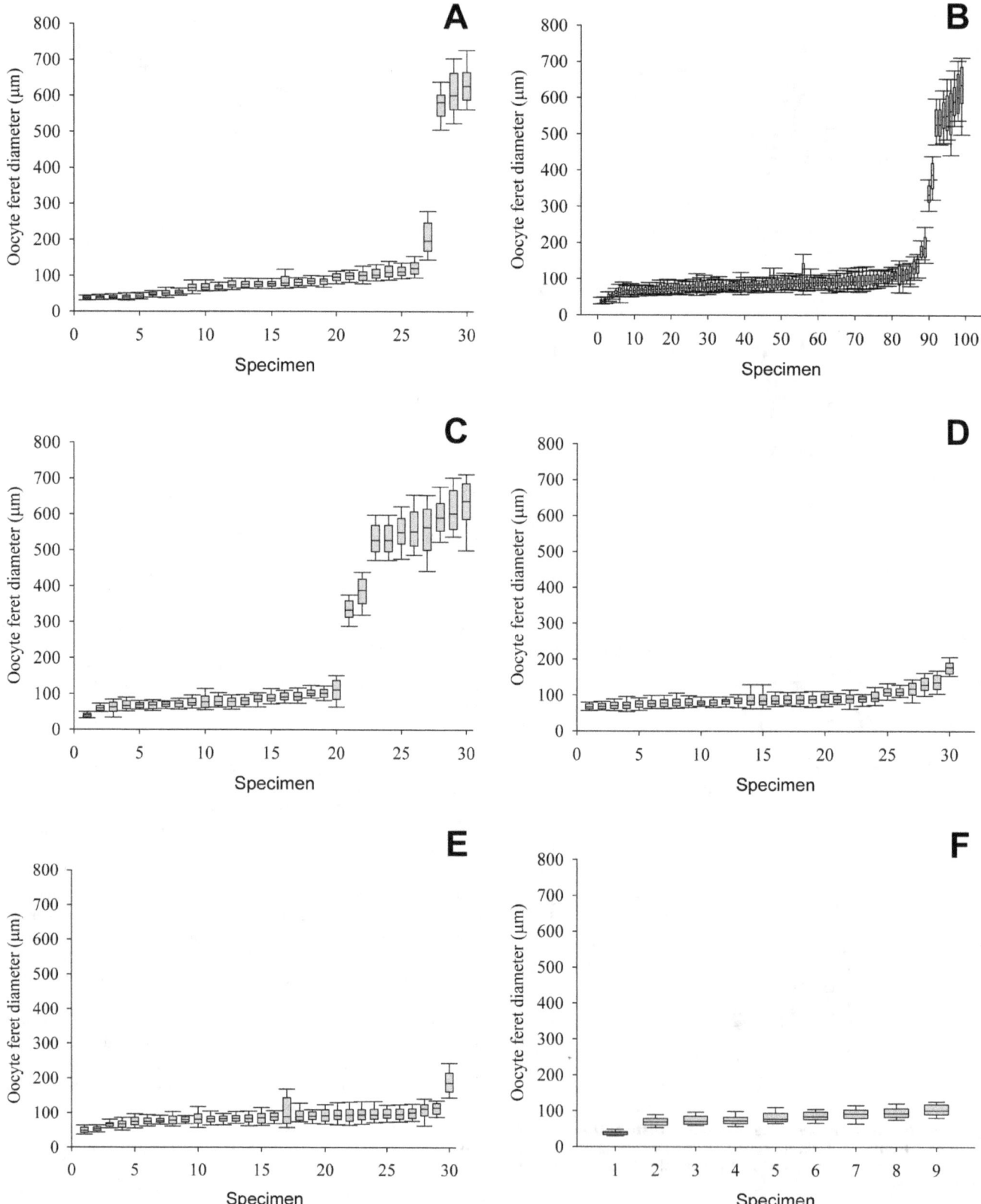

Figure 7. *Rimicaris hybisae*: **Spatial variation in oocyte size-frequency distribution at the Beebe (B–F) and Von Damm (A) vent fields, January 2012.** (A) Von Damm Vent Field; (B) Beebe Vent Field (all samples); (C) sample J2-613-24; (D) sample J2-619-15; (E) sample J2-613-19 (F) sample J2-620-32.

mental conditions. Further experimental studies in the laboratory and *in situ* would be required to elucidate these knowledge gaps.

Tyler & Young [2] reported that a complete life cycle had yet to be elucidated for a single vent or seep species. Following their template for a generalised marine invertebrate life cycle, what is now known, inferred, or unknown in gametogenesis, copulation/spawning, embryo development, larval ecology, dispersal and recruitment are summarised for *Rimicaris hybisae* in Figure 8.

Embryos of *Rimicaris hybisae* were variable in size but there was no spatial variation in embryo size. Embryos ranged in size from 0.42–0.76 mm (greatest diameter) with a mean size that was

consistent with those reported in other alvinocaridid species to date (Table 2). The small size of embryos revealed in all the alvinocaridids for which data are available is indicative of planktotrophic larvae with extended development [27]. This hypothesis is supported by biochemical and experimental studies for other species [30,31,32,33,34,35,36,37,38,39]. A long larval duration could facilitate extended dispersal between patchily distributed chemosynthetic environments and promote genetic diversity and colonisation of new vents [40,41,42]. Biochemical studies would be required to confirm the trophic ecology of *R. hybisae* during the larval (and adult) phase.

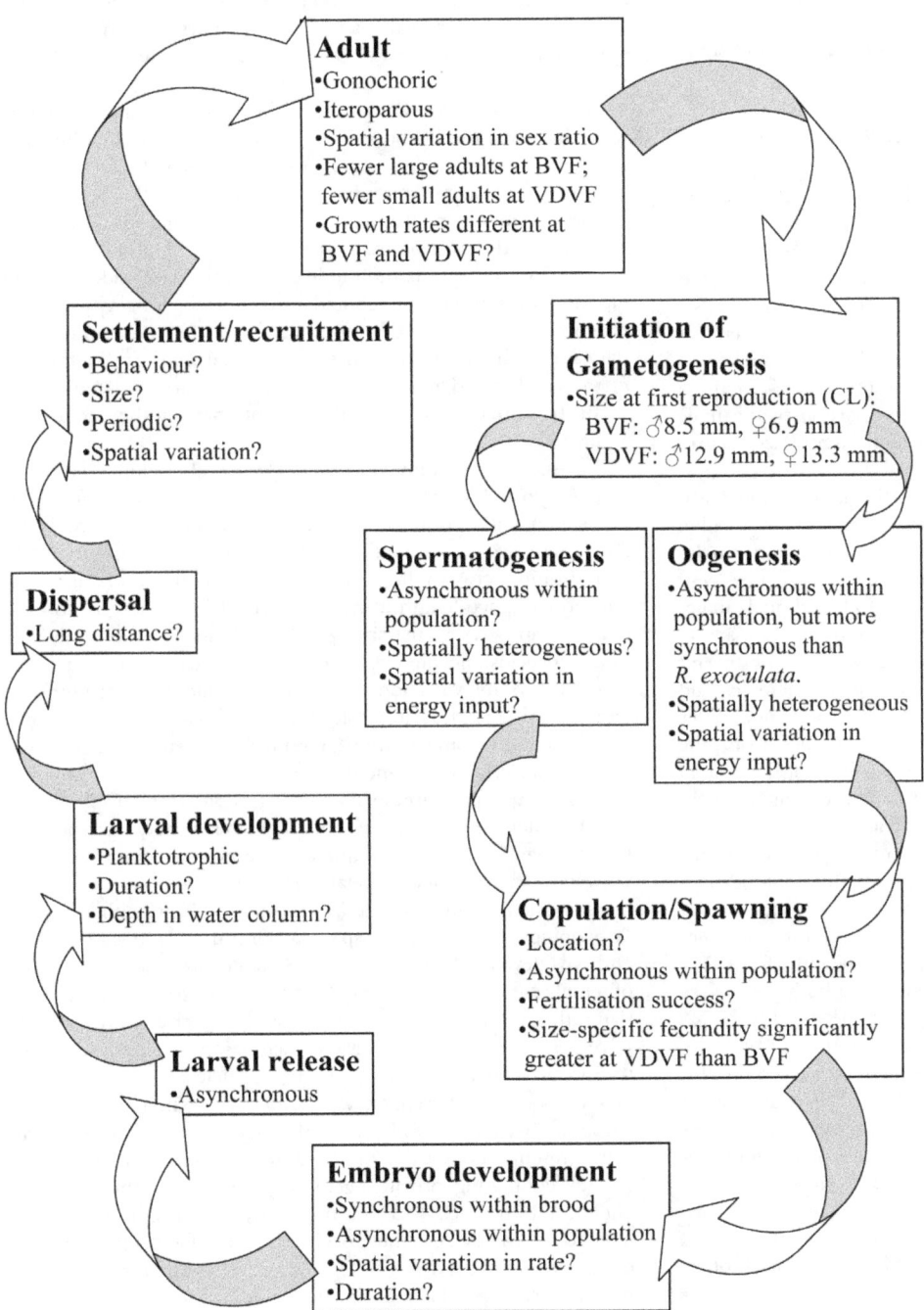

Figure 8. *Rimicaris hybisae*: **Inferred life-history (sensu Tyler & Young, 1999).** Stages are in bold. BVF, Beebe Vent Field; CL, carapace length; VDVF, Von Damm Vent Field.

It has been suggested that upward vertical movement during the larval phase could explain the broad geographic distribution of certain species with planktotrophic larvae entrained in deep-water currents [4]. Vent shrimp postlarvae have been found to be present in midwater above MAR vents and to extend great distances laterally from known chemosynthetic sites [43]. However, the period and depth of planktotrophic development in the water column has not been specified for a single alvinocaridid species to date. The larval phase remains one of the least known stages in the life-cycle of most deep-sea species [6], yet elucidating the larval ecology of a species is a prerequisite to understand patterns of reproduction and recruitment. Studies of larval tolerances, lifespan and mortality during dispersal combined with hydrographic and phylogeographic data are required to address this significant gap in the life-cycle of *Rimicaris hybisae* and other alvinocaridid species.

Spatial variation in population structure and reproductive features

In January 2012, the sampled population of *Rimicaris hybisae* at the Von Damm Vent Field was dominated by large females and a lesser proportion of slightly smaller males. The sex ratio was significantly female-biased but the ratio of males to non-ovigerous females did not differ significantly from 1:1. However, there was a significantly smaller proportion of large non-ovigerous females compared with males. Large ovigerous females accounted for a high proportion of the female population. Males with a spermatophore represented quite a high proportion of the male population but were only present in the larger size classes (CL>12 mm).

Overall, the sampled population at the Beebe Vent Field was similar to that at the Von Damm Vent field (see above), but with a significantly greater proportion of ovigerous and brooding females and males with spermatophores. The sex ratio at the Beebe Vent Field was also significantly female-biased overall, although males represented a greater proportion of the population in comparison with the Von Damm Vent Field. There was no significant difference in size between males and non-ovigerous females and the ratio of males to non-ovigerous females did not differ significantly from 1:1. However, individual samples from the Beebe Vent Field were heterogeneous. Recognition of this spatial heterogeneity is crucial to consider in any attempts to infer temporal patterns from limited spatial samples.

Ovigerous female *Rimicaris hybisae* and males with spermatophores were present with the greatest frequency in sample J2-613-24. This sample was dominated by ovigerous females and the sex ratio was significantly female-biased. However, the ratio of males to non-ovigerous females did not differ significantly from 1:1. Females were significantly larger than males and both sexes were significantly larger than those in samples J2-613-19 and J2-620-32.

Large male and female *Rimicaris hybisae* were present with the greatest frequency in sample J2-619-15. The sex ratio in this sample did not differ significantly from 1:1. Ovigerous females were present with the second greatest frequency here and females were significantly bigger than males. Males with spermatophores were present with the lowest frequency in this sample.

Ovigerous female *Rimicaris hybisae* were present with the lowest frequency in samples J2-613-19 and J2-620-32 and the size of males and females were not significantly different from each other in these samples. The sex ratio in sample J2-620-32 was significantly male-biased, whereas it did not differ significantly from 1:1 in J2-613-19.

The population of *Rimicaris hybisae* showed a significant clear bias towards females in the sample from the Von Damm Vent Field and sample J2-613-24 from Beebe Woods. These samples also contained the greatest proportion of ovigerous females. Both samples were collected from large, high-density aggregations of shrimp. Ramirez-Llodra & Segonzac [16] described a clear bias towards females in *Alvinocaris muricola* in the cold-seep site north of Regab in the Congo Basin. Copley & Young [15] identified a specific distribution of males and females within the Brine Pool mussel bed, with ovigerous females avoiding the sulphidic or anoxic extremes in the environment. Hydrothermal vents exhibit fine-scale heterogeneity in physico-chemical conditions [44]. Environmental variables within shrimp aggregations may influence the distribution of ovigerous females, resulting in a spatially heterogeneous pattern of reproductive development in *R. hybisae*, as found in other vent taxa [45,46]. Further sampling and collection of environmental data would be required to test this hypothesis.

Shrimp at the Von Damm Vent Field were significantly larger than shrimp at the Beebe Vent Field both overall and within each population category. The smallest ovigerous female (CL 6.9 mm) and male with a spermatophore (CL 8.5 mm) at the Beebe Vent Field were much smaller than the smallest ovigerous female (CL 13.3) and male with spermatophore at the Von Damm Vent Field (CL 12.9 mm). Size at the onset of maturity is considered a key life-history parameter that should also reflect the longevity and lifetime investment in reproduction of a species [26]. Assuming these data represent the minimum size of sexual maturity in *Rimicaris hybisae*, males and females at the Beebe Vent Field may reach maturity at smaller sizes than their counterparts at the Von Damm Vent Field.

Size-specific fecundity in *Rimicaris hybisae* falls within the range of values reported in other alvinocaridid species (Table 2). Fecundity was significantly positively correlated with body size in *R. hybisae* and there was no significant difference in the slope of the relationship between $\log_e$-transformed fecundity and $\log_e$-transformed carapace length at the two vent fields. There is no clear relationship between the size-specific investments in reproduction in alvinocarids and phylogeny, depth or environment (vent versus seep) [16]. Some variation in measured fecundity may result from females losing eggs during collection [3] but variation in fecundity may also result from variations in reproductive success, which may be affected by environmental conditions [15]. Environmental factors such as pressure, availability of photosynthetically-derived nutrition, temperature, and/or fluid chemistry may vary between the two vent fields which are considerably different in depth. However, a non-linear relationship between fecundity and carapace length may result from volume effects either for the developing ovary or in the space around the pleopods on which embryos develop. Consequently, the greater size-specific fecundity of females at Von Damm may be attributable to their larger body size rather than environmental conditions, given the homogeneity of regression between $\log_e$-transformed carapace lengths and $\log_e$-transformed fecundities across both vent fields.

The oocyte size-frequency distributions exhibited significant variation between samples and all stages of developing oocytes were present amongst females in January 2012. Most females had gonads containing previtellogenic oocytes (<100 μm) and early vitellogenic oocytes (>100 μm). However, several large non-ovigerous females from the Von Damm Vent Field and sample J2-613-24 contained large vitellogenic oocytes. These data indicate iteroparous reproduction in *Rimicaris hybisae*. However, the variation observed was less than that recorded among individual female *R. exoculata* from TAG [17], suggesting a greater degree of synchrony in the oocyte development of *R. hybisae* than reported for *R. exoculata*.

Embryo development was clearly asynchronous between females of *Rimicaris hybisae*, indicating that larval release may also be asynchronous for the population as a whole. Environmental variability may potentially affect every reproductive process, resulting in embryo development proceeding at different rates within a vent field.

Video from a previous cruise (NOAA 'Okeanos Explorer') to the Von Damm Vent Field in August 2011 showed no evidence of brooding *Rimicaris hybisae*, whilst individuals in video collected during January 2012 showed clear evidence of embryo carrying, suggesting possible periodic reproduction for *R. hybisae*. The collection and analyses of additional samples spanning several seasons of the year are a prerequisite to determine the potential periodicity or even seasonality in reproduction and possibly recruitment of *R. hybisae*.

Periodic reproduction has also been suggested for *Rimicaris exoculata* based on the almost complete absence of ovigerous females in collections from summer months (when most samples have been collected) [47]. However, observations of oocyte size-frequency distributions of females collected in different times of the year indicate iteroparous, asynchronous reproduction and lack of seasonal reproduction in *R. exoculata* from MAR vents [13,17]. An alternative hypothesis that has been proposed is that ovigerous females stay outside the main populations [13]. However, recent observations of ovigerous females of *R. exoculata* within the main aggregations around high temperature zones at Logatchev in March 2007 [48] refute the latter hypothesis.

Periodic production of eggs has also been proposed for *Mirocaris fortunata* [13], whereas continuous egg production with periodic spawning was suggested for *Alvinocaris muricola* [16]. In contrast, a seasonal pattern of reproduction has been revealed in *A. stactophila* at the Brine Pool cold seep [15] and a few other species with planktotrophic larvae from vent environments (e.g., [14,49,50]). Surface productivity may therefore be important for the nutrition of planktotrophic larvae of some vent and seep species, particularly at shallower depths.

Conclusions

Reproductive features examined in *Rimicaris hybisae* at the Beebe and Von Damm vent fields are consistent with those described previously for other alvinocaridid species, consistent with phylogenetic constraint of such features in vent species. Several gaps remain, however, in understanding the life cycle of this and other alvinocaridid species.

Samples collected from the Von Damm and Beebe vent fields in January 2012 revealed spatial variation in the population structure and reproductive features of *Rimicaris hybisae*. These data highlight a high degree of spatial variability in the reproductive features of a mobile species in the vent environment. The sample from the Von Damm Vent Field and sample J2-613-24 from Beebe Woods exhibited the highest frequencies of ovigerous females and significantly female-biased sex ratios. Any bias in sex ratio was not the result of immature males being misidentified as females. Nevertheless, when the generally large ovigerous females were excluded, there was no significant deviation from a 1:1 sex ratio. Environmental variables within shrimp aggregations may influence the distribution of ovigerous females, resulting in a spatially heterogeneous pattern of reproductive development in *R. hybisae*, as found in other vent taxa. This hypothesis would require testing with the collection and analysis of further samples with environmental data (temperature, fluid chemistry).

Reproduction in *Rimicaris hybisae* is iteroparous. The oocyte development of *R. hybisae* appears to exhibit a greater degree of synchrony than reported previously for *R. exoculata*. However, embryo development and larval release may be asynchronous for the population as a whole. Analysis of video from a previous cruise to the Von Damm Vent Field in a different season revealed no evidence of ovigerous females. These results suggest a lack of synchrony and possible periodicity in the reproductive development of *R. hybisae*. Any subsequent investigation of temporal variation in the reproductive development of *R. hybisae*, however, needs to take into account the spatial variation also revealed by this study.

Specimens of *Rimicaris hybisae* from the Von Damm Vent Field were significantly larger than specimens from the Beebe Vent Field and may reach maturity at larger sizes than their counterparts at the Beebe Vent Field. Given a possible non-linear relationship between fecundity and carapace length, the larger body sizes of brooding females at Von Damm may result in a greater size-specific fecundity compared with females at the Beebe Vent Field.

Acknowledgments

The authors thank the Master and ship's company of the RV 'Atlantis', the crew of the ROV 'Jason II' and fellow scientists on the 18th voyage (16th leg) of RV 'Atlantis', especially C.L. Van Dover, S. Plouviez and J.W. Clarke; C. German, chief scientist. The helpful comments of two reviewers are gratefully acknowledged.

Author Contributions

Particpants on Atlantis18_16: VN PAT. Conceived and designed the experiments: VN JTC PAT. Performed the experiments: VN. Analyzed the data: VN JTC. Contributed reagents/materials/analysis tools: VN JTC PAT. Wrote the paper: VN JTC PAT.

References

1. Young CM, Vazquez E, Metaxas A, Tyler PA (1996) Embryology of vestimentiferan tubeworms from deep-sea methane/sulphide seeps. Nature 381: 514–516.
2. Tyler PA, Young CM (1999) Reproduction and dispersal at vents and cold seeps. J Mar Biol Assoc U.K. 79: 193–208.
3. Ramirez-Llodra E (2002) Fecundity and life-history strategies in marine invertebrates. Adv Mar Biol 43: 87–170.
4. Van Dover CL, German CR, Speer KG, Parson LM, Vrijenhoek RC (2002) Evolution and Biogeography of Deep-Sea Vent and Seep Invertebrates. Science 295: 1253–1257.
5. Desbruyères D, Segonzac M, Bright M (2006) Handbook of deep-sea hydrothermal vent fauna. Vienna: Biologiezentrum der Oberosterreichische Landesmuseen. 544 p.
6. Young CM (2003) Reproduction, development and life-history traits. In: Tyler PA, editor. Ecosystems of the deep oceans.Amsterdam: Elsevier. pp 381–426.
7. Connelly DP, Copley J, Murton BJ, Stansfield K, Tyler PA, et al. (2012) Hydrothermal vents on the world's deepest seafloor spreading centre. Nat Commun 3: doi: 10.1038/ncomm s1636.
8. Nye V, Copley J, Plouviez S (2012) A new species of *Rimicaris* (Crustacea: Decapoda: Caridea: Alvinocarididae) from hydrothermal vent fields on the Mid-Cayman Spreading Centre, Caribbean. J Mar Biol Assoc U.K. 92: 1057–1072. doi: 10.1017/S0025315411002001.
9. Nye V, Copley J, Linse K, Plouviez S (2012) *Iheyaspira bathycodon* new species (Vetigastropoda: Trochoidea: Turbinidae: Skeneinae) from the Von Damm Vent Field, Mid-Cayman Spreading Centre, Caribbean. J Mar Biol Assoc U.K. doi: 10.1017/S0025315412000823.
10. Nye V, Copley J, Plouviez S, Van Dover C (2012) A new species of *Lebbeus* (Crustacea: Decapoda: Caridea: Hippolytidae) from the Von Damm Vent Field, Caribbean Sea. J Mar Biol Assoc U.K. doi: 10.1017/S0025315412000884.
11. Martin JW, Haney TA (2005) Decapod crustaceans from hydrothermal vents and cold seeps: a review through 2005. Zool J Linn Soc 145: 445–522.
12. Van Dover CL, Factor JR, Williams AB, Berg CJ Jr (1985) Reproductive strategies of hydrothermal vent decapod crustaceans. Bull Biol Soc Wash 6: 223–227.
13. Ramirez-Llodra E, Tyler PA, Copley JTP (2000) Reproductive biology of three caridean shrimp, *Rimicaris exoculata*, *Chorocaris chacei* and *Mirocaris fortunata*

(Caridea: Decapoda), from hydrothermal vents. J Mar Biol Assoc U.K. 80: 473–484.

14. Perovich GM, Epifanio CE, Dittel AI, Tyler PA (2003) Spatial and temporal patterns in the vent crab *Bythograea thermydron*. Mar Ecol Prog Ser 251: 211–220.

15. Copley J, Young CM (2006) Seasonality and zonation in the reproductive biology and population structure of the shrimp *Alvinocaris stactophila* (Caridea: Alvinocarididae) at a Louisiana Slope cold seep. Mar Ecol Prog Ser 315: 199–209.

16. Ramirez-Llodra E, Segonzac M (2006) Reproductive biology of *Alvinocaris muricola* (Decapoda: Caridea: Alvinocarididae) from cold seeps in the Congo Basin. J Mar Biol Assoc U.K. 86: 1347–1356.

17. Copley JTP, Jorgensen PBK, Sohn RA (2007) Assessment of decadal-scale ecological change at a deep Mid-Atlantic hydrothermal vent and reproductive time-series in the shrimp *Rimicaris exoculata*. J Mar Biol Assoc U.K. 87: 859–867.

18. Hilário A, Vilar S, Cunhar MR, Tyler P (2009) Reproductive aspects of two bythograeid crab species from hydrothermal vents in the Pacific-Antarctic Ridge. Mar Ecol Prog Ser 378: 153–160.

19. Christoffersen ML (1986) Phylogenetic relationships between Oplophoridae, Atyidae, Pasiphaeidae, Alvinocarididae fam. n., Bresiliidae, Psalidopodidae and Disciadidae (Crustacea: Caridea: Atyoidea). Bol Zool 10: 273–281.

20. Williams AB (1988) New marine decapod crustaceans from waters influenced by hydrothermal discharge, brine and hydrocarbon seepage. Fish Bull 86: 263–287.

21. Sheader M, Van Dover CL (2007) Temporal and spatial variation in the reproductive ecology of the vent-endemic amphipod *Ventiella sulphuris* in the eastern Pacific. Mar Ecol Prog Ser 331: 181–194.

22. Rogers AD, Tyler PA, Connelly DP, Copley JT, James R, et al. (2012) The Discovery of New Deep-Sea Hydrothermal Vent Communities in the Southern Ocean and Implications for Biogeography. PLoS Biol 10: e1001234. doi:10.1371/journal.pbio.1001234.

23. Williams AB, Rona PA (1986) Two new caridean shrimps (Bresiliidae) from a hydrothermal field on the Mid-Atlantic Ridge. J Crustacean Biol 6: 446–462.

24. Gebruk AV, Galkin SV, Vereshchaka AL, Moskalev LI, Southward AJ (1997) Ecology and biogeography of the hydrothermal vent fauna of the Mid-Atlantic Ridge. Adv Mar Biol 32: 93–144.

25. Clarke A. (1993) Reproductive Trade-Offs in Caridean Shrimps. Funct Ecol 7: 411–419.

26. Anger K, Moreira GS (1998) Morphometric and reproductive traits of tropical caridean shrimps. J Crustacean Biol 18: 823–838.

27. Bauer RT (2004) Remarkable shrimps: adaptations and natural history of Carideans. Norman: University of Oklahoma Press. 282 p.

28. Company J-B, Sarda F (2002) Growth parameters of deep-water decapod crustaceans in the northwestern Mediterranean Sea: a comparative approach. Mar Biol 136: 79–90.

29. Corey S, Reid DM (1991) Comparative fecundity of decapod crustaceans. I. The fecundity of thirty-five species of nine families of caridean shrimp. Crustaceana 60: 270–294.

30. Koyama S, Nagahama T, Ootsu N, Takayama T, Horri M, et al. (2005) Survival of deep-sea shrimp (*Alvinocaris* sp.) during decompression and larval hatching at atmospheric pressure. Mar Biotechnol 7: 272–278.

31. Creasey S, Rogers AD, Tyler PA (1996) Genetic comparison of two populations of the deep-sea vent shrimp Rimicaris exoculata (Decapoda: Bresiliidae) from the Mid-Atlantic Ridge. Mar Biol 125: 473–482.

32. Pond D, Dixon D, Sargent J (1997) Wax-ester reserves facilitate dispersal of hydrothermal vent shrimps. Mar Ecol Prog Ser 146: 289–290.

33. Pond D, Dixon DR, Bell MV, Fallick AE, Sargent JR (1997) Occurrence of 16:2(n-4) and 18:2(n-4) fatty acids in the lipids of the hydrothermal vent shrimps *Rimicaris exoculata* and *Alvinocaris markensis*: nutritional and trophic implications. Mar Ecol Prog Ser 156: 167–174.

34. Pond D, Segonzac M, Bell MV, Dixon DR, Fallick AE, et al. (1997) Lipid and lipid carbon stable isotope composition of the hydrothermal vent shrimp *Mirocaris fortunata*: evidence for nutritional dependence on photosynthetically fixed carbon. Mar Ecol Prog Ser 157: 221–231.

35. Allen CE (1998) Lipid profiles in deep-sea organisms. PhD dissertation: University of Southampton.

36. Allen Copley CE, Tyler PA, Varney MS (1998) Lipid profiles of hydrothermal vent shrimps. Cah Biol Mar 39: 229–231.

37. Gebruk A, Southward EC, Kennedy H, Southward AJ (2000) Food sources, behaviour, and distribution of hydrothermal vent shrimps at the Mid-Atlantic Ridge. J Mar Biol Assoc U.K. 80: 485–499.

38. Pond D, Gebruk A, Southward EC, Southward AJ (2000) Unusual fatty acid composition of storage lipids in the bresillioid shrimp *Rimicaris exoculata* couples the photic zone with MAR hydrothermal vent sites. Mar Ecol Prog Ser 198: 171–179.

39. Stevens CJ, Limen H, Pond DW, Gelina Y, Juniper SK (2008) Ontogenetic shifts in the trophic ecology of two alvinocaridid shrimp species at hydrothermal vents on the Mariana Arc, western Pacific Ocean. Mar Ecol Prog Ser 356:225–237.

40. Shank TM, Lutz RA, Vrijenhoek RC (1998) Molecular systematics of shrimp (Decapoda: Bresiliidae) from deep-sea hydrothermal vents, I: Enigmatic "small orange" shrimp from the Mid-Atlantic ridge are juvenile *Rimicaris exoculata*. Mol Mar Biol Biotechnol 7: 88–96.

41. Tyler PA, Young CM (2003) Dispersal at hydrothermal vents: a summary of recent progress. Hydrobiologia 503: 9–19.

42. Teixeira S, Serrao EA, Arnaud-Haond S (2012) Panmixia in a Fragmented and Unstable Environment: The Hydrothermal Shrimp *Rimicaris exoculata* Disperses Extensively along the Mid-Atlantic Ridge. PLoS One 7:e38521. doi: 10.1371/journal.pone.003852 1.

43. Herring PJ, Dixon DR (1998) Extensive deep-sea dispersal of postlarval shrimp from a hydrothermal vent. Deep Sea Res Part 1 Oceanogr Res Pap 45: 2105–2118.

44. Luther GW, Rozan TF, Taillefert M, Nuzzio DB, et al. (2001) Chemical speciation drives hydrothermal vent ecology. Nature 410: 813–816.

45. Copley JTP, Tyler PA, Van Dover CL, Philp SJ (2003) Spatial variation in the reproductive biology of *Paralvinella palmiformis* (Polychaeta: Alvinellidae) from a vent field on the Juan de Fuca Ridge. Mar Ecol Prog Ser 255: 171–181.

46. Pradillon F, Gaill F (2007) Adaptation to deep-sea hydrothermal vents: Some molecular and developmental aspects. J Mar Sci Tech-Taiw 15: 37–53.

47. Herring PJ (1998) North Atlantic midwater distribution of the juvenile stages of hydrothermal vent shrimps (Decapoda: Bresiliidae). Cah Biol Mar 39: 387–390.

48. Gebruk A, Fabri M-C, Briand P, Desbruyères D (2010) Community dynamics over a decadal scale at Logatchev, 14° 42′N, Mid-Atlantic Ridge. Cah Biol Mar 51: 383–388.

49. Colaço A, Martins I, Laranjo M, Pires L, Leal C, et al (2006) Annual spawning of the hydrothermal vent mussel *Bathymodiolus azoricus* under controlled aquarium conditions at atmospheric pressure. J Exp Mar Biol Ecol 333: 166–171.

50. Dixon DR, Lowe DM, Miller PI, Villemin GR Colaço A, et al. (2006) Evidence of seasonal reproduction in the Atlantic vent mussel *Bathymodiolus azoricus*, and an apparent link with the timing of photosynthetic primary production. J Mar Biol Assoc U.K. 86: 1363–1371.

Capturing Ecosystem Services, Stakeholders' Preferences and Trade-Offs in Coastal Aquaculture Decisions: A Bayesian Belief Network Application

Laetitia Helene Marie Schmitt[1], Cecile Brugere[2]*

1 Centre for Health Economics/Department of Economics and Related Studies, University of York, Heslington, York, United Kingdom, **2** Stockholm Environment Institute, University of York, Heslington, York, United Kingdom

Abstract

Aquaculture activities are embedded in complex social-ecological systems. However, aquaculture development decisions have tended to be driven by revenue generation, failing to account for interactions with the environment and the full value of the benefits derived from services provided by local ecosystems. Trade-offs resulting from changes in ecosystem services provision and associated impacts on livelihoods are also often overlooked. This paper proposes an innovative application of Bayesian belief networks - influence diagrams - as a decision support system for mediating trade-offs arising from the development of shrimp aquaculture in Thailand. Senior experts were consulted (n = 12) and primary farm data on the economics of shrimp farming (n = 20) were collected alongside secondary information on ecosystem services, in order to construct and populate the network. Trade-offs were quantitatively assessed through the generation of a probabilistic impact matrix. This matrix captures nonlinearity and uncertainty and describes the relative performance and impacts of shrimp farming management scenarios on local livelihoods. It also incorporates export revenues and provision and value of ecosystem services such as coastal protection and biodiversity. This research shows that Bayesian belief modeling can support complex decision-making on pathways for sustainable coastal aquaculture development and thus contributes to the debate on the role of aquaculture in social-ecological resilience and economic development.

Editor: Jose Luis Balcazar, Catalan Institute for Water Research (ICRA), Spain

Funding: This research was supported by the Marie Curie Intra European Fellowship project ECOLIVA "Sustainable ECOsystem services and LIVelihoods through Aquaculture development" (Project No: PIEF-GA-2009- 235835, 2010–11) funded under the European Community 7th Framework Programme. The funders had no role in study design, data collection and analysis, decision to publish, or preparation of the manuscript.

Competing Interests: The authors have declared that no competing interests exist.

* E-mail: cecile.brugere@sei-international.org

Introduction

The shrimp aquaculture industry of Thailand, which accounted for 15.4% of the global aquaculture production of shrimps in 2009 [1], is a major contributor to the country's economy, bringing in over US$ 2 billion of foreign revenues annually [2] and employing more than one million people [3]. The majority of the industry (98%) is based on the production of the exotic Pacific white-leg shrimp (*Litopenaeus vannamei*), which has replaced the indigenous Giant black tiger shrimp (*Penaeus monodon*) since the early 2000s [4].

The introduction of intensive farming techniques in the 1980s led to Thailand becoming a world-leading shrimp producer. However, this shift was associated with water pollution and acidity build-up in shrimp ponds. In conjunction with poor husbandry methods, this resulted in the onset of catastrophic viral diseases and a collapse in production [5,6], prompting farmers to abandon their ponds and dig new ones further into mangrove-forested areas. Mangroves swamps were considered of low economic value as the ecosystem services they deliver (e.g. pollutant sink, storm protection, provision of wood and fish [7,8]) were not taken into account. As a result, Thailand's mangrove forests were halved between 1961 and 1993, although the extent to which shrimp farming alone was responsible for the loss of mangrove cover in the country remains debatable [5].

The Ecosystem Approach to Aquaculture (EAA) has recently emerged in response to calls for increased sustainability in the aquaculture sector and recognition that aquaculture is often embedded in sensitive social-ecological systems [9]. This approach advocates the consideration of biophysical and human dimensions of ecosystems in order to achieve the dual aims of sustainable production and human wellbeing [10]. At the heart of the EAA lies the notion of social-ecological resilience, defined as "the capacity to maintain integrity when responding to external changes and feedbacks" [11]. Decision-making in the context of coastal social-ecological systems requires considering several different issues: i) interconnectivity, i.e. the fact that "disruption in one system is likely to cause disruption in the other" [10], (ii) complexity and uncertainty, as coastal ecosystem responses are often non-linear [12] and the minimum level of ecosystem structure needed to maintain a constant flow of services is unknown [13], and (iii) stakeholder conflicts stemming from competing uses over coastal resources and institutional failures [14].

However, a gap in the methodology exists as these issues are not simultaneously considered in decision-making or when the EAA is implemented in coastal areas. This calls for a decision support system that can holistically assess and quantify the trade-offs arising from coastal aquaculture management decisions. A prior review of the range and uses of decision support tools currently available (e.g. extended cost-benefit analysis, multi-criteria decision analysis) showed that Bayesian Belief Networks (BBN) – also called influence diagrams – stand out for their ability to handle uncertainty, complex non-linear relationships as well as qualitative and quantitative data of an ecological and economic nature. BBNs also have the advantage of remaining straightforward enough to allow clear communication of complex decision outcomes to policy makers [15].

BBNs have been applied to natural resource management to model human pressures on diverse ecosystems [16–18], to assist in the assessment of the social, ecological and economic dimensions of coastal water and catchment management [19–21] and to evaluate and compare fisheries management plans [22]. Applications of BBNs to aquaculture have been limited to modeling the impacts of water management options on the simultaneous production of rice, fish, crab and shrimp in Viet Nam [23] and to quantifying the human drivers behind the adoption of sea cucumber aquaculture as an alternative livelihood strategy in Tanzania [15].

The purpose of the present study, and the novelty of this research, is to broaden the use of Bayesian Belief Networks by incorporating indicators of social-ecological resilience and refining their application to aquaculture. The specific objective of this work is to show how such a decision support tool can help mediate complex and multiple trade-offs in the context of shrimp aquaculture development in Thailand through: (i) the articulation of available knowledge on the impacts of coastal aquaculture on social, ecological and economic systems via a network structure, and (ii) the generation of a probabilistic impact matrix that highlights trade-offs – or "social conflicts over interest and values" [24] – and uncertainties in management decision outcomes. The underlying goal is to provide policy-makers with a full picture of the potential consequences of their aquaculture development decisions on specific characteristics of the system. Section 2 of this paper presents the methodology used to develop a BBN applied to shrimp aquaculture. Section 3 shows modeling results for six scenarios of aquaculture management and assesses their implications in terms of trade-offs between ecosystem services provision, local livelihoods, financial profits and economic development. Finally, section 4 discusses the findings and the role of BBN as a decision-support system for sustainable aquaculture development, in line with the EAA.

Materials and Methods

1. Study area and data collection

In Thailand, the shrimp farming sector is characterized by the prevalence of small production units (<2 ha), which coexist alongside a minority of very large farms [25]. As of 2007, more than 90% of shrimp farms were intensive, i.e. with high stocking densities of juvenile shrimp in ponds [26]. Production predominantly takes place in semi-closed systems characterized by minimal exchange of water with the outside environment to limit the introduction of viruses and pathogens and the release of pollution [27–29]. However, pond water and sludge are still partially discharged in adjacent waterways, notably at the time of harvest. In addition, *L. vannamei* shrimps have been reported in the vicinity of ponds, despite measures aimed at preventing their accidental release in the environment [30]. Fully closed systems also exist but their requirements of a water reservoir, sedimentation pond and continuous aeration make them capital-intensive operations [28]. Whilst shrimp aquaculture has increasingly taken place in supra-tidal areas, i.e. behind the mangrove fringe (S. Funge-Smith,

personal communication, 2011), a large number of shrimp farms are still suspected to be operating in inter-tidal areas due to lack of law enforcement and legislation loopholes [5,31].

Nevertheless, awareness of Better Management Practices (BMPs), framed by the International Principles for Responsible Shrimp Farming [32], has been growing. BMPs aspire to reduce the impact of shrimp farming on the environment and improve the livelihoods of farmers by implementing measures such as storage ponds and increasing efficiency in production by optimizing the shrimp feeding process [33]. In Thailand, the implementation of BMPs has been promoted, in particular among small producers (e.g. [34]), but their adoption remains limited in comparison to other countries such as India and Viet Nam [35,36].

At the other end of the production spectrum is aquasilviculture, an extensive system that promotes harmonious aquatic production alongside mangrove forestation [37,38]. In the context of the present research, aquasilviculture is defined as a mangrove-shrimp system where 70% of the pond surface is planted with mangroves and 30% is dedicated to shrimp production. Although aquasilviculture remains marginal in Thailand in comparison to other Southeast Asian countries, it has enabled farmers to return to traditional low input-low output shrimp farming after crop failures [39]. It is also recognised that, under the right conditions, it can be used as a suitable management strategy to rehabilitate disused shrimp ponds, enhance the provision of mangrove ecosystem services, and provide a complementary source of income to farmers [40,41].

The characteristics of the development of shrimp farming in Thailand, its history and impacts have been extensively documented and general information and data on the sector were collected from the literature. In line with the EAA, Figure 1 captures the multiple interactions of the sector with the environment (including services provided by coastal ecosystems), the economy and livelihoods. Four groups of key stakeholders have been identified (underlined in Figure 1): (i) farmers themselves, organized in associations and seeking income from their activity, (ii) the Government of Thailand, seeking export revenues and employment through the entire shrimp value chain, (iii) members of coastal communities who can simultaneously benefit (e.g. employment) and suffer (e.g. loss of mangrove-dependent livelihoods) from the conversion of mangroves to ponds, and (iv) future generations, representing the stakes of long-term sustainable coastal development.

The bulk of the literature on the economics of shrimp farming relates to *P. monodon* and not *L. vannamei*, which is associated with different yields, costs and profits [4]. To fill this gap and provide up-to-date data on exploitation costs and revenues for *L. vannamei*, onsite surveys were implemented. Ten intensive shrimp farmers were sampled in Surat Thani, Surat Thani Province (sample stratified by farm size: 3 very small farms (<2 ha), 3 small farms (2.1–5 ha), 3 medium farms (5–50 ha) and 1 large farm (>50 ha)). Ten smaller-scale intensive farmers were also sampled in Samroyiot, Prachuap Kiri Khan Province (Figure 2). Of the Samroyiot farmers, six were implementing Better Management Practices (BMPs) and also provided economic and production data. These six farmers were gathered in a "cluster" and working together towards the implementation of BMPs.

Ethical clearance of the research was provided by the European Commission under Framework Programme 7 (PIEF-GA-2009-235835) as a condition of the award. Questionnaires were designed and tested to ensure an optimal balance between cultural sensitivity and achievement of the study objectives. Particular attention was given to gender and to the sensitiveness of the questions asked when interviewing women, minority or vulnerable

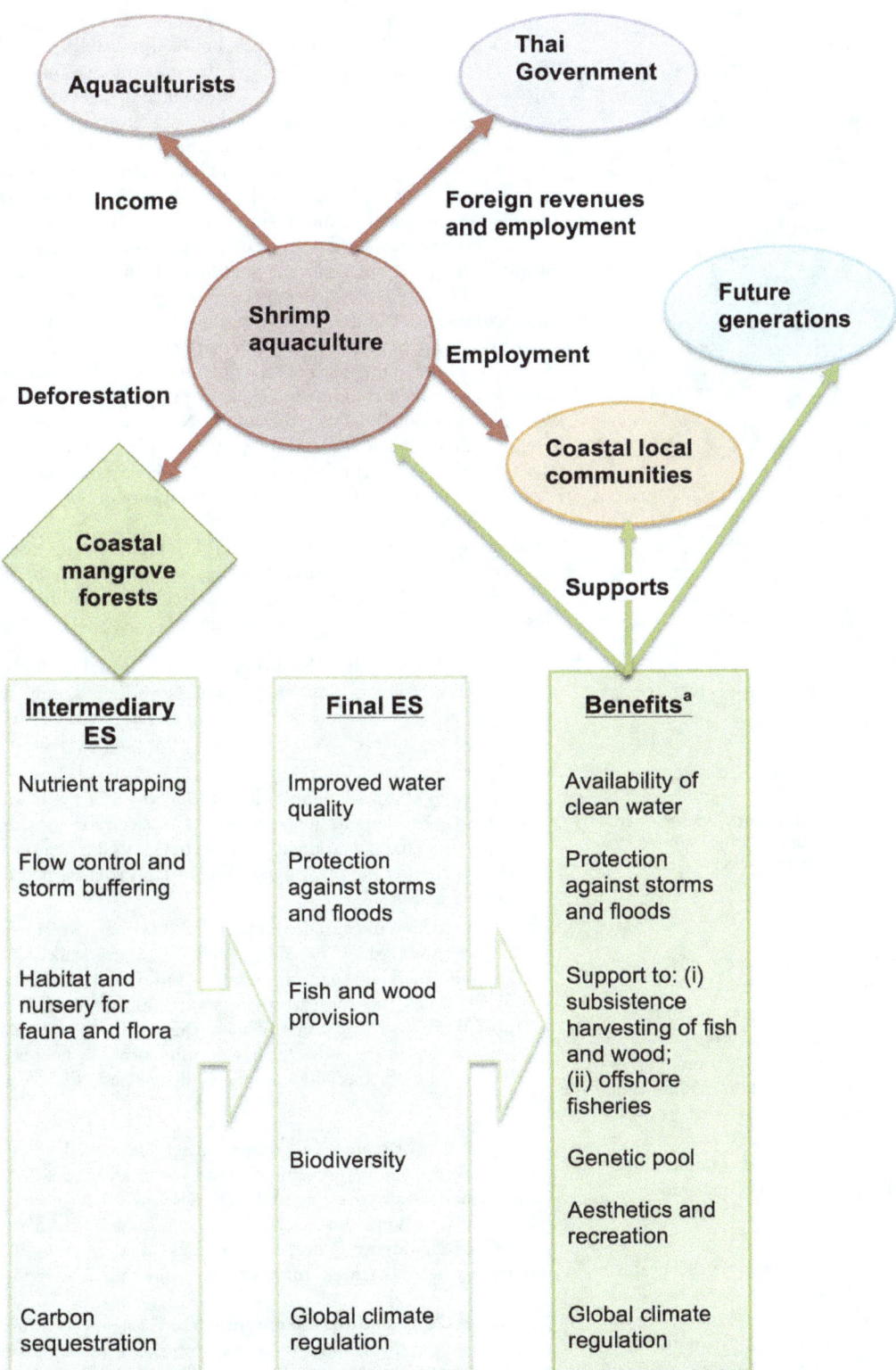

Figure 1. Aquaculture's main impacts on the environment, the economy and livelihoods and related user-conflicts. ES = Ecosystem Services. [a] Benefits should be distinguished from final ES since they are often a product of final ES and human inputs (e.g. fishing gear) and economic valuation of ecosystem should apply to ecosystem benefits only [13,54,55]. Figure based on information from [40] and [8], relying on the ecosystem services classification by [13,55].

Figure 2. Locations of onsite surveys in Thailand. 1: Surat Thani (Surat Thani Province) representing conventional shrimp farming. 2: Samroyiot (Prachap Kiri Khan Province) representing small-scale shrimp farming.

people (e.g. elders, disabled, illiterate, poor), in order to ensure the free expression of their views. Informed verbal consent was provided by each interviewee in light of guarantees of confidentiality, anonymity, privacy, data protection and the possibility to withdraw from the survey at any time. Ethical training was provided to all partners involved in data collection. Monitoring and evaluation of participation and of the research process occurred according to agreed codes of conduct and standards of research practice, including anonymizing the questionnaire recording forms and destroying these at the end of the study.

2. BBN development

2.1. BBN definition. A BBN is a directed acyclic graph, which parameters are treated as random variables and modeled via nodes. The typology of nodes used in BBN modeling is provided in Table 1. Nodes are interconnected in "parent-child" (i.e. predictive) relationships. A "child" node has a conditional probability distribution (CPD) for each possible combination of states of its "parent" nodes and the width of the distributions indicates the level of uncertainty in the cause-effect relation [42]. Unlike hierarchical Bayesian networks, which rely extensively on model simulations, the CPDs that populate BBN chance nodes are generally obtained analytically, with expert opinion used to overcome data gaps [43]. Scenario analysis allows the updating of chance nodes' CPDs into posterior probability distributions (PDs) via Bayes' rule of joint probabilities [44].

In addition to explicitly handling uncertainty via the notion of conditional probabilistic dependence between variables [45], a BBN can incorporate both qualitative and quantitative information by connecting a decision node with chance nodes and utility nodes (Figure 3). Finally, thanks to its network structure, a BBN provides a visual representation of the causal relationships that underpin complex systems to which management decisions are applied, as well as straightforward probabilistic information on uncertain outcomes of management actions [46,47].

2.2. Network construction. Information on the impacts of shrimp farming collected from the literature was incorporated in a network via chance nodes, whereby cause-effect relationships were made explicit. Leaf nodes, which are the nodes of focus for our scenario analysis, were specifically chosen to represent the diversity of interests of the four groups of stakeholders previously identified, so that user-conflicts could be made explicit in output results. The BBN was developed via the software GeNIe from the Decision Systems Laboratory of the University of Pittsburgh, USA (http://dsl.sis.pitt.edu). This software is freely available and has the advantage over other free commonly used software of not limiting the network size.

Following Marcot et al.'s guidance [48], the number of parent nodes per child node was limited to three so that the number of CPDs of child nodes would remain within reasonable limits. As a consequence, three modeling choices were made:

(i) The BBN was built at the shrimp farm level and thus did not model the whole shrimp value chain. Nevertheless, employment generated by the sector was indirectly dealt with via the node "long-term contribution to the country's shrimp exports".

(ii) The issue of abandoned ponds was addressed in the model via the node "long term contribution to the country's shrimp exports" (Thai government's perspective) and the computation of the net present value (NPV) of the profit earned by aquaculturists.

(iii) The modeling of coastal communities' resilience, which was inspired by Ashley et al.'s livelihoods framework [49], focused solely on natural, human and financial capital, which were the most straightforward to handle in the present analysis. Modeling of these three sources of capital was made via the nodes "fish/wood for locals", "health of locals" and "potential for income diversification" respectively.

The modeling of mangroves' ecosystem services called upon a value judgment on how the value of these services would be best communicated. Mangroves' ecosystem services were all modeled qualitatively via chance nodes. However, coastal protection was also modeled quantitatively in monetary units (utility node) as it is estimated to be mangroves' most economically valuable service [12].

2.3. Case study and management scenarios. A hypothetical case study representing the range of pond management possibilities was defined to show how the constructed BBN could support complex decision-making on coastal aquaculture development pathways. This case study used six mutually exclusive aquaculture and land management scenarios (Table 2), underpinned by a set of assumptions defined in Table 3.

2.4. Expert involvement. A group of 42 senior experts in coastal aquaculture from academia, research institutions and the industry were contacted by email to validate the network and the case study and to elicit the CPDs that populate chance nodes. They were sent conditional probability tables to fill for each

Table 1. Typology of nodes in BBN modeling, according to their modeling role and position in the network.

Modeling role	
Chance node	Node modeling a random variable over discrete states and defined by a joint conditional probability distribution
Utility node	Node populated with utility values that express preferences over outcomes
Decision node	Node modeling choices that can be made by the decision maker, comprising one state for each scenario
Position in network	
Parent node	Node predicting one (or more) child node(s)
Child node	Node linked to one (or more) parent node(s) via a predictive relationship. A child node has a conditional probability distribution for each combination of state of its parent nodes
Root node	A node with no parent, defined by a probability distribution
Leaf node	A node with no child

chance node, alongside extensive explanations about the constructed BBN, the case study and the management scenarios considered. Of these experts, twelve contributed to the fine-tuning of the network variables and causal relationships, and to the refinement of the assumptions underpinning the case study. CPDs were elicited by four of the twelve experts through a number of iterations and an in-depth dialogue.

2.5. BBN parameterisation. *Chance nodes.* Because biophysical modeling was outside the scope of this research and due to lack of available datasets, all the CPDs of the model's chance nodes were elicited by experts. As the BBN comprises a large number of chance nodes (24), all chance nodes were attributed the states of "high", "medium" and "low" to facilitate the CPD elicitation exercise. Although applying such a restriction on the states of variables clearly represents a simplification of reality, it ultimately enables information on trade-offs to be presented in an easily understandable form to decision-makers. Normalized average values of elicited CPDs were then calculated.

Utility node for coastal protection. Barbier at al. found that mangroves' level of wave attenuation was a quadratic function of habitat size, which allowed computation of a set of estimates for the monetary value of coastal protection services [12]. In order to link these estimates to the qualitative level of coastal protection quality modeled in the BBN, each coastal protection level ("high",

"middle", "low") was arbitrarily associated to a non-linear change in wave height and the monetary value estimates were rescaled to rai units. This procedure, which enabled coastal protection values to be integrated into the model, is summarized in Figure 4.

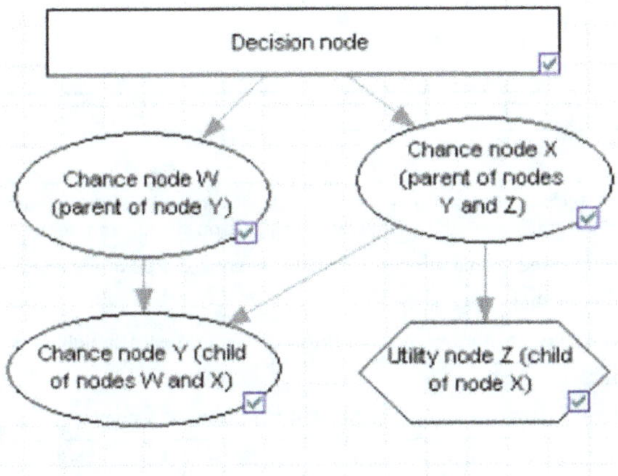

Utility nodes for aquaculturists' annual profit. The computation of the profit earned by the group of aquaculturists encompassed only the costs and revenues expected to be substantially impacted by the type of farm management. Therefore, the term "profit" here is an indicator of the financial performance of each management option, and not of the accounting profit earned. Annual profit was expressed in US$/rai so that it could be easily rescaled to any number of farms. Table S1, in supporting information, details how economic and production data from onsite surveys in Surat Thani and Samroyiot were integrated in the profit computations, in combination with secondary data and expert opinion.

For each management scenario*s*, aquaculturists' annual profit was computed in GeNIe, via several utility nodes, based on formulas (1) and (2). The net present value (NPV) was computed in Excel according to formula (3).

$$R_s = Q_s(1 - CL_s)P_s \qquad (1)$$

$$\pi_s = R_s - VC_s \qquad (2)$$

$$NPV(\pi_s) = -I_s + \sum_{t=1}^{T} \frac{\pi_{s,t}}{(1+r)^t} \qquad (3)$$

R_s is aquaculturists' annual gross revenue (US$/rai, 1 rai = 0.16 ha); Q_s is the total annual shrimp production (kg/rai); CL_s is the annual crop loss (%); P_s is the shrimp farmgate price (US$/kg); π_s is the annual profit (US$/rai); VC_s is the annual production variable cost (US$/rai); I_s is the investment cost (US$/rai); T is the farm lifespan (years) and r is the discount rate.

Results

1. Network and impact matrix

The BBN in Figure 5 provides a graphical representation of the knowledge and belief of causal relationships between social, ecological and economic impacts associated with land and farm management scenarios relative to shrimp farming. Modeling output for the five leaf nodes (aquaculturists' annual profit; long-term contribution to the country's shrimp exports; value of coastal

Figure 3. Convention adopted to graphically represent each type of node in a Bayesian Belief Network.

Table 2. Mutually exclusive aquaculture and land management scenarios used in BBN modeling.

Scenario name	Description
BAU - Business as usual	Conventional intensive farming as currently practised
BMP	Introduce Better Management Practices
Restore forest+closed-system	Fully restore the existing farm as a mangrove forest and build a closed-system in the supra-tidal zone, behind the mangrove fringe
Replant 20%	Replant mangroves on 20% of the farm pond area
Replant 40%	Replant mangroves on 40% of the farm pond area
Aquasilviculture	Replant mangroves on 70% of the pond area and integrate the culture of mangroves with low-density shrimp aquaculture

Notes:
- The last three scenarios combine intertidal land conversion with mangroves replanting and conservation and were designed to capture non-linearity in mangroves provision of ecosystem services [12].
- The "aquasilviculture" and "restore forest+closed-system" scenarios were associated with a 100% probability of "high" level of coastal protection, while "BAU" and "BMP" scenarios were associated with a 100% probability of "low" level of coastal protection.

protection; biodiversity; resilience of local community), which were specifically chosen to reflect trade-offs and user-conflicts, are presented in the form of a probabilistic impact matrix (Figure 6). The scenario "restore forest+closed-system" scored the highest on each criterion modeled via the leaf nodes and thus represents the best-case scenario.

2. Interpretation and sensitivity analyses

2.1 Profit. Whilst levels of farm production drove the difference in annual profit between management scenarios, production loss due to mangrove replanting in ponds was partially compensated by a lower risk of shrimp loss from disease outbreak or typhoons and a higher likelihood of securing a price premium via production certification. For instance, the annual profit under scenarios "replant 20%" and "replant 40%" was lower than under "BAU" by only 15% and 32% respectively. This was due to a lower expected crop loss (10.3% and 8% of annual crop respectively, vs. 12.6% for "BAU") and a greater likelihood of certification (12.7% and 19.7% respectively, vs. 8% for "BAU"). The better financial performance of scenario "BMP" against "BAU" stemmed from lower feeding costs following optimization of the shrimp feeding process.

Sensitivity to assumptions underpinning the NPV computations was tested. As a result of compounded discounting, sensitivity to

discount rate was especially acute for scenarios "aquasilviculture" and "restore forest+closed-system", since these management systems were assumed to be operating over the longest periods (50 years and 30 years respectively). Conversely, sensitivity to farm lifespan was stronger for scenarios "BAU" and "BMP" since they were associated with a shorter lifespan (10 and 13 years respectively). Under the assumption that all systems would operate for only 10 years, scenario "restore forest+closed-system" was no longer associated with the greatest NPV of profits due to the high investment cost incurred in year 0 (US$ 4,249/rai).

2.2 Coastal protection value. The larger the replanted pond surface was, the higher the expected value of coastal protection. Following assumptions pertaining to the probability of coastal protection levels associated with each management scenario (see Table 2), the BBN encompassed uncertainty about coastal protection quality only for the scenarios which involved replanting 20% and 40% of the pond surface. Economic values of coastal protection were underpinned by the assumed association between qualitative levels of protection and quantitative provision of wave-attenuation service (see Figure 3). Sensitivity analysis showed that the stronger the assumption of non-linearity, the more spread out across management scenarios coastal protection values were.

2.3 Long-term contribution to the country's exports. The performance of scenarios "aquasilviculture" and

Table 3. Assumptions behind the BBN modeling case study.

Assumption	Description
A1	Small-scale intensive shrimp farms relying on semi-closed systems are operating as a group and cultivate and export the non-native species *L. vannamei* (using specific pathogen free broodstock), thus generating cumulative impacts on the environment
A2	Farms are located in Thailand's intertidal area (i.e. formerly forested by mangroves), and have recently started operating
A3	The farm owner fully undertakes the management option of his/her choice at t = 0[1]
A4	The rate of survival of replanted mangroves is high and replanted mangroves provide their ecosystem service straightaway[1] [2]
A5	All production, including from aquasilviculture, is exported[3]
A6	To the exception of aquasilviculture, stocking densities remain identical among the different management scenarios

Notes:
[1]A3 and A4 were motivated by the difficulty to integrate time dynamics in BBNs.
[2]Mangroves are expected to deliver their services after 3 years of normal growth (J. Primavera, personal communication, 2011).
[3]While the quality and size of shrimps may greatly determine whether shrimps are exported or not, this assumption reflects the fact that 85% of the Thai shrimp production is exported [3].

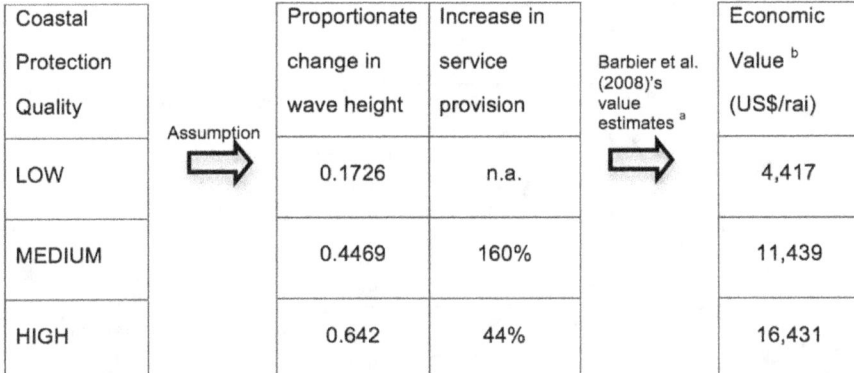

Coastal Protection Quality		Proportionate change in wave height	Increase in service provision		Economic Value [b] (US$/rai)
	Assumption			Barbier et al. (2008)'s value estimates [a]	
LOW		0.1726	n.a.		4,417
MEDIUM		0.4469	160%		11,439
HIGH		0.642	44%		16,431

Figure 4. Parametrisation of the utility node "Value of coastal protection". [a] Estimates computed by [12] based on the expected damage cost method combined with a quadratic wave attenuation function and the assumption that each km^2 of mangroves represents a mangrove forest area of 100 m inshore along a 10 km coastline. Estimates were converted in (US$/rai) with 1 rai = 0.0016 km^2. [b] Net present values computed over a 20-year time-horizon with a 10% discount rate.

"restore forest+closed-system" on the long-term contribution to the country's exports differed sharply from the "BAU" scenario. It was expected to be high with a 62% probability under scenario "restore forest+closed-system" (vs. 37% under "BAU"), while it was expected to be low with a 69% probability under scenario "aquasilviculture" (vs. 28% under "BAU"). By contrast, in the other scenarios, the level of contribution to exports did not differ substantially from the "BAU" scenario as the decrease in production due to replanting was partially offset by an increase in farm lifespan (lower bioaccumulation of pollutants). For example, the likelihood of high contribution to exports under scenario "replant 40%" was only 4 percentage points lower than under "BAU".

2.4 Biodiversity. Whilst biodiversity levels ("high", "medium", "low") were not specifically defined here, they were meant to reflect the degree of variety of living organisms and genetic variability encapsulated in the definition of biodiversity adopted in Article 2 of the Convention on Biological Diversity in 1993. The likelihood of a high level of biodiversity was shown to increase from 7% under "BAU" to 8% under "BMP", 17% under "replant 20%", 25% under "replant 40%", 58% under "aquasilviculture" and 75% under "restore forest+closed-system". Interestingly, the likelihood of high biodiversity increased by 50% between scenarios "replant 20%" and "replant 40%" and by 130% between scenarios "replant 40%" and "aquasilviculture" (i.e. replant 70%), highlighting non-linearity in ecosystem service provision. Sensitivity analysis consisted of evaluating the separate impacts on biodiversity levels under each sensitivity scenario of: (i) a high risk of spread of diseases to native species; (ii) high water quality and (iii) low pollution from other activities, against the base case simulation (run 1) where the root node "pollution from other activities" was populated with a uniform PD. For these sensitivity analyses, the nodes "risk of spread of diseases to native species", "surface water quality" and "pollution from other activities" were separately controlled, in three successive runs, to be with a 100% probability in a given state of interest (e.g. "low" or "high"). Results, presented in Figure 7, showed that water quality, itself partially driven by levels of external pollution, is expected to have a greater influence on biodiversity levels than the risk of spread of disease to native species.

2.5 Resilience of local community. Scenarios "restore forest+closed-system" and "aquasilviculture" stood out as very likely to bring high coastal community resilience, with a 93% and

82% probability respectively, vs. 27% under "BAU". By contrast, the spread of posterior PDs under replanting scenarios showed much greater uncertainty in the expected level of resilience. Sensitivity analysis focused on the node "locals' subsistence capacity" as it strongly influenced the node "resilience of local community" (see Figure 5). Each parent node of the node "locals' subsistence capacity" (i.e. nodes "quantity of wood for locals", "quantity of fish for locals" and "local employment") was separately controlled, in three successive runs, to be in the state "low" with a 100% probability. Results against run 1, presented in Figure 8, showed that uncertainty in the level of restoration of productive services was the greatest contributor to overall uncertainty in subsistence capacity and thus, of resilience. Additionally, under the replanting and aquasilviculture scenarios, the likelihood of high subsistence capacity was found to decrease more under the "low wood" sensitivity run than under the "low fish" sensitivity run. However, when the mangrove forest was restored to its original state (scenario "restore forest+closed-system"), fish and wood provision were found to have an equal influence on locals' subsistence capacity.

Discussion

1. Highlights and contextualisation of the modeling results

Firstly, economic results aim solely to provide an indication of the difference in scenarios' financial performance and of the economic impact of improving coastal protection by replanting mangroves. As such, they should not be used for predictive purposes.

Secondly, the differential performance of scenarios in terms of contribution to Thailand's shrimp exports underlines substantial differences in production capacity and reliability between systems. Indeed, the productive capacity of aquasilviculture is small compared to closed systems, which are themselves more reliable than conventional systems ("BAU") due to better disease risk management. Additionally, results suggest that, in the long run, the loss in production capacity from mangroves replanting in ponds is at least partially offset by the increase in farm lifespan due to a lower bioaccumulation of pollutants.

Thirdly, the non-linearity in biodiversity provision highlighted in the results is of particular relevance for the identification of an optimal level of mangrove replanting. The impact of management

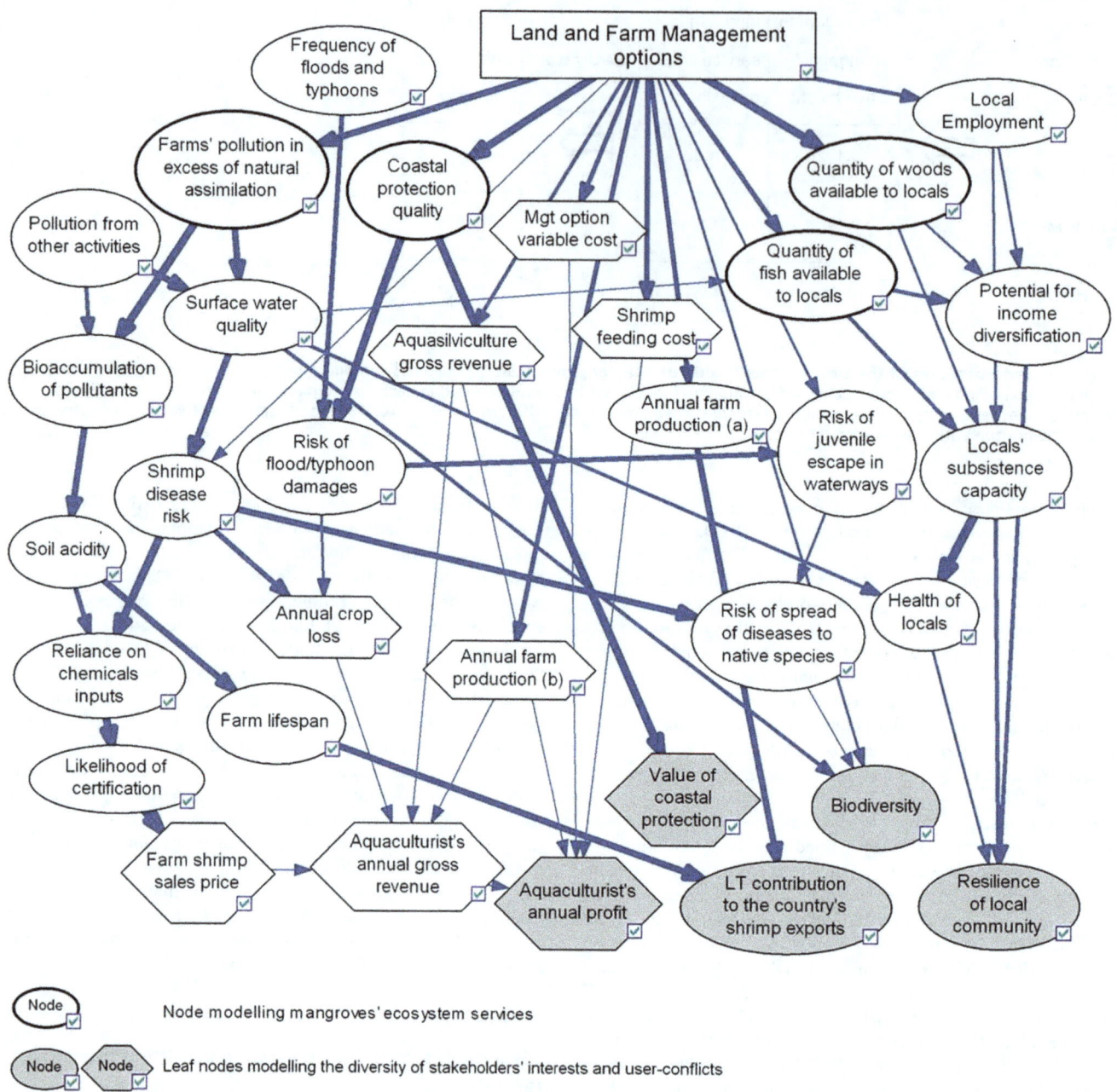

Figure 5. Bayesian belief network and strength of influence between variables. The width of the arrows linking nodes depicts the strength of influence between parameters ciphered by the conditional probability distributions. [a] Chance node defining farm production qualitatively over the states "high", "medium" and low". [b] Utility node defining farm production quantitatively (kg/rai).

scenarios on biodiversity is, however, expected to be more complex than our results suggest and should be validated by field data.

Fourthly, systems where farming is mixed with mangroves may differ in the way they control organisms entering and exiting the shrimp pond (M. Troell, personal communication, 2011). Therefore, when considering local community resilience, uncertainty about the level of restoration of productive services is likely to be larger for the provision of nursery habitat and fish seed than for wood. The greater uncertainty in the restoration of fish provision following mangrove replanting may explain why wood availability appears as a stronger contributor to locals' subsistence capacity

than fish provision. This hypothesis should, however, be tested empirically.

2. Role of this BBN as a decision support tool for sustainable coastal resource use

The BBN presented here provided a holistic representation of the main user conflicts and trade-offs associated with various forms of shrimp farming development. Although not a predictive tool, this BBN is a means of articulating existing knowledge and beliefs about the multiple interactions of shrimp farming with the environment, the economy and livelihoods. It also enables to

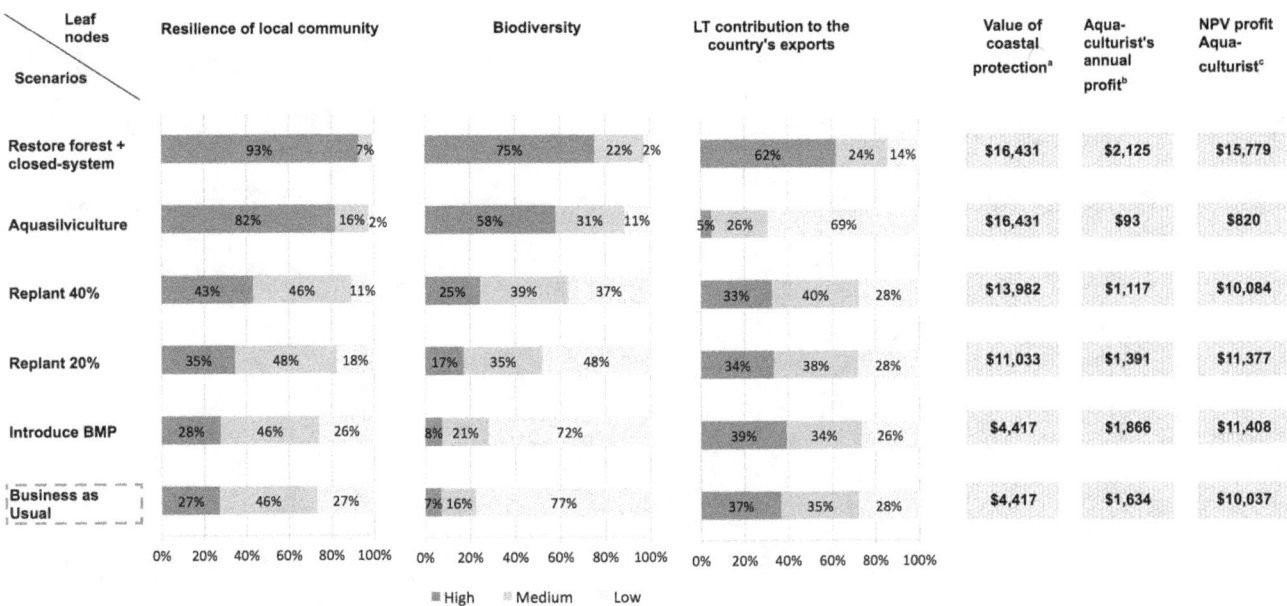

Figure 6. Impact matrix. Following belief propagation, posterior probability distributions and expected utility values for each leaf node were obtained for each management scenario. The matrix summarizes the performance of the five management scenarios against the BAU scenario on the criteria modeled via the leaf nodes. [a] NPV values (US$/rai) computed over a 20 year-time horizon, using a 10% discount rate. [b] Encompasses only exploitation costs and revenues impacted upon by the management option. [c] NPV values (US$/rai) computed using a 10% discount rate and based on estimates of farm lifespan (see Table S1). 1 ha = 6.25 rai.

quantitatively compare the trade-offs associated with a range of aquaculture development strategies. In a context of renewed calls for the improved governance of the sector, especially in countries where the social-ecological resilience of coastal areas has been eroded in the past, this approach could help policy-makers

understand the potential consequences of their decisions and increase the transparency of their policy choices by prompting them to explicitly value pre-defined sustainability criteria.

Sustainable land and farm management practices should be in line with the principles of the EAA and adaptable to local

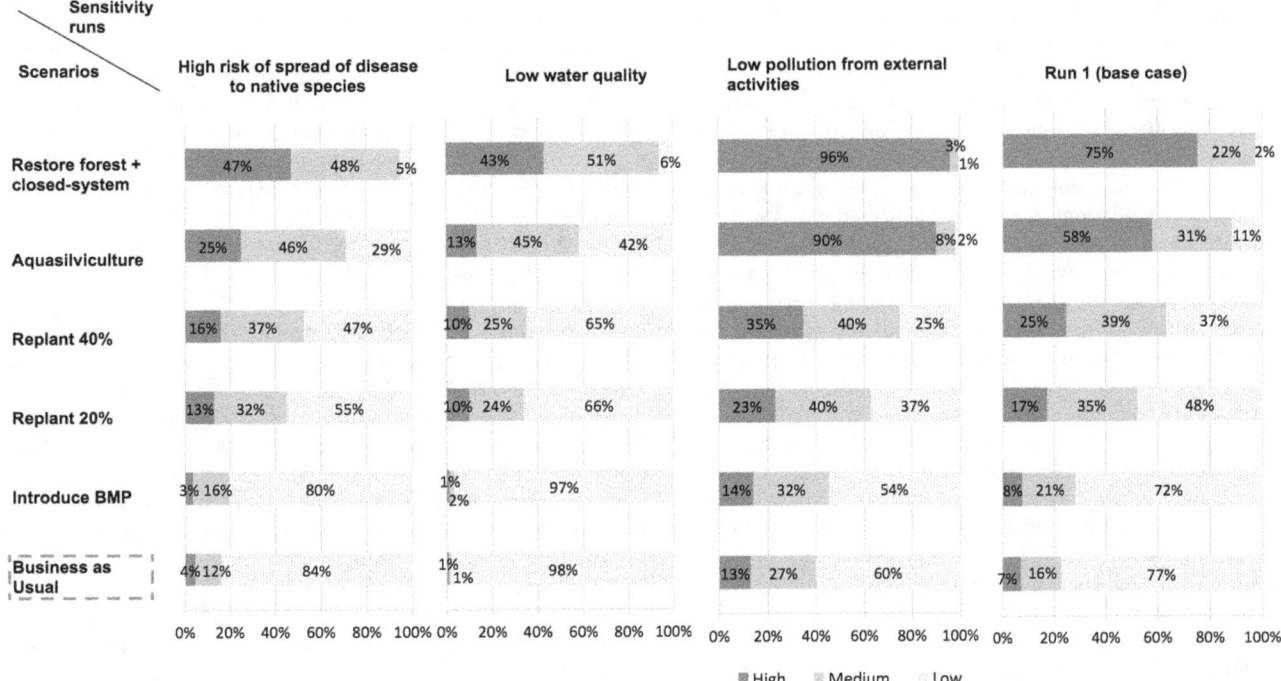

Figure 7. Sensitivity analyses for the node "Biodiversity". Assessment of the predictive influence of the variables "risk of spread of diseases to native species", "surface water quality" and "pollution from other activities" on the posterior probability distribution of the node "biodiversity".

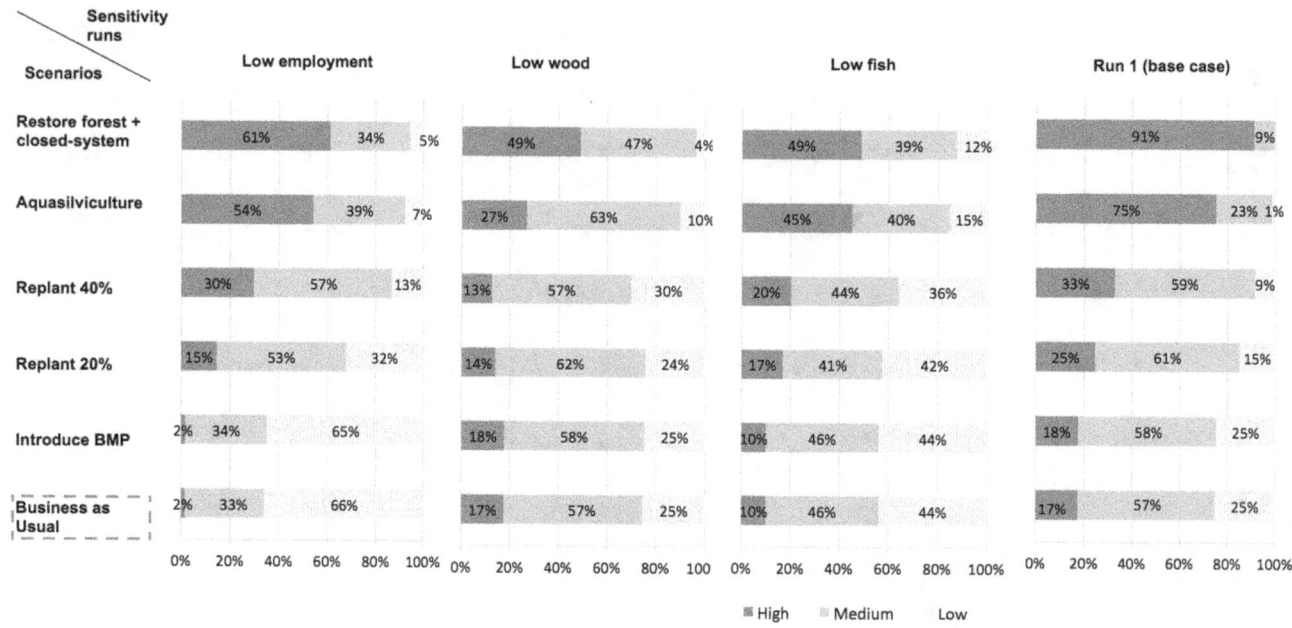

Figure 8. Sensitivity analyses for the node "Locals' subsistence capacity." Assessment of the predictive influence of the variables "quantity of wood for locals", "quantity of fish for locals" and "local employment" on the posterior probability distribution of the node "locals' subsistence capacity".

constraints as part of adaptive management [10]. Consequently, the objective of Thai policy-makers should not be to identify the best aquaculture scenario to promote to all coastal areas, but to define an appropriate diversified portfolio of environmentally-friendly and socially-acceptable practices, where aquasilviculture and closed-systems are only the extremes of a spectrum of possible options [11]. By making explicit how different alternatives may mediate the major trade-offs associated with shrimp farming, the BBN developed here can support the construction of such a diversified portfolio of management practices and inform policy choices.

Furthermore, since conventional intensive shrimp farmers have traditionally focussed on short-term profitability instead of sustainability [39], economic incentives are expected to play a key role in the successful diffusion and adoption of more sustainable management practices among aquaculturists. Potential governmental levers include the improvement of farmers' access to certified markets that drive a price premium, the set-up of microfinance schemes (e.g. to help fund capital-intensive closed-systems), and the implementation of schemes for payments for ecosystem services. Regarding the latter, and subject to property rights and farmers' acceptance, modeling findings on the estimation of the difference in financial performance between the "BAU" scenario and alternatives involving the restoration of mangroves ecosystem services could help in defining a level of compensation to provide to aquaculturists. However, as yields, costs and revenues can vary widely between sites [50,51], our data on aquaculturists' profits would need to be complemented by other data on costs and revenues to generate reliable estimates of the impacts of management scenarios on aquaculturists' profits at local, regional or national scales [50,51].

Although adjustments would be required, the model developed in this paper could be replicated for capturing trade-offs between aquaculture development and environmental and livelihood protection objectives in other countries of Southeast Asia (particularly the Philippines and Indonesia) and Latin America

(e.g. Mexico, Honduras, Venezuela), where shrimp farming has displaced mangroves. The issue of scale, however, should be at the core of potential replications of the model. Local land and farm management practices should also be encompassed in wider integrated coastal assessments since "mangroves destruction goes beyond the shrimp industry alone" [52]. Additionally, in areas where shrimp pond abandonment has been a widespread phenomenon following disease outbreaks [5], the model could provide a basis for modeling the rehabilitation of disused shrimp ponds and restoration of mangrove ecosystem services. To transfer the application of the BBN to such cases, the extent to which the ecosystem has been altered, e.g. acidity levels, tidal hydrology and soil alteration, and the objective of ecosystem services restoration, e.g. coastline protection, supporting community livelihoods through restoration of coastal fisheries, or aquasilviculture development, should be incorporated [53]. Pond rehabilitation options for other commercial purposes such as salt production and coconut plantations could also be considered.

3. Suggestions for methodological improvements

Given the size of the network and for practical reasons, our elicitation exercise left what constituted "high", "middle" and "low" levels for each variable open to experts' interpretation. Ideally, the discrete states of all variables should have been characterised by a wider consultation with them.

Furthermore, interactions with experts at each stage of the network development should be complemented with broader consultations with key stakeholders, such as aquaculturists, local communities, environmental organisations and government representatives. Not only this is likely to strengthen the robustness of the model [23], it should also ease the implementation of policy measures stemming from the modeling findings [46].

Finally, although it was decided to focus only on ecosystem services that had a substantial economic value or that could be considered in terms of their qualitative contribution to local communities' resilience, the BBN could be further refined by incorporating other ecosystem services such as cultural services and carbon sequestration. The integration of profits from the latter, potentially secured via payments for ecosystem services, could enable to more accurately model the financial performance associated with each land management scenario.

Conclusion

This paper aimed to develop a Bayesian belief network (BBN) as a decision support system for mediating trade-offs between economic development, protection of natural ecosystems and coastal livelihoods, piloted in the case of the Thai coastal shrimp aquaculture. Modeling insights consisted of identifying for each land and aquaculture management scenario: (i) the expected magnitude of trade-offs due to user-conflicts and (ii) the level of uncertainty surrounding scenarios' performance on criteria reflecting stakeholders' diverse interests. Further analyses enabled quantitative measurement of the sensitivity of the model outputs to pre-defined assumptions (e.g. farm lifespan), input values (e.g. percentage of crop loss, pollution from external activities) and conditional probabilistic dependencies between the network's variables.

Whilst the BBN was developed for coastal shrimp farming in Thailand, suggestions were provided on how to apply this decision tool to other coastal aquaculture contexts. The presently developed BBN can therefore support the implementation of the Ecosystem Approach for Aquaculture in three ways: (i) by articulating available knowledge and beliefs on aquaculture's multiple interactions with the environment, the economy and livelihoods, (ii) by promoting comprehensiveness, explicit handling of uncertainty and transparency in the valuation of pre-defined

sustainability criteria and (iii) by supporting innovative policy measures. Examples of such measures include the design of a diversified portfolio of sustainable farm management practices and of schemes of payments for ecosystem services. Finally, from a wider perspective, this research underlines the potential of BBNs to help frame the sustainable development of productive industries that interfere with the provision of ecosystem services.

Acknowledgments

The authors are grateful to project partners in Thailand (Coastal Resources Institute (CORIN) – Asia, and Seafresh Industry Public Co. Ltd) and to all the experts consulted during the study. They would like to acknowledge in particular the contributions of Dr. Dominique Gautier (Aquastar Ltd, UK); Dr. Malcolm Beveridge (WorldFish Center, Africa); Dr. Stuart Bunting (Essex University, UK) and Dr. Neil Ridler (University of Brunswick, Canada) in the probability elicitation exercise. The authors are also grateful to Drs. Murray Rudd and Richard Cookson at the University of York, UK, for their useful comments on an earlier draft, and to Sarah West for her proofreading of the manuscript.

Author Contributions

Conceived and designed the experiments: LS CB. Performed the experiments: LS. Analyzed the data: LS CB. Contributed reagents/materials/analysis tools: CB. Wrote the paper: LS CB.

References

1. FAO (2011) Aquaculture production (quantities and values) 1950–2009. Fishstats. Rome, Italy, Food and Agriculture Organization.

2. Thailand Customs Department (2007) Statistics on Shrimp Import and Export (in Thai). http:\\www.customs.go.th (Accessed 24 September 2013)

3. Fisheries Department (2008) Thailand experience and opportunities for aquaculture certification: Thai quality shrimp for international markets. Bangkok, Thailand: Department of Fisheries, Government of Thailand.

4. Wyban J (2007) Thailand's shrimp revolution. AQUA Culture Asia Pacific Magazine: 16–18.

5. Huitric M, Folke C, Kautsky N (2002) Development and government policies of the shrimp farming industry in Thailand in relation to mangrove ecosystems. Ecol Econ 40: 441–455.

6. Szuster B (2006) Coastal shrimp farming in Thailand: searching for sustainability. In: Hoanh CT, Tuong TP, Gowing JW, Hardy B, editors. Environment and Livelihoods in Tropical Coastal Zones. Oxford, UK: CAB International, pp. 86–98.

7. Sathirathai S (1998) Economic valuation of mangroves and the roles of local communities in the conservation of natural resources: case study of Surat Thani, South of Thailand. Singapore: Environment and Economics Program for South East Asia, International Development Research Centre.

8. Brander LM, Florax R, Vermaat JE (2006) The empirics of wetland valuation: A comprehensive summary and a meta-analysis of the literature. Environmental & Resource Economics 33: 223–250.

9. Soto D, Aguilar-Manjarrez J, Brugere C, Angel D, Bailey C, et al. (2008) Applying an ecosystem-based approach to aquaculture: principles, scales and some management measures. In: Soto D, Aguilar-Manjarrez J, Hishamunda N, editors. Building an ecosystem approach to aquaculture. FAO/Universitat de les Illes Balears Expert Workshop, 7–11 May 2007, Palma de Mallorca, Spain. FAO Fisheries and Aquaculture Proceedings. No. 14. Rome, Italy: Food and Agriculture Organization, pp. 15–35.

10. Bailey C. (2008) Human dimension of an ecosystem approach to aquaculture. In: Soto D, Aguilar-Manjarrez J, Hishamunda N, editors. Building an ecosystem approach to aquaculture. FAO/Universitat de les Illes Balears Expert Workshop, 7–11 May 2007, Palma de Mallorca, Spain. FAO Fisheries and Aquaculture Proceedings. No. 14. Rome, Italy: Food and Agriculture Organization, pp. 37–42.

11. Bush SR, van Zwieten PAM, Visser L, van Dijk H, Bosma R, et al. (2010) Scenarios for Resilient Shrimp Aquaculture in Tropical Coastal Areas. Ecology and Society 15.

12. Barbier EB, Koch EW, Silliman BR, Hacker SD, Wolanski E, et al. (2008) Coastal ecosystem-based management with nonlinear ecological functions and values. Science 319: 321–323.

13. Fisher B, Turner K, Zylstra M, Brouwer R, de Groot R, et al. (2008) Ecosystems services and economic theory: integration for policy-relevant research. Ecol Appl 18: 2050–2067.

14. Brugere C (2006) Can integrated coastal management solve agriculture-fisheries-aquaculture conflicts at the land-water interface? A perspective from New Institutional Economics. In: Hoanh CT, Tuong TP, Gowing JW, Hardy B, editors. Environment and Livelihoods in Tropical Coastal Zones. Oxford, UK: CAB International, pp. 258–273.

15. Slater MJ, Mgaya YD, Mill AC, Rushton SP, Stead SM (2013) Effect of social and economic drivers on choosing aquaculture as a coastal livelihood. Ocean Coast Manag 73: 22–30.

16. Borsuk ME, Stow CA, Reckhow KH (2004) A Bayesian network of eutrophication models for synthesis, prediction, and uncertainty analysis. Ecol Modell 173: 219–239.

17. Marcot BG, Holthausen RS, Raphael MG, Rowland MM, Wisdom MJ (2001) Using Bayesian belief networks to evaluate fish and wildlife population viability under land management alternatives from an environmental impact statement. For Ecol Manage 153: 29–42.

18. McNay RS, Marcot BG, Brumovsky V, Ellis R (2006) A Bayesian approach to evaluating habitat for woodland caribou in north-central British Columbia. Can J For Res 36: 3117–3133.

19. Sadoddin A, Kelly R, Letcher RA, Jakeman T, Croke B, et al. (2012) A bayesian decision support system model: dryland salinity management application. The International Journal of Environmental Resources Research 1: 1–17.

20. Kragt ME, Bennett J (2009) Integrated hydro-economic modelling: challenges and experiences in an Australian catchment. Canberra, Australia: Environmental Economics Research Hub, Crawford School of Economics and Government.

21. Ticehurst JL, Newham LTH, Rissik D, Letcher RA, Jakeman AJ (2007) A Bayesian network approach for assessing the sustainability of coastal lakes in New South Wales, Australia. Environ Model Softw 22: 1129–1139.

22. Levontin P, Kulmala S, Haapasaari P, Kuikka S (2011) Integration of biological, economic, and sociological knowledge by Bayesian belief networks: the interdisciplinary evaluation of potential management plans for Baltic salmon. ICES J Mar Sci 68: 632–638.

23. Baran E, Jantunen T, Cheng P (2006) Developing a consultative Bayesian model for integrated management of aquatic resources: an inland coastal zone case study. In: Hoanh CT, Tuong TP, Gowing JW, Hardy B, editors. Environment

and Livelihoods in Tropical Coastal Zones. Oxford, UK: CAB International, pp. 206–218.

24. Martinez-Alier J (2001) Ecological conflicts and valuation: mangroves versus shrimps in the late 1990s. Environ Plann C-Gov Policy 19: 713–728.

25. Davy FB, De Silva SS (2011) Success Stories in Asian Aquaculture. Ottawa, Canada: Springer.

26. Loawapong A (2010) Background report fisheries value chain: Kingdom of Thailand. Under project "A value-chain analysis of international fish trade and food security with an impact assessment of the small scale sector". Bangkok, Thailand: Department of Fisheries (DoF), Government of Thailand.

27. (ADB) Asian Development Bank, (NACA) Network of Aquaculture Centres in Asia-Pacific (1997) Shrimp and carp aquaculture sustainability. Beijing, China, October 1995. Bangkok, Thailand: NACA.

28. Kongkeo H (1997) Comparison of intensive shrimp farming systems in Indonesia, Philippines, Taiwan and Thailand. Aquaculture Research 28: 789–796.

29. Funge-Smith SJ, Aeron-Thomas M (1995) The economic factors and risks influencing the sustainability of Thai intensive shrimp farms. Stirling, UK: Institute of Aquaculture, University of Stirling.

30. Senanan W, Tangkrock-Olan N, Panutrakul S, Barnette P, Wongwiwatanawut C, et al. (2007) The presence of the Pacific whiteleg shrimp (Litopenaeus vannamei, Boone, 1931) in the wild in Thailand. J Shellfish Res 26: 1187–1192.

31. Hishamunda N, Ridler NB, Bueno P, Yap WG (2009) Commercial aquaculture in Southeast Asia: Some policy lessons. Food Policy 34: 102–107.

32. FAO NACA, UNEP WB (2006) International principles for responsible shrimp farming. Bangkok, Thailand: NACA.

33. NACA (2011) Network of Aquaculture Centers in Asia-Pacific. Bangkok, Thailand. http://www.enaca.org

34. NACA, WWF (2010) Supporting and demonstrating small-scale shrimp farmer group to access international and market through certification schemes. Bangkok, Thailand: NACA.

35. Umesh NR (2007) Development and adoption of BMPs by self-help farmer groups. Aquaculture Asia Magazine XII: 8–10.

36. Umesh NR, Chandra Mohan AB, Ravibadu G, Padiyar PA, Phillips MJ, et al. (2010) Shrimp farmers in India: Empowering small-scale farmers through a cluster-based approach. In: De Silva SS, Davy FB, editors. Success Stories in Asian Aquaculture. Ottawa, Canada: Springer, pp. 41–66.

37. Binh C, Phillips M, Demaine H (1997) Integrated shrimp-mangrove farming systems in the Mekong delta of Vietnam. Aquaculture Research 28: 599–610.

38. Primavera J (2000) Integrated mangrove-aquaculture systems in Asia. Integrated Coastal Zone Management, Autumn ed: 121–130.

39. Stevenson NJ (1997) Disused shrimp ponds: options for redevelopment of mangrove. Coast Manage 25: 423–425.

40. Troell M (2009) Integrated marine and brackishwater aquaculture in tropical regions; research, implementation and prospects. In: Soto S, editor. Integrated mariculture: a global review. FAO Fisheries and Aquaculture Technical Paper. No. 529. Rome, Italy: Food and Agriculture Organization, pp. 47–131.

41. Bunting SW, Bosma RH, van Zwieten PA, Sidik A (2013) Bioeconomic modeling of shrimp aquaculture strategies for the mahakam delta, Indonesia. Aquaculture Economics and Management 17: 51–70.

42. Varis O, Kuikka S (1999) Learning Bayesian decision analysis by doing: lessons from environmental and natural resources management. Ecol Model 119: 117–195.

43. Uusitalo L (2007) Advantages and challenges of Bayesian networks in environmental modelling. Ecol Model 203: 312–318.

44. Farmani R, Henriksen HJ, Savic D (2009) An evolutionary Bayesian belief network methodology for optimum management of groundwater contamination. Environ Model Softw 24: 303–310.

45. Castelletti A, Soncini-Sessa R (2007) Bayesian networks in water resource modelling and management. Environ Model Softw 22: 1073–1074.

46. Cain J (2001) Planning improvements in natural resources management. Wallingford, Oxon, UK: Centre for Ecology and Hydrology.

47. McCann RK, Marcot BG, Ellis R (2006) Bayesian belief networks: applications in ecology and natural resource management. Can J For Res 36: 3053–3062.

48. Marcot BG, Steventon JD, Sutherland GD, McCann RK (2006) Guidelines for developing and updating Bayesian belief networks applied to ecological modeling and conservation. Can J For Res 36: 3063–3074.

49. Ashley C, Carney D, Britain G (1999) Sustainable livelihoods: Lessons from early experience. London, UK: Department for International Development.

50. Kasai C, Nitiratsuwan T, Baba O, Kurokura H (2005) Incentive for shifts in water management systems by shrimp culturists in southern Thailand. Fish Sci 71: 791–798.

51. Raux P, Bailly D (2001) Literature review on world shrimp farming. Brussels, Belgium: European Commission Framework Programme 5.

52. Neiland AE, Soley N, Varley JB, Whitmarsh DJ (2001) Shrimp aquaculture: economic perspectives for policy development. Marine Policy 25: 265–279.

53. Lewis RR (2001) Mangrove restoration - costs and benefits of successful ecological restoration. University Sains, Penang, Malaysia, 4–8 April 2001. Stockholm, Sweden: Beijer Institute of Ecological Economics.

54. Boyd J, Banzhaf S (2007) What are ecosystem services? The need for standardized environmental accounting units. Ecol Econ 63: 616–626.

55. Fisher B, Turner RK, Morling P (2009) Defining and classifying ecosystem services for decision making. Ecol Econ 68: 643–653.

Spatial Variability of Benthic-Pelagic Coupling in an Estuary Ecosystem: Consequences for Microphytobenthos Resuspension Phenomenon

Martin Ubertini[1,2], Sébastien Lefebvre[4], Aline Gangnery[3], Karine Grangeré[1,2], Romain Le Gendre[3], Francis Orvain[1,2,5]*

1 Université de Caen Basse-Normandie, FRE3484 BioMEA, Caen, France, **2** CNRS INEE, FRE3484 BioMEA, Caen, France, **3** IFREMER, LERN, Port en Bessin, France, **4** Université de Lille1, UMR CNRS 8187 LOG "Laboratoire d'océanologie et geosciences", Station Marine de Wimereux, Wimereux, France, **5** CNRS, UMR 7208 BOREA, Muséum d'histoire naturelle, CRESCO, Dinard, France

Abstract

The high degree of physical factors in intertidal estuarine ecosystem increases material processing between benthic and pelagic compartments. In these ecosystems, microphytobenthos resuspension is a major phenomenon since its contribution to higher trophic levels can be highly significant. Understanding the sediment and associated microphytobenthos resuspension and its fate in the water column is indispensable for measuring the food available to benthic and pelagic food webs. To identify and hierarchize the physical/biological factors potentially involved in MPB resuspension, the entire intertidal area and surrounding water column of an estuarine ecosystem, the Bay des Veys, was sampled during ebb tide. A wide range of physical parameters (hydrodynamic regime, grain size of the sediment, and suspended matter) and biological parameters (flora and fauna assemblages, chlorophyll) were analyzed to characterize benthic-pelagic coupling at the bay scale. Samples were collected in two contrasted periods, spring and late summer, to assess the impact of forcing variables on benthic-pelagic coupling. A mapping approach using kriging interpolation enabled us to overlay benthic and pelagic maps of physical and biological variables, for both hydrological conditions and trophic indicators. Pelagic Chl *a* concentration was the best predictor explaining the suspension-feeders spatial distribution. Our results also suggest a perennial spatio-temporal structure of both benthic and pelagic compartments in the ecosystem, at least when the system is not imposed to intense wind, with MPB distribution controlled by both grain size and bathymetry. The benthic component appeared to control the pelagic one via resuspension phenomena at the scale of the bay. Co-inertia analysis showed closer benthic-pelagic coupling between the variables in spring. The higher MPB biomass observed in summer suggests a higher contribution to filter-feeders diets, indicating a higher resuspension effect in summer than in spring, in turn suggesting an important role of macrofauna bioturbation and filter feeding (*Cerastoderma edule*).

Editor: Simon Thrush, National Institute of Water & Atmospheric Research, New Zealand

Funding: This work was funded by the Regional council of Basse-Normandie. The funders had no role in study design, data collection and analysis, decision to publish, or preparation of the manuscript.

Competing Interests: The authors have declared that no competing interests exist.

* E-mail: francis.orvain@unicaen.fr

Introduction

Estuaries are known to be among the most productive systems in the biosphere [1]. Their high productivity is mainly due to the presence of nutrients and of multiple food resources for the trophic web, coming from both riverine, marine planktonic and benthic compartments [2]. Moreover, in most of these shallow water environments, the intensity of the physical factors reinforces the connections between benthic and pelagic environments by increasing material processing, nutrient cycling and erosion/deposition exchanges. Among all these processes, microphytobenthos (MPB) resuspension is a major phenomenon involved in benthic-pelagic coupling since MPB can contribute up to 50% or more of the primary production for such ecosystems [3]. Consequently, MPB resuspension has major implications both for the food web and for ecosystem stability [4], [5][6].

Benthic-pelagic coupling and especially MPB resuspension are controlled by a complex set of interactions (Fig. 1) between biological, physical, and chemical components or processes [7]. Physical processes such as waves and tidal currents are responsible for erosion of the sediment, leading to sediment resuspension in the water column [8], and hence modifying the properties of the sediment. These mechanisms directly control sediment erodibility, especially sediment composition and compaction [9]. The associated MPB is resuspended at the same time, with wind effect being the major physical factor controlling its resuspension [4]. Even if MPB resuspension is directly controlled by bulk sediment properties related to erodibility, MPB is partly able to control its own resuspension behavior by producing exopolymeric substances (EPS), which reinforces the surface cohesion [10], [11], [12]. This biofilm structure may also cause physical armoring of the sediment, thus limiting its erosion [7]. Macrofauna may also affect the resuspension of MPB by bioturbation, affecting sediment

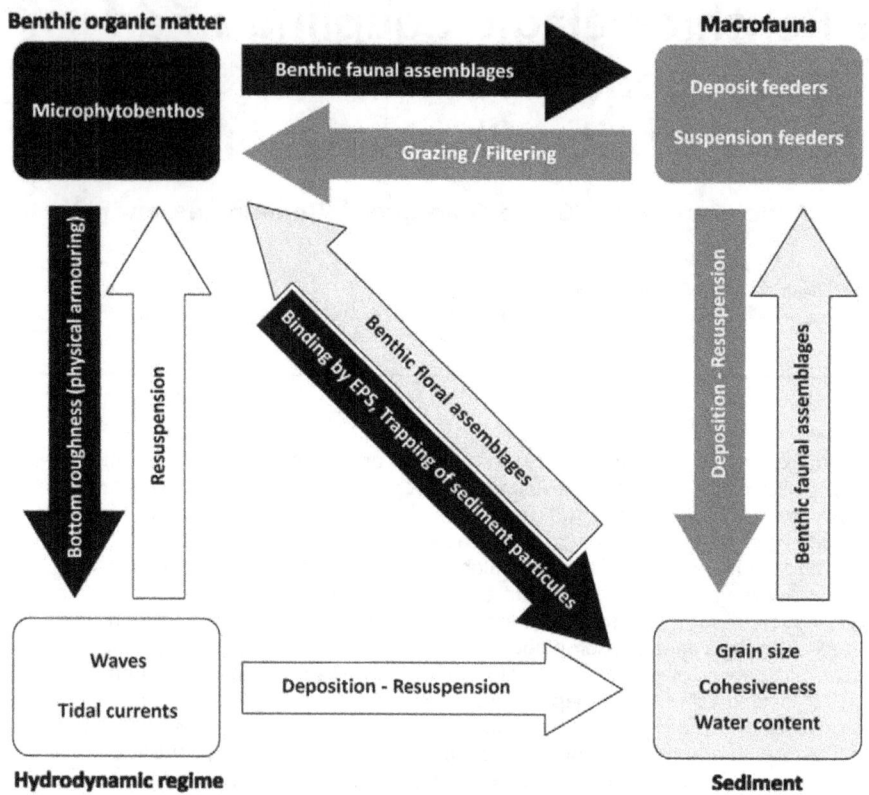

Figure 1. Factors involved in sediment resuspension and the associated microphytobenthos.

erodibility by 1) releasing a material with a high concentration of microphytobenthos [13] and 2) reducing MPB biomass due to nutrition [14]. As a consequence of trophic interactions, MPB can influence long-term trends in benthic macrofauna composition [15], which in turn influence differently MPB resuspension by bioturbation.

MPB biomass always varies in space and over time at all scales in the sedimentary landscape. For instance, surface MPB biomass can double at a given site within one day [16], and MPB biofilms also oscillate in response to the tidal 14-day cycle [17], [18]. In intertidal areas, MPB biomass also varies with the season, and the lowest and highest biomass are found in winter and summer, respectively [16], [19]. Variation in MPB biomass from one year to the next appears to be low [19]. Physical variables such as light irradiance [20], temperature [21], nutrient concentration [22] or wind intensity [19], may be responsible for these seasonal variations in MPB biomass, and many of these time forcing variables can also cause variations at different spatial scales. Even if irradiance varies over the year, light availability is closely related to the bathymetry, and thus influences benthic production [18], [19]. Grain-size is not homogenous within the intertidal area, leading to differential distribution of sediments with different degrees of erodibility, and sediment composition could be the main factor that regulates the spatial patterns of MPB biomass [17]. These differences between sediment types lead to different diatom assemblages with epipsamic or epipelic diatoms causing related variation in MPB biomass. If the previously described physical variables act as a bottom-up control of MPB biomass [17], biological phenomena such as grazing can act as a top-down control [23]. Even if the spatio-temporal dynamics of microphytbenthos production and biomasses are now better understood,

the extent to which the MPB biomass supplies the water column is poorly described and quantified.

Different approaches have been used to characterize benthic-pelagic exchanges caused by MPB resuspension phenomenon. This phenomenon has been widely studied using flume experiments, which enable quantification of the relationship between bed erodibility and sediment properties [24]. However, flume studies focus on the initial point of the erosion phenomenon, without tracking the source distribution of particles along Lagrangian movements of water bodies. Most of the time those data are used for model parameterization for a further evaluation of its fate in the water column. Although flume and small mesocosms experiments are useful to quantify resuspension rates at small scales, they do not enable assessment of the implications of resuspension processes for benthic-pelagic coupling and trophic redistribution at the ecosystem scale [25]. Bivalve farmings have often been recognized as habitats where microphytobenthic communities colonize rapidly the sediments in relation to deposit and bed flow properties mediation by the effect of farming structures and alimentary behavior of animals [17], [25].

Different proxies have been used to study the benthic-pelagic coupling and they can be used as well to better define the trophic routes of resuspended microphytobenthos within an ecosystem. Chlorophyll a (Chl a) biomass, which is often used as a proxy for phytoplankton biomass [26], can also be used as a proxy for resuspension [27], but this variable include both benthic and pelagic sources of Chl a. The taxonomic ratio of benthic to pelagic microalgae can be used as a quantitative indicator for resuspension phenomena [28], thus refining the Chl a concentration indicator. However, differences between benthic and pelagic diatoms are not that obvious since some species are tychopelagic, i.e. live in both environments. Like for Chl a concentration, particulate suspended inorganic matter (SPiM) can be a good indicator of resuspension if both benthic and pelagic compartments are studied at the same time, but the time lag is difficult to avoid *in situ*, particularly when a whole ecosystem is being studied. Some authors used isotopic signatures with $\delta^{13}C$ and $\delta^{15}N$ values of suspension feeders to determine the MPB contribution to their diets [29], [30], [31]. In fact, they could be indirectly used as a proxy of amounts of resuspended MPB, but isotopic studies focus on the final point, i.e. consumption, without knowing whether the initial MPB primary production was autochthonous or allochthonous. Such information could be very useful to consider for coastal management and ecological implications in terms of habitat connection and trophic interaction. The use of phaeopigments as a grazing indicator has been discussed by several authors and judged to be useful for studies of the water column [32], [33]. Because *in situ* studies include many parameters and all these indices provide substantial information concerning different aspects of benthic-pelagic coupling, the combination of them is the best way to assess the implication of MPB resuspension and its redistribution in the pelagic ecosystem and along the trophic chain.

Understanding the set of multifactorial interactions at the ecosystem scale is of critical importance to quantify exports of MPB to the water column, its relative importance compared to the phytoplankton communities and to hierarchize the physical and biological factors potentially involved in MPB exportation. To our knowledge, field experiments have never included both benthic and pelagic compartments at a large scale to explore MPB resuspension phenomenon even though they are complementary and very difficult to separate in estuaries. Because MPB is simultaneously consumed and exported to the water column, in this study we overlaid benthic and pelagic maps of physical and biological variables, for both hydrological conditions and trophic indicators.

The multiple criteria approach we used to study the indices at all scales enabled us to explain the resuspension within the whole ecosystem approach and to cope with the absence of flux measurements (i.e. erosion as well as trophic fluxes). This study also included a spatial survey of MPB distribution, the factors explaining its resuspension and finally its consumption by filter feeders. To better assess the temporal variations in benthic-pelagic coupling, benthic and pelagic compartments were studied simultaneously at two contrasted seasons in terms of forcing variables and MPB and phytoplankton biomass within a temperate macrotidal and exploited coastal ecosystem, the "Baie des Veys" (BDV, France). The whole intertidal area was sampled to account for the spatial heterogeneity within the Bay including different spatial patterns of forcing factors (presence/absence of shellfish farmings, sediment composition, macrofauna distribution, bed shear stress, salinity). Concerning temporal variability, MPB production is normally low in early spring and high in late summer, but the spring phytoplankton bloom is normally higher than the late summer bloom, so resuspension and its relative contribution as a trophic resource in the water column is expected to be higher in late summer. Bioturbation activities that could lead to the resuspension of microphytobenthos from intertidal sediments are also expected to be amplified at the end of summer because of the high levels of biomass but also because of the positive effects of temperature.

Materials and Methods

1. Study Area

The *Baie des Veys* (BDV, Fig. 2) is an estuarine bay located in the western part of the Bay of Seine in the eastern English Channel.

It is characterized by an intertidal area covering 37 km^2 and a macrotidal regime that reaches 7 m maximum tidal amplitude during spring tides and 2.5 m during neap tides [34]. The bay is quite well protected from the prevailing wind by the Cotentin peninsula. Current velocity can reach 3 m.s^{-1} during flood tides and 1.5 m.s^{-1} during ebb tides [29]. Four rivers flow into the BDV through two channels, the Isigny channel in the east and the Carentan channel in the west. Freshwater runoff is low in summer and high in winter, with flows ranging from 3.7 to 26.4 m^3.s^{-1} in the Carentan Channel and from 23.9 to 40.4 m^3.s^{-1} for the Isigny Channel. The oyster farming area extends into the north-eastern part of the bay.

2. Sampling Strategy

Both benthic and pelagic variables were sampled during spring tides to better assess the contribution of resuspended MPB to the total Chl *a* content in the water column [4]. Benthic samples were collected within a week between March 29 and April 2, 2010, and water samples were collected on April 29 and 30, 2010. For the 2 sampling periods, the tidal amplitude was approximately 5.5 m. The same strategy was applied at the end of summer, to assess the impact of the increased number and activity of mollusks on the resuspension phenomenon. Benthic samples were collected from September 8 to 12, 2010, and water samples on the 13 and 14 of September. Because farming structures are located in the north-eastern part of the bay, they may benefit from both autochtonous and allochtonous resources coming from the south of the bay. As a consequence, water was sampled during spring ebb tides, to account for the flux of MPB from the southern part to the north part of the bay, potentially feeding the suspension-feeders reared in the bay. A systematic grid of 88 points was extended over the entire map of the intertidal area with a sampling interval of 500 m (Fig. 2). Heterogeneity in soil organism distribution occurs at nested scales, and is shaped by a spatial hierarchy of environmental factors, intrinsic population processes and disturbances [35]. To explore smaller scale distributions, a nested sampling design was applied [36], [17]. The intertidal area was divided into three sub-domains that were considered as distinct areas due to their separation by the Isigny and Carentan Channels. In each sub-domain, a semi-cross sampling design was applied with an interval of 100 m between each point (Fig. 2). Each semi-cross was placed on high gradient areas previously observed on the field [17].

3. Field Measurements

Sediment and benthos sampling. At low tide, four 20 cm diameter cores were collected at each sampling site. The first cm of each core was removed and placed in a separate plastic bag. At each site, macrofauna were harvested within a square of 0.25 m^2. The choice of a surface of 0.25 m^2 is appropriate when referring to several studies carried out on the benthos in the intertidal zone showing that this surface is sufficiently large and well suited for estimating the abundance of bivalves [37]. This surface allows a suitable and satisfactory sampling of the fauna whatever its distribution (contagious, regular, or random), even when populations are small in number [38].

The entire sediment was sampled to a depth of 10 cm by hand, and then sieved directly on the spot using a 1 mm mesh size sieve [39]. The depth of 10 cm was chosen in order to take in account most of the mollusk biomass potentially involved in resuspension phenomenon, assuming the fact that some bivalve species like *Mya arenaria* that was observed on the field live under this depth [40]. Sieved samples were placed in plastic bags for transport to the lab.

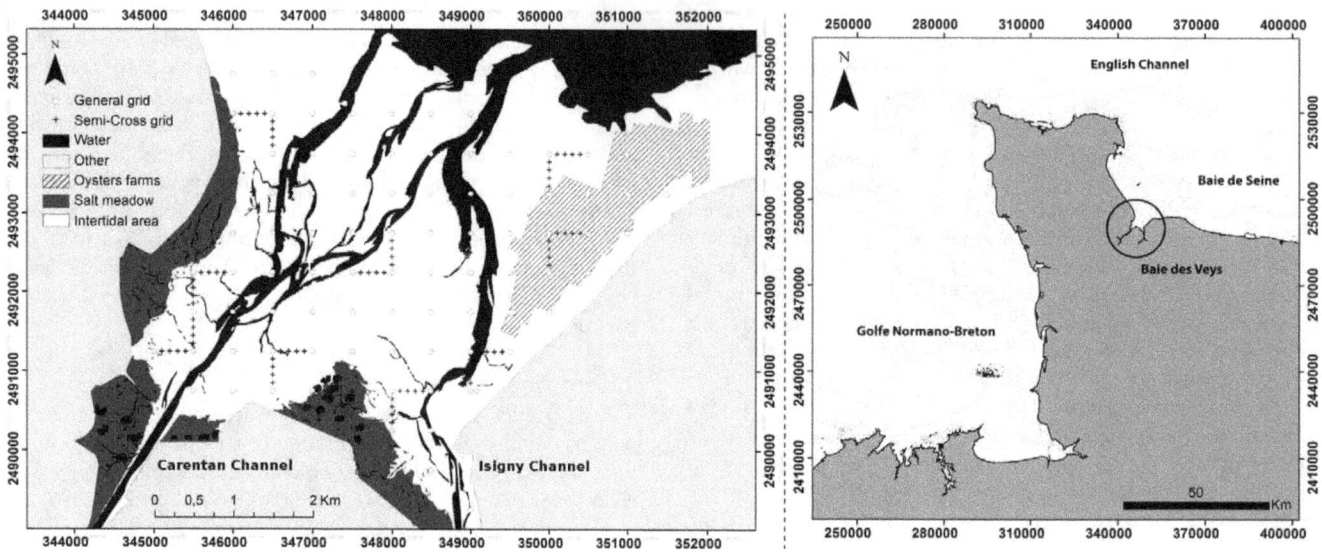

Figure 2. Location of the Baie des Veys and sampling grid.

Animal sampling for isotopy. Six sampling sites were selected within the farming structures to assess the spatial variability of the suspension-feeder *Crassostrea gigas* diet by sampling the widest extent of the structures. The bathymetry of all sites was between +1.65 and +3.50 m. Sites 1 and 2 were on hard substrata, and other sites were on soft-bottom. At each sampling site, five oysters were sampled in the two seasons one month after the spring and late summer samplings, in order to investigate their diets by stable isotopic ratios of carbon and nitrogen.

Water column sampling. Surveys were carried out during ebb tide at all grid points, from one hour after high tide to one hour before low tide. For the 2 periods, the wind velocities were found to be relatively low and similar (3.44 and 4.67 m s^{-1} for April and September respectively), as well as the wind direction with dominant west-northwestern winds (213.33 and 251.46° in a 360° compass rose for April and September respectively). At each point, 5 L water samples were collected by pumping water at a height of 1 m above the seafloor, to ensure access to resuspended MPB. Water samples passed directly through a home-made device equipped with a multi-parameter sensor YSI 6600 (YSI, Yellow Springs, Ohio, USA), before being stored for laboratory measurements. Water subsamples were immediately preserved in Lugol solution for the determination of flora.

4. Laboratory Analyses

Sediment. Back at the laboratory, sediment samples were pooled and mixed thoroughly, and a 1.5 ml subsample was removed and stored at −20°C in the dark until Chl *a* analyses. The remaining sediment was also stored at −20°C until grain-size measurements. Each subsample of sediment was freeze-dried and then weighed to determine the sediment water content. Chl *a* was measured on freeze-dried subsamples using a fluorometric method to estimate algal biomass (µg.g^{-1} sediment). The Chl *a* content of the sediment was extracted in 90% acetone at 4°C for 18 h in the dark. The chlorophyll extracts were measured after centrifugation on a Turner Designs TD 700 fluorimeter (USA) following the method of Welschmeyer [41]. Analysis of particle size distribution was performed by using a grain-size laser method. Sediment samples were dried at 60°C for 3 days and sieved (for coarse-grained particles >2000 µm). Organic matter was removed from

the samples with H_2O_2, followed by soil dispersion with sodium hexametaphosphate. Then, grain-size analysis was performed using a laser granulometer (Coulter, LS200, USA). For the sake of simplicity, the size fractions obtained using the Wenworth scale were then classified in two groups: mud (0–63 µm) and sand (63–2000 µm).

Macrofauna. Samples were fixed in a 10% formaldehyde solution for 24 h and transferred to 70% ethanol for storage until further analyses. All samples were carefully sorted to separate organisms and the remaining sediment. The mollusk species were then determined [42]. Mollusk flesh was separated from the shell, dried at 60°C for 3 days and weighed without the shell. Small specimens with a tough shell (e.g. *Peringia ulvae*) were treated with a drop of 33% hydrochloric acid solution for a few minutes to dissolve the shell. The organisms were then dried in an oven at 450°C for 4 hours to obtain the ash free dry weight.

Freeze–dried, powdered, and homogenized oyster samples were analyzed using a CHN elemental analyzer (EuroVector, Milan, Italy) for particulate organic carbon (POC) and particulate nitrogen (PN) in order to calculate their C/N atomic ratio (Cat/Nat). The analytical precision of the experimental procedure was estimated to be less than 2% DW for POC and 6% DW for PN. The gas resulting from the elemental analyses was introduced online into an isotope ratio mass spectrometer (IRMS) (GV IsoPrime, UK) to determine carbon and nitrogen isotopes. Stable isotopic data are expressed as the relative per mil (‰) differences between the samples and the conventional standards, Pee Dee Belemnite (PDB) for carbon and atmospheric N_2 for nitrogen, according to the following equation:

$$\delta(‰) = \left[\left(\frac{R_{sample}}{R_{standard}} \right) - 1 \right] \times 1000$$

where δ is ^{13}C or ^{15}N abundance and R is the ^{13}C: ^{12}C or ^{15}N:^{14}N ratio. The internal standard was the USGS 40 of the International Atomic Energy Agency ($\delta^{13}C = -26.2$; $\delta^{15}N = -4.5$). The typical analytical precision was ±0.05‰ for carbon and ±0.19‰ for nitrogen. The Phillips and Gregg mixing model [43] was used to estimate spatio-temporal variations in the contribution of suspended organic matter (OMS), including particulate organic

Table 1. Variogram models with their parameter values and cross-validation results.

Spring sampling

Variable	Benthic Chl *a*	Mud fraction	Macrofauna biomass	Pelagic Chl *a*	SPiM	Salinity
Kriging type	Ordinary	Ordinary	Ordinary	Universal	Universal	Ordinary
Detrending order	None	None	None	First	First	None
Transformation	Log	Log	Log	Log	Log	None
Variogram model	Spherical	Gaussian	Exponential	Circular	Circular	Gaussian
Anisotropy	True	True	False	False	True	True
Nugget	0.132	0.469	0.080	0.136	0.163	1.655
Sill	0.894	1.460	1.232	0.419	0.489	6.953
Range	3122.358	2937.798	968.969	1273.001	423.666	3078.753
R^2 (variogram)	0.991	0.987	0.881	0.925	0,732	0,857
Mean std.	0.016	0.000	−0.024	−0.025	−0.046	-0.005
RMSS	0.951	1.198	0.968	1.122	0.946	1.054

Summer sampling

Variable	Benthic Chl *a*	Mud fraction	Macrofauna biomass	Pelagic Chl *a*	SPiM	Salinity
Kriging type	Ordinary	Ordinary	Ordinary	Ordinary	Ordinary	Ordinary
Detrending order	None	None	None	None	None	None
Transformation	Log	None	Log	Log	None	Log
Variogram model	Gaussian	Circular	Exponential	Circular	Exponential	Spherical
Anisotropy	False	True	False	False	False	False
Nugget	0.319	37.4	0.040	0.037	0.865	0.002
Sill	0.677	101	1.46	0.174	18.0	0.347
Range	2242	3456	1101	2675	2714	2568
R^2 (variogram)	0.841	0.730	0.924	0.965	0.847	0.934
Mean std.	−0.002	−0.017	−0.050	−0.023	0.001	−0.022
RMSS	1.06	1.067	1.178	1.068	1.172	0.943

Mean std = Mean standardized; RMSS = Root Mean Square standardized.

matter (POM), MPB, resuspended POM (rPOM), and macroalgae (ULV), to the suspension-feeders' diets, following the protocol of Lefebvre et al. [7] but with fractionation values of 1.85‰ for $\delta^{13}C$ and 3.79‰ for $\delta^{15}N$, obtained from Dubois et al. [44].

Water samples. To measure the concentration of suspended particulate matter, two subsamples (1L) were sieved and passed through weighed and dried glass-fiber filters (Whatman GF-C), washed with distilled water to avoid errors due to salt, packed in petrislides (Millipore, USA), and immediately stored at −20°C until analyses. The filters were dried in an oven at 60°C for 72 hours. For Chl *a* concentration measurements, two subsamples were sieved and passed through a glass-fiber filter (Whatman GF-C), folded and placed in a tube at −20°C before analyses. The Chl *a* content was extracted in 90% acetone for 18 h at 4°C in the dark. After short centrifugation (3500 G), the chlorophyll extracts were measured on a Turner Designs TD 700 fluorimeter (USA) following the method of Welschmeyer [41] and expressed as chlorophyll content ($\mu g.L^{-1}$) in the spring samples. The summer samples were analyzed using Lorenzen's method [45] in order to examine the phaeopigment content. Calibration was performed between the two methods to compare the result of the two samplings (y = 0.9624x+1.5399, $R^2 = 0.999$). Each sample preserved in Lugol was observed for quantitative/qualitative determination of microalgae flora, following the Utermohl method

described in [46] using light microscopy on Sedgewick-Rafter cells. In some samples, 400 individual cells were counted whatever the total number of cells, following the European standard for phytoplankton counting (NF EN 15204, 2006). Finally, a list of diatoms and the ratio of benthic to pelagic diatom species were established for each site following the protocol of Kasim and Mukaï [28]. Actually, the quantity of larger species is underestimated using abundances, while of the smaller species is underestimated using biomass [47]. To get round this problem, log-transformed abundance scores were used to calculate this ratio [48].

5. Statistical Analyses

Geostatistical analyses were performed with the ArcGIS extension Geostatistical Analyst (ESRI, USA) in order to map the different variables measured on the field. Since there was a high spatial dependency in all the variables measured, kriging was chosen as the best interpolation method to predict values for the whole intertidal area. Normal distribution was checked before each analysis and log-transformation was applied as a function of the variable concerned. Global trends were also examined, to enable removal of the possible effect of the tidal circulation on the water column. If necessary, detrending was applied using a polynomial algorithm of chosen order. Each variable was studied to find the best semivariogram model fitting for data, between circular, spherical, exponential and gaussian models (Table 1). Cross-validation enabled us to check the validity of the semi-variogram models we selected. Nugget effect was always small and never reached up to 1/3 of the sill value (Table 1), confirming the validity of the sampling scale and chosen nested design scale. If the prediction errors are not biased, the mean prediction error should be near zero. However, this value depends on the scale of the data; to standardize these, the standardized prediction errors give the prediction errors divided by their prediction standard errors. The mean of these, called "mean standardized", should also be near zero. If the prediction standard errors are valid, the root mean squared standardized error should be close to 1. If it is greater than 1, the variability of the predictions has been underestimated, and inversely. The MARS-3D hydrodynamic model [49] was used to obtain the mean bottom current velocities at the two sampling periods. The results were plotted using the ArcgiS Toolbox "MGET" [50].

Multivariate analysis using the R package ADE4 (R-project) were used to better identify spatial and seasonal effects and explore the benthic-pelagic coupling through correlations between the variables. Principal Components Analyses (PCA) were performed on benthic and pelagic log-transformed datasets for both seasons, completed with estimated data from the kriging matrices for the few numbers of points where there were some missing values. For these analyses, bathymetry was considered as an auxiliary variable because it can play a role in both benthic and pelagic compartments. Co-inertia analysis was used to explore the relationships between the benthic and pelagic compartments by coupling the previous PCA, and its validity was checked by performing a Monte-Carlo test on the sum of eigenvalues of the analysis [51]. Frequency distribution of the RV values for 100 random co-inertia simulations was tested to check the validity of the co-inertia analyses.

Regression analyses were performed using Minitab (Minitab inc., USA) in order to find the best model predicting variable distribution. Stepwise regression was used to identify the best subsets of predictors in sampled variables. A linear regression was then applied on the most appropriate subset of data, corresponding to the best Awaike Information Criterion (AIC), meaning the lowest values when comparing the different regression models.

Results

Benthos

Data for each measured variable were analyzed using a PCA (Fig. 3). Only the two first components were kept, and these explained 93.7% of the total variation. The correlation circle (Fig. 3A) showed a clear relationship between the mud fraction, Chl a content and the bathymetry of the intertidal area. These three variables were well represented in the 1st axis and explained 61.4% of the total variation, confirming that the distribution patterns for both the mud fraction and the Chl a concentration remained stable between the two seasons. Mollusk biomass distribution was not correlated with the above variables; it was well represented on the 2nd axis and explained 32.3% of the total variance.

The scatter plot of individuals (Fig. 3B) showed a clear spatial structure, and sampling points were merged into three groups, corresponding to three areas of the bay: the eastern part located on the east side of the Isigny channel, and the northern and southern areas to the west (Fig. 4H). Individual distribution was explained by the correlation circle, with a Chl a concentration gradient from north to south, and an eastern area with a lower mollusk biomass.

Geostatistical analysis and kriged maps confirmed that benthic Chl a concentration (Fig. 4A, B) was characterized by the same distribution patterns at the two seasons, with higher concentration close to the salt meadow particularly at the southern borders of the bay, and also under the farming structures in the east. Less concentration was found in the central and northern part of the bay, resulting in a decreasing gradient from the coast to the center of the bay. Regarding to the three areas determined by the PCA, the southern and eastern areas were characterized by relatively high Chl a concentration compared to the northern area. The mud fraction (Fig. 4C, D) was correlated with the previous parameter with a gradient from the southern part of the bay with muddy to mixed sediment to sandy areas in the north and east. There was a slight increase in the mud fraction in two patches in the eastern part of the bay sampled during summer. The Chl a concentration and mud fraction were both clearly linked to the bathymetry of the bay (Fig. 4G), with the shallower parts located close to the salt meadow and along the eastern coast. Ordinary kriging was required for the Chl a concentration and mud fraction (Table 1), and the variogram structure was close considering the range (ca. 3000 m), reflecting a similar patch size for these two variables.

In contrast, there was a change in mollusk biomass between the two sampling periods (Fig. 4E,F). Five major species were identified at each season, with the cockle *Cerastoderma edule* as the dominant species (Table 2). Only one of the major species changed between the two seasons: *S. plana* was present in spring but replaced by *Abra tenuis* in late summer (Table 2). Mean mollusks biomass increased 20-fold between the two seasons (Table 2). Despite these differences, the distribution type of mollusks remained the same for the two seasons since the variogram structure was similar in terms of kriging method (Ordinary), in terms of variogram model (Exponential) and range values (ca. 1 km). Maximum biomass increased 3-fold between the two sampling dates, from 87.6 g.m^2 in spring to 216.3 g.m^2 in summer. The spring map shows the three-parted intertidal area, with a high mollusk biomass in the south, a lower biomass in the north and a very low biomass in the east. The summer map shows a larger high biomass area, and a contrast between the eastern part with low biomass and the northern and southern areas characterized by high biomass. Two high biomass patches were present in spring, and were still present but far bigger in summer. Very low mollusk biomass was found under the farming structures in the east at both sampling dates.

Regression analysis (Table 3) revealed that benthic Chl a concentration can be predicted by the whole set of benthic and pelagic variables in spring and mud fraction, bathymetry and water Chl a concentration in summer. For both seasons, mollusk biomass was best predicted by the association of chl a water and bathymetry, with a better model adjustment for spring (R^2 = 0.48).

Pelagos

A PCA was applied to the pelagic variables and only the two first components were retained, which explained 81.6% of the total variation. The correlation circle (Fig. 5A) shows that SPiM was anti-correlated with the bathymetry and was well represented on the 1st axis, where it explains 57.3% of the total variation. There was also a good correlation between the pelagic Chl a concentration and the concentration of SPiM, even if the former was partly represented on the 2nd axis. Salinity was not correlated with the pelagic Chl a concentration, and was represented to the same extent on both axes, but poorly anti-correlated with the concentration of SPiM. This low or null relationship between salinity and both pelagic Chl a concentration and SPiM

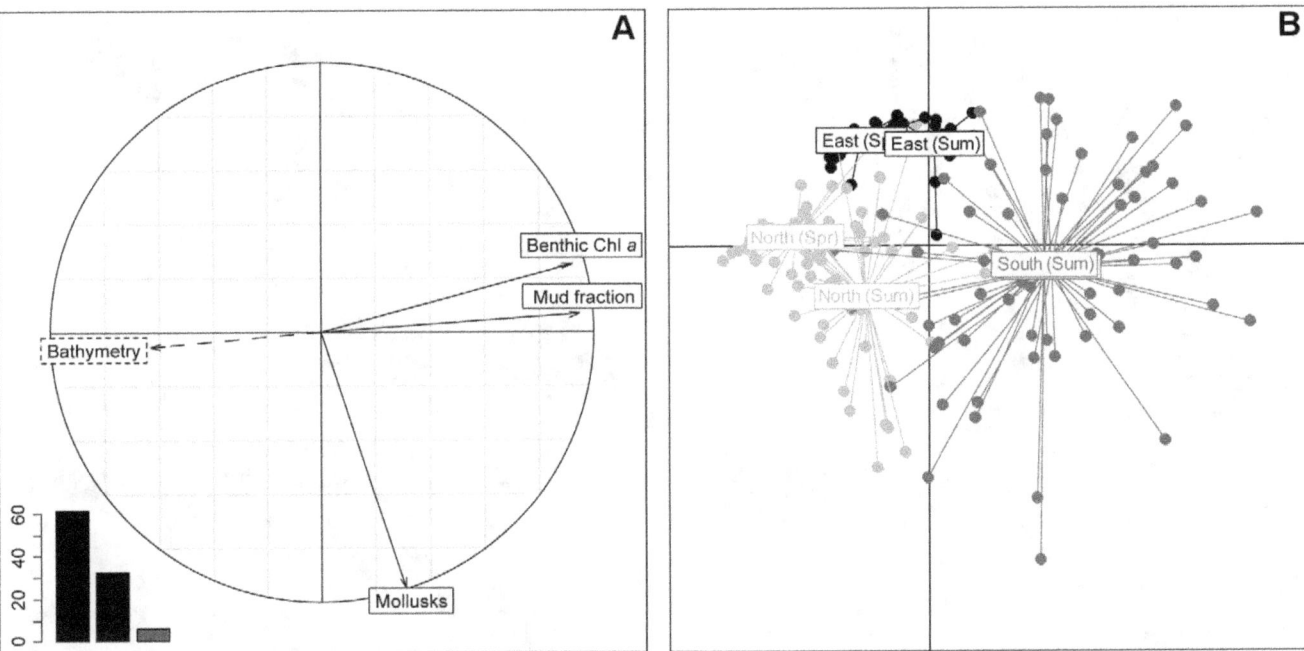

Figure 3. PCA results of the benthic log-transformed variables for the 2 seasons. Bathymetry (m), Chl a concentration (μg.g^{-1}), mud fraction (% of total sediment) and mollusk biomass (g AFDW.m^{-2}). Data used for the PCA resulted from the extraction of the corresponding kriged maps on the general sampling grid. Bathymetry was used as an auxiliary variable. A: Correlation circle; B: Scatter plot of individuals, "South (Spr)" and "South (Sum)" captions are confounded.

concentration showed that these two variables were not entirely related to the river inlets.

In line with the results for benthos, the scatterplot of individuals (Fig. 5B) showed a clear spatial structure in the data: the sites were merged into three groups, corresponding to three spatial areas within the bay: eastern, northern and southern areas. The drift observed in the scatterplot of individuals between the two seasons was the same in all parts of the bay and appeared to be related to salinity. The southern area was characterized by the highest pelagic Chl a concentration while the eastern area had the lowest.

Kriged salinity maps (not shown) showed a common structure with a south to north gradient from low to high salinity. Salinity was twice lower in spring with stronger river inputs, particularly from the eastern channel. The southern part of the bay was characterized by high Chl a concentration and SPiM (Fig. 6), whereas the northern area showed lower concentrations. Both sampling periods were characterized by a depletion observed in the eastern area, which was stronger in spring. In late summer, Chl a concentration ranged from 2.78 to 18.8 μg.L^{-1}, and the area was smaller than that found in April (from 0.64 to 26.1 μg.L^{-1}). Like on the spring map, on the summer map, a large area at the north-east was characterized by a limited depletion of Chl a concentration in the water column. The quantity of SPiM was higher in the southern and north-western part of the bay than in the eastern part. Except in the area with the farming structures where SPiM was low, concentrations were related to the bathymetry of the bay, with lower concentrations in areas with deeper water. Mean currents at the bottom showed velocities of between 0.05 and 0.40 m.s^{-1}, with higher velocities along the two channels. The two sampling periods showed similar hydrodynamic conditions with a general field of current vectors oriented towards the north-north-west of the bay during this ebb tide causing the bay to empty. A small area with lower velocities was observed in the center of the bay.

Benthic-Pelagic Coupling

In order to examine the benthic-pelagic coupling at the bay scale, co-inertia analysis was performed on both benthic and pelagic variables. Correlation circles revealed the close link between the co-structure described by co-inertia axis F1/F2, and the structure of each dataset described by the respective components in each PCA. In fact, projected variances on axis F1/F2 of the co-inertia analysis were close to the values of maximum projected variances on the 1st and 2nd axis of the PCA (Table 4). Comparison between the co-inertia coefficient RV and its empirical distribution during the Monte-Carlo test showed a strong co-structure between the two tables (RV = 0.239). Next, the test procedure was run on the two seasons separately, to check for a seasonal impact on the co-structure between benthos and pelagos. Results were more significant in spring (RV = 0.500), even if the summer RV (RV = 0.110) remained good (p<0.01 for the 3 tests).

The cross-table resulting from the co-inertia (Fig. 7) confirmed the strong impact of the season on benthos-pelagos coupling. In spring, mollusks exhibited a negative correlation with salinity, and a positive correlation with SPiM and water Chl a concentration. The levels of correlations were much lower than in spring data. Spring benthic Chl a concentration showed a negative correlation with salinity, whereas it showed a positive correlation with SPiM in both seasons. The mud fraction was positively correlated with water Chl a concentration in spring, and negatively correlated with salinity at the same period. The same relationship was found in summer, but to a lesser degree. Finally, in both season, the mud fraction was negatively correlated with bathymetry and positively correlated with SPiM.

Diatom taxonomic analysis revealed a number of taxa originating from different environments - marine, brackish and benthic - and characterized by different shapes and sizes (Table 5). Even if the species composition was almost the same at the two

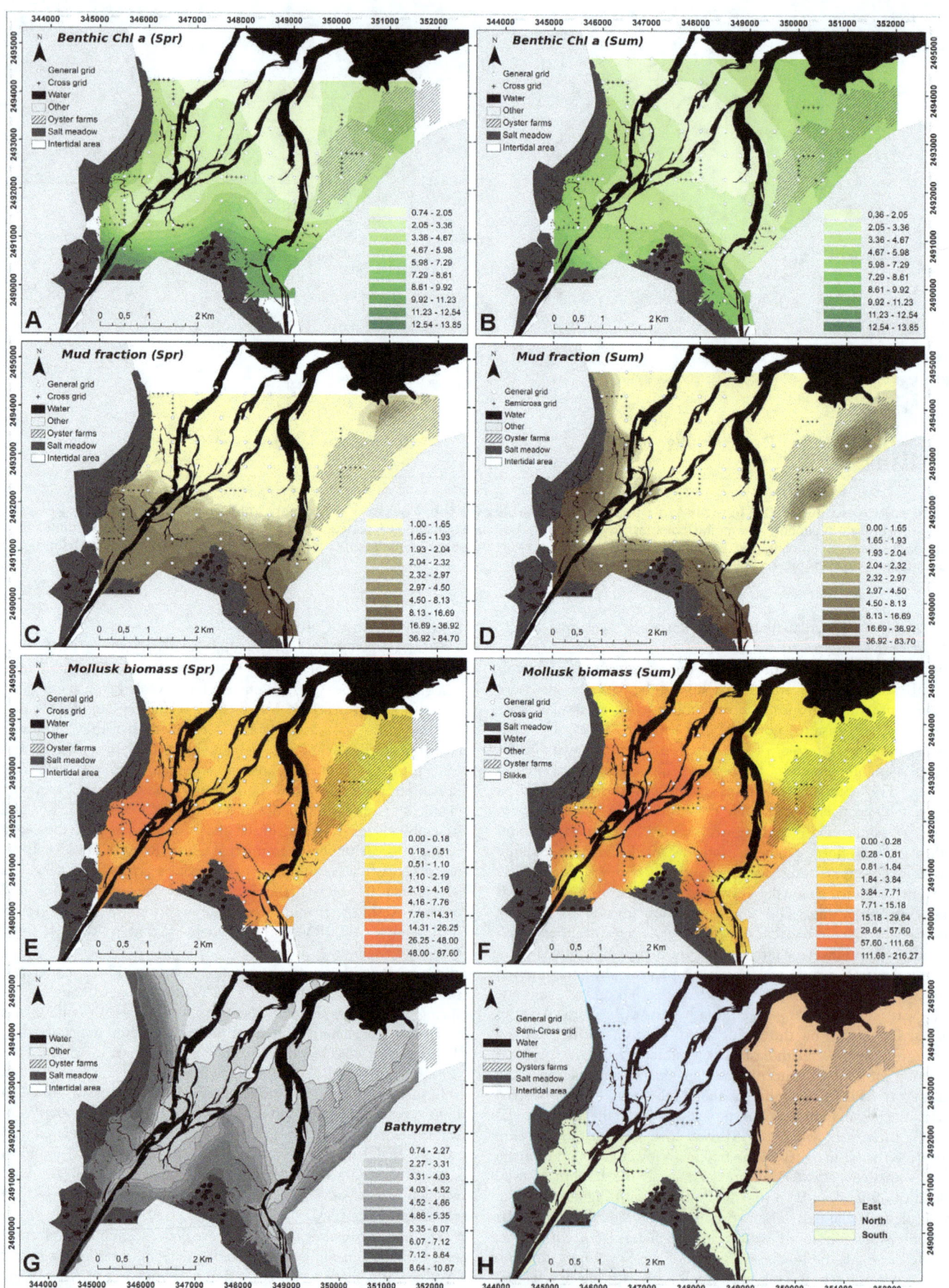

Figure 4. BDV spring and summer kriged maps of benthic variables for the 2 seasons. All variables were kriged on different variogram models depending on the data (Table 1). Geometrical scales were used to maximize the visualization of both gradients and the patchiness of the different variables. Mollusk maps are at different scales to account for the discrepancy in the data between the 2 sampling campaigns. **A, B**: Chl a concentration (μg.g^{-1}); **C, D**: Mud fraction (%<63 μm of total sediment); **E, F**: Mollusks biomass (g AFDW g.m^2). **G**: Bathymetry of the BDV, from low to high tide spring tide levels (m). **H**: Representation of the 3 subdomains defined by the PCA.

sampling periods, the relative proportions of the different species differed. The long-chain diatom *Asterionellopsis glacialis*, a brackish species observed at both sampling dates, was the dominant species in the Bay during the spring sampling (92.4%). The marine genus *Chaetoceros sp.* was dominant during the autumn sampling (73.4%) but was not identified in the spring samples. Benthic-pelagic ratios were calculated at several points distributed throughout the intertidal area, and represented on the Chl a concentration maps for both seasons (Fig. 5). At the first sampling date, the benthic-pelagic ratios were low throughout the Bay. However, benthic species reached 40% of the diatom community at two sampling points in the late summer sampling. Results showed that the benthic-pelagic ratios supported pelagic species over the entire map (Fig. 6), with percentages ranging between 60.2% and 100% in spring and between 59.6% and 100% except at two sites in summer. In fact, at these two sampling points, benthic diatoms species reached 47.0% and 47.2% of the diatom community. These two points corresponded to the area where the highest water Chl a concentration and SPiM were observed.

The benthic phaeopigment percentage map (Fig. 8) showed a higher concentration close to the channels, and a lower concentration under the farming structures and in the central area between the two channels. The water column phaeopigment map showed a negative relationship with the water column Chl a concentration map (Fig. 6, 8), whereas no relationship was observed with the benthic phaeopigments map. The two depletion areas previously seen for the pelagic Chl a concentration were the two areas with the maximum phaeopigment percentages, ranging from 21.50% to 30.39% of the total pigments.

Oysters sampled a month after each field campaign showed significant differences in isotopic signature (Fig. 9A). The δ_{13}C values ranged between -20 and -19‰, and δ_{15}N values between 9 and 10‰ in spring. These values increased in summer, ranging from -19 to -18‰ for δ_{13}C and from 9.5 to 11‰. After correction of the trophic step fractionation (Fig. 9B), these values were clearly distributed between particulate organic matter (POM) and MPB. There was an increased contribution of MPB to the oyster diets (paired t-test, P-value $= 0.027$), which increased from 18.0% in spring to 39.2% in summer on average (Fig. 9C). The spatial pattern of this contribution differed in the two seasons, with a decreasing south-to-north gradient in spring, whereas the maximum contribution was found at the OYST3 and OYST4 located the middle of the farming structures in summer (Fig. 9D). This area was found to be enriched in both mud and benthic Chl a concentration at the same season.

Discussion

MPB Spatial Distribution: Strong Effect of Mud Fraction and Bed Elevation

As shown in Orvain et al. [17] for this ecosystem, there was a clear relationship between MPB biomass and the grain-size of the sediment. Chl a concentration appeared to increase as a function of the mud fraction, in agreement to other studies ([52], [53]) which found higher Chl a content when expressed per mass unit. Moreover, both Chl a biomass and mud fraction were closely correlated with the bathymetry of the intertidal area, especially in area located on the west side of the Carentan channel. In fact, shallower water in areas with less hydrodynamic stress favored the silting up of these areas, and increased sunlight intensity, all of which favored MPB production [54]. Thus, in the present study, MPB biomass was well correlated to both mud fraction and bathymetry, as shown for other temperate estuarine ecosystems ([19], [55], [56]). In spite of the strong contrast between the two sampling periods in terms of temperature or light and nutrients availability, results revealed a perennial spatial structure of the intertidal sediments and MPB biomass in the bay (Fig. 4A, B, C, D) regarding to the stability of patterns between seasons at the year scale. The southern area close to the salt meadow was characterized by shallower waters, resulting in a muddy area because of the combination of direct river inputs and lower hydrodynamic conditions. Conversely, the northern part was under marine influence, with higher hydrodynamic conditions leading to sandy sediments. Finally, the eastern area appeared to be mainly influenced by the farming structures. The limited seasonal effect on the ranges of benthic Chl a concentration found in the BDV underlines the predominant effect of grain size and bathymetry on MPB distribution and biomass. As this bay is mostly made up of sandy sediments, these results correspond to those observed by van der Wal et al. [19], showing lower variability in sandy sediments than in muddy sediments. MPB distribution patterns were in close agreement with the results observed in April 2003 [17] even if the biomass levels were higher, certainly due to the much higher solar radiation observed during that exceptionally hot year.

Mollusk Spatial Distribution: Direct Linkage with Water Chl a Spatial Patterns

Distribution patterns of mollusk biomass, mainly represented by *Cerastoderma edule*, appeared to be related to the water Chl a concentration for the two seasons. Distribution of macrozoo-benthos in response to microphytobenthos and sediment has been

Table 2. Mean weight and number of mollusks per m^2 in the 2 samplings.

Species	C. edule		M. balthica		A. tenuis		H. ulvae		S. plana	
Mean	ind.m^2	g. m^2	ind. m^2	g. m^2	ind. m^2	g. m^2	ind. m^2	g. m^2	ind. m^2	g. m^2
Spring	16.6	2.99	2.172	0.064			102	0.077	10.7	0.035
Summer	169	65.6	4.331	0.632	6.55	0.067	148	3.18		

Table 3. Response of selected variables to log-transformed benthic and pelagic variables.

Variable	Linear predictor	R^2_{adj}	AIC
Benthic Chl a (SPR)	−2.60 −0.145 mollusks*** +0.546 mud fraction*** −0.513 bathymetry*** +0.319 SPiM* −0.235 chl a water** +2.26 salinity*	0.78	−211
Benthic Chl a (SUM)	0.943+0.325 mud fraction*** - 0.143 bathymetry* −0.346 chl a water***	0.76	−257
Mollusks (SPR)	0.339+0.725 chl a water*** −0.624 bathymetry***	0.48	−35.3
Mollusks (SUM)	−1.46+1.81 chl a water*** +0.826 bathymetry***	0.13	64.5

All variables included in linear predictor are significant (*p<0.05, **p<0.01, ***p<0.001). R^2 represents the percentage of response variable variation that is explained by its relationship with one or more predictor variables, adjusted for the number of predictors in the model. AIC (Akaike information criterion) is a measure of the relative goodness of fit of the models, best models (lower values) were kept.

studied by van der Wal [57] at an intertidal area scale, finding good model predictors to describe surface and deep deposit-feeders biomass using benthic variables such as MPB biomass or median grain-size, but the model found for suspension feeders was less satisfying with no predictor terms of the model significant. Honkoop et al. [58] also found low relationship between abiotic factors and the distribution patterns of benthos, suggesting that they could be influenced or determined by biotic interactions which may be more important than the assumed abiotic structuring they measured.

Our results confirm their suggestions about including pelagic variables to improve comprehension and modeling of macrofauna abundance or biomass distribution. If benthic variables may explain biomass and/or distribution of deposit-feeders, both benthic and pelagic variables must be assessed to explain better suspension-feeders biomass and/or distribution, underlining the necessity of including food compartment associated with studied communities. As a consequence the reverse reasoning has to be considered too when studying Chl a concentration in the water

column, with higher cockle densities leading to higher bioturbation and consequently higher MPB resuspension.

Benthic-pelagic Coupling: Impact of Resuspension Phenomenon

The same three part structure as for the benthic sampling was observed in the water column at two sampling seasons (Fig. 4G), showing on the one hand the fundamental influence of physical factors on benthic-pelagic coupling, and, on the other hand, its robustness over time in terms of both structure and resuspension phenomena. The pelagic structure is not perennial since water bodies are highly variable over time in terms of phytoplankton abundance and composition. As a consequence, the strong spatial structure of the benthic compartment influences the pelagic compartment through a domino effect controlled mainly by hydrodynamics and currents. Wind effect is well recognized as one of the first factor implicated in the temporal variation of resuspension phenomenon [4]. The results by de Jonge and van Beusekom focused on temporal dynamics in terms of resuspension

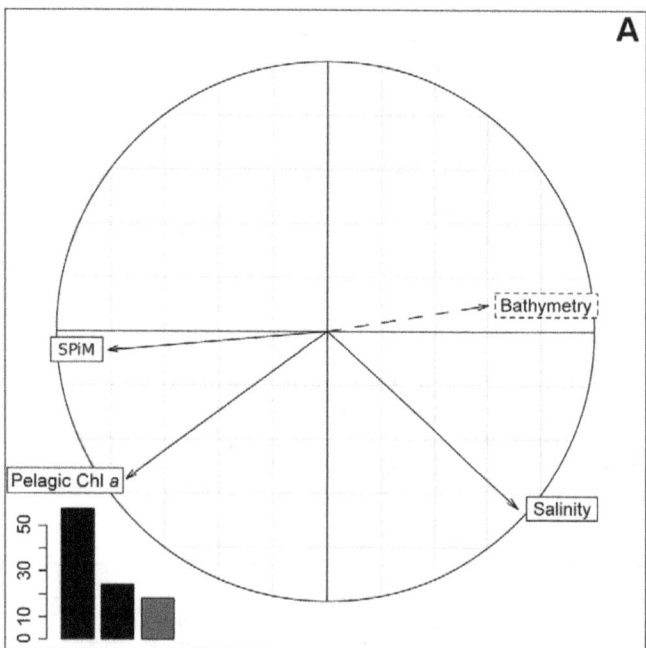
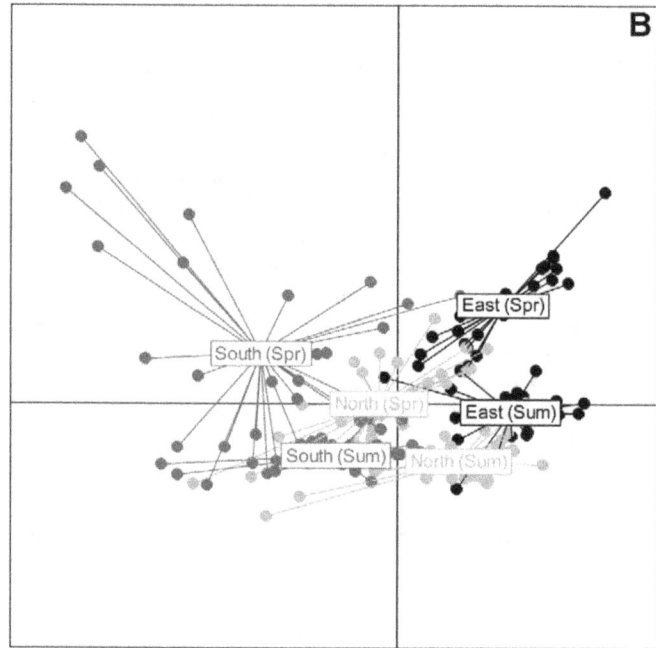

Figure 5. PCA results of the pelagic log-transformed variables for the 2 seasons. Bathymetry (m), Chl a concentration (µg.L⁻¹), SPIM (mg.L⁻¹) and salinity. Data used for the PCA resulted from the extraction of the corresponding kriged maps at the location of the general sampling grid. A: Correlation circle; B: Scatter plot of individuals (Spr = Spring; Sum = Summer).

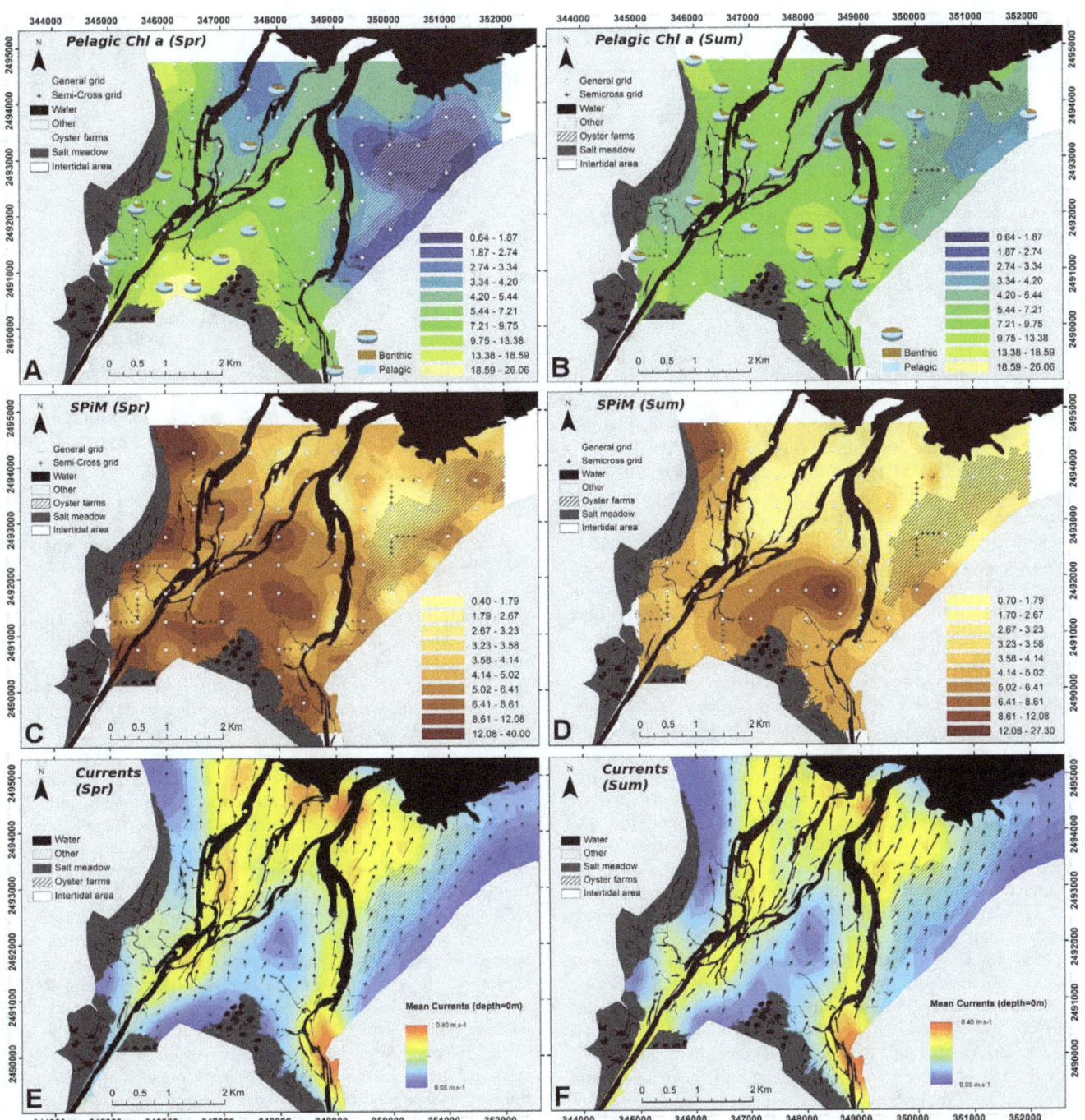

Figure 6. BDV spring and summer kriged maps of pelagic variables for the 2 seasons. All variables were kriged on different variogram models depending on the data (Table 1). Geometrical scales were used to maximize the visualization of both gradients and the patchiness of the different variables. Mollusk maps are at different scales to account for the discrepancy in the data between the 2 samplings. **A, B**: Chl *a* concentration (µg.L⁻¹); **C, D**: SPiM amount (mg.L⁻¹). **E, F**: Bottom mean current velocities and direction at the 2 sampling periods, calculated by the MARS-3D hydrodynamic model.

phenomenon. By contrast, the present study was based on two samplings with similar hydrodynamic conditions but with a comprehensive number of stations to examine spatial patterns. Temporal detailed dynamics of the resuspension of microphytobenthos as well as wind effects were out of the scope of the present study which aims at describing spatial patterns of the benthic and pelagic variables without considering the wind. The results presented here and particularly the difference observed between

the two samplings focus on other phenomena implicated in resuspension phenomenon. We must mention that contrary to more open estuarine ecosystems, BDV is relatively protected of wind effects by the geographical configuration of this basin, which is protected by southern and/or western dominant winds by the Cotentin Peninsula. Only northern (and especially north-eastern) winds can have an impact on the general functioning of this bay in terms of erosion.

Table 4. Comparison of inertia resulting from the separate analyses of each dataset.

Axis	InerBen	InerPel	InermaxBen	InermaxPel
F1	1.78	1.63	1.84	1.72
F2	1.02	0.738	0.967	0.731

Two co-inertia axes (F1 and F2) were selected. InerBen = inertia of the benthic table projected on co-inertia axes; InerPel = inertia of the pelagic table projected on co-inertia axes; InermaxBen = maximal projected inertia of the benthic table (1st and 2nd eigenvalue of the PCA); InermaxPel = maximum projected inertia of the pelagic table (1st and 2nd eigenvalue of the PCA).

The pelagic Chl *a* concentration was closely correlated with the concentration of the SPiM, but the low or null relationship with salinity revealed the influence of resuspension events rather than river inputs. In fact, the two channels were characterized by low SPiM concentrations (Fig. 6E and F). Moreover, both Chl *a* concentration and SPiM concentration were inversely correlated with bed elevation (Fig. 5A), reinforcing the hypothesis of resuspension events from the muddier sediments with a higher impact in these shallow waters [59]. Similarly, good levels of correlation between SPiM and Chl a in the water column were obtained by Guarini et al. during large-scale [60] and long-term [48] samplings in another estuarine bay (Marennes-Oléron) where microphytobenthic communities are more developed than in BDV.

The significant co-structure found between the benthic and pelagic compartments confirms the hypothesis of a strong coupling, maintaining the 3-part structure in both compartments and at both seasons. The dominance of *Asterionelopsis glacialis* and *Chaetoceros spp.* during the respective sampling periods in this ecosystem has already been reported in the literature [61]. It reflects changes in the estuarine microalgae communities over the year, with dominance of brackish species in early spring and of marine species in late summer. However, resuspension phenomena appear to be relatively stable, given the range of the MPB ratios revealed by taxonomic identification. Only two locations showed a high MPB ratio during the late summer sampling, corresponding to a higher SPiM and an area of lower current velocities observed at the same period. According to the benthic:pelagic ratio of this patch and to microscopic observations, a part of this resuspended MPB is probably the result of inputs from the eastern channel. However, taxonomic analyses have to be interpreted with caution. The distinction between plankton and benthos is not perfectly clear because some microalgae are tychopelagic, i.e. they live in both environments [62]. Actually, in the surf zone (the zone extending from the outermost line of breakers to the limit of wave uprush) communities dominated by long chain diatoms like *Asterionellopsis glacialis* can be deposited on the sediment by the ebb tide, because mucus and particles attached to the cells increase their density, hence increasing sedimentation [63]. As a consequence, the number of living *A. glacialis* cells per sediment area behind the surf zone can be on average four orders of magnitude higher than the concentrations found in the respective water column [64].

The mollusk biomass was 20-fold higher in summer than in spring and macrofaunal activity is also known to increase between spring and summer because of the effects of temperature, so higher biological activity at the bay scale resulted in a higher phytoplankton consumption and subsequently in an increase in biodeposition. This high consumption rate was confirmed by

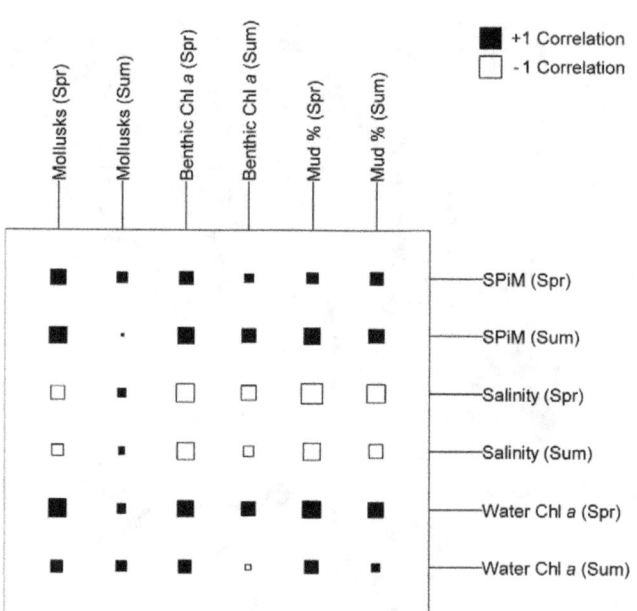

Figure 7. Cross-table resulting from the co-inertia analysis. Represents the correlation between the benthic and pelagic datasets.

phaeopigments released in the water column (Fig. 8B). Finally, the isotopic signature of the diatom *Asterionellopsis glacialis* needs to be investigated since it could regulate its buoyancy in order to stay in the bay [65], since it belongs both to benthic and pelagic environment. Its presence, highest in spring, could help explain the good overlay of the benthic and pelagic maps.

In spring, resuspension phenomena were mainly under hydrodynamic influences resulting in an almost complete overlaying of benthic and pelagic compartment map, reflecting the close coupling between them. In summer, the increase in mollusk biomass increased the effects of bioturbation that could be involved in different chl *a* fluxes like consumption, biodeposition and bioresuspension processes. The role of bioturbators is well known as an important factor controlling microphytobenthos resuspension [13] [66]. Macrofauna activities and especially those of the cockle *Cerastoderma edule* must be better explored and evaluated because our results suggest that this is, when not considering wind effect, the prime factor controlling resuspension rates of microphytobenthos at the scale of the bay. This process must be thoroughly involved in the good relationship between suspension-feeder biomass and concentrations of resuspended chl *a*.

Impact of Cultivated Oysters in the Benthic-pelagic Coupling

The intertidal area is divided into two parts with respect to the Isigny channel, the two parts being clearly separated by the presence/absence of oysters farming structures. The spatial distribution of studied variables in the eastern part of the bay allows deciphering of the oyster impact on the benthic-pelagic coupling. Even if this area was characterized by sandy sediments with a very small mud fraction and a high depth, MPB biomass was high. Several explanations are possible and/or a combination of them.

First, microalgae communities can benefit from biological phenomena such as biodeposition under the farming structures, providing a favorable habitat for MPB assemblages [67]. Increased oyster filtration activity in late summer [68] led to

Table 5. List of determined microalgal taxa from Lugol fixed water samples.

Diatom species	Lifestyle	Shape	Size classes	Spring	Automn
Asterionellopsis glacialis	Tychopelagic	Pennate	Small (<15.10³ µm³.cell⁻¹)	■	■
Chaetoceros spp.	Pelagic	Centric		□	■
Cyclotella spp.	Pelagic	Centric		■	■
Navicula spp.	Benthic	Pennate		■	■
Diploneis spp.	Benthic	Pennate	Medium (15.10³ µm³.cell⁻¹ < x<150.10³ µm³.cell⁻¹)	■	□
Gyrosigma fasciola	Benthic	Pennate		■	■
Gyrosigma hippocampus	Benthic	Pennate		■	■
Nitzschia longissima	Benthic	Pennate		■	■
Pleurosigma spp.	Benthic	Pennate		■	■
Pseudo-nitzschia spp.	Benthic	Pennate		■	■
Paralia marina	Benthic	Centric		■	■
Thalassionema nitzschioides	Pelagic	Pennate		□	■
Thalassiosira rotula	Pelagic	Centric		■	■
Coscinodiscus wailesii	Pelagic	Centric	Large (>150.10³ µm³.cell⁻¹)	■	■
Guinardia delicatula	Pelagic	Centric		■	■
Guinardia striata	Pelagic	Centric		■	■
Lauderia annulata	Pelagic	Centric		■	■
Odontella regia	Pelagic	Centric		■	■
Rhizosolenia imbricata	Pelagic	Centric		■	■

Each species is classified by living, shape, size class and presence/absence during the two samplings.

a higher biodeposition, explaining two little patches with a higher mud fraction in the east in late summer (Fig. 4D) and the high phaeopigment percentages in the water column (Fig. 8B). Nevertheless, biodeposition was a local phenomenon which was mostly visible under the farming structures, and may be insufficient to significantly increase the mud fraction of the sediment at the scale of the eastern area. The two small patches could be explained by a combination of cultivated stocks, currents or bathymetry difference over farming structures [69]. Secondly, the high clearance rate of oysters significantly affected

light availability by reducing water turbidity [18], thus enhancing MPB production.

Thirdly, the eastern area was characterized by very low concentrations of wild mollusks and biomass, in contrast to the western part of the bay. The exclusion of wild suspension-feeders under farming structures has already been observed at this site by Dubois et al. [70], showing that there was a shift in the trophic chain to high levels, with a predominance of predators, especially under farming structure. The primary consumption rate of MPB by grazers must therefore be low under the farming structures.

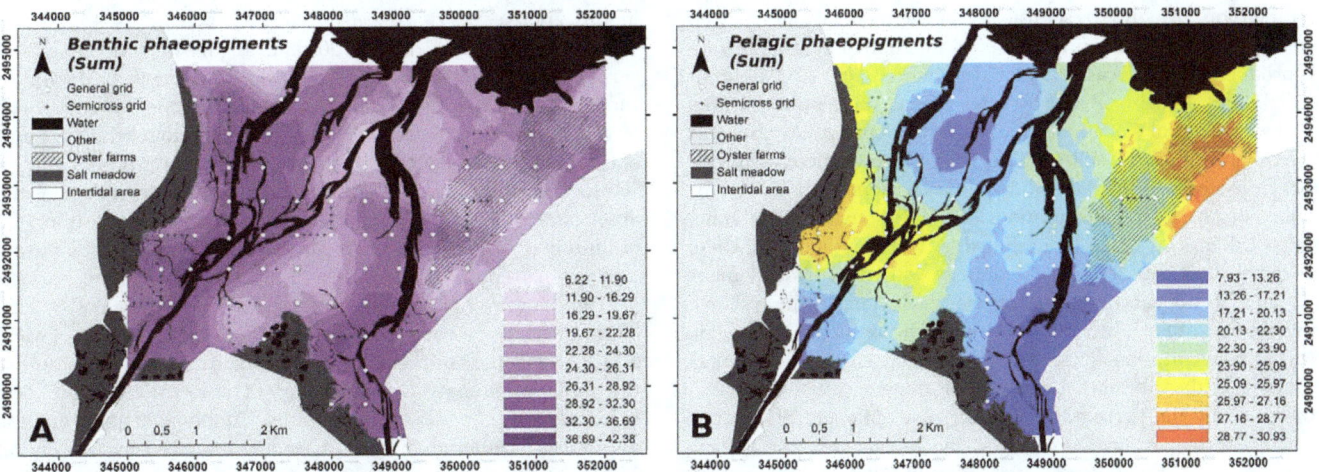

Figure 8. BDV summer kriged maps of both benthic (A) and pelagic (B) phaeopigments. Results are presented as % of total pigments.

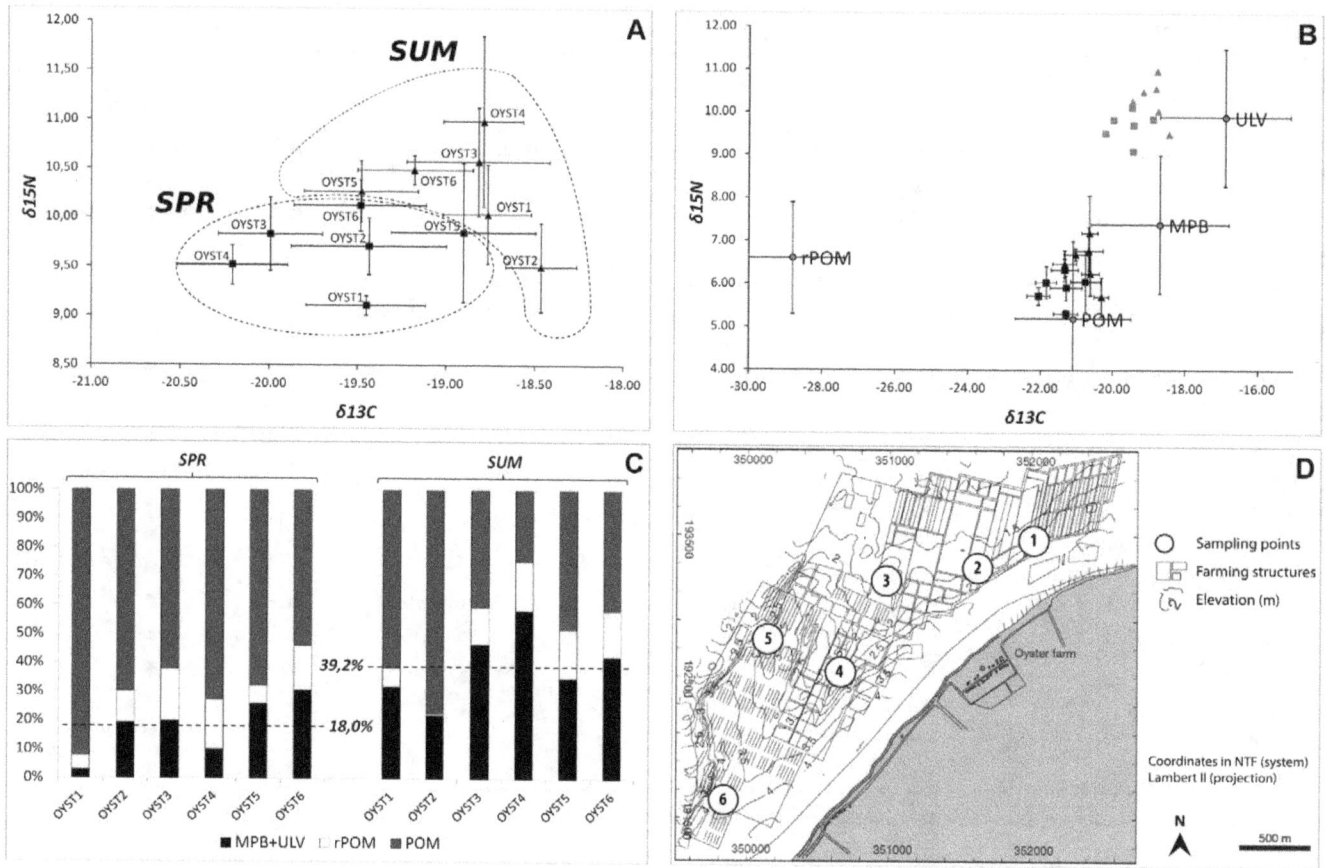

Figure 9. Temporal variations of δ13C and δ15N for *C. gigas* at 6 locations in the BDV. Isotopic signature (A), Isotopic signature before (gray) and after (black) fractionation (B), contribution of sources to oyster diets (C), location of oysters within the farming structures in the north-western part of the bay (D). Organic matter sources are plotted with standard deviations (see Lefebvre et al., 2009) to distinguish their relative contribution to the diets in the 2 sampling campaigns (B). Horizontal bars indicate the ±SD of the mean for n = 5.

Finally, this part of the bay could be dominated by epipsammic species, explaining the contrast between the null or low mud fraction and the high Chl *a* concentration in both spring and summer. Sandy sediments have been reported to show more diverse assemblages than muddy sediments, including epipsammic diatoms, euglenids and cyanobacteria [71].

Porter et al. [25] found that a low degree of tidal resuspension is also responsible for a general shift from phytoplankton primary production to microphytobenthic primary production (by a cascade of effects where light and nutrient availabilities are also involved). Such effects must be implicated in the ecological functioning of this farming zone and this general shift must be still reinforced by the biodeposition fluxes due to oysters. Among the 4 hypotheses mentioned above for explaining the high concentrations of MPB biomass in this zone, none of them can be really excluded. A combination of all these processes must interfere in interactions with tidal hydrodynamics, and it appears very delicate to disentangle the relative contribution of each of these processes. The reduction in pelagic Chl *a* concentration observed above the farming structures highlighted the filtration efficiency of oysters, which was also confirmed by the percentage of water phaeopigment. The latter variable provides an argument in favour of the direct production of pseudofeces after consumption of microalgae and/or resuspension events of easily erodible sediments with high biodeposits under the farming structures [66]. However,

the lack of match between benthic phaeopigments and both pelagic Chl *a* concentration and water phaeopigments suggests that it resulted mostly from the direct consumption of microalgae. These observations confirm the adequacy of the model proposed by Grangeré et al. [69], which represent ecosystem functioning with or without the presence of oysters, and revealed the prevailing effect of their top-down regulation in this area.

Since mollusk biomass was 20 times higher in summer than in spring, primary consumption would be expected to be higher in summer. However, the relative constancy of the MPB biomass levels between the two seasons suggests that the primary consumption is balanced by a higher primary production in summer. The eastern part of the bay was characterized by the absence of wild mollusks under farming structures, confirming the exclusion of suspension feeders already observed by Dubois et al. [70]. Suspension feeders may be disturbed by both biodeposition [72] and/or overconsumption of organic matter by the oysters. Thus, the lack of a correlation between the spatial patterns of MPB and macrofauna can be mainly attributed to the fact that most of the mollusk biomass was made up of suspension feeders and especially of *C. edule*, which widely dominated the mollusk assemblage, rather than deposit feeders that feed exclusively on MPB.

Impact of Resuspension for Higher Trophic Levels: Evidence for Allochtonous Feeding

Isotopic signatures showed that oysters consumed 2 times more MPB in summer than in spring, leading us to formulate three hypotheses: i) local feeding of autochtonous MPB directly associated with resuspended biodeposits under farming structures, ii) reduced phytoplankton abundance in late summer compared to spring, leading to a higher relative abundance of MPB in the potential food pool, iii) a higher resuspension at the bay scale and especially from the adjacent area in the south, that consequently supplies trophic resources to the cultivated oysters [73]. Regarding the limited differences between the two seasons in terms of benthic Chl a concentration, the first one can be ignored. Moreover, fluxes would not be expected to be very different since similar hydrodynamics and mollusk densities were found at both seasons under the farming structures, supporting the hypothesis of an allochtonous feeding of oysters. At 1 m depth Chl a concentration levels are similar in April and September, so that the second hypothesis seems unlikely. Taxonomic analyses revealed higher benthic-pelagic ratios for the 2 samplings in the adjacent area of the forming zone at the south/west, reinforcing the third hypothesis, which seems to be the most reliable one, when merging all data. A higher resuspension rate at the bay scale results from a change in the forcing variables, meaning one of the following compartments: hydrodynamics, sediment properties and/or macrofauna. Both samplings campaigns were conducted during spring tides with similar hydrodynamic conditions, and sediment properties revealed only slight differences between the two seasons. Higher resuspension could be explained by the huge increase in mollusk biomass between the two seasons. In fact, the cockles *Cerastoderma edule* have been shown to increase resuspension phenomena via bioturbation [74]. Therefore, the better coupling between benthic and pelagic compartments in spring can probably be explained by a predominance of physical factors. The increase in mollusk biomass/activity in summer drastically altered the balance between benthic and pelagic Chl a, with an increased role for biological factors in resuspension phenomena.

Conclusion

To better assess resuspension phenomenon at an ecosystem scale, a special effort was made to study benthic and pelagic variables at the same time, to better unravel causes of resuspension between biotic and abiotic factors. This *in situ* study is the first to analyze benthic-pelagic coupling at a bay scale in terms of body masses advection and trophic routes. The spatial heterogeneity of this ecosystem enabled the predominant physical and/or biological processes to be highlighted as a function of the area and/or season. The perennial structure observed at the scale of the whole bay provides evidence for the significant involvement of resuspension phenomena at the bay scale. Although physical factors appeared to predominate during winter/spring, in summer, biological factors can significantly increase exchanges between benthic and pelagic compartments when not considering the wind.

The use of a multicriteria approach (robust approach plus unusual indicators) to trophic/taxonomic indicators made it possible to strongly suggest a role for resuspension and benthic zonation in the spatial distribution of Chl a concentration in the water columns in two contrasting seasons and also that mollusks and particularly the cockle *Cerastoderma edule* play a role in microphytobenthic resuspension and its availability for oysters (*Crassostrea gigas*). When the biomass of these mollusks increases too much, this positive effect is masked by a high consumption rate leading to local depletion of Chl a concentration and SPiM. In fact, mollusk spatial distribution has a direct linkage with water Chl a concentration spatial patterns, which might have a structuring role on suspension-feeders. This highlights the fact that it is of critical importance to consider the connection between adjacent areas in terms of trophic relationships and microphytobenthos advection for farming structure management. These results underline the importance of taking biological phenomena into account in benthic-pelagic coupling to better evaluate the impact of resuspension for higher trophic levels.

Our study clearly suggests that there is not only a direct resuspension of microphytobenthos from the south of the bay but also an exportation from the water body of this habitat to another one at the north, then supplying food items to the cultivated oysters of the bay. Such trophic connections between adjacent habitat is of prime importance to consider because ecosystem models must consider these processes from the primary benthic production to the final consumption by suspension-feeders by including resuspension and advection in order to evaluate the real contribution of these areas as potential sink/sources of carbon [60].

Acknowledgments

We would particularly like to thank Jean-Paul Lehodey both for his help with innovative equipment during sampling and for his help in the field. We are grateful to Frédéric Guyon and Christophe Roger for their help in the field, and to Frank Maheux for his valuable help with organizing the water sampling. We thank Emmanuel Karakachian and Sébastien Lemaire for their help in the field and for their significant contribution to mollusk determination and grain-size analysis. We would also like to thank Pascal Claquin, Juliette Fauchot, Anne-Marie Russig, Isabelle Mussio, Clothilde Heudes, Olivier Desmur, Vincent Justome, Olivier Pierre-Duplessix, Emilie Rabiller, and Benjamin Simon as well as interns Olivier Goetz, Maxime Lafont for their help in the field. We are very grateful to Sandra Sritharan for her help with isotopic analysis, and to the GEOPHEN laboratory and especially to Laëtitia Birée for their help with grain-size analysis. We also thank METEOFRANCE for wind data. We would like to thank the two anonymous reviewers for their insightful and constructive comments.

Author Contributions

Conceived and designed the experiments: MU FO SL. Performed the experiments: MU FO AG KG. Analyzed the data: MU FO SL. Contributed reagents/materials/analysis tools: MU AG FO RL. Wrote the paper: MU.

References

1. Schelske CL, Odum EP (1962) Mechanisms maintaining high productivity in Georgia estuaries. Proc Gulf Caribb Fish Inst: 75–80.
2. Malet N, Sauriau P, Ryckaert M, Malestroit P, Guillou G (2008) Dynamics and sources of suspended particulate organic matter in the Marennes-Oléron oyster farming bay: Insights from stable isotopes and microalgae ecology. Estuar Coast Shelf Sci 78: 576–586.
3. Underwood G, Kromkamp J (1999) Primary production by phytoplankton and microphytobenthos in estuaries. Adv Ecol Res 29: 93–153.
4. De Jonge VN, Van Beusekom JEE (1995) Wind and tide induced resuspension of sediment and microphytobenthos in the Ems estuary. Limnol Oceanogr 40: 766–778.
5. Perissinotto R, Nozais C, Kibirige I, Anandraj A (2003) Planktonic food webs and benthic-pelagic coupling in three South African temporarily-open estuaries. Acta Oecol 24: 307–316.
6. Kang C, Lee Y, Choy E, Shin J, Seo I, et al. (2006) Microphytobenthos seasonality determines growth and reproduction in intertidal bivalves. Mar Ecol Prog Ser 315: 113–127.

7. Tolhurst TJ, Jesus B, Brotas V, Paterson DM (2003) Diatom migration and sediment armouring – an example from the Tagus Estuary, Portugal. Mar Ecol 503: 183–193.

8. Lucas CH, Widdows J, Brinsley MD, Salkeld PN, Herman PMJ (2000) Benthic-pelagic exchange of microalgae at a tidal flat: 1. Pigment analysis. Mar Ecol Prog Ser 196: 59–73.

9. Amos C, Feeney T, Sutherland T, Luternauer J (1997) The stability of fine-grained sediments from the Fraser River Delta. Estuar Coast Shelf Sci 45: 507–524.

10. Tolhurst TJ, Gust GM, Paterson D (2002) The influence of an extracellular polymeric substance (EPS) on cohesive sediment stability. Fine Sediment Dynamics in the Marine Environment. Elsevier, Vol. 5. 409–425.

11. Consalvey M, Jesus B, Perkins RG, Brotas V, Underwood GJC, et al. (2004) Monitoring Migration and Measuring Biomass in Benthic Biofilms: The Effects of Dark/far-red Adaptation and Vertical Migration on Fluorescence Measurements. Photosynth Res 81: 91–101.

12. Spears B, Saunders J, Davidson I, Paterson DM (2008) Microalgal sediment biostabilisation along a salinity gradient in the Eden Estuary, Scotland: unravelling a paradox. Mar Freshwater Res 59: 313–321.

13. Orvain F, Sauriau PG, Sygut A, Joassard L, Le Hir P (2004) Interacting effects of Hydrobia ulvae bioturbation and microphytobenthos on the erodibility of mudflat sediments. Mar Ecol Prog Ser 278: 205–223.

14. Austen I, Andersen TJ, Edelvang K (1999) The Influence of Benthic Diatoms and Invertebrates on the Erodibility of an Intertidal Mudflat, the Danish Wadden Sea. Estuar Coast Shelf Sci 49: 99–111.

15. Pillay D, Branch GM, Forbes AT (2007) Effects of Callianassa kraussi on microbial biofilms and recruitment of macrofauna : a novel hypothesis for adult – juvenile interactions. Mar Ecol Prog Ser 347: 1–14.

16. Koh CH, Khim JS, Araki H, Yamanishi H, Koga K (2007) Within-day and seasonal patterns of microphytobenthos biomass determined by co-measurement of sediment and water column chlorophylls in the intertidal mudflat of Nanaura, Saga, Ariake Sea, Japan. Estuar Coast Shelf Sci 72: 42–52.

17. Orvain F, Lefebvre S, Montepini J, Sébire M, Gangnery A, et al. (2012) Spatial and temporal interaction between sediment and microphytobenthos in a temperate estuarine macro-intertidal bay. Mar Ecol Prog Ser 458: 53–68.

18. Newell RIE, Cornwell JC, Owens MS (2002) Influence of simulated bivalve biodeposition and microphytobenthos on sediment nitrogen dynamics: A laboratory study. Limnol Oceanogr 47: 1367–1379.

19. van der Wal D, Wielemaker-van den Dool A, Herman PMJ (2010) Spatial Synchrony in Intertidal Benthic Algal Biomass in Temperate Coastal and Estuarine Ecosystems. Ecosystems 13: 338–351.

20. Serodio J, Coelho H, Vieira S, Cruz S (2006) Microphytobenthos vertical migratory photoresponse as characterised by light-response curves of surface biomass. Estuar Coast Shelf Sci 68: 547–556.

21. Colijn F, De Jonge V (1984) Primary production of microphytobenthos in the Ems-Dollard Estuary. Mar Ecol Prog Ser 14: 185–196.

22. Cibic T, Blasutto O, Falconi C, Fondaumani S (2007) Microphytobenthic biomass, species composition and nutrient availability in sublittoral sediments of the Gulf of Trieste (northern Adriatic Sea). Estuar Coast Shelf Sci 75: 50–62.

23. Blanchard GF, Guarini J-M, Orvain F, Sauriau P-G (2001) Dynamic behaviour of benthic microalgal biomass in intertidal mudflats. J Exp Mar Biol Ecol 264: 85–100.

24. Lucas C, Widdows J, Wall L (2003) Relating spatial and temporal variability in sediment chlorophyll a and carbohydrate distribution with erodibility of a tidal flat. Estuar Coast 26: 885–893.

25. Porter E, Mason R, Sanford L (2010) Effect of tidal resuspension on benthic-pelagic coupling in an experimental ecosystem study. Mar Ecol Prog Ser 413: 33–53.

26. Huot Y, Babin M, Bruyant F (2007) Does chlorophyll a provide the best index of phytoplankton biomass for primary productivity studies? Biogeosciences 4: 707–745.

27. de Jonge VD, van Beusekom JEE (1995) Wind-and tide-induced resuspension of sediment and microphytobenthos from tidal flats in the Ems estuary. Limnol Oceanogr 40: 766–778.

28. Kasim M, Mukai H (2006) Contribution of benthic and epiphytic diatoms to clam and oyster production in the Akkeshi-ko estuary. J Oceanogr 62: 267–281.

29. Dubois S, Orvain F, Marin-léal JC, Ropert M, Lefebvre S (2007) Small-scale spatial variability of food partitioning between cultivated oysters and associated suspension- feeding species, as revealed by stable isotopes. Mar Ecol Prog Ser 336: 151–160.

30. Lefebvre S, Marín Leal JC, Dubois S, Orvain F, Blin J-L, et al. (2009) Seasonal dynamics of trophic relationships among co-occurring suspension-feeders in two shellfish culture dominated ecosystems. Estuar Coast Shelf Sci 82: 415–425.

31. Choy EJ, Richard P, Kim K-R, Kang C-K (2009) Quantifying the trophic base for benthic secondary production in the Nakdong River estuary of Korea using stable C and N isotopes. J Exp Mar Biol Ecol 382: 18–26.

32. Carpenter SR, Elser MM, Elser JJ (1986) Chlorophyll production, degradation, and sedimentation: Implications for paleolimnology. Limnol Oceanogr 31: 112–124.

33. Spooner N, Harvey HR, Pearce GES, Eckardt CB, Maxwell JR (1994) Biological defunctionalisation of chlorophyll in the aquatic environment. II: Action of endogenous algal enzymes and aerobic bacteria. Org Geochem 22: 773–780.

34. Deroin J-P (2012) Combining ALOS and ERS-2 SAR data for the characterization of tidal flats. Case study from the Baie des Veys, Normandy, France. Int J Appl Earth Obs 18: 183–194.

35. Ettema CH, Wardle DA (2002) Spatial soil ecology. Trends Ecol Evol 17: 177–183.

36. Webster R, Welham SJ, Potts JM, Oliver MA (2006) Estimating the spatial scales of regionalized variables by nested sampling, hierarchical analysis of variance and residual maximum likelihood. Comput Geosci 32: 1320–1333.

37. Eleftheriou A, Holme NA (1984) Macrofauna techniques. N.A. Holme & A.D. McIntyre (Eds.), Methods for the study of marine benthos, Oxford, Blackwell Scientific. 140–216.

38. Underwood AJ, Chapman M (1996) Scales of spatial patterns of distribution of intertidal invertebrates. Oecologia 1996: 212–224.

39. Hammerstrom KK, Ranasinghe JA, Weisberg SB, Oliver JS, Fairey WR, et al. (2010) Effect of sample area and sieve size on benthic macrofaunal community condition assessments in California enclosed bays and estuaries. Integrated Environ Assess Manag: 1–10.

40. Hansen K, King GM, Kristensen E (1996) Impact of the soft-shell clam Mya arenaria on sulfate reduction in an intertidal sediment. Aquat Microb Ecol 10: 181–194.

41. Welschmeyer NA (1994) Fluorometric of chlorophyll a in the presence of analysis b and pheopigments chlorophyll. Limnol Oceanogr 39: 1985–1992.

42. Hayward PJ, Ryland JS (1990) The marine fauna of the British Isles and north West Wales. Clarendon Press, Oxford 1: 1–627.

43. Phillips DL, Gregg JW (2003) Source partitioning using stable isotopes: coping with too many sources. Oecologia 136: 261–269.

44. Dubois S, Jean-louis B, Bertrand C, Lefebvre S (2007) Isotope trophic-step fractionation of suspension-feeding species : Implications for food partitioning in coastal ecosystems. J Exp Mar Biol Ecol 351: 121–128.

45. Lorenzen CJ (1967) Determination of chlorophyll and phaeopigments: spectrophotometric equations. Limnol Oceanogr: 343–346.

46. Utermöhl von H (1931) Neue Wege in der quantitativen Erfassung des Planktons. (Mit besonderer Beriicksichtigung des Ultraplanktons). Verh Int Verein Theor Angew Limnol 5: 567–595.

47. Snoeijs P, Busse S, Potapova M (2002) The importance of diatom cell size in community analysis. J Phycol 38: 265–281.

48. Guarini J-M, Gros P, Blanchard G, Richard P, Fillon A (2004) Benthic contribution to pelagic microalgal communities in two semi-enclosed, European-type littoral ecosystems (Marennes-Oléron Bay and Aiguillon Bay, France). J Sea Res 52: 241–258.

49. Lazure P, Dumas F (2008) An external–internal mode coupling for a 3D hydrodynamical model for applications at regional scale (MARS). Adv Water Resour 31: 233–250.

50. Roberts JJ, Best BD, Dunn DC, Treml EA, Halpin PN (2010) Environmental Modelling & Software Marine Geospatial Ecology Tools : An integrated framework for ecological geoprocessing with ArcGIS, Python, R, MATLAB, and C++. Environ Modell Softw 25: 1197–1207.

51. Dolédec S, Chessel D (1994) Co-inertia analysis: an alternative method for studying species-environment relationships. Freshwater Biol 31: 277–294.

52. Perkins R (2003) Changes in microphytobenthic chlorophyll a and EPS resulting from sediment compaction due to de-watering: opposing patterns in concentration and content. Cont Shelf Res 23: 575–586.

53. Cartaxana P, Mendes C, Vanleeuwe M, Brotas V (2006) Comparative study on microphytobenthic pigments of muddy and sandy intertidal sediments of the Tagus estuary. Estuar Coast Shelf Sci 66: 225–230.

54. Ni Longphuirt S, Clavier J, Grall J, Chauvaud L, Leloch F, et al. (2007) Primary production and spatial distribution of subtidal microphytobenthos in a temperate coastal system, the Bay of Brest, France. Estuar Coast Shelf Sci 74: 367–380.

55. Brotas V, Cabrita T, Portugal A, Serodio J, Catarino F (1995) Spatio-temporal distribution of the microphytobenthic biomass in intertidal flats of Tagus Estuary (Portugal). Hydrobiologia 300–301: 93–104.

56. Saburova M, Polikarpov I (1995) Spatial structure of an intertidal sandflat microphytobenthic community as related to different spatial scales. Mar Ecol Prog Ser 129: 229–239.

57. van der Wal D, Herman P, Forster R, Ysebaert T, Rossi F, et al. (2008) Distribution and dynamics of intertidal macrobenthos predicted from remote sensing: response to microphytobenthos and environment. Mar Ecol Prog Ser 367: 57–72.

58. Honkoop PJC, Pearson GB, Lavaleye MSS, Piersma T (2006) Spatial variation of the intertidal sediments and macrozoo-benthic assemblages along Eighty-mile Beach, North-western Australia. J Sea Res 55: 278–291.

59. Lund-Hansen LC, Petersson M, Nurjaya W (1999) Vertical sediment fluxes and wave-induced sediment resuspension in a shallow-water coastal lagoon. Estuaries 22: 39–46.

60. Guarini J-M, Sari N, Moritz C (2008) Modelling the dynamics of the microalgal biomass in semi-enclosed shallow-water ecosystems. Ecol Model 211: 267–278.

61. Pannard A, Bormans M, Lagadeuc Y (2008) Phytoplankton species turnover controlled by physical forcing at different time scales. Can J Fish Aquat Sci 65: 47–60.

62. Safi KA (2003) Microalgal populations of three New Zealand coastal locations: forcing functions and benthic-pelagic links. Mar Ecol Prog Ser 259: 67–78.

63. Odebrecht C, Abreu PC, Fugita C, Bergesch B (2003) The Impact of Mud Deposition on the Long Term Variability of the Surf-Zone Diatom Asterionellopsis glacialis (Castracane) Round at Cassino Beach, Brazil. J Coastal Res 35: 486–491.

64. Rörig LR, Garcia VMT (2003) Accumulations of the surf-zone diatom *Asterionellopsis glacialis* (CASTRACANE) ROUND in Cassino Beach, Southern Brazil, and its Relationship with Environmental Factors. J Coastal Res 35: 167–177.

65. Waite AM, Thompson PA, Harrisson PJ (1992) Does energy control the sinking rates of marine diatoms? Limnol Oceanogr 37: 468–477.

66. Orvain F, Sauriau PG, Bacher C, Prineau M (2006). The influence of sediment cohesiveness on bioturbation effects due to *Hydrobia ulvae* on the initial erosion of intertidal sediments: a study combining flume and model approaches. J Sea Res 55: 54–73.

67. Cognie B, Barille L (1998) Does bivalve mucus favour the growth of their main food source, microalgae? Oceanol Acta 22: 441–450.

68. Bougrier S, Geairon P, Deslous-Paoli J, Bacher C, Jonquières G (1995) Allometric relationships and effects of temperature on clearance and oxygen consumption rates of *Crassostrea gigas* (Thunberg). Aquaculture 134: 143–154.

69. Grangeré K, Lefebvre S, Bacher C, Cugier P, Ménesguen A (2010) Modelling the spatial heterogeneity of ecological processes in an intertidal estuarine bay: dynamic interactions between bivalves and phytoplankton. Mar Ecol Prog Ser 415: 141–158.

70. Dubois S, Marin-Léal JC, Ropert M, Lefebvre S (2007) Effects of oyster farming on macrofaunal assemblages associated with *Lanice conchilega* tubeworm populations : A trophic analysis using natural stable isotopes. Aquaculture 271: 336–349.

71. Underwood GJC, Barnett M (2006) What determines species composition in microphytobenthic biofilms? Functioning of microphytobenthos in estuaries. Kromkamp J, editor. Microphytobenthos symposium. Amsterdam, The Netherlands: Royal Netherlands Academy of Arts and Sciences. 121–138.

72. Callier MD, Richard M, McKindsey CW, Archambault P, Desrosiers G (2009) Responses of benthic macrofauna and biogeochemical fluxes to various levels of mussel biodeposition: an in situ "benthocosm" experiment. Mar Pollut Bull 58: 1544–1553.

73. Kon K, Hoshino Y, Kanou K, Okazaki D, Nakayama S, et al. (2012) Importance of allochthonous material in benthic macrofaunal community functioning in estuarine salt marshes. Estuar Coast Shelf Sci 96: 236–244.

74. Ciutat A, Widdows J, Readman JW (2006) Influence of cockle *Cerastoderma edule* bioturbation and tidal-current cycles on resuspension of sediment and polycyclic aromatic hydrocarbons. Mar Ecol Prog Ser 346: 114–126.

Assessing the Health of the U.S. West Coast with a Regional-Scale Application of the Ocean Health Index

Benjamin S. Halpern[1,2,3*⁹], **Catherine Longo**[1⁹], **Courtney Scarborough**[1⁹], **Darren Hardy**[1,9⁹], **Benjamin D. Best**[1], **Scott C. Doney**[4], **Steven K. Katona**[5], **Karen L. McLeod**[6], **Andrew A. Rosenberg**[7], **Jameal F. Samhouri**[8]

1 National Center for Ecological Analysis and Synthesis, Santa Barbara, California, United States of America, 2 Bren School of Environmental Science and Management, University of California, Santa Barbara, California, United States of America, 3 Imperial College London, Silwood Park Campus, Ascot, United Kingdom, 4 Marine Chemistry & Geochemistry Department, Woods Hole Oceanographic Institution, Woods Hole, Massachusetts, United States of America, 5 Conservation International, Arlington, Virginia, United States of America, 6 COMPASS, Oregon State University, Department of Zoology, Corvallis, Oregon, United States of America, 7 Union of Concerned Scientists, Cambridge, Massachusetts, United States of America, 8 Conservation Biology Division, Northwest Fisheries Science Center, National Marine Fisheries Service, National Oceanic and Atmospheric Administration, Seattle, Washington United States of America, 9 Digital Library Systems & Services, Stanford University, Stanford, California, United States of America

Abstract

Management of marine ecosystems increasingly demands comprehensive and quantitative assessments of ocean health, but lacks a tool to do so. We applied the recently developed Ocean Health Index to assess ocean health in the relatively data-rich US west coast region. The overall region scored 71 out of 100, with sub-regions scoring from 65 (Washington) to 74 (Oregon). Highest scoring goals included tourism and recreation (99) and clean waters (87), while the lowest scoring goals were sense of place (48) and artisanal fishing opportunities (57). Surprisingly, even in this well-studied area data limitations precluded robust assessments of past trends in overall ocean health. Nonetheless, retrospective calculation of current status showed that many goals have declined, by up to 20%. In contrast, near-term future scores were on average 6% greater than current status across all goals and sub-regions. Application of hypothetical but realistic management scenarios illustrate how the Index can be used to predict and understand the tradeoffs among goals and consequences for overall ocean health. We illustrate and discuss how this index can be used to vet underlying assumptions and decisions with local stakeholders and decision-makers so that scores reflect regional knowledge, priorities and values. We also highlight the importance of ongoing and future monitoring that will provide robust data relevant to ocean health assessment.

Editor: James P. Meador, Northwest Fisheries Science Center, NOAA Fisheries, United States of America

Funding: Beau and Heather Wrigley generously provided the founding grant. Additional financial and in-kind support was provided by the Pacific Life Foundation, Thomas W. Haas Fund of the New Hampshire Charitable Foundation, the Oak Foundation, Akiko Shiraki Dynner Fund for Ocean Exploration and Conservation, Darden Restaurants Inc. Foundation, Conservation International, New England Aquarium, National Geographic, and the University of California Santa Barbara's National Center for Ecological Analysis and Synthesis, which supported the Ecosystem Health Working Group as part of the Science of Ecosystem-Based Management project funded by the David and Lucile Packard Foundation. Individual authors also acknowledge support from the U.S. National Science Foundation. The funders had no role in study design, data collection and analysis, decision to publish, or preparation of the manuscript.

Competing Interests: The authors have declared that no competing interests exist.

* Email: halpern@bren.ucsb.edu

⑨ These authors contributed equally to this work.

Introduction

As decision-makers shift towards more comprehensive approaches to managing ecosystems [1–3], management goals and targets increasingly focus on overall ecosystem health rather than on single sectors or stressors. This trend is particularly apparent for marine systems where efforts to implement ecosystem-based management (EBM) often have the stated objective of improving ocean health [3–6]. Along the United States west coast this emphasis exists in the regional governing body (West Coast Governor's Alliance on Ocean Health; [2]), NOAA's National Marine Sanctuaries' regular assessments of condition [7], ecosystem based approaches to fisheries management plans [8], and state-level and local efforts such as the Marine Life Protection Act, Puget Sound Action Agenda, and the west coast EBM network [9–11]. Until recently a standard tool to measure and track changes in

ocean health in a repeatable, transparent, quantitative and goal-driven manner was lacking, although it is key to informing management and policy [3,12–13]. Within the United States, both federal and state agencies must make decisions regarding changing ocean uses, new regulations, balancing needs of multiple stakeholders, and supporting coastal economies, along with many other issues. We developed the Ocean Health Index (hereafter, the Index) in part to help address these needs [14].

Public policy necessarily serves multiple interests and goals, such as biodiversity conservation, food production and many others, and thus relies on assessments of ecosystem health through the human lens of meeting societal goals and delivering desired benefits (see File S1). This perspective on ecosystem health is a departure from traditional conservationist views that focus on health as a measure of pristineness (recently debated by [15–17]). Assessments of ocean health are thus measured and bounded by

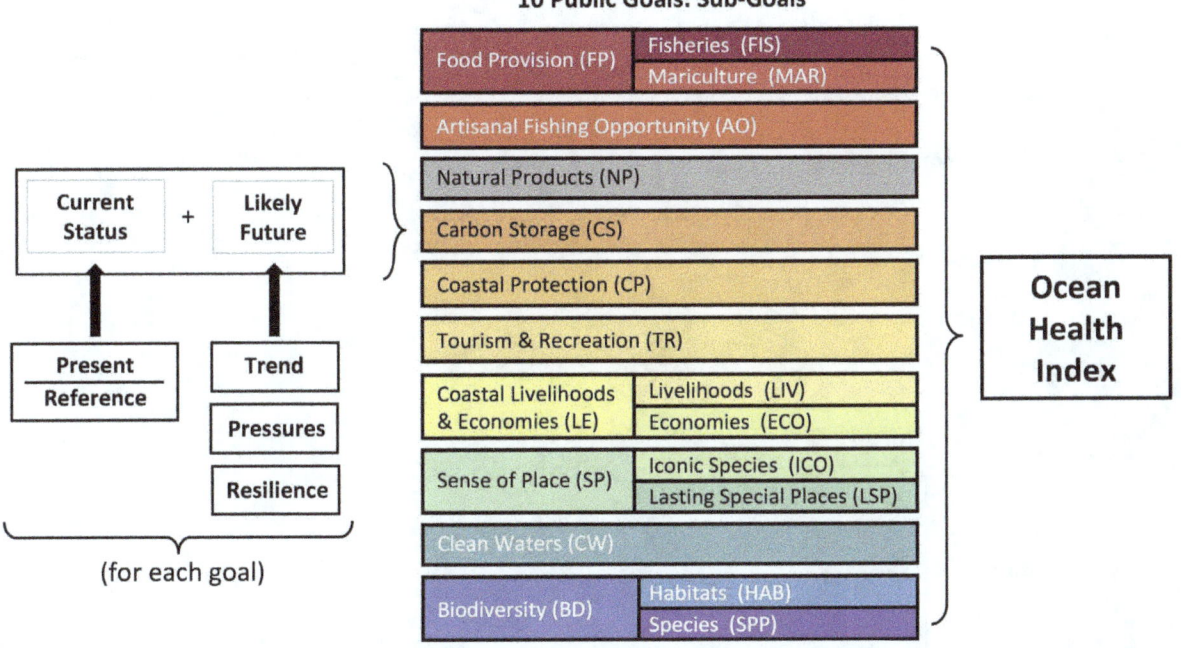

Figure 1. Schematic of the Ocean Health Index showing the 10 goals that comprise it, some with sub-goals, and broadly how each goal is calculated. Natural products is grey to indicate that it was not relevant in the U.S. west coast and thus not assessed.

human interactions with the ecosystem, not solely by the state of the natural ecosystem. A growing scientific literature also focuses on ecosystems as coupled social-ecological systems (e.g., [18–20]). We thus define a healthy ocean as one that sustainably delivers a range of benefits to people now and in the future (Fig. 1; [14]).

As a consequence of this human focus, individuals may perceive and understand ecosystems differently and have different views and sets of values that influence their assessment of ecosystem health. These differences require addressing questions such as: which aspects of ecosystem health are more or less important, how does one set reference points used to quantify health (e.g., ambitious versus practical), and which proxy measures can and cannot be used to estimate the status of each dimension of ecosystem health. Answering these kinds of questions exemplifies the types of inherently subjective decisions that must be made when developing an indicator of ecosystem health. The role of subjectivity cannot be ignored or avoided, but instead it should be made fully transparent, allowing subjective decisions to be modified to suit specific applications in a way that reflects regional values. The Ocean Health Index was designed in part to address this challenge with respect to measuring the health of marine ecosystems.

The Index has many potential applications. One, comparison of performance among regions, requires data consistency across regions. A second, comparison of ocean health within a region over time, makes using the best regionally (as opposed to globally) available data paramount. Halpern et al. [14] exemplifies the first type of assessment. Here we demonstrate an application of both types of assessment, exploring how well the Index performs at a regional scale in a relatively data-rich setting and how sub-regional comparisons might inform local and regional management decisions along the west coast of the United States.

We addressed three core questions: 1) How can the Index framework be adapted to a regional setting and make the best use of locally-available information, i.e., calculating components based on new and different data? 2) What is the health of the U.S. west coast and how is it changing? 3) How can the Index and its underlying framework tie into regional policy decisions? In addressing these questions we also hope to provide guidance and insight to other scientists and practitioners who may want to apply the Index to a new region.

Methods

Case study region

We calculated Index scores for five coastal sub-regions of the U.S. west coast –Washington, Oregon, Northern California, Central California, and Southern California – as well as an area-weighted average of these sub-regions to produce an overall regional score. These sub-regions were chosen based on a compromise between political and ecological boundaries. Our focus here is primarily on political boundaries, as most data are gathered and reported by agencies and organizations based on jurisdictional boundaries. However, because California's large geographical extent and ecological, social and economic diversity justified further subdivision, we used county boundaries, which closely align with biogeographic boundaries (Fig. 2), to define three sub-region boundaries. We did this because many data are reported at the county level, which facilitated calculating the Index scores within these sub-regions. The Index could be applied to ecological regions (such as marine ecoregions; [21]), but insufficient data are currently reported at these scales to make this feasible.

From a socio-economic perspective, the five sub-regions differ substantially (e.g., [22]). Southern California contains the heavily urbanized and densely populated coasts of San Diego and Los Angeles but also includes eight large, nearly uninhabited offshore islands. Central California is more sparsely populated but also includes the densely-populated San Francisco Bay. Northern California is even less densely populated, except in Sonoma county bordering San Francisco Bay. The coast of Oregon is uniformly

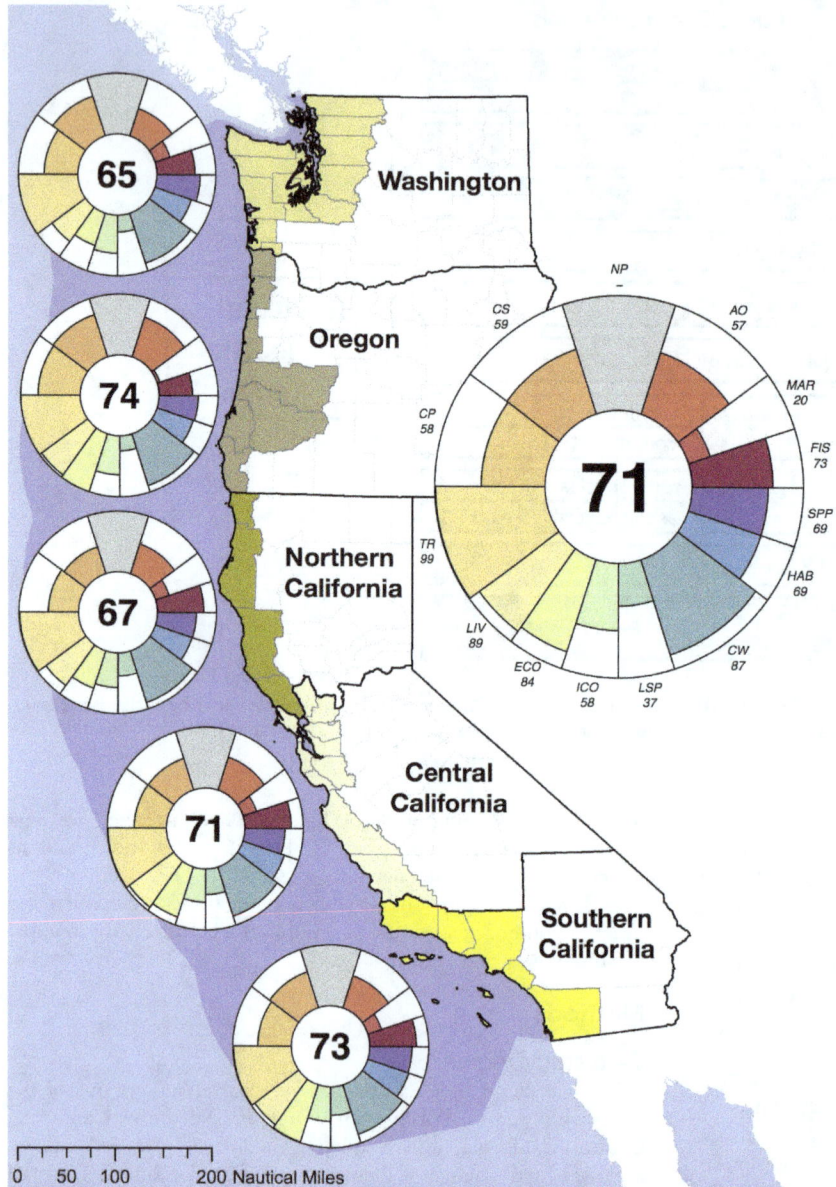

Figure 2. Map of the study region with each goal score per sub-region (left) and for the overall U.S. west coast (right). Each petal in the plots represents the score (radius) and weight (width) for the goal or sub-goal; see Fig. 1 for color legend and goal names. The number in the center is the overall Index score. Natural products is not assessed. Regions are depicted with coastal counties and the 200 nm exclusive economic zone is shaded in darker blue for reference only; regional scores are the area-weighted average of sub-region scores.

sparsely populated, with just 16,000 inhabitants in the largest town (Coos Bay). Washington state includes two distinct regions – the outer coast, which is similar to Oregon's coast, and Puget Sound, which includes many urbanized areas, most notably Seattle and Tacoma.

Index calculations

The Index is comprised of ten widely-held goals for healthy oceans that capture the different aspects of how people use, benefit from and value ocean ecosystems: food provision, artisanal fishing opportunities, natural products, carbon storage, coastal protection, sense of place, tourism & recreation, coastal livelihoods & economies, clean waters, and biodiversity (see Table 1 for goal definitions). These goals broadly map to the set of ecosystem services described by others (e.g., [23]), but have important

differences (such as inclusion of coastal livelihoods and economies, which is not an ecosystem service) that motivate calling them public goals rather than ecosystem services.

An overall Index score, I, is calculated as the weighted sum of the scores for each goal assessed in the Index ([14] and Fig. 1), such that:

$$I = \sum_{i=1}^{N} \alpha_i I_i, \tag{1}$$

where α is the importance (i.e. weight) placed on each goal i. For the U.S. west coast analysis we assumed equal weights for all N goals. In a previous application of the Index at the global scale, when a goal was not relevant to a location (for example, extractive goals for uninhabited islands or assessments of coral health in

Table 1. Details for the current status calculation of goals and sub-goals that comprise the Ocean Health Index.

Goal	Sub-goal	Definition	Reference point type	Reference point	Data used
Food provision	**Fisheries**	Harvest of sustainably caught wild seafood	Functional Relationship	Single species biomass at maximum sustainable yield (BMSY) and single species fishing mortality at maximum sustainable yield (FMSY)	B/BMSY and F/FMSY estimates from stock assessments; mean annual commercial catch per species
	Mariculture	Production of sustainably cultured seafood	Established Target	350% increase in production from 2005 levels, distributed evenly among farmable areas in all sub-regions	Tons of shellfish produced; areas deemed safe for mariculture farming by NOAA
Artisanal fishing opportunity		Opportunity to engage in artisanal-scale fishing for subsistence or and/or recreation	Established Target	**Physical Access:** One coastal access points per mile of coastline	**Physical Access:** Number of coastal access points per mile
				Economic Access: No increase in the ratio of fuel price to median income over a five-year period	**Economic Access:** Change in gas price over time
				Resource Access: Perfect sustainability score for all fish stocks	**Resource Access:** Condition of fish stocks as measured by the NOAA fish Stock Sustainability Index (FSSI)
Natural products		Sustainable harvest of natural products, such as shells, algae, and fish oil used for reasons other than food provision	N/A	N/A	N/A
Carbon storage		Conservation status of natural habitats affording long-lasting carbon storage	Temporal Comparison (historical benchmark)	**Salt Marshes:** 50% of historical areal extent (roughly since 1850s)	**Salt Marshes:** Areal extent in 2006, 2002, 1996, and the 1850s
				Seagrasses: Zero pressure to coastal areas from nutrient input	**Seagrasses:** Nutrient input model applied within the 100 m depth contour
Coastal protection		Conservation status of natural habitats affording protection of the coast from inundation and erosion	Temporal Comparison (historical benchmark)	**Salt Marshes:** 50% of historical areal extent (roughly since 1850s)	**Salt Marshes:** Areal extent in 2006, 2002, 1996, and the 1850s
				Sand Dunes: 100% of the areal extent in 1960	**Sand Dunes:** Areal extent in 2006, 2002, 1996, and 1960
				Seagrasses: Zero pressure to coastal areas from nutrient input	**Seagrasses:** Nutrient input model applied within the 100 m depth contour
Tourism & recreation		Opportunity to enjoy coastal areas for recreation and tourism	Temporal Comparison (moving target)	No net loss in participation in marine-related activities over a 10 year period	Model of per capita participation rates in 19 marine-related activities based on demographic variables
Coastal livelihoods & economies	**Coastal livelihoods**	Jobs and wages from marine-related sectors	Temporal Comparison (historical benchmark)+Spatial Comparison	**Jobs:** No net loss in the number of jobs in marine-related sectors relative to all job sectors in each region over a five-year period	Jobs and wages data for 20 marine-related sectors; total jobs (marine and non-marine sectors)
				Wages: Highest per capita average annual wages across all regions and marine sectors	
	Coastal economies	Revenues from marine-related sectors	Temporal Comparison (moving target)	No net loss in revenue in marine-related sectors relative to all economy sectors over a five-year period	Revenue data for 20 marine-related sectors; total revenue (marine and non-marine sectors)

Table 1. Cont.

Goal	Sub-goal	Definition	Reference point type	Reference point	Data used
Sense of place	Iconic species	Cultural, spiritual, or aesthetic connection to the environment afforded by iconic species	Established Target	All assessed species coservation status classified as of least concern	Species conservation status as determined by NatureServe criteria
	Lasting special places	Cultural, spiritual, or aesthetic connection to the environment afforded by coastal and marine places of significance	Established Target	30% of all marine and terrestrial areas protected	Marine and terrestrial areas protected and managed for conservation
Clean waters		Clean waters that are free of nutrient and chemical pollution, marine debris and pathogens	Established Target	Zero marine debris, nutrient run-off, beach closures due to pathogens, and chemical contaminants in sediments and bivalve tissue	Nutrient plume models; beach closure data; beach clean-up data; concentration of chemicals in sediment and bivalve tissue samples
Biodiversity	Habitats	The existence value of biodiversity measured through the conservation status of habitats	Temporal Comparison (historical benchmark)	**Salt Marshes:** 50% of historical areal extent (roughly since 1850s)	**Salt Marshes:** Areal extent in 2006, 2002, 1996, and the 1850s
				Sand Dunes: 100% of the areal extent in 1960	**Sand Dunes:** Areal extent in 2006, 2002, 1996, and 1960
				Seagrasses: Zero pressure to coastal areas from nutrient input	**Seagrasses:** Nutrient input model applied within the 100 m depth contour
				Soft-bottom: Zero pressure from bottom trawl fishing	**Soft-bottom:** Amount of fish caught using bottom-trawl methods and location of current soft-bottom habitats
	Species	The existence value of biodiversity measured through the conservation status of marine-associated species	Established Target	All assessed species extinction risk status classified as of least concern	Species extinction risk status as determined by IUCN criteria

countries that do not have coral reefs), it was dropped from the assessment as it was deemed not applicable. In other words, in these cases the community assigns no value to the goal (i.e., the goal weight is zero; [14]), thus resulting in its irrelevance to the overall assessment. In this U.S. west coast application, we exclude the natural products goal because for most products there is no recorded trade within the region, even though it likely occurs at small scales, and for kelp, limited commercial harvest existed in Southern and Central California but no longer occurs (for unknown reasons). We further explain our rationale for excluding this goal in File S1.

Goal scores were calculated as the average of current (x_i) and likely future status ($\hat{X}_{i,F}$) Current status was measured as the present value (X_i) relative to a reference value ($X_{i,R}$), such that $x_i = X_i / X_{i,R}$. Likely future status was measured as current status modified by the recent trend (T), cumulative pressures (p), and resilience (r), such that:

$$\hat{x}_{i,F} = (1+\delta)^{-1}[1 + \beta T_i + (1-\beta)(r_i - p_i)] \cdot x_i, \qquad (2)$$

where δ is the discount rate ($\delta = 0$) and β is the relative importance of trend versus the difference between pressures and resilience in determining the likely future status (we assumed $\beta = 0.67$, following Halpern et al. [14]). Note that the likely future status

does not predict the future, but only estimates what the status score is likely to be approximately 5 yr hence, given what is known today about recent trends and the counterbalance of pressure versus resilience metrics. Reference points for each goal are described in Table 1 (and further detailed in File S1). These reference points are similar but not equivalent to management targets; a reference point as we define it here is the maximum sustainable level of production of each goal. In some cases management may choose a target different than these reference points for practical or sociopolitical reasons [24]. Resilience measures focused on the presence of relevant institutions, but in general could not evaluate their effectiveness due to lack of such data. Details of goal models and parameters are provided in Table 2.

Changes to Index calculations for regional application

As compared with the global application of the Index [14], higher resolution data and longer time series, along with a better understanding of the regional context, allowed for improved approaches to modeling and setting reference points for many goals. Briefly, the following changes were made (see also File S1):

(1) **Higher resolution data.** We used regional-scale data wherever possible, relying on national data used in Halpern et

Table 2. Models and parameter used to calculate each goal and sub-goal.

Goal	Sub-Goal	Status Model Equations	Variables
Food Provision (x_{FP})		$x_{FP} = w_{FP} \cdot x_{FIS} + (1 - w_{FP}) \cdot x_{MAR}$	
		$w_{FP} = \frac{C_T}{C_T + Y_r}$	Y_r = Total sustainable mariculture harvest; C_T = Total current wild-caught fishing yield; w_{FP} = weight per seafood sector
		$x_{FIS} = \sum_i w_i \cdot \frac{F'_i + B'_i}{2}$	w_i = weight per stock i; F = current fishing mortality of stock i; B = current biomass of stock i
		$w_i = \frac{\bar{C_i}}{\sum \bar{C}}$	$\bar{C_i}$ = mean catch of stock i throughout the time-series
	Fisheries (x_{FIS})	$B' = \begin{cases} \dfrac{\frac{B}{Bmsy}}{0.8} & \text{when } \frac{B}{Bmsy} < 0.8 \\ 1 & \text{when } 0.8 \le \frac{B}{Bmsy} < 1.5 \\ \dfrac{3.35 - \frac{B}{Bmsy}}{1.8} & \text{when } \frac{B}{Bmsy} \ge 1.5 \end{cases}$	$Bmsy$ = biomass of stock i producing maximum sustainable yield
		$F' = \begin{cases} 0 & \text{when } \frac{B}{Bmsy} < 0.8 \text{ and } \frac{F}{Fmsy} > \frac{B}{Bmsy} + 1.5 \\ \dfrac{\frac{F}{Fmsy}}{\frac{B}{Bmsy} - 0.2} & \text{when } \frac{B}{Bmsy} < 0.8 \text{ and } \frac{F}{Fmsy} < \frac{B}{Bmsy} - 0.2 \\ \dfrac{\frac{B}{Bmsy} + 1.5 - \frac{F}{Fmsy}}{1.5} & \text{when } \frac{B}{Bmsy} < 0.8 \text{ and } \frac{B}{Bmsy} + 0.2 < \frac{F}{Fmsy} < \frac{B}{Bmsy} + 1.5 \\ 1 & \text{when } \frac{B}{Bmsy} < 0.8 \text{ and } \frac{B}{Bmsy} - 0.2 \le \frac{F}{Fmsy} < \frac{B}{Bmsy} + 0.2 \\ 1 & \text{when } \frac{B}{Bmsy} \ge 0.8 \text{ and } 0.8 \le \frac{F}{Fmsy} < 1.2 \\ \dfrac{\frac{F}{Fmsy}}{0.8} & \text{when } \frac{B}{Bmsy} \ge 0.8 \text{ and } \frac{F}{Fmsy} < 0.8 \\ \dfrac{2.5 - \frac{F}{Fmsy}}{1.3} & \text{when } \frac{B}{Bmsy} \ge 0.8 \text{ and } \frac{F}{Fmsy} \ge 1.2 \end{cases}$	$Fmsy$ = Fishing mortality that yields long term maximum sustainable yield of stock i
	Mariculture (x_{MAR})	$x_{MAR} = \frac{Y_C}{Y_r}$	Y_C = Current total sustainable mariculture harvest
		$Y_{C_k} = \sum_i Y_{k,i} \cdot S_i$	S = Sustainability coefficient for species i; k = sub-region
		$Y_r = \frac{FA_k}{FA_T} \cdot 3.5 \cdot Y_{2005}$	FA = farmable area; Y_{2005} = yield in 2005
Artisanal Fishing Opportunities (x_{AO})		$x_{AO} = \dfrac{\left(\frac{AP_c}{AP_r}\right) + \left(\frac{FSSI}{FSSI_r}\right) + AE_i}{3}$	AP = average distance between coastal access points; c = current; r = reference; $FSSI$ = catch-weighted average NOAA Fish Stock Sustainability Index score
		$AE_i = \left(\frac{G_r}{I_r} - \frac{G_c}{I_c}\right)$	AE = economic access; G = gas price per gallon; I = median income
Natural Products (x_{NP})		Not assessed for this region	
Carbon Storage (x_{CS})		$x_{CS} = \sum_j \left(\frac{C_{c,j}}{C_{r,j}} \cdot \frac{A_j}{A_T}\right)$	A_k = area covered by habitat j; A_T = total area covered by all habitats; C_j = condition of habitat j
Coastal Protection (x_{CP})		$x_{CP} = \sum_{j=1}^{J} \left(\alpha_j \cdot \frac{C_{c,j}}{C_{r,j}}\right)$	r_j = protective ability rank of habitat j
		$\alpha_j = \frac{w_j \cdot A_j}{\sum_j (w_j \cdot A_j)}$	
		$w_j = \frac{r_j}{\sum_j r_j}$	
Coastal Livelihoods & Economies (x_{LE})		$x_{LE} = \frac{x_{LIV} + x_{ECO}}{2}$	

Table 2. Cont.

Goal	Sub-Goal	Status Model Equations	Variables
	Livelihoods (x_{LIV})	$x_{LIV} = \dfrac{\left(\dfrac{\sum_z j_{c,z}}{\sum_z j_{r,z}} + \dfrac{\sum_z g_{k,z}}{\sum_z g_{r,z}}\right)}{2}$	j = adjusted number of direct and indirect jobs within sector z; g = average PcPPP-adjusted wages within sector z; c = current year; r = reference year (j, e) or reference location (for g)
	Economies (x_{ECO})	$x_{ECO} = \sum_z \dfrac{e_{c,z}}{e_{r,z}}$	e = total adjusted revenue generated directly and indirectly from sector z
Tourism and Recreation (x_{TR})		$x_{TR} = \dfrac{\sum_{i=1}^{19} P_{c,i}}{\sum_{i=1}^{19} P_{r,i}}$	P_c = current predicted participation in each recreation activity i (of 19); P_r = observed participation in recreation activity i in year 2000
Sense of Place (x_{SP})		$x_{SP} = \dfrac{x_{ICO} + x_{LSP}}{2}$	
	Iconic Species (x_{ICO})	$x_{ICO} = \dfrac{\sum_{l=1}^{6} S_l \cdot w_l}{\sum_{l=1}^{6} S_l}$	l = IUCN threat category; S_m = number of assessed iconic species in category l; w_l = weight per threat category l
	Lasting Special Places (x_{LSP})	$x_{LSP} = \dfrac{\dfrac{MPA_{3\,nm}}{0.3*A_{3\,nm}} + \dfrac{MPA_{EEZ}}{0.3*A_{EEZ}} + \dfrac{TA_{PA}}{0.3*TA}}{3}$	MPA = fully protected marine area; EEZ = offshore waters (3–200 nm); $3\ nm$ = coastal waters (0–3 nm); TA = area on coastal land (0–1 mi); TA_{PA} = protected area on coastal land
Clean Waters (x_{CW})		$x_{CW} = \sqrt[4]{a \cdot u \cdot l \cdot d}$	a = population without access to sanitation relative to global maximum; $u = 1 -$ nutrient inputs; $l = 1 -$ chemical inputs; $d = 1 -$ marine debris
Biodiversity (x_{BD})		$x_{BD} = \dfrac{x_{SPP} + x_{HAB}}{2}$	
	Species (x_{SPP})	$x_{SPP} = \dfrac{\sum_M \left(\dfrac{\sum_N w_{i,k}}{N}\right) \cdot A_c}{A_T}$	n = number species per grid cell c; m = number of grid cells in the assessment region; A_c = total area of grid cell c; A_T = total area of the assessment region
	Habitats (x_{HAB})	$x_{HAB} = \sum_j \left(\dfrac{C_{c,j}}{C_{r,j}} \cdot \dfrac{A_j}{A_T}\right)$	See variables above

See File S1 for details on data and rationales for each.

al. [14] in only 20% (10 of 49) of data layers. In particular, we were able to use higher resolution data from local sources for the status and trend calculation of all goals. Most of the pressure layers, and all of the resilience layers, were also calculated using local data sets.

(2) **Models adapted to better represent regional goals for ocean health**. For several goals we modified the approach to assessing current status based on higher-quality regional data. (a) For food provision derived from wild-caught fisheries, we were able to use formal stock assessments routinely employed for local fisheries management to capture the status of major commercially-caught species. These estimates are derived from complex models developed by working groups of experienced local experts. In contrast, the global analysis had to rely on models requiring many simplifying assumptions, leading to higher uncertainty. (b) For food provision derived from mariculture, we improved our estimate of potential sustainable productivity by assuming cultivation could only increase in areas already under production. In contrast, the global analysis assumed potential

productivity scaled to total coastal area and highest observed production density, i.e. China. (c) For the tourism & recreation goal, we were able to use information on participation rates in a range of coastal and marine tourism and recreational activities. Participation rates more closely match the intent of this goal and are a more robust proxy than the international tourist arrivals data used in the 2012 global study [14] and are a more direct measure than the tourism employment proxy used in the 2013 global study [25]. We also changed the reference point from spatial (used in the global analysis) to temporal, because adequate time-series data were available.

(3) **Reference points based on U.S. west coast priorities.** (a) For the mariculture sub-goal, the reference point was based on regional projections of nationally-desired economic and food security targets. (b) For the habitat health scores used in coastal protection, carbon storage, and the habitats sub-goal of biodiversity, we used reconstructions of historic extents, rather than recently recorded trends, to set targets that were more ambitious than in the global analysis. (c) For the lasting

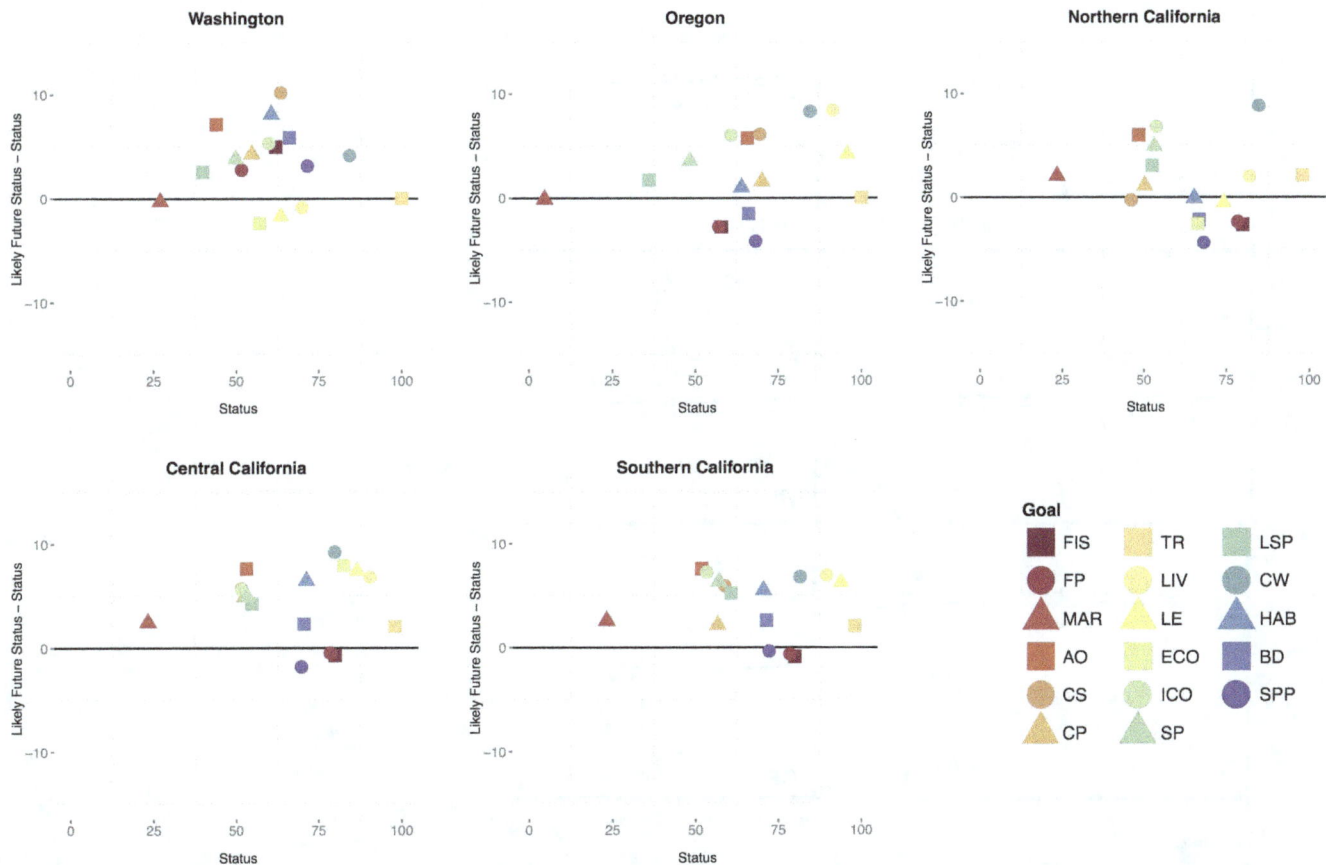

Figure 3. Current status versus the difference between likely future and current status for each goal and sub-goal within each sub-region. Values above the y-axis indicate the likely future status is greater than the current status. Note that y-axis is scaled −10 to 10.

special places sub-goal, in addition to evaluating protected areas inland and within 3 nm of the coast, we included a third zone (3 to 200 nm offshore) because we assume assessment of places offshore as well as nearshore is important to people in this region.

We conducted a number of analyses to assess how results were affected by various assumptions and data constraints (File S1), three of which we focus on here. First, to assess how results changed when goals were modeled differently, we calculated the regionally-modified goals using methods from the global study for comparison, when possible. Second, we assessed how alternate reference points modified goal scores, in particular for the mariculture sub-goal (see File S1). Finally, we assessed the consequences of using empirically-derived unequal weights, elicited from regional experts representing a diverse cross-section of stakeholders [26], for combining goals into a single Index score (Table S11 in File S1). Experts weighted sense of place and clean waters goals highest, and three to four times more heavily than the tourism & recreation and coastal livelihoods & economies goals, the two lowest-weighted goals (Table S11 in File S1).

Although we were able to estimate past status values for all goals and sub-goals except tourism and recreation, iconic species, and species diversity (which had insufficient time series; see File S1), most pressure and resilience metrics were not available for past time periods, precluding calculation of the 'likely future status', and thus the overall Index scores, for past years.

Scenario analyses

To further explore how the Index could be used within typical regional-scale decision contexts and to illustrate how the Index responds to typical management actions, we simulated several scenarios and recalculated the overall Index. The intent of this analysis was not to model precise changes but rather to illustrate expected types and relative magnitudes of change across goals. Rather than being prescriptive, these scenarios were chosen to illustrate how one can use the Index to explore consequences of management decisions. We recognize that realistic implementation would require engagement with decision-makers, normative decisions about management goals, fine-tuning of assumptions, and model-based simulations of future conditions. Our heuristic scenarios assessed what scores would be if 1) regulations had been adopted 5 years ago that successfully reduced land-based runoff of nutrients and pollutants each by 25%, 2) habitat restoration activities had been successfully implemented such that coastal wetlands and sand dunes were increased in extent by 10%, and 3) the Marine Life Protection Act (MLPA) process in California had not occurred and the currently-existing network of MPAs therefore had not been established within the state. For this third scenario we implemented three successive versions intended to measure changes over time in how the system, and thus the Index, would respond (for details see File S1).

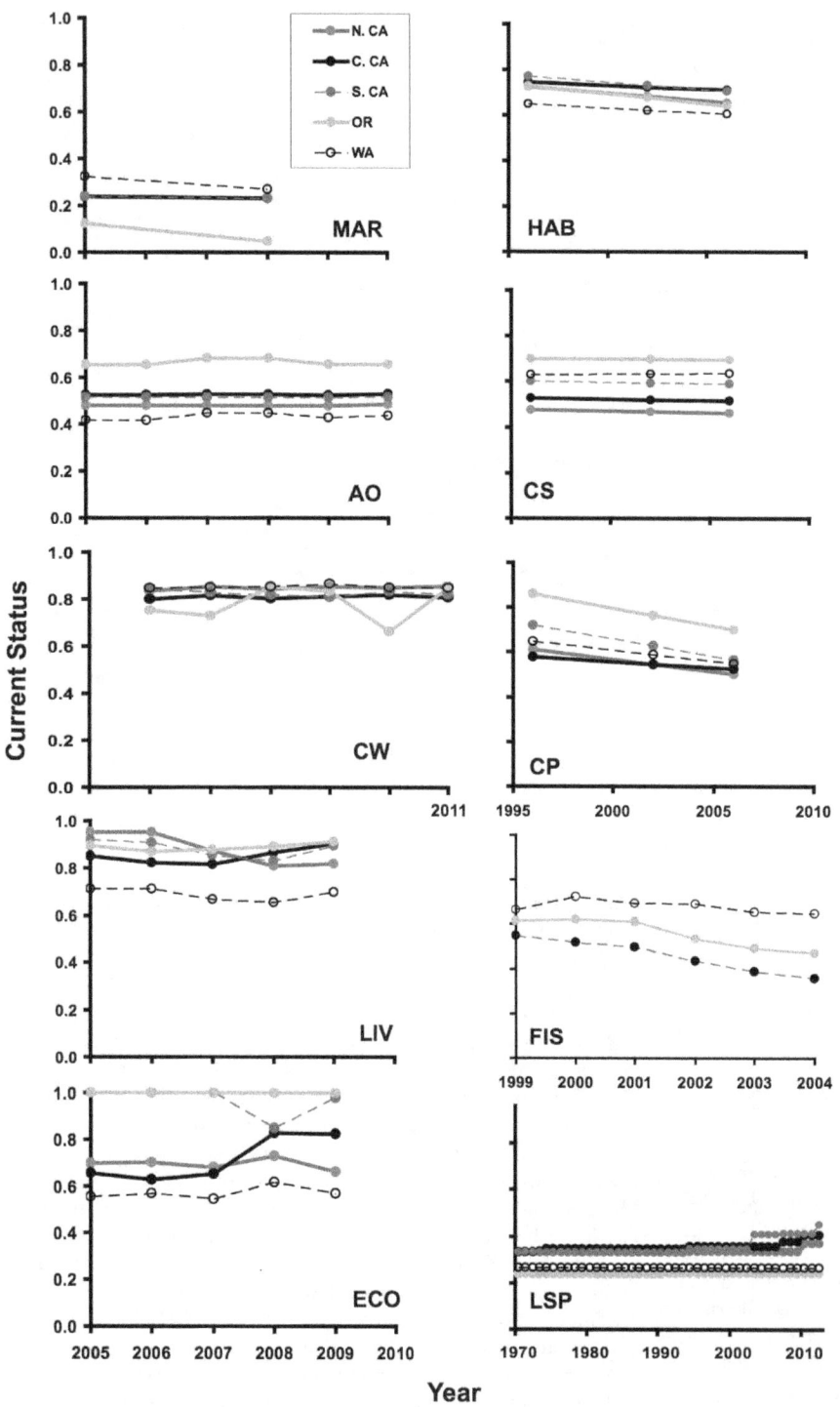

Figure 4. Time series of current status scores for goals and sub-goals with available historical data. Note different time scales on x-axes in right-hand plots. Plots are for the habitat sub-goal (HAB), carbon storage (CS), coastal protection (CP), artisanal fishing opportunities (AO), mariculture sub-goal (MAR), clean waters (CW), fisheries sub-goal (FIS), coastal livelihoods sub-goal (LIV), coastal economies sub-goal (ECO), and the lasting special places sub-goal (LSP). FIS could not be assessed for sub-regions within California and so a single state-level result is presented in that case.

Results

Overall the U.S. west coast scored 71 out of 100. Washington scored lowest of all sub-regions (65), with increasingly higher scores in Northern California (67), Central California (71), Southern California (73), and Oregon (74; Fig. 2). Goal scores

varied from 22 (mariculture) to 99 (tourism and recreation), with tourism and recreation and clean waters scoring highest, and carbon storage, coastal protection, lasting special places and mariculture scoring lowest for all 5 sub-regions (see also Table S33 in File S1). Despite biophysical and socioeconomic differences

Table 3. Changes in goal scores for which sensitivity analyses were conducted using alternate methods for calculating goal status.

Region	Fisheries		Mariculture					Tourism and Recreation		Artisanal Fishing Opportunity	
	Orig without data poor stocks	Alt with data poor stocks	Orig federal target	Alt1 spatial ref; global target	Alt2 spatial ref; national target	Alt3 temporal ref	Alt4 production function	Orig temporal ref	Alt spatial ref	Orig regional model	Alt global model
Washington	64	55	27	1	81	80	10	100	55	47	61
Oregon	56	54	5	0.1	6	0	1	100	41	69	61
California	79	65		0.1	9	96	19	99	91		
N. California			24							51	59
C. California			24							57	58
S. California			25							55	59

Results are reported for each sub-region separately when possible. 'Orig' is the original approach used for reporting main results; 'Alt' is the alternative approach used for sensitivity analyses. Separate analyses for each sub-region within California were only possible for the artisanal fishing opportunity goal and for the original mariculture goal; results for other cases are reported for all of California as a single value. For mariculture we tested four different alternate reference points. See File S1 for details.

among sub-regions, overall Index scores for the sub-regions were within a 7 point range. Differences in scale, available data, and goal methodologies for several goals preclude direct quantitative comparison of these results to global scores [14].

For nearly every goal in each of the five sub-regions, likely near-term future scores were greater than those for the present (Fig. 3). Although the likely future state is strongly influenced by the recent trend (see Eq. 2), and the recent trend for many goals was (slightly) negative, the likely future status is also influenced by the balance between resilience measures and cumulative pressures, and in many cases resilience was greater than pressures (Table S34 in File S1). In only 24% of cases (11 of 45) were likely future scores worse, namely coastal livelihoods & economies in Washington; fisheries and species biodiversity in Oregon; fisheries, species biodiversity, carbon storage, and coastal livelihoods & economies in Northern California; and species biodiversity and fisheries in Central and Southern California. The potential future declines in the health of the fisheries (food provision) and species (biodiversity) sub-goals, despite significant resources being committed to their improvement in the region, is largely due to recent declines creating a negative trend (see below, Fig. 4, and Table S34 in File S1).

The likely future status does not tell the whole story, however. For goals and sub-goals with sufficient data to calculate past values of the current status, different patterns emerged. Habitat-based goals, notably habitat diversity and coastal protection, showed declines (8–17% respectively) across all sub-regions over the past ten years (Fig. 4). Coastal livelihoods and economies showed initial small declines in some sub-regions but recent recovery in many cases (Fig. 4). Because these values are standardized to remove broader economic patterns, this result suggests stronger effects of the global recession that began in 2008 and slower economic recovery in marine sectors compared to other sectors. Lasting special places showed recent improvements, in large part because of California's MPA initiatives, while remaining goals showed little recent change (Fig. 4). Because risk status of most species is rarely assessed more than once, we were not able to calculate past status scores for species diversity or iconic species sub-goals.

Comparisons of results obtained for several goals when assessed with the previous global approaches versus the refined regional approaches showed important differences. For the tourism and recreation goal, scores from the regional analysis were considerably higher than those obtained applying the global model, most likely reflecting both use of more informative data on participation rates instead of international tourist arrival data and the choice of a local temporal reference point instead of an across-region spatial one (Table 3). For large countries with coasts spanning sizable

Table 4. Changes in Index scores for each subregion and the U.S. West coast with goals weighted equally or unequally based on regionally-specific, empirically-derived preferences (Halpern et al. 2013b).

Region	Equal	Unequal
Washington	65	66
Oregon	74	74
N. California	67	66
C. California	71	69
S. California	73	71
U.S. west coast	71	70

biophysical gradients or bordering different oceans, such as the United States, sub-national assessments such as the one here are likely to produce Index scores that differ from those derived from the national-level global assessment. In contrast, the choice of a very different approach to modeling the artisanal fishing opportunity goal had relatively small effects on resulting scores (Table 3). Changes in the mariculture reference point significantly increased scores relative to the global approach, with highly variable results using other methods for setting reference points (Table 3). Unequal goal weighting [26], which represents one example of how people value goals differently, produced lower Index scores for some sub-regions and higher scores for others (Table 4).

The three management scenarios showed goal score changes from about −12% up to +11% depending on the goal and scenario (Fig. 5). For example, because land-based pollution is a pressure on nearly every goal, hypothetical decreases in this

stressor led to modest increases in most goals (scenario 1). Simulated habitat restoration had a relatively large effect on habitat-based goals, a result that was influenced by choice of habitat reference points (see File S1, scenario analysis section), but not on other goals (scenario 2). The three versions of scenario 3 illustrate how an initial action (or in this case hypothetical lack thereof) could have cascading effects across multiple goals that may lead to increases in some goals and decreases in others. In the example here, the hypothetical removal of MPAs decreased the lasting special places score, increased the food provision score (through increased fishing), and decreased scores for other goals as a result of increased fishing pressure.

Discussion

By calculating the Index for the U.S. West coast, we were able to take advantage of regional data and knowledge of the system to

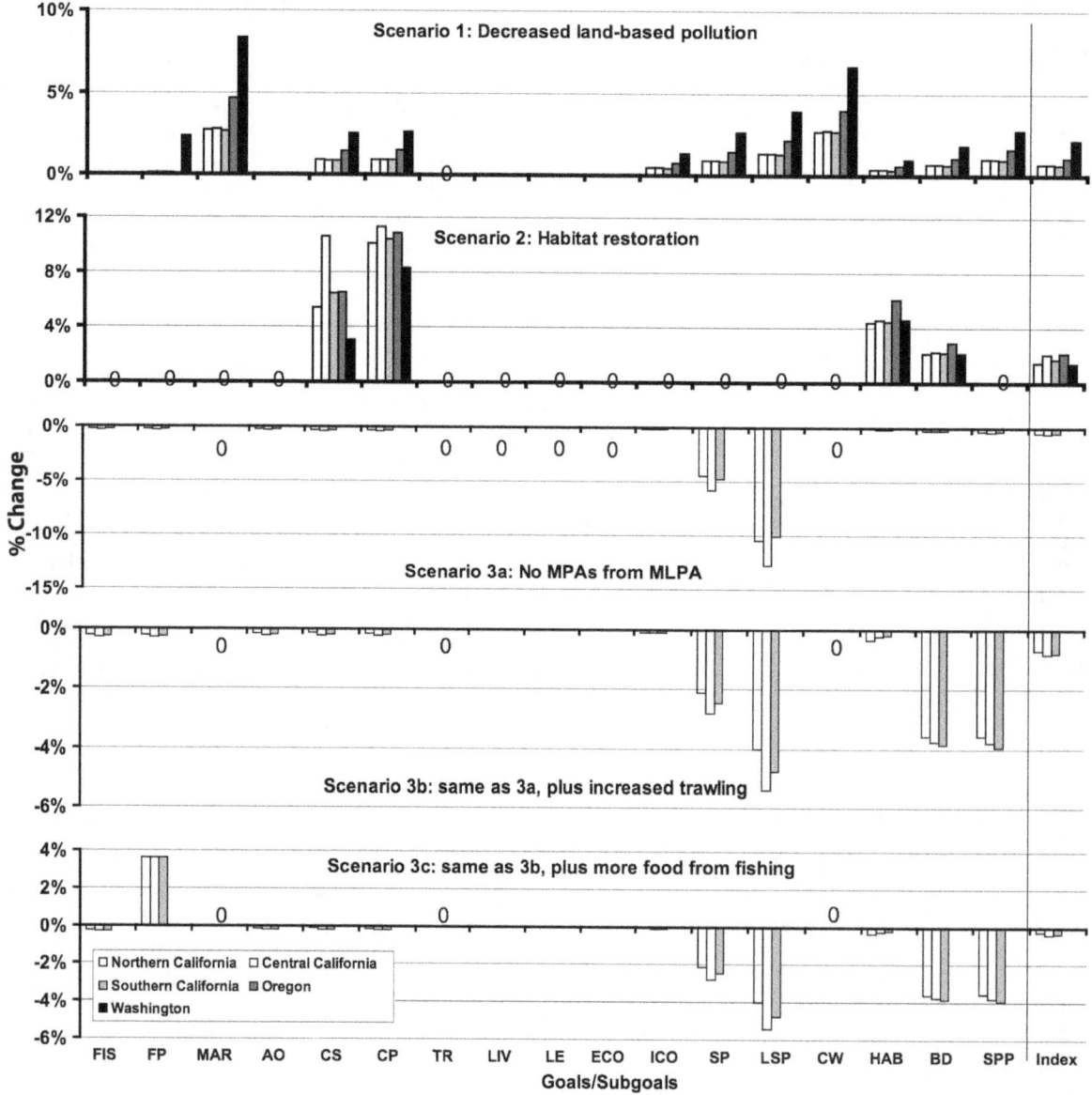

Figure 5. Scenario results as percent change in goal and Index scores for each sub-region. Goals with no change are indicated with a zero. Overall Index scores are on the far right, separated by the horizontal gray line. Note different scales on y-axes.

assess how particular goals and overall ocean health are faring at a regional scale, and whether conditions are getting better or worse. We found that current status scores for individual goals have gotten worse in the past decade or so (with the exception of recent improvements in lasting special places an livelihoods and economies for some sub-regions), but in most cases the near future looks better than the present (Figs. 3, 4). Assessment of the likely future status rewards the presence of regulatory and management measures; however, as data do not often exist on effectiveness of these measures, future estimates may be overly optimistic. In addition, the likely future status makes incorporates the potential impacts of climate change only as current climate-related pressures and not estimates of future conditions. Regardless, these differences highlight the importance of having time-series and maintaining on-going records of both ecological and governance information in order to understand likely future changes. The Index was designed explicitly to capture and quantify these different temporal components.

Spatial comparisons of sub-regional scores offered additional insights. Sub-regional scores had relatively small differences compared to the large range of scores globally in Halpern et al. [14]. This is understandable because biophysical and socioeconomic characteristics within the study region vary much less than among countries across the globe. Regional scores from this assessment were all higher than the score for the entire U.S. calculated from the global assessment. We cannot distinguish whether this difference stems from the use of different methods and data sources or if it suggests support for the widely-held view that the U.S. west coast is a relatively healthy and sustainably managed system. The overall U.S. score derived in the global study integrates scores from diverse coastal regions (i.e., Alaska, Hawaii, Gulf of Mexico, and east and west coasts of the U.S.) that vary historically, ecologically and in resource management actions.

Individual goal scores showed bigger sub-regional differences, but these differences were dampened when averaged with other goal scores to create overall Index scores (assuming equal weights for all goals; see File S1). For example, the fisheries sub-goal scored considerably higher in California (79; the Californian sub-regions could not be assessed separately for this goal due to the resolution of reported data) than in Oregon (56) and Washington (64; Fig. 2, Table S33 in File S1). This difference is in part due to differences in the dominant stock in the different regions; hake historically constituted roughly 30% of total catch in Oregon and Washington and are currently at low biomass and fishing effort levels (giving a score of 0.39 for the stock), whereas in California yellowfin tuna historically represented roughly 40% of total catch and are currently at ideal biomass and fishing effort levels (a score of 1.0).

The coastal livelihoods and economies sub-goals had very high scores for Central and Southern California and Oregon yet much lower scores for Northern California and Washington (Fig. 2, Table S33 in File S1). This goal uses a 'moving window' reference point, comparing each region to itself five years prior (while correcting for broader economic patterns, such as the global recession, that are independent of the condition of marine and coastal systems), based on the assumption that people mostly care about how they are doing economically relative to recent and local economic conditions. This technique avoids direct comparison, for example, of the absolute size of Southern California's coastal economy to that of Oregon. Consequently, Washington and Northern California scored lower because the largest sectors had significant declines in the last five years (in Washington, large declines occurred for jobs in tourism and transportation sectors and for revenue in tourism and living resources sectors; in Northern California, declines occurred for jobs and revenue in

tourism), while such declines generally did not occur in the other regions.

Variation among sub-regions for the carbon storage goal (as well as the generally low scores for this goal for all sub-regions) is primarily due to the status of salt marsh habitats. Although salt marsh habitat loss has occurred throughout the U.S. west coast, this has been particularly severe in Central and Northern California. The exact values of the scores for this goal are highly dependent on the choice of reference point, which in this case is challenged by both practical and philosophical issues. Practically, few spatial data exist prior to the 1990s. Therefore, to set an ambitious yet realistic reference point, we relied on estimates of historical loss of these habitats from pre-industrial times and set the reference point to a fraction (50%) of this original extent. Philosophically, one must (subjectively) decide what serves as an ambitious yet realistic target (cf. [24]) for restoring this habitat, given that a great deal of potential ecosystems services were lost but also considering that the massive alteration driven by urbanization of estuarine systems is unlikely to be completely reversed.

Finally, differences among sub-region scores for the artisanal fishing opportunity goal were driven primarily by differences in coastal access, which together with economic factors and fish stock status determined this goal's score. Public access is provided along Oregon's entire coastline, leading to a higher score, whereas Washington and California both allow privately-owned access to the coast and have large stretches of restricted-access coastline. The choice of a common reference point for all sub-regions allows for direct comparison, but ultimately may not reflect sub-regional differences in management objectives, such as how local people want artisanal fishing to occur. If such differences are significant enough and the objective of local managers is primarily on managing at this finer scale, then a separate Index score, calculated with sub-regional best available data and locally-determined reference points, would be more appropriate.

Other goals are consistently low or high across all sub-regions but with results that may not immediately seem intuitive. For example, all sub-regions would like to increase coastal tourism, yet all scored nearly perfectly on this goal. One might hope that a different type of reference point could resolve this paradox. A functional relationship between people's values and the effects on the ecosystem caused by different levels of participation in coastal recreation would be ideal, as it could indicate what absolute levels of coastal tourism are both wanted by local communities and sustainable for local ecosystems. Unfortunately we do not have the information to construct one and it could differ in the different sub-regions. A spatial comparison reference point would probably not be appropriate as coastal communities in Oregon, for example, are very different biogeographically than those in Southern California, with local population density, weather and beach access being some of the place-specific factors affecting the number and frequency of people recreating in and around the ocean. Consequently, spatial comparison using a region-wide reference point would unfairly penalize one of the locations and would not be a useful indicator of ocean health. Currently no stated objectives for desired levels of tourism exist that could be used for an 'established target' reference point. This leaves temporal reference points as the best choice in this specific case, and, for reasons similar to economically-based temporal reference points for the livelihoods & economies goal, we used a moving-window reference point (i.e., that conditions are as good or better than they were 5 years ago). For all sub-regions, participation in coastal recreational activities has remained the same or increased in the past 5 years.

A similar issue of choice of reference point affected the scores for the mariculture sub-goal. We used an established target (increase mariculture by 350% from 2005 to 2020; [27]), which was based on socio-economic projections of seafood demand. This target produced relatively low scores, but our assessment of a range of other types of reference points showed that these scores are strongly dependent on choice of reference point (Table 3). Ideally we would have used a functional relationship reference point based on biophysical variables and societal preferences for how much available ocean space should be allocated to mariculture versus all the many other uses that currently exist and that will emerge in the near future. Unfortunately we currently have very little of these data, and so we relied on rough estimates of socially and ecologically desirable 'farmable area' in each sub-region. Although the estimate from Nash [27] is potentially arbitrary, a 350% increase in production is not unreasonable from an environmental perspective (i.e., very little area is currently dedicated to shellfish farms and the production of shellfish species has a minimal environmental impact), such that the socially/economically desirable reference point reported by federal managers seems reasonable using SMART principles for setting reference points [24]. A production function based on these parameters would likely lead to higher mariculture scores, as the reference points (i.e., targets) would likely be lower. Uncertainty in what these target values should be remains an important gap in our current understanding.

Lessons learned for regional assessments

Although from a global perspective the U.S. west coast is a relatively data-rich location, data from the recent past were largely lacking, and historical data even more so. Most notably these gaps include habitat condition (current quality, and historical and current extent), conservation status of most species, fisheries stock assessments, historical levels of human pressures on ecosystems, and the nature and effectiveness of regulatory measures – data gaps common nearly everywhere [28–29]. The process of pulling together the information necessary to calculate the Index serves as a means to systematically evaluate where key gaps remain. Such gaps are a perpetual challenge for managers and policy makers. Prioritizing efforts to fill those gaps remains critical. The assessment here also highlights the need for new or continued assessment of pressures and resilience measures, not only of status variables, for effective assessment of overall ocean health. Our study offers a valuable starting point, or baseline, for future assessments in this region, but only by filling key data gaps will we gain the ability to determine trends in overall ocean health, a critical need for ecosystem-based management [30]. The sooner that other regions can begin comprehensive and repeatable assessments of ocean health, the better equipped they will be to make fully informed and strategic resource management decisions.

A key challenge for any assessment of ecosystem health is to detect meaningful and significant change in condition. Ideally one can then attribute that change to natural versus anthropogenic drivers of change, although such attribution is notoriously difficult. Along the U.S. west coast efforts to detect and attribute change face the challenge of distinguishing between broad shifts in the ecosystem due to natural climate variability on timescales of interannual (such as El Nino-driven changes) to multi-decadal (such as Pacific Decadal Oscillations; PDO) from longer-term trends driven by human impacts such as climate change. In the Index most goals document the cumulative effect of natural and anthropogenic change but do not explicitly attribute the underlying cause. For example, the biodiversity, carbon storage, coastal protection and clean water goals all have fixed targets that are independent of the cause of change, and the human-focused goals of coastal livelihoods & economies and tourism & recreation are more indirectly affected by such natural variation and are also independent of the cause of change. The fisheries sub-goal of food provision is one case where attribution of change is important, as fisheries management benefits from knowing the cause(s) of mortality. In this case, stock assessment models are usually refined to adjust the reference points and assessments in consideration of known changes in key oceanographic conditions (i.e., adjusting B_{MSY} to account for what is sustainable for a given oceanographic regime). By using information from local stock assessments, the index is adjusted for these effects whenever local information and modeling tools allow.

Importantly, and perhaps unsettling to some, is the reality that assessing something as diverse and comprehensive as 'ocean health' requires accommodating regional values and perspectives of the people that are part of the coastal ecosystems being assessed. Within the Ocean Health Index, this means that some conceptual aspects of implementing the Index are inherently subjective. Although the Index framework provides guidelines that can help adapt models to available regional data, using simpler models or different proxies when necessary [14], it cannot prescribe which available regional data sets are preferred. Nor does the framework dictate the most appropriate models to use or choices of reference points, proxy data or goal weights. For example, the way we modified models or reference points used in the global assessment [14] for regional use in the food provision, tourism & recreation, and artisanal fishing opportunities goals (see File S1) has important consequences for the resulting goal scores (Table 3). These adaptations highlight the flexibility of the Index to incorporate different perspectives on how goals should be assessed, as reflected by debates on how fisheries were modeled [31–34]. The Index can also accommodate a different set of goals if they better reflect what local communities value, although we posit that the ten goals currently defined within the Index are sufficiently broad to capture a vast majority of values. To some this flexibility may seem to come at the cost of comparability or objectivity, but we argue that any local assessment of ecosystem health faces similar challenges of accounting for local variables and community values. No indicator is exempt from such subjective decisions. The Index's framework, however, requires one to identify, justify and track such assumptions explicitly and thus fosters careful and well-documented assessment of the sensitivity of results to such decisions.

Policy implications

The Index was explicitly designed to help inform decision-making by providing a comprehensive, comparable, and quantitative assessment of the range of components that drive overall ocean health. As with any decision support tool, the scale of assessment should match the scale of decision-making [35]. Our assessment here is thus most valuable to regional-scale (e.g., West Coast Governors Alliance, [2]; California Current LME, [36] and state-level (e.g., California's Marine Life Protection Act, [37]) decision-making. Decisions at smaller scales (such as Puget Sound or San Francisco Bay) would be best informed by recalculating the Index using best available local-scale information where possible. To help support application of the Index in these (and other) processes in the future, we have developed a software tool (www.ohi-science.org) that allows people to explore Index results as well as recalculate (or calculate anew) scores as new data become available. A key strength, and challenge, of the Index is that it requires an explicit statement of all assumptions and assignment of specific targets for each ocean health goal. The strength lies in providing stakeholders and decision-makers a forum to articulate

Table 5. Comparative summary of assessment tools and methods that have been applied to regions of the US West coast.

	OHI	IEA	PSP	FEP	CalCOFI
Ecological system assessed explicitly	no	yes	yes	yes	yes
Social system assessed explicitly	no	yes	no	no	no
Integrated assessment of socio-ecological systems	yes	no	no	no	no
Scalable to sub-regional level	yes	yes	yes	no	yes
Includes scenario analyses	yes	yes	no	no	no
Part of PFMC process	no	yes	no	yes	yes
Part of WCGA process	no	yes	no	no	no
Addresses most/all sectors	yes	yes	yes	yes	no
Combines all sectors into an overall quantitative assessment	yes	no	no	no	no
Reference points are explicitly delineated	yes	no	no	no	no

OHI, IEA and FEP methods have been applied to the entire west coast; PSP and CalCOFI are sub-regional assessments but are included for comparative purposes. Attributes which all or none of the methods achieve are not included in this table. See legend below for definition of acronyms.
Legend: OHI = Ocean Health Index; IEA = Integrated Ecosystem Assessment; PSP = Puget Sound Partnership; FEP = PFMC Fisheries Ecosystem Plan annual reports; PFMC = Pacific Fisheries Management Council; WCGA = West Coast Governors Agreement; CalCOFI = California Cooperative Oceanic Fisheries Investigations.

their reference points and assumptions, while leveraging their values and knowledge, and a means to disentangle and clearly define their multiple, interacting objectives. The challenge arises from the practical (e.g., data constraints) and political (e.g., managing expectations, achieving consensus) process of making these important decisions, and the inherent sensitivity of Index scores to these choices [24]. The Index offers a tool to engage stakeholders and decision-makers in these difficult but necessary discussions, while also helping agencies fulfill their mandates.

For example, the ability to use scenarios to evaluate the likely consequences of any particular management action for overall ocean health provides a powerful decision-support tool, but requires additional assumptions and decisions about how things will likely change in the future. We illustrated such a process with several heuristic scenarios (Fig. 5) intended to show how the Index could inform regional-scale decision-making on issues such as land-use regulations and MPA creation. Scenarios intended to inform decision-making at these or smaller scales in the future would benefit from vetting model assumptions through a planning process, and require that the Index be applied at the relevant spatial scale. Although hypothetical, the scenarios demonstrate several key aspects of the Index relevant to decision makers: 1) it responds quickly to management actions, giving initial 'credit' for those actions, and then further responds over time as the system (social, economic, and ecological) changes; 2) tradeoffs inherent in many decisions are captured by the Index (either explicitly as they are built into the Index or implicitly as they would emerge after management actions); and 3) the Index allows one to compare very different management actions in a transparent and quantitative way across different sub-regions, thus supporting strategic decision-making. The magnitude of expected change in the Index will necessarily be related to the scale of management action relative to the scale of assessment.

Such scenario analyses are also a key way that the Index can be used to explore potential implications of climate change on ocean health. As with the other scenario examples, because the Index does not model the future it cannot predict future ocean health. Instead, dynamic process models can be used to simulate ecological and social conditions, and then these results can be fed in as input parameters for calculating an alternate Index score. In this case, the Index can be used to indicate the likely overall ocean health in the future under status quo conditions and a

changing climate. Additional management scenarios could then be layered on top of those outputs to better understand the likely effect of climate change on future ocean health.

Scenario analyses also illustrate how the Index can be used to identify and understand tradeoffs among goals. Some of these known tradeoffs are built into the architecture of the Index, for example in how increased (sustainable) fishing produces higher scores for food provision but lowers other goals due to its negative pressure on them. Other more complex, emergent tradeoffs become visible only when the Index is measured over time and one can track how goal scores change in similar or opposite directions. Because of the complexity of ecosystem responses, full attribution of a change in one goal causing a change in another goal is difficult, but such patterns can provide insight on where to direct further exploration of such possible tradeoffs. The ability to calculate past status scores, and then correlate changes in the Index with past management actions, illustrates a key way it can be used to assess management effectiveness. If the Index were adopted as a management tool, recalculating scores regularly could reveal whether management actions had the intended effect on both overall ocean health and particular goals. This objective demonstrates the power (and necessity) of having a quantitative, repeatable, transparent and comprehensive method for assessment.

The process of adapting the Index to finer geographic scales highlights its flexibility but also the limits to comparability of Index scores across scales. Most decision making focuses on optimizing outcomes for a region of interest (e.g., a particular country, or a state within a country), regardless of how other regions are performing, such that adapting the Index to the best available regional information is appropriate and ideal. However, it is human nature to ask how one is doing relative to others, and that desire for comparability can lead to misunderstanding of Index results if the comparisons are made across assessments at different scales (e.g., global version regional). Here we focus on results within and among U.S. west coast regions and minimize comparisons to global results for the U.S. for these exact reasons.

Many other assessment frameworks and tools have been applied to the U.S. west coast to evaluate different aspects of its health (see Table 5 for a summary of several prominent ones). Although it is instructive to compare the approaches to understand their strengths and weaknesses, it is important to note that each method was developed and applied for specific purposes, such that direct comparison among them is not always appropriate. Integrated

Ecosystem Assessments (IEAs), Fisheries Ecosystem Plans (FEPs) and CalCOFI reports are all part of the Pacific Fisheries Management Council's (PFMC) decision process, such that those assessments are directly affecting and assessing management actions, however the Ocean Health Index is too new to have had a chance to be vetted and potentially included in the PFMC process. All of the other methods directly and explicitly assess the ecological and biophysical aspects of the system, whereas in the Ocean Health Index these assessments are not separately available because they are combined within integrated socio-ecological indicators. On the other hand, for this reason the Index is currently the only method to offer a fully integrated assessment. Most of the methods assess the full range of sectors active in the region, but the Ocean Health Index generally combines them together into overall goal measures rather than tracking individual sectors separately. Finally, the Ocean Health Index makes explicit the process of defining and setting quantitative reference points that establish when goals are fully achieved, whereas the other methods tend to rely on expert judgment and informal evaluations.

Another important policy implication of applying the Index is to help prioritize data collection and primary research efforts. Most monitoring focuses on biological impacts without connecting them explicitly to benefits that people want and need. The Index framework, by explicitly showing the connection between societal goals and the ability of the system to provide those goals, highlights the importance of collecting ecological, social, institutional, and economic data to monitor and inform management, and motivates all stakeholders to strive for a more sustainable human-ocean system.

Application of the Ocean Health Index to the US west coast not only provided an assessment of ocean health for the region but also guidance on the opportunities and challenges in applying and adapting the general Index framework to a regional setting. In the relatively data-rich US west coast, we were able to take advantage of the best available knowledge and information and make sub-regional assessments, sub-regions that share some ecological and socio-economic aspects but also show many differences that are important for defining management strategies. Such sub-regional assessments are likely to be important in most regions of the world. In particular, this downscaled, regional application of the Index offers a means and a medium for conversations among disparate marine use sectors by providing measures of diverse aspects of ocean health in a common currency.

Supporting Information

File S1 Complete set of supplementary information, including supplementary methods, tables S1–S38, and figures S1–S2.

Acknowledgments

Thanks to Dan Ovando for help exploring options for calculating the status of data poor stocks in the region and Judith Kildow and Charles Colgan for help with livelihoods data. Dave Fluharty and an anonymous reviewer provided very helpful comments.

Author Contributions

Conceived and designed the experiments: BSH CL SKK KLM JFS. Performed the experiments: BSH CL DH BDB. Analyzed the data: BSH CL CS DH BDB. Wrote the paper: BSH CL CS DH BDB SCD SKK KLM AAR JFS.

References

1. MSFD (2008) Marine Strategy Framework Directive. Directive 2008/56/EC of the European Parliament and Council establishing a framework for community action in the field of marine environmental policy. http://eurlex.europa.eu/LexUriServ/LexUriServ.do?uri=OJ:L:2008:2164:0019:0040:EN:PDF.
2. WCGA (2006) West coast governors' agreement on ocean health. Agreement on Ocean Health Action Plan.
3. McLeod KL, Leslie HM (2009) Ecosystem-based management for the oceans. Island Press, Washington, DC.
4. COA (1998) Commonweath of Australia. Australia's Oceans Policy.
5. DFO (2002) Directorate, Fisheries and Oceans Canada. Canada's Oceans Strategy, Ottawa, Ontario.
6. Obama B (2010) Executive Order 13547: Stewardship of the ocean, our coasts, and the great lakes, Washington DC.
7. NMS (2004) National Marine Sanctuary Program. A Monitoring Framework for the National Marine Sanctuary System.
8. PFMC (2013) Pacific Fishery Management Council: *Pacific Coast Fishery Ecosystem Plan for the U.S. Portion of the California Current Large Marine Ecosystem – Public Review Draft, February 2013.* (Document prepared for the Council and its advisory entities.) Pacific Fishery Management Council, 7700 NE Ambassador Place, Suite 101, Portland, OR, 97220-1384.
9. CDFG (1999) California Department of Fish and Game. Marine Life Protection Act. Fish and Game Code Sections 2850-2863.
10. PSP (2008) Puget Sound Partnership. Puget Sound Action Agenda: Protecting and restoring the Puget Sound ecosystem by 2020, Olympia, Washington.
11. WCEBM (2010) West Coast EBM Network. Community-Based Management of Coastal Ecosystems: Highlights and Lessons of Success from the West Coast Ecosystem-Based Management Network. www.westcoastebm.org/WestCoastEBMNetwork_EBMGuide_June2010.pdf
12. Halpern BS, Diamond J, Gaines S, Gelcich S, Gleason M, et al. (2012a) Near-term priorities for the science, policy and practice of Coastal and Marine Spatial Planning (CMSP). Marine Policy 36:198-205.
13. Tallis H, Levin PS, Ruckelshaus M, Lester SE, McLeod K, et al. (2010) The many faces of ecosystem-based management: Making the process work today in real places. Marine Policy 34:340-348.
14. Halpern BS, Longo C, Hardy D, McLeod KL, Samhouri JF, et al. (2012b) An index to assess the health and benefits of the global ocean. Nature 488:615-620.
15. Karieva P, Marvier M, Lalasz R (2011) Conservation in the Anthropocene. Breakthrough Journal 2.
16. Karieva P, Marvier M (2012) What is conservation science? BioScience 62: 962-969.
17. Soulé M (2013) The "new conservation". Conservation Biology 27: 895-897.
18. Folke C, Carpenter SR, Walker BH, Scheffer M, Chapin III FS, et al. (2011) Reconnecting to the biosphere. Ambio doi:10.1007/s13280-011-0184-y;
19. Liu J, Dietz T, Carpenter SR, Alberti M, Folke C, et al. (2007) Complexity of coupled human and natural systems. Science 317: 1513–1516.
20. Walker B, Salt D (2006) Resilience thinking: sustaining ecosystems and people in a changing world. Washington, DC: Island Press.
21. Spalding M, Fox HE, Allen GR, Davidson N, Ferdana ZA, et al (2007) Marine ecoregions of the world: a bioregionalization of coastal and shelf areas. BioScience 57: 573–583.
22. Samhouri JF, Kim S, Zhang CI, Fogarty M (2014) Marine ecosystem-based management in temperate systems. In: Marine ecosystem-based management, Fogarty MJ, McCarthy JJ, editors. Vol. 16 of *The Sea.* Harvard University Press, Boston. pp. 325–367
23. MEA (2005) Millennium Ecosystem Assessment. Ecosystems and human well-being: synthesis report.
24. Samhouri JF, Lester SE, Selig ER, Halpern BS, Fogarty MJ, et al. (2012) Sea sick? Setting targets to assess ocean health and ecosystem services. Ecosphere 3: article 41.
25. Halpern BS, Longo C, Stewart Lowndes JS, Best BD, Frazier M, et al (in review) Patterns and emerging trends in global ocean health. PLoS ONE
26. Halpern BS, Longo C, McLeod KL, Cooke RG, Fischhoff B, et al. (2013b) Elicited preferences for components of ocean health in the California Current. Marine Policy 42:68–73.
27. Nash CE (2004) Achieving policy objectives to increase the value of the seafood industry in the United States: the technical feasibility and associated constraints. Food Policy 29:621–641.
28. Pereira HM, Cooper DH (2006) Towards the global monitoring of biodiversity change. Trends In Ecology & Evolution 21:123–129.
29. Rands MRW, Adams WM, Bennun L, Butchart SHM, Clements A, et al. (2010) Biodiversity conservation: challenges beyond 2010. Science 329:1298–1303.
30. Francis T B, Levin PS, Harvey CJ (2011) The perils and promise of future analysis in marine ecosystem-based management. Marine Policy 35:675–681.
31. Branch TA, Hively DJ, Hilborn R (2013) Fisheries assessments are biased in Ocean Health Index. Nature 495: E5–D6.
32. Halpern BS, Gaines S, Kleisner K, Longo C, Pauly D, et al. (2013a) Reply to Branch et al. Nature 495: E7.
33. Kleisner KM, Longo C, Coll M, Halpern BS, Hardy D, et al. (2013) Exploring patterns of seafood provision revealed in the global Ocean Health Index. Ambio DOI 10.1007/s13280-013-0447-x

34. Pauly D, Hilborn R, Branch TA (2013) Fisheries: does catch reflect abundance? Nature 494:303–306.

35. Cash DW, Moser SC (2000) Linking global and local scales: Designing dynamic assessment and management processes. Global Environmental Change 10: 109–120.

36. Aquarone MC, Adams S (2008) XIV-44 California Current: LME#3. In: K. Sherman K, Hempel G, editors. The UNEP large marine ecosystem report: A perspective on changing conditions in LMEs of the world's regional seas, UNEP Regional Seas Report and Studies No. 182. United Nations Environment Programme, Nairobi, Kenya. pp. 593–603.

37. CCC (1997) California Coastal Commission, California Coastal Commission Strategic Plan. Available: http://www.coastal.ca.gov/strategy.html. Accessed 2014 Feb. 10.

Fish Oil Replacement in Current Aquaculture Feed: Is Cholesterol a Hidden Treasure for Fish Nutrition?

Fernando Norambuena[1]**, Michael Lewis**[1]**, Noor Khalidah Abdul Hamid**[2]**, Karen Hermon**[1]**, John A. Donald**[2]**, Giovanni M. Turchini**[1]*

1 School of Life and Environmental Sciences, Deakin University, Warrnambool, Victoria, Australia, 2 School of Life and Environmental Sciences, Deakin University, Waurn Ponds, Geelong, Victoria, Australia

Abstract

Teleost fish, as with all vertebrates, are capable of synthesizing cholesterol and as such have no dietary requirement for it. Thus, limited research has addressed the potential effects of dietary cholesterol in fish, even if fish meal and fish oil are increasingly replaced by vegetable alternatives in modern aquafeeds, resulting in progressively reduced dietary cholesterol content. The objective of this study was to determine if dietary cholesterol fortification in a vegetable oil-based diet can manifest any effects on growth and feed utilization performance in the salmonid fish, the rainbow trout. In addition, given a series of studies in mammals have shown that dietary cholesterol can directly affect the fatty acid metabolism, the apparent *in vivo* fatty acid metabolism of fish fed the experimental diets was assessed. Triplicate groups of juvenile fish were fed one of two identical vegetable oil-based diets, with additional cholesterol fortification (high cholesterol; H-Chol) or without (low cholesterol; L-Chol), for 12 weeks. No effects were observed on growth and feed efficiency, however, in fish fed H-Col no biosynthesis of cholesterol, and a remarkably decreased apparent *in vivo* fatty acid β-oxidation were recorded, whilst in L-Chol fed fish, cholesterol was abundantly biosynthesised and an increased apparent *in vivo* fatty acid β-oxidation was observed. Only minor effects were observed on the activity of stearyl-CoA desaturase, but a significant increase was observed for both the transcription rate in liver and the apparent *in vivo* activity of the fatty acid Δ-6 desaturase and elongase, with increasing dietary cholesterol. This study showed that the possible effects of reduced dietary cholesterol in current aquafeeds can be significant and warrant future investigations.

Editor: Yoav Gothilf, Tel Aviv University, Israel

Funding: This research was supported under the Australian Research Council's Discovery Projects funding scheme (Project DP1093570). The views expressed herein are those of the authors and are not necessarily those of the Australian Research Council. The funding agency had no role in study design, data collection and analysis, decision to publish, or preparation of the manuscript.

Competing Interests: The authors have declared that no competing interests exist.

* E-mail: giovanni.turchini@deakin.edu.au

Introduction

Cholesterol modulates the fluidity of cell membranes, is essential for cell membranogenesis, growth and differentiation, is a key structural component of muscle, brain and the nervous system, and is the precursor for many physiologically active compounds, including sex and molting hormones, adrenal corticoids, bile acids and vitamin D [1–4]. Because of the role of cholesterol in cardiovascular diseases, it has been extensively studied in human and mammalian models since the early 1900s [5]. In aquatic animals, most of the information about cholesterol and its nutritional requirement/roles has focused on marine crustaceans, since invertebrates lack the enzymes necessary for its biosynthesis [6,7]. Teleost fish have no dietary requirement for cholesterol as they are capable of synthesizing it [8], and as such the topic of the potential effects of dietary cholesterol in cultured fish feed (aquafeed) has received scarce research interest. Marine-derived raw materials traditionally used in aquafeed (namely fish meal and fish oil) are rich sources of cholesterol [9]. However, in present-day aquaculture, because of environmental and economic consideration, there is an ever increasing utilization of alternatives to substitute marine ingredients, with fish oil commonly being replaced by vegetable oils [10]. The remarkably different fatty

acid composition of vegetable oils compared to fish oils has fuelled an intensive global research effort focusing on fatty acid nutrition in cultured fish [11]. Yet, another important difference between these oils is that vegetable oils contain high levels of phytosterols [12] and very little cholesterol, compared to fish oils, which contain large amount (from 3.5 to 7.7 g kg^{-1}) of cholesterol [9,13,14]. Therefore, the increasing level of substitution of fish oil in aquafeed is not only responsible for the modification of the fatty acid composition in feed, but is also simultaneously responsible for progressively reduced levels of cholesterol.

In mammals, the availability of dietary cholesterol has been shown to affect the fatty acid metabolism [15–18], however, there have been only a few studies investigating dietary supplementation of cholesterol in fish feed [19–26]. These were all implemented using diets containing abundant levels of fish oil, and thus even the non-fortified (control) treatments were providing large amounts of dietary cholesterol. In a study focusing on fatty acid metabolism in Atlantic salmon (*Salmo salar*) fed a fish oil or a vegetable oil based diet, and implementing the gene microarray technique, it was observed that, in liver of fish fed the vegetable oil diet, the genes of the cholesterol biosynthesis pathway were up-regulated [8]. This study confirmed that in fish, as was observed in other vertebrates

[27], the rate of cholesterol biosynthesis is highly responsive to the amount of available dietary cholesterol, which is known to inhibit biosynthesis by regulating HMG-CoA reductase. This mechanism makes sure that all the vital cholesterol-dependant metabolic pathways are preserved, independently from dietary cholesterol availability.

It can be argued that current aquafeeds, being low in cholesterol content, are forcing cultured fish to constantly produce cholesterol; but an important consideration is that cholesterol biosynthesis is a rather expensive metabolic exercise, requiring 18 acetyl-CoA, 18 ATP, 16 NADPH and 4 O_2 molecules per molecule of cholesterol produced [27]. The possible effects of limited dietary cholesterol on fish energy expenditures, and also on fatty acid metabolism and tissue content, are unknown. The aim of this study was to determine if dietary cholesterol fortification in a vegetable oil based-diet can manifest any effect on growth, feed utilization performance, and apparent *in vivo* fatty acid metabolism in cultured fish. The selected target species was the popular cultured salmonid rainbow trout (*Oncorhynchus mykiss*).

Materials and Methods

Fish husbandry and experimental diets

All procedures implemented during this experiment were approved by the Deakin University Animal Ethics Committee (AEC ref A41/2011). All possible efforts to minimize animal suffering were taken. Rainbow trout (*Oncorhynchus mykiss*) were hatched and reared at the Department of Primary Industries facility (Snobs Creek, Victoria, Australia), and were then translocated and acclimatised to experimental conditions at Deakin University (Warrnambool, Australia) and fed on a commercial diet (Ridley Aquafeed, Australia) for two weeks. At the start of the experiment, ten fish were euthanized with an overdose of anaesthetic (AQUI-S, New Zealand) and samples from fillet, liver or the whole body were collected, weighed and stored at $-20°C$. One hundred and twenty six fish (body weight 17.3±1 g) were anaesthetised, weighed and then randomly distributed into six tanks (21 fish per tank) of 1000 L capacity within a fully controlled multi-tank recirculation system (RAS). Fish were held at 15°C, under 12:12 light:dark cycle. Total ammonia and nitrite levels were monitored using Aquamerck test kits (Merck, Germany) and were maintained at optimal levels. Three tanks were randomly assigned to one of two dietary treatments. Fish were fed twice a day to apparent satiation for 84 days. Then, after a 24 h gut evacuation period, all fish were anaesthetised and weighed. A final sample of 11 fish per tank (33 per treatment) was randomly collected and euthanized, and samples of fillet, liver or the whole body were collected and stored at $-20°C$ for further chemical analysis. In addition, samples of liver were also snap frozen by immersion in liquid nitrogen and then stored at $-80°C$ for further biomolecular analysis. Growth and feed utilisation parameters over the experimental period were calculated as previously described [28]; these included initial and final average weight (g), average feed consumption (g fish^{-1}), gain in weight (g and %), food conversion ratio (FCR), specific growth rate (SGR, % day^{-1}), feed ratio (% of body weight), dress-out percentage (DP%), fillet yield percentage (FY%), hepatosomatic index (%) and condition factor (K). Two experimental vegetable oil based diets were formulated with or without the addition of dietary cholesterol, and named L-Chol (low cholesterol) and H-Chol (high cholesterol), respectively (Table 1). The low cholesterol diet (L-Chol) contained only the cholesterol originating from the raw materials used in the formulation, whilst the high cholesterol diet (H-Chol) was fortified with an additional 1 g kg^{-1} of cholesterol

(as free cholesterol), mimicking the amount of cholesterol that would have been present if fish oil was used as the added lipid source. The experimental diets were formulated to be iso-proteic (450 mg g^{-1}), iso-lipidic (200 mg g^{-1}) and iso-energetic (22 kJ g^{-1}) and were manufactured as previously described [29]. A blend of linseed and canola (rapeseed) oils was used as the lipid source while fish meal, poultry meal, soy protein concentrate and blood meal were used for protein sources. Alpha-cellulose was used as an inert filler to compensate for different cholesterol addition.

Chemical analysis

The chemical composition of the experimental diets, faeces and fish samples were determined via proximate composition analysis according to standard procedures [30]. Briefly, moisture content was determined by drying samples in an oven at 100°C to constant weight, and protein (Kjeldahl nitrogen; N×6.25) in an automated Kjeltech (Tecator, Sweden). Lipids were determined by chloroform:methanol extraction (2:1) [31] with the substitution of chloroform with dichloromethane for safety reasons, and the addition of butylated hydroxytoluene (BHT) (50 mg L^{-1}) to

Table 1. Formulation and proximate composition of the two experimental diets with (H-Chol) or without (L-Chol) cholesterol fortification.

	Experimental Diets[1]	
	L-Chol	H-Chol
Diet formulation (g kg^{-1})		
Protein sources[2]	614	614
Vegetable oil[3]	147	147
Starch[4]	149	149
Min. & Vit.[5]	50	50
Others[6]	10	10
a-cellulose[7]	30	29
Cholesterol[8]	0	1
Proximate composition (mg g^{-1})		
Protein	442.3	457.4
Lipid	200	201.7
Moisture	54.1	49.5
Ash	87.4	82.1
NFE[9]	216.2	209.3
Energy (MJ Kg^{-1})[10]	22.1	22.4
Total cholesterol (mg g^{-1})	1.2	2.35

[1]Experimental diet nomenclature: L-Chol diet contained no added cholesterol, H-Chol diet containing 1 g Kg^{-1}. added cholesterol.
[2]Basal diet composition (g Kg^{-1}): poultry meal 211, soy protein concentrate 144, fish meal 87, blood meal 66, soybean meal 58, wheat gluten 57, whey protein 40; Ridley Agriproducts, Narangba, Queensland, Australia.
[3]Vegetable oil: 70% linseed oil, Sceney Chemical Pty., Ltd., Sunshine, VIC, Australia and 30% Canola oil, Black and Gold, Tooronga, VIC. Australia.
[4]Starch: Pre-gel starch, Ridley Agriproducts, Narangba, Queensland, Australia.
[5]Min & Vit.: Complete minerals and vitamins mix supplement; Sigma-Aldrich, Inc. St. Louis, MO, USA.
[6]Others (g Kg^{-1}): Amino acid mix (L-Methionine, L-Lysine, glutamic acid) 3, Celite® 7, Sigma-Aldrich, Inc. St. Louis, MO, USA.
[7]a-cellulose: alpha cellulose, Sigma-Aldrich, Inc. St. Louis, MO, USA.
[8]Cholesterol: Sigma-Aldrich, Inc. St. Louis, MO, USA.
[9]NFE: Nitrogen free extract calculated by difference.
[10]Calculated on the basis of 23.6, 39.5 and 17.2 KJ g^{-1} of protein, fat and carbohydrate, respectively.

reduce lipid oxidation during processing. Ash was determined by incinerating samples in a muffle furnace (Wit, C & L, Australia) at 550°C for 18 h. Nitrogen free extract (NFE) was calculated by difference and total energy was computed on the basis of 23.6, 39.5 and 17.2 kJ g^{-1} of protein, lipid and carbohydrate, respectively.

Fatty acid (FA) analysis was performed in duplicate for each fish tissue sample and experimental diet with the exception of the initial whole body sample, that was performed in triplicate. After lipid extraction, a known amount of tricosanoate (23:0) was added as internal standard, and FA were esterified into methyl esters using the acid catalysed methylation method [32]. The identification of FA methyl esters was determined using an Agilent Technologies GC 7890A gas chromatograph (Agilent Technologies, USA) equipped with an BPX70 capillary column (120 m, 0.25 mm internal diameter, 0.25 µm film thickness; SGE Analytical Science Pty Ltd, Ringwood, Vic, Australia), a flame ionisation detector (FID), an Agilent Technologies 7693 auto-sampler injector, and a split injection system. The injection volume was 1 µl (split ratio 50:1), with the injector and detector temperature set at 300°C and 270°C, respectively. The oven temperature program was: 60°C held for 2 min, from 60 to 150°C at 20°C min^{-1}, held at 150°C for 2 min, from 150 to 205 at 1.5°C min^{-1}, from 205 to 240 at 4°C min^{-1}, and held at 240°C for 24 min. The carrier gas is helium at 1.5 mL min^{-1}, at a constant flow. Each of the fatty acids was identified relative to known external standards (Sigma-Aldrich, Inc., St. Louis, MO, USA, and Nu-Chek Prep, Elysian, MN, USA), and the resulting peaks were then corrected by the theoretical relative FID response factors [33] and quantified relative to the internal standard.

The total cholesterol content of diets, faeces and whole body was analysed by an external accredited laboratory (The National Measurement Institute, Melbourne, Australia), with blind samples provided to the laboratory.

Digestibility, cholesterol mass balance and FA metabolism estimation

During days 53 to 84, faeces were collected from each individual tank using a previously described method [34]. Nutrients apparent digestibility was determined by assessing acid insoluble ash (AIA) as inert marker in diets and faeces, as described by Van Keulen and Young [35] and adapted to rainbow trout [36].

A simple mass balance for estimating overall cholesterol metabolism was implemented. The possible appearance (cholesterol *de novo* production/biosynthesis) or disappearance (cholesterol catabolism) was estimated at whole body level using the following equation: Cholesterol Appearance/Disappearance = (Cholesterol in final fish)−(Cholesterol in initial fish)−(Total cholesterol net intake); where: (Cholesterol in final fish) = (Cholesterol content in final fish, mg g^{-1})×(Final fish weight, g); (Cholesterol in initial fish) = (Cholesterol content in initial fish, mg g^{-1})×(Initial fish weight, g); and (Total cholesterol net intake) = (Cholesterol content in diet, mg g^{-1})×(Total weight of feed consumed, g fish^{-1})× (Apparent cholesterol digestibility, %).

The estimation of the apparent *in vivo* fatty acid metabolism (i.e., fatty acid *de novo* production, β-oxidation, elongation and desaturation) was calculated via the implementation of the whole body fatty acid mass balance method [37], with subsequent developments [37,38], and it is suggested to refer to these references for full details and computations.

Tissue RNA extraction and polymerase chain reaction (RT-PCR)

In order to study the expression of fatty acyl Δ-6 desaturase (*D6fad*) and elongase (*Elovl5*) in the liver, which is the main organ where fatty acid biosynthesis occurs, total RNA was extracted from approximately 10 mg of tissue by organic solvent (Tri-reagent), according to the manufacturer's instructions (Sigma Aldrich, USA), followed by phase separation with chloroform, then precipitation with isopropanol. RNA quality and quantity were assessed by gel electrophoresis and spectrophotometry (NanoDrop) (Thermo Scientific, USA). One µg of total RNA per sample was reverse-transcribed into cDNA using a Superscript III Reverse Transcriptase (Invitrogen, USA) according to the manufacturer's protocol. The synthesized first-strand cDNA (40 µL) was diluted to 80 µL using nuclease-free water and stored at −20°C. The concentration of single-stranded cDNA was quantified against an oligonucleotide standard in an assay using Oligreen reagent (Invitrogen, USA). The mRNA expression of *D6fad* and *Elovl5* were measured by RT-PCR. Semi quantitative real-time PCR was performed in duplicate using a Rotor Gene RG 3000 (Qiagen, Germany) in a 25 µL reaction containing 1 ng cDNA, primer pair (100 nM of forward and reverse) and 12.5 µL of SYBR® Premix Ex Taq™ (TaKaRa, Japan). The real time reaction cycle was as two-step thermal profile (94°C for 5 seconds, 60°C for 20 seconds), acquiring to a SYBR green fluorescence channel between the annealing and extension steps. The melting curve analysis was performed at the end of 40 cycles to determine the specificity of the reaction. The expression of the genes was normalized by the ratio of the threshold cycle (Ct) value to the concentration of the single stranded cDNA and is given by $2^{-\Delta\Delta Ct}$. Specific primer pairs were designed for rainbow trout based on the gene sequences available in GenBank (http://www.ncbi.nlm.nih. gov): *D6fad* (accession no. NM001124287) forward; 5′-ACTAGTGGCTCCTCTGGTC-3′, reverse; 5′-CAGATCCCCTGACTTCTTCA-3′) and *Elovl5* (accession no. AY605100) forward; 5′- TCAACATCTGGTGGTTCGTCAT-3′, reverse; 5′-TGTTCAGGGAGGCACCAAAG-3′) using Primer Express (ver. 3, Applied Biosystems, USA). The sequences were confirmed by sequencing the amplicons.

Statistical analysis

All data were reported as mean ± standard error ($n = 3$; $N = 6$). The compliance of data with normality and homogeneity of variance were tested using the Kolmogorov–Smirnov and Bartlett (Chi-Sqr) tests and when necessary, log-transformation was carried out. Data interpretation was based on independent *T-test* at a significance level of 0.05. All statistical analyses were computed by using SPSS version 17.0 (SPSS, Inc., Chicago, IL. USA).

Results

Experimental diets were iso-proteic, iso-lipidic and iso-energetic, and differed only for their total cholesterol content, which varied from 1.20 to 2.35 mg g^{-1}, for L-Chol and H-Chol, respectively (Table 1). Both diets were readily accepted by fish, and by the end of the experiment fish achieved 8.5-fold growth, with no mortality recorded during the experimental period (Table 2). No significant differences were noted in any growth and feed utilization parameters. A reduction in the coefficient of variation in fish fed H-Chol was noted (Table 2). Dress-out percentage (DP%) showed a statistically significant increase in H-Chol group (P<0.05). No differences were noted for any of the nutrient apparent digestibility coefficients, with the only exception recorded for total cholesterol digestibility which was significantly lower in L-Chol experimental diet (Table 2). The total lipid content was significantly higher in the fillets of the H-Chol fed fish compared to L-Chol fed fish (P<0.05), with the same trend (though not statistically significant) observed in fish whole body (Table 3). Protein, ash, moisture and total cholesterol in fillet and whole body were not different between the two groups (Table 3). The cholesterol mass balance showed that dietary cholesterol net intake (mg fish^{-1}) was higher (P<0.05) in fish fed H-Chol, whilst total cholesterol content of fish at the end of the experiment was significantly higher in L-Chol fed fish (P<0.05). L-Chol fed fish recorded a cholesterol appearance (positive final balance) of 19 mg fish^{-1}, whilst H-Chol fed fish recorded a net disappearance (negative final balance) of -208 mg fish^{-1}, during the duration of the experimental trial (Table 4).

The fatty acid composition in the two experimental diets was almost identical, and was characterised by relatively high content of 18:1n-9, 18:3n-3, 18:2n-6 and 16:0 (Table 5). At the end of the feeding trial, the trout liver showed no differences in all, except for one fatty acid (20:3n-6), this being significantly higher in H-Chol fed fish. In fish fillet, few statistically significant differences were noted, with total polyunsaturated fatty acids (PUFA), total n-3 PUFA, 18:3n-3 and 20:0 being higher in L-Chol fed fish, and total saturated fatty acids (SFA), 14:0, 16:0 and 16:1n-7 being higher in H-Chol fed fish (Table 5). Similar trends (though not statistically significant) were observed in whole body fatty acid composition. Statistically significant differences were recorded for 20:4n-3, 22:6n-3, total long chain PUFA (LC-PUFA) and n-3 LC-PUFA, being higher in H-Chol fed fish compared to L-Chol fed fish (Table 5). Several statistically significant differences were observed in the apparent in vivo fatty acids β-oxidation (Table 6), with the only exceptions being a few quantitatively minor FA (12:0, 14:0, 14:1n-5, 16:1n-7 and 22:4n-6), which recorded no differences. In all instances where statistically significant differences were recorded, the β-oxidation of FA was higher in fish fed the L-Chol diet, compared to H-Chol fed fish. This resulted in the total FA β-oxidation being significantly higher (1.4 fold higher) in L-Chol group, compare with H-Chol (P<0.05).

The desaturase D6fad and elongase Elovl5 gene expression increased in fish fed H-Chol in comparison to the L-Chol diet group (Fig. 1). Levels of D6fad transcripts were significantly higher (1.9-fold) in the liver of H-Chol fish groups compared with L-Chol, and a similar trend, albeit not statistically significant, was observed in Elovl5 transcript gene expression.

Several statistically significantly (P<0.05) higher activities were recorded in the apparent in vivo FA bioconversion (desaturation, elongation and peroxisomal chain shortening) in H-Chol fed fish, compared to the L-Chol group (Table 7). Accordingly, the apparent activity of stearyl-CoA desaturase or Δ-9 fatty acid desaturase (D9fad) was higher in H-Chol, with a significantly 2.2-fold higher bioconversion of 20:0 to 20:1n-11 in H-Chol (P<0.01). Likewise, D6fad apparent activity observed in the H-Chol group was higher (P<0.05), with a higher bioconversion of 18:2n-6 to 18:3n-3, and 24:5n-3 to 24:6n-3, being 1.3-fold and 1.2-fold higher, respectively. The same trend was recorded for Δ-5 fatty acid desaturase (D5fad) apparent activity, with higher bioconversion of 20:4n-3 to 20:5n-3 in the H-Chol group. The apparent in vivo activity of FA elongase (Elovl2 and Elovl5) was also higher in the H-Chol group compared with the L-Group (P<0.05), particularly in the elongation of 20:0 (2.2 fold), 18:1n-9 (2.2 fold), 18:4n-3 (1.7 fold) and 18:3n-6 (1.3 fold). Similarly, the apparent in vivo fatty acid peroxisomal chain chartering of 24:6n-3 to 22-6n-3 was higher in H-Chol (P<0.05). No difference was noted in FA neogenesis (Table 7).

Table 2. Growth performance, feed utilization and nutrient digestibility in rainbow trout fed with the two experimental diets with (H-Chol) or without (L-Chol) cholesterol fortification.

	Experimental Diets[1]		P-value	CV%	
	L-Chol	H-Chol		L-Chol	H-Chol
Growth and feed utilization					
Initial weight (g)	17.8±0.3	17.8±0.3	ns	5.4	5.5
Final weight (g)	149.6±11	153.8±4.4	ns	22.0	14.0
Feed consumption (g fish^{-1})[2]	191.9±2.0	185.6±5.9	ns	1.5	4.5
Weight gain (g)[3]	131.8±11.8	136.1±4.2	ns	12.6	4.4
Weight gain (%)[4]	742.3±71.8	766.0±18.1	ns	13.7	3.3
FCR[5]	1.48±0.1	1.36±0.02	ns	11.8	2.1
SGR[6]	2.66±0.1	2.70±0.03	ns	5.5	1.4
Feed Ratio%[7]	2.89±0.2	2.70±0.1	ns	9.3	2.5
Biometrical parameters					
DP%[8]	82.4±0.1[b]	83.8±0.5[a]	*	0.1	0.8
FY%[9]	48.1±0.5	49.2±1.9	ns	1.6	5.5
HSI%[10]	1.54±0.1	1.63±0.1	ns	8.0	4.4
K[11]	1.72±0.1	1.69±0.1	ns	3.0	1.8
Apparent digestibility (%)					
Dry matter	77.0±1.7	77.8±0.7	ns		
Lipid	94.7±0.8	93.5±0.8	ns		
Protein	86.7±1.7	86.9±0.7	ns		
Cholesterol	70.8±2.4	82.7±1.1	*		
Fatty acid digestibility (%)					
ADC[12] (14:0)	91.6±0.6	89.9±0.7	ns		
ADC (18:1n-9)	96.7±0.8	95.9±0.7	ns		
ADC (18:2n-6)	97.3±0.8	96.5±0.6	ns		
ADC (18:3n-3)	97.7±0.8	97.3±0.4	ns		
ADC (20:4n-6)	92.8±1.3	89.3±2.1	ns		
ADC (20:5-n3)	84.8±3.6	76.9±6.7	ns		
ADC (22:6-n3)	94.2±1.1	92.2±1.2	ns		

Data are presented as a mean ± s.e.m., n=3, N=6. P>0.05 = ns (not significant),
* =P<0.05,
** =P<0.01,
CV% = Coefficient of variance in percentage.
[1]See Table 1 for experimental diet abbreviations.
[2]Feed consumption (g fish^{-1}) = dry feed consumed per fish over the 84 day experimental period.
[3]Weight gain (g) = (final weight) − (initial weight).
[4]Weight gain% = (final weight−initial weight)×(initial weight)$^{-1}$×100.
[5]FCR (Food Conversion Ratio) = (dry feed fed)×(wet weight gain)$^{-1}$.
[6]SGR (Specific Growth Rate) = [Ln (final weight)−Ln (initial weight)]×(number of days)$^{-1}$×100.
[7]Feed Ration (% day^{1}) = (dry food fed per day)×(average weight)$^{-1}$×100.
[8]DP% (Dressed−out percentage) = (gutted fish weight)×(total fish weight)$^{-1}$×100.
[9]FY% (Fillet yield percentage) = (fillet weight)×(total weight)$^{-1}$×100.
[10]HSI% (Hepatosomatic Index) = (weight of liver)×(total weight fish)$^{-1}$×100.
[11]K (g cm^{-3}) (Condition Factor) = (total fish weight)×(total fish length)$^{-3}$.
[12]ADC = Apparent Digestibility Coefficients.

Table 3. Proximate composition and total cholesterol (mg g^{-1} on wet basis) of fillet and whole body of rainbow trout fed with the two experimental diets with (H-Chol) or without (L-Chol) cholesterol fortification.

	Experimental Diets[1]		
	L-Chol	H-Chol	P-value
Fillet (mg g^{-1})			
Moisture	716.0±4.1	705.0±5.3	*ns*
Protein	198.9±4.0	202.2±3.9	*ns*
Lipid	72.8±3.0	80.4±6.2	*
Ash	12.3±0.5	12.3±0.3	*ns*
Cholesterol	0.7±0.1	0.7±0.0	*ns*
Whole Body (mg g^{-1})			
Moisture	657.9±5.2	653.6±8.3	*ns*
Protein	172.7±3.9	170.3±3.5	*ns*
Lipid	154.3±4.8	161.2±5.1	*ns*
Ash	15.1±0.6	15.0±0.5	*ns*
Cholesterol	1.4±0.1	1.2±0.0	*ns*

Data are presented as a mean ± s.e.m., $n = 3$, $N = 6$. P>0.05 = ns (not significant),
* = P<0.05,
** = P<0.01.
[1]See Table 1 for experimental diet abbreviations.

Discussion

In this study, it was shown that juvenile rainbow trout fed the non-fortified vegetable oil based-diet (L-Chol) were actively producing cholesterol and were fully capable of compensating for the reduced dietary intake. This is in agreement with the notion that decreased dietary availability of cholesterol would stimulate *de novo* biosynthesis to maintain overall cholesterol homeostasis [8,27]. By contrast, fish fed the cholesterol fortified diet (H-Chol) did not need to biosynthesise cholesterol, as the dietary supply was sufficient, and in fact, there was evidence that some of the dietary cholesterol was catabolised. Similarly to what has been reported for mammals [39], it seems evident that fish cells face the dual requirement of providing sufficient cholesterol for membrane growth and replenishment and, at the same time, avoiding excessive accumulation, via cholesterol catabolism. Accordingly, the main hypothesis of the present study was that a vegetable oil-based diet, containing limited dietary cholesterol, would be responsible for increased energy expenditure for *de novo* cholesterol biosynthesis, and consequently would have an impact on fish performance. To validate this hypothesis, one would expect improved fish performance in fish fed the diet fortified with the additional cholesterol (H-Chol). However, in the present study, juvenile rainbow trout fed for 12 weeks with either diet showed no differences in any of the measured growth and feed efficiency parameters. In a recent study on the same species [26], but using diets containing fish oil and primarily focusing on the issue of soybean inclusion in aquafeed and its hypocholesterolemic effect [40], a positive effect of dietary cholesterol supplementation on fish performances was observed. Similarly, mixed results are currently available for other teleost species. No effects of dietary cholesterol fortification on fish performances were observed in Atlantic salmon fed high fish meal and fish oil based diets [20] and in channel catfish (*Ictalurus punctatus*) fed casein based diets [22].

Table 4. Cholesterol mass balance in rainbow trout fed with the two experimental diets with (H-Chol) or without (L-Chol) cholesterol fortification.

	Experimental Diets[1]		
	L-Chol	H-Chol	P-value
Cholesterol balance (mg fish^{-1})			
Cholesterol in initial fish	38.0±1.4	37.3±0.6	*ns*
Total cholesterol net intake (absorbed)	164±4	361±15	**
Cholesterol in final fish	221±2	190±11	*
Cholesterol Appearance/Disappearance	19±0	−208±12	**

Data are presented as a mean ± s.e.m., $n = 3$, $N = 6$. P>0.05 = ns (not significant),
* = P<0.05,
** = P<0.01.
[1]See Table 1 for experimental diet abbreviations.

However, improved growth performance in response to dietary cholesterol fortification was recorded when channel catfish were fed soybean based diets [22]. In hybrid striped bass (*Morone chrysops* x *M. saxatilis*) fed diets containing abundant fish meal and fish oil no effect of cholesterol fortification on growth was recorded [23]. In Japanese flounder (*Paralichthys olivaceus*) contrasting effects were recorded, with increasing performances in fish fed soybean protein isolate based diets, but decreasing in fish fed fish protein concentrate based diet, as a result of cholesterol fortification [24]. However, it should also be reported that in the latter trial all treatments showed growth retardation, when compared to the control treatment containing fish meal and no added cholesterol. In juvenile turbot (*Scophthalmus maximus*) fed plant protein based diets (soybean meal and wheat gluten) growth improvement was recorded when cholesterol was added up to 1% of total diet, but then growth reduction was observed for higher level of cholesterol fortification [25].

Despite the lack of effect on fish performance in the present study, there was evidence of decreased perivisceral fat deposit (higher DP%), reduced lipid content in fillet and whole body, and significantly decreased apparent *in vivo* fatty acid β-oxidation, in H-Chol fed fish compared to L-Chol fed fish. This may suggest that increased dietary cholesterol was indeed responsible for reduced energy expenditures. Higher weight dispersion (larger coefficient of variation, CV%) in fish fed L-Chol (CV = 22%) compared with H-Chol (CV = 14%), was observed. This might suggest that, within the same population, individual differences in adapting to low dietary cholesterol availability (i.e., cholesterol biosynthesis capability) could be present. Interestingly, in a recent study focusing on dietary cholesterol fortification in rainbow trout fed a diet containing fish oil, but in which the fishmeal fraction was abundantly replaced by soybean meal, it was also shown that increased dietary cholesterol availability contributed to improved disease resistance and overall immune status of the fish [41]. Overall, these may be considered as important, albeit preliminary evidence, clearly warranting future experimental trials over a longer period of time (i.e., the full growing cycle for cultured fish), where the possible effects of reduced dietary cholesterol on performance and health status may be fully manifested.

A remarkably reduced fatty acid β-oxidation was recorded in fish fed H-Chol, compared to L-Chol. Consistently, it has been documented that in rats, dietary cholesterol is directly reducing the overall fatty acid β-oxidation via the inhibition of the activity of carnitine palmitoyltransferase (CPT) [42]. Additionally, it can be

Table 5. The total fatty acid (TFA) content (mg g^{-1} of lipid) and the fatty acid composition (% TFA) of the two experimental diets and of the liver, fillet and whole body of rainbow trout fed with the two experimental diets with (H-Chol) or without (L-Chol) cholesterol fortification.

| Fatty acids | Diets[1] | | Tissues | | | | | | | | |
| | | | Liver | | | Fillet | | | Whole body | | |
	L-Chol	H-Chol	L-Chol	H-Chol	P	L-Chol	H-Chol	P	L-Chol	H-Chol	P
TFA[2] (mg g^{-1} L)	782	780	751±17	743±5	ns	781±13	827±15	ns	798±4	792±6	ns
(% of TFA)											
14:0	0.5	0.5	0.6±0.0	0.6±0.1	ns	0.7±0.0	0.8±0.0	**	0.8±0.1	0.8±0.0	ns
16:0	10.8	10.9	12.6±0.8	13.2±0.3	ns	12.9±0.1	14.0±0.1	**	12.6±0.4	13.0±0.1	ns
18:0	4.3	4.3	6.3±0.1	6.7±0.2	ns	4.6±0.0	4.7±0.1	ns	4.6±0.1	4.6±0.0	ns
20:0	0.2	0.2	0.2±0.0	0.1±0.0	ns	0.2±0.0	0.1±0.0	*	0.2±0.0	0.2±0.0	ns
Total SFA[3]	15.9	16.0	19.7±0.9	20.7±0.2	ns	19.0±0.1	20.2±0.2	**	18.1±0.4	18.6±0.2	ns
16:1n-7	1.9	1.9	1.9±0.2	2.2±0.2	ns	2.6±0.1	3.2±0.1	**	3.0±0.4	3.1±0.1	ns
18:1n-9	33.4	33.3	24.2±1.8	23.5±0.7	ns	33.2±0.5	33.4±0.2	ns	34.4±0.4	34.1±0.3	ns
18:1n-7	1.9	1.9	2.0±0.1	1.9±0.0	ns	2.1±0.0	2.0±0.0	ns	2.1±0.0	2.0±0.0	ns
20:1n-9	0.6	0.6	2.0±0.2	2.0±0.1	ns	1.0±0.0	1.0±0.0	ns	1.0±0.1	1.0±0.0	ns
22:1n-9	-[7]	-	0.2±0.0	0.2±0.0	ns	0.2±0.0	0.2±0.0	ns	0.2±0.0	0.2±0.0	ns
Total MUFA[4]	37.9	37.9	31.4±2.0	30.8±0.9	ns	39.3±0.6	40.1±0.3	ns	40.8±0.7	40.6±0.4	ns
18:2n-6	15.0	15.2	6.2±0.3	6.0±0.1	ns	11.9±0.1	12.0±0.2	ns	12.1±0.4	12.6±0.2	ns
20:2n-6	-	-	1.2±0.1	1.2±0.0	ns	0.6±0.0	0.6±0.0	ns	0.5±0.0	0.5±0.0	ns
20:3n-6	0.1	0.1	1.2±0.0	1.4±0.0	*	0.5±0.0	0.5±0.0	ns	0.4±0.0	0.5±0.0	ns
20:4n-6	0.2	0.2	2.9±0.3	2.9±0.1	ns	0.5±0.0	0.5±0.0	ns	0.4±0.0	0.4±0.0	ns
22:4n-6	0.1	0.1	0.2±0.0	0.2±0.0	ns	0.2±0.1	0.2±0.0	ns	0.1±0.0	0.1±0.0	ns
Total n-6 PUFA[5]	15.8	15.9	12.9±0.2	13.0±0.1	ns	14.1±0.1	14.1±0.1	ns	14.0±0.3	14.5±0.3	ns
18:3n-3	29.3	29.3	5.7±0.7	4.7±0.2	ns	17.7±0.2	16.5±0.4	*	18.1±1.0	17.2±0.4	ns
18:4n-3	0.1	0.1	0.8±0.1	0.7±0.0	ns	2.2±0.1	2.0±0.0	ns	2.4±0.3	2.2±0.1	ns
20:3n-3	0.1	0.1	0.9±0.0	0.8±0.0	ns	0.8±0.0	0.8±0.0	ns	0.7±0.1	0.7±0.0	ns
20:4n-3	-	-	1.0±0.1	1.0±0.0	ns	0.9±0.0	0.9±0.0	ns	0.9±0.0	1.0±0.1	*
20:5n-3	0.1	0.1	4.9±0.2	4.9±0.4	ns	1.3±0.1	1.3±0.0	ns	1.1±0.0	1.1±0.1	ns
22:6n-3	0.5	0.5	21.3±1.4	21.8±0.3	ns	4.6±0.3	4.1±0.3	ns	3.3±0.0	3.5±0.1	*
Total n-3 PUFA[5]	30.3	30.3	36.1±1.2	35.6±0.9	ns	27.7±0.4	25.6±0.4	*	27.1±0.8	26.3±0.3	ns
Total PUFA	46.1	46.2	48.9±1.4	48.5±1.0	ns	41.8±0.6	39.7±0.5	*	41.1±1.1	40.8±0.5	ns
n-6 LC-PUFA[6]	0.8	0.6	6.5±0.5	6.8±0.1	ns	1.8±0.2	1.8±0.1	ns	1.6±0.1	1.6±0.1	ns
n-3 LC-PUFA	0.9	0.9	29.6±1.7	30.1±0.7	ns	7.8±0.4	7.3±0.4	ns	6.5±0.1	6.9±0.0	**
Total LC-PUFA	1.8	1.5	36.0±2.1	36.8±0.7	ns	9.6±0.6	9.0±0.4	ns	8.1±0.1	8.5±0.0	**

Data are presented as a percentage of total fatty acid ± s.e.m., $n=3$, $N=6$. $P>0.05=$ ns (not significant),
* = $P<0.05$,
** = $P<0.01$.
[1]See Table 1 for experimental diet abbreviations.
[2]TFA = Total fatty acid (mg g^{-1} lipid).
[3]SFA = Saturated fatty acids.
[4]MUFA = Monounsaturated fatty acids.
[5]PUFA = Polyunsaturated fatty acids.
[6]LC-PUFA = Long chain polyunsaturated fatty acids.
[7]- = not detected.

argued that the increased FA β-oxidation resulting from a reduced dietary cholesterol supply may be due to the increased energy and acetyl-CoA demands that are required for the increased cholesterol synthesis [43,44].

A second objective of the present study was to assess if dietary cholesterol had any effect on fatty acid metabolism, as the key enzymes involved this pathway are known to be affected by several physiological and nutritional factors [45], including dietary fatty acid composition and cholesterol [17]. Specifically, in rats, dietary supplementation of cholesterol has been shown to increase the activity of stearyl-CoA desaturase (*D9fad*), leading to accumulation of MUFA [17,42], whereas the activities of *D5fad* and *D6fad* were both shown to be reduced by dietary cholesterol [15,16,18]. Contrary to these findings, in the present study on a teleost species, minor effects were observed on the activity of *D9fad* and final MUFA content, but a direct and positive effect was shown on the

Table 6. The apparent *in vivo* fatty acid β-oxidation (nmol g^{-1} day^{-1}) in rainbow trout fed with the two experimental diets with (H-Chol) or without (L-Chol) cholesterol fortification.

Fatty acids	Experimental Diets[1]		
	L-Chol	H-Chol	P-value
12:0	1.4±0.7	0.0±0.0	ns
14:0	2.9±2.7	0.0±0.0	ns
16:0	451±79	170±48	*
18:0	213±15	131±11	*
20:0	16.3±0.2	9.2±1.4	*
Total SFA[2]	684±95	311±60	*
14:1n-5	1.2±0.1	1.0±0.1	ns
16:1n-7	35.0±20.9	0.0±0.0	ns
18:1n-7	95.0±8.5	65.3±3.4	*
18:1n-9	1,955±132	1,372±96	*
22:1n-11	8.4±0.1	7.3±0.1	**
Total MUFA[2]	2,095±158	1,445±99	*
18:2n-6	1,122±37	841±22	**
22:2n-6	39.3±0.9	11.6±0.8	**
22:4n-6	3.5±1.8	3.5±0.3	ns
22:5n-6	12.3±1.1	6.7±0.8	*
18:3n-3	2,329±58	1,890±65	**
Total PUFA[2]	3,506±98	2,752±88	**
Total n-6 PUFA	1,177±40	863±22	**
Total n-3 PUFA	2,329±58	1,890±65	**
Total â-Oxidation	6,285±336	4,508±240	*

Data are presented as a mean ± s.e.m., n = 3, N = 6. P>0.05 = ns (not significant),
* = P<0.05,
** = P<0.01.
[1]See Table 1 for experimental diet abbreviations.
[2]See Table 5 for fatty acid class abbreviations.

Table 7. Apparent *in vivo* activity (nmol g^{-1} day^{-1}) of the key enzymes in fatty acid biosynthesis pathways in rainbow trout fed with the two experimental diets with (H-Chol) or without (L-Chol) cholesterol fortification.

	Experimental Diets[1]		
	L-Chol	H-Chol	P-value
D9fad[2]	14.2±13.8	39.4±15.7	ns
16:0 to16:1n-7	13.8±13.8	38.3±15.6	ns
20:0 to 20:1n-11	0.5±0.1	1.1±0.1	**
D6fad[3]	840.0±51.3	942.2±39.6	ns
18:2n-6 to18:3n-6	59.5±6.7	75.6±0.2	*
18:3n-3 to18:4n-3	586.4±36.5	640.4±28.4	ns
24:5n-3 to24:6n-3	194.1±8.9	226.3±11.0	*
D5fad[4]	294.1±16.3	338.0±19.9	ns
20:3n-6 to 20:4n-6	9.4±1.5	12.8±0.9	ns
20:4n-3 to 20:5n-3	284.7±15.0	325.2±19.4	*
Elovl5 & Elovl2[5]	932.9±52.7	1,119.7±45.0	*
12:0 to 14:0	5.1±5.1	16.6±2.3	ns
20:0 to 22:0	0.8±0.4	1.4±0.1	ns
18:1n-9 to 20:1n-9	14.0±4.3	32.0±2.8	*
18:2n-6 to 20:2n-6	39.5±1.4	45.7±0.3	*
18:3n-6 to 20:3n-6	35.1±3.6	46.4±0.6	*
18:3n-3 to 20:3n-3	62.3±3.6	65.8±0.2	ns
18:4n-3 to 20:4n-3	357.5±18.3	418.3±16.6	*
20:1n-9 to 22:1n-9	12.6±1.1	15.8±0.8	*
20:3n-3 to 22:3n-3	2.7±0.6	2.8±0.1	ns
20:5n-3 to 22:5n-3	205.6±9.7	243.9±13.3	*
22:1n-9 to 24:1n-9	1.9±0.6	3.6±0.2	ns
22:5n-3 to 24:5n-3	194.1±8.9	226.3±11.0	*
Peroxisomal chains shortening			
24:6n-3 to 22:6n-3	194.1±8.9	226.3±11.0	*
Neogenesis	4.6±4.6	15.4±2.5	ns

Data are presented as a mean ± s.e.m., n = 3, N = 6. P value: ns = not significant,
* = P<0.05,
** = P<0.01.
[1]See Table 1 for experimental diet abbreviations.
[2]D9fad = fatty acid △-9 desaturase.
[3]D6fad = fatty acid △-6 desaturase.
[4]D5fad = fatty acid △-5 desaturase.
[5]Elovl5 & Elovl2 = fatty acid elongase (−5 and −2).

transcription rate and activity of *D6fad*. FA elongase was also positively stimulated by the addition of dietary cholesterol, and this ultimately resulted in the modification of the whole body fatty acid

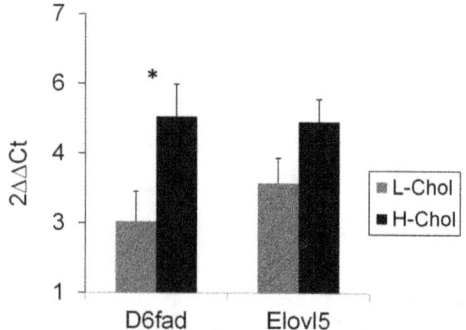

Figure 1. Differential gene expression of fatty acid △-6 desaturase (D6fad) and fatty acid elongase 5 (Elovl5) in liver of rainbow trout, fed with the two experimental diets with (H-Chol) or without (L-Chol) cholesterol fortification. (* significant differences between L-Chol and H-Chol, P<0.05, n = 3).

composition in fish fed H-Chol, which recorded higher content of n-3 LC-PUFA.

The direct effect of dietary cholesterol on MUFA metabolism in rats has been suggested to be an adaptive mechanism in order to provide more oleoyl-CoA as a substrate for cholesterol esterification and storage [46], whilst no explanation regarding the negative effects of cholesterol on PUFA metabolism in rats has been made. Available information regarding the possible effects of cholesterol on fatty acid bioconversion are somewhat mixed, as *in vitro* studies using hamster ovary have shown the opposite effect, with *D9fad* gene expression and its enzyme activity being repressed by increased available cholesterol [47]. Additionally, when it comes to comparing results obtained in different species, it should be noted that it has been recently shown that generally, the fatty acid metabolism and its regulation in teleost fish can be markedly different from that of mammals [48,49]. Therefore, it is not surprising that fish fatty acid metabolism may respond differently to dietary cholesterol, when compared to mammals.

In general agreement with what was previously reported in rats, one study on Atlantic salmon reported an apparent increase in *D9fad* activity, accompanied by a minor lowering effect on *D5fad* and *D6fad* activities, when fish were fed with high dietary cholesterol (~14 mg g^{-1}) compared to fish fed low dietary cholesterol (~4 mg g^{-1}) [20]. However, it should be noted that this study used fish oil-based diets, and thus all diets contained much higher cholesterol content, and had abundant LC-PUFA. As such, it is almost impossible to compare these findings with those of the present study, where diets contained limited LC-PUFA (and thus allowing for their biosynthesis) and much lower cholesterol content. Additionally, the involvement of cholesterol in the membrane bilayer, and the subsequent potential effect on the fluidity of the membrane, is likely to be extremely important for fish, particularly for those inhabiting very cold water. The fish studied in present experiment were rainbow trout, and reared in freshwater at temperate conditions (15°C) and thus likely differed in their requirements and/or cholesterol utilisation, compared with the Atlantic salmon (reared at 7–9°C) of the previous study.

Peroxisomal chain shortening (β-oxidation) in H-Chol fed fish was observed, which resulted in higher 22:6n-3 (DHA) production and deposition. It is important to highlight that peroxisomal β-oxidation is a gender- related process in mammals [50], along with cholesterol regulation [46], and fatty acid bioconversion [51,52]. Recently, it was shown that, in a marine teleost fish (*Solea*

senegalensis), *Elovl5* and Δ-4 fatty acid desaturase (*D4fad*) displayed significant differences in gene expression between males and females [53]. In the present study, juvenile, mixed (and un-sexed) fish were used, which may have contributed in the relatively large variability observed in the extensive number parameters that were recorded. Thus, the possible effects of gender on cholesterol and fatty acid metabolism and their modulation at different temperatures should be an important factor to be considered in future studies.

In conclusion, the possible effects of reduced dietary cholesterol in diets formulated with a substantial replacement of fish meal and fish oil with vegetable products can be quite significant. The regulation of cholesterol in fish clearly requires further research, towards the optimisation of the formulation of future aquafeed.

Acknowledgments

The author are grateful to Dr. Ramez Alhazzaa, Prof. Andy Sinclair (Deakin University, Australia) and Dr. Gunveen Kaur (Victoria University, Australia) for their generous technical assistance and Dr. Richard Smullen (Ridley Aquafeed, Ridley AgriProducts, Australia) for kindly providing feedstuff ingredients.

Author Contributions

Conceived and designed the experiments: GMT. Performed the experiments: ML KH. Analyzed the data: FN ML KH NKAH JAD. Contributed reagents/materials/analysis tools: JAD GMT. Wrote the paper: FN GMT.

References

1. Vlahcevic ZR, Hylemon PB, Chiang JYL (1994) Hepatic cholesterol metabolism. In: Arias I, editor. The Liver Biology and Pathobiology. 3 ed. Raven Press, New York, NY. pp. 379–389.
2. Kritchevsky D (2008) Fats and oils in human health. In: Min CCADB, editor. Food Lipids: chemistry, nutrition, and biotechnology. CRC Press Boca Raton, FL. pp. 499–511.
3. Parish EJ, Li S, Bell AD (2008) Chemistry of waxes and sterols In Food Lipids: chemistry, nutrition, and biotechnology; Min CCADB, editor: CRC Press Boca Raton, FL., USA.
4. Zhang W, Mai K, Yao C, Xu W, Wang W, et al. (2009) Effects of dietary cholesterol on growth, survival and body composition of juvenile abalone Haliotis discus hannai Ino. Aquaculture 295: 271–274.
5. Aschoff L (1932) Observations concerning the relationship between cholesterol metabolism and vascular disease. Br Med J 2: 1131–1134.
6. Coutteau P, Geurden I, Camara MR, Bergot P, Sorgeloos P (1997) Review on the dietary effects of phospholipids in fish and crustacean larviculture. Aquaculture 155: 149–164.
7. Sheen S-S (2000) Dietary cholesterol requirement of juvenile mud crab Scylla serrata. Aquaculture 189: 277–285.
8. Leaver M, Villeneuve L, Obach A, Jensen L, Bron J, et al. (2008) Functional genomics reveals increases in cholesterol biosynthetic genes and highly unsaturated fatty acid biosynthesis after dietary substitution of fish oil with vegetable oils in Atlantic salmon (Salmo salar). BMC Genomics 9: 299.
9. Tocher DR, Bendiksen EÅ, Campbell PJ, Bell JG (2008) The role of phospholipids in nutrition and metabolism of teleost fish. Aquaculture 280: 21–34.
10. Olsen Y (2011) Resources for fish feed in future mariculture. Aquacult Environ Interact 1: 187–200.
11. Turchini GM, Torstensen BE, Ng W-K (2009) Fish oil replacement in finfish nutrition. Rev Aquaculture 1: 10–57.
12. Liland NS, Espe M, Rosenlund G, Waagbo R, Hjelle JI, et al. (2013) High levels of dietary phytosterols affect lipid metabolism and increase liver and plasma TAG in Atlantic salmon (Salmo salar L.). Br J Nutr 1: 1–10.
13. Bell JG, Koppe W (2008) Lipids in aquafeed. In: Giovanni Turchini W-KN, Douglas R. Tocher, editor. Fish oil replacemnet and alternative lipid sources in aquaculture feeds. SRC Press Boca raton, FL. pp. 61–98.
14. Moffat C (2009) Fish oils: the chemical building blocks. In: Rossell B, editor. Fish oils. Chichester, UK.: Blackwell. pp. 51–79.
15. Garg ML, Sebokova E, Thomson AB, Clandinin MT (1988) Delta 6-desaturase activity in liver microsomes of rats fed diets enriched with cholesterol and/or omega 3 fatty acids. Biochem J 249: 351–356.
16. Garg ML, Thomson AB, Clandinin MT (1988) Effect of dietary cholesterol and/or omega 3 fatty acids on lipid composition and delta 5-desaturase activity of rat liver microsomes. J Nutr 118: 661–668.
17. Garg ML, Wierzbicki AA, Thomson AB, Clandinin MT (1988) Dietary cholesterol and/or n-3 fatty acid modulate delta 9-desaturase activity in rat liver microsomes. Biochim Biophys Acta 962: 330–336.
18. Muriana FJG, Ruiz-Gutierrez V, Vazquez CM (1992) Influence of dietary cholesterol on polyunsaturated fatty acid composition, fluidity and membrane-bound enzymes in liver microsomes of rats fed olive and fish oil. Biochimie 74: 551–556.
19. Farrell AP, Saunders RL, Freeman HC, Mommsen TP (1986) Arteriosclerosis in Atlantic salmon. Effects of dietary cholesterol and maturation. Arterioscl Throm Vas 6: 453–461.
20. Bjerkeng Storebakken, Wathne (1999) Cholesterol and short-chain fatty acids in diets for Atlantic salmon Salmo salar (L.): effects on growth, organ indices, macronutrient digestibility, and fatty acid composition. Aquacult Nutr 5: 181–191.
21. Liland NS, Roselund G, Hjelle JI, Torstensn B (2012) Metabolic regulation of cholesterol levels in Atlantic salmon (salmo salar); Molde, Norway, 4-7 June 2012. pp. 29–30.
22. Twibell RG, Wilson RP (2004) Preliminary evidence that cholesterol improves growth and feed intake of soybean meal-based diets in aquaria studies with juvenile channel catfish, Ictalurus punctatus. Aquaculture 236: 539–546.
23. Sealey WM, Craig SR, Gatlin DM (2001) Dietary cholesterol and lecithin have limited effects on growth and body composition of hybrid striped bass (Morone chrysops x M-saxatilis). Aquaculture Nutrition 7: 25–31.
24. Deng J, Mai K, Ai Q, Zhang W, Wang X, et al. (2010) Interactive effects of dietary cholesterol and protein sources on growth performance and cholesterol metabolism of Japanese flounder (Paralichthys olivaceus). Aquaculture Nutrition 16: 419–429.
25. Yun BA, Mai KS, Zhang WB, Xu W (2011) Effects of dietary cholesterol on growth performance, feed intake and cholesterol metabolism in juvenile turbot (Scophthalmus maximus L.) fed high plant protein diets. Aquaculture 319: 105–110.
26. Deng JM, Bi BL, Kang B, Kong LF, Wang QJ, et al. (2013) Improving the growth performance and cholesterol metabolism of rainbow trout (Oncorhynchus mykiss) fed soyabean meal-based diets using dietary cholesterol supplementation. British Journal of Nutrition 110: 29–39.
27. Parish EJ, Li S, Bell AD (2008) Chemistry of waxes and sterols. In: Casimir C Akoh, David B Min, editors. Food Lipids: chemistry, nutrition, and biotechnology Third ed. Boca Raton, FL: CRC Press pp. 99–123.
28. Francis DS, Turchini GM, Jones PL, De Silva SS (2007) Dietary lipid source modulates in vivo fatty acid metabolism in the freshwater fish, Murray cod (Maccullochella peelii peelii). J Agric Food Chem 55: 1582–1591.
29. Brown TD, Francis DS, Turchini GM (2010) Can dietary lipid source circadian alternation improve omega-3 deposition in rainbow trout? Aquaculture 300: 148–155.

13

Effect of Fish Oil Supplementation on Fasting Vascular Endothelial Function in Humans: A Meta-Analysis of Randomized Controlled Trials

Wei Xin[1], Wei Wei[2], Xiaoying Li[1]*

1 First Department of Geriatric Cardiology, Chinese PLA General Hospital, Beijing, PR China, **2** Medical College of Nankai University, Tianjin, PR China

Abstract

Background: Effect of fish oil supplementation on flow-mediated dilation, an index of endothelial function in humans, remains controversial. We performed a meta-analysis to determine whether fish oil supplementation could improve endothelial function.

Methods: Human intervention studies were identified by systematic searches of Medline, Embase, Cochrane's library and references of related reviews and studies. A random-effect model was applied to estimate the pooled results. Meta-regression and subgroup analyses were performed to evaluate the impact of study characteristics on the effect of fish oil supplementation on flow-mediated dilation.

Results: A total of sixteen records with 1,385 subjects were reviewed. The results of the pooled analysis showed that fish oil supplementation significantly improved flow-mediated dilation (weighed mean difference: 1.49%, 95% confidence interval 0.48% to 2.50%, p = 0.004). Meta-regression and subgroup analysis suggested that the quality of included studies were inversely related to the overall effect (regression coefficient = −1.60, p = 0.04), and the significance of the effect was mainly driven by the studies with relatively poor quality. Sensitivity analysis including only double-blind, placebo-controlled studies indicated fish oil supplementation has no significant effect on endothelial function (weighed mean difference: 0.54%, 95% confidence interval −0.25% to 1.33%, p = 0.18). Besides, normoglycemic subjects or participants with lower diastolic blood pressure seemed to be associated with remarkable improvement of endothelial function after fish oil supplementation.

Conclusions: Although current evidence suggested a possible role of fish oil in improving endothelial function, large-scale and high-quality clinical trials are needed to evaluate these effects before we can come to a definite conclusion.

Editor: Rudolf Kirchmair, Medical University Innsbruck, Austria

Funding: No current external funding sources for this study.

Competing Interests: The authors have declared that no competing interests exist.

.* E-mail: lixy301@163.com

Introduction

Accumulating evidence from epidemiological studies and clinical trials have suggested that increased intake of non-fried fish or supplementation with fish oil is associated with lower risk of cardiovascular mortality, indicating a potential role of fish oil supplementation in the primary and secondary prevention of cardiovascular diseases (CVD) [1,2]. Fish oil – mainly consisting of two categories of marine omega 3 polyunsaturated fatty acids (PUFAs) – eicosapentaenoic acid (EPA) and docosahexaenoic acid (DHA) – may exert its cardioprotective effects via many mechanisms including lowering blood pressure, regulation of blood lipids, lowering heart rate, anti-inflammation, anti-arrhythmia, and possible improvement of endothelial dysfunction et al. [3,4].

Endothelial dysfunction is considered to be an early pathophysiologic feature of many CVD including atherosclerosis and hypertension et al., and is also an independent predictor and prognostic factor for these CVD [5–8]. Clinically, endothelial function can be determined by measuring flow-mediated dilation (FMD), which represents the ability of the brachial artery to dilate in response to ischemia-induced hyperemia and reflects of the local bioavailability of endothelium-derived vasodilator, mainly including nitric oxide (NO) [9–11]. Although a cross-sectional study has suggested that dietary fish or fish oil consumption may be associated with enhanced FMD in women [12], results of prospective randomized controlled trials evaluating the effect of fish oil supplementation on FMD, a surrogate of endothelial function, are generally controversial [13–28], partly due to the small number of included participants. Therefore, we performed a meta-analysis to systematically evaluate the effect of fish oil supplementation on FMD in humans. More importantly, we tried to explore the influence of the participant and study characteristics, especially those of established risk factors, to the effect of fish oil on endothelial function.

Methods

This systematic review and meta-analysis was performed according to PRISMA (Preferred Reporting Items for Systematic

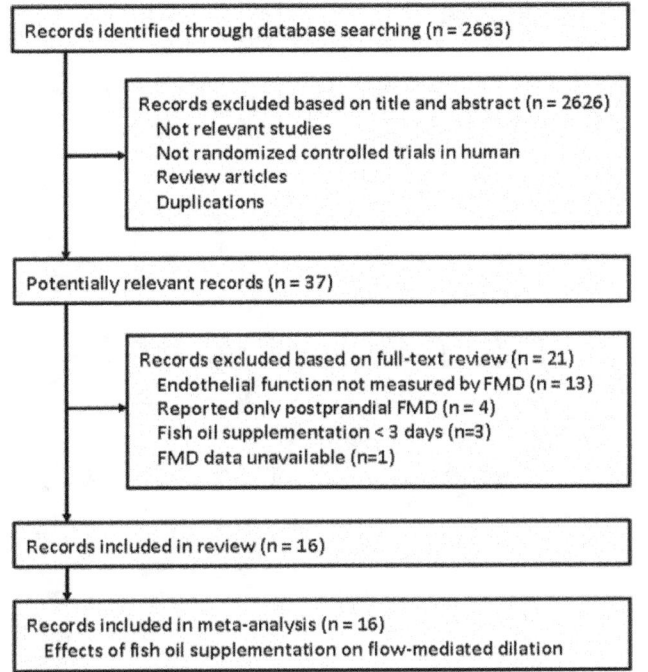

Records identified through database searching (n = 2663)

Records excluded based on title and abstract (n = 2626)
Not relevant studies
Not randomized controlled trials in human
Review articles
Duplications

Potentially relevant records (n = 37)

Records excluded based on full-text review (n = 21)
Endothelial function not measured by FMD (n = 13)
Reported only postprandial FMD (n = 4)
Fish oil supplementation < 3 days (n=3)
FMD data unavailable (n=1)

Records included in review (n = 16)

Records included in meta-analysis (n = 16)
Effects of fish oil supplementation on flow-mediated dilation

Figure 1. Flow diagram of the study selection procedure.

Reviews and Meta-Analyses) statement [29] and Cochrane Handbook guidelines [30].

Search strategy

We systematically searched Pubmed (from 1950 to February, 2012), Embase (from 1966 to February, 2012) and the Cochrane Library (Cochrane Center Register of Controlled Trials) for relevant records, using the term "omega-3 fatty acids", "fish oil", "fish-oil", "marine oil", "eicosapentaenoic acid", "EPA", "docosahexaenoic acid", "DHA", "dietary therapy" paired with the following: "endothelial", "endothelium", "coronary heart disease", "cardiovascular disease", "stroke", "cerebrovascular disease", and "trial". The search was limited to studies in humans. We also analyzed reference lists of original and review articles using a manual approach.

Study selection

Original studies were included if they met the following criteria: 1) published as full-length articles in English; 2) reported as a prospective, randomized, and controlled trial with either a parallel or a crossover design (regardless of sample size); 3) analyzed human subjects who were assigned to oral fish oil supplementation or a control group for ≥3 days; 4) evaluated endothelial function by measuring of fasting FMD in the brachial artery; 5) data [means and standard deviations (SDs)] concerning changes in FMD from baseline were reported or could be estimated. Review articles, nonhuman studies, observational studies without longitudinal follow-up, cross-sectional studies, duplicate publications, and studies in which changes in FMD were not reported or could not be estimated were excluded.

Data extraction and quality assessment

Two authors (WX and WW) independently performed the literature searching, data extraction, and quality assessment according to inclusion criteria. Discrepancies were resolved by

consensus. Extracted data included study design characteristics (parallel or crossover), patient characteristics [e.g., number, age, sex, body mass index (BMI), general healthy status, percentages of the smokers and patients with type 2 diabetes (T2DM), baseline systolic blood pressure (SBP), diastolic blood pressure (DBP), serum triglyceride (TG) and total cholesterol (TC) levels, usage of statins, angiotensin-converting enzyme inhibitors (ACEIs) or angiotensin receptor blockers (ARBs), and baseline FMD of the participants], intervention strategies (total dose of fish oil, dose of EPA and DHA, and the treatment in control groups), follow-up duration, and means and SDs for changes of FMD from baseline. Because previous studies indicated that technical aspects of FMD measurement may influence the results [10], characteristics of FMD measurement (occlusion position and occlusion duration) were also extracted for analysis. If these data were missing or not reported in the content of the paper, corresponding authors were contacted to ask if the unpublished data were available. For trials in which fish oil was supplied of more than one dose or treatment duration (e.g., FMD data were repeatedly measured at more than one time point) multiple studies were considered.

The quality of the studies was judged by Jadad Score, which evaluates the quality of randomization, generation of random numbers, concealment of treatment allocation, blinding, and reporting of withdrawals [31]. Trials scored one point for each area addressed, with a possible score between 0 and 5, where 5 represented the highest level of quality.

Statistical analysis

The primary outcome of this meta-analysis was the change in FMD between baseline and endpoint in response to oral fish oil intervention. The pooled effect was presented as weighted mean difference (WMD) with 95% confidence intervals (CI). Inter-study heterogeneity was formally tested using Cochrane's test, and significant heterogeneity was considered existing if p value was <0.10. The I^2 statistic, which describes the percentage of total variation across studies that is due to heterogeneity rather than chance, was also examined, and a value of $I^2 > 50\%$ indicated significant heterogeneity [32]. We used a random-effect model to estimate the overall effect instead of a fixed-effect model, considering that this is a more conservative method that takes into account that study heterogeneity can vary beyond chance, thus providing a more generalizable results. Univariate meta-regression analysis was performed to identify the possible source of heterogeneity, using the following variables: number of subjects in each study; mean age; percentage of males; BMI; percentage of smokers; SBP, DBP, percentage of patients with T2DM; mean TG level; mean TC level; baseline FMD; fish oil dose; DHA dose; EPA dose; follow-up duration and Jadad Score. Predefined subgroup analyses were also performed to further explore the possible influence of the above study characteristics on the pooled outcome. Median values of continuous variables were used as cutoff values for grouping studies. Sensitivity analysis by excluding certain studies was performed to test the stability of the results. Furthermore, potential publication bias was assessed with funnel plot, Egger regression asymmetry test [33], as well as fail-safe N test [30]; p values were two-tailed and statistical significance was set at 0.05. Meta-analysis and statistical analysis was performed with Stata software (version 12.0; Stata Corporation, College Station, TX).

Table 1. Baseline characteristics of participants of included studies.

Study	No. of subjects	Participants	Mean age (years)	Male (%)	BMI (kg/m²)	T2DM (%)	Smokers (%)	Mean SBP (mmHg)	Mean DBP (mmHg)	Mean TG (mmol/L)	Mean TC (mmol/L)	Statins used	ACEI/ARB used	Baseline FMD (%)
Woodman 2003a	20	Hypertensive T2DM	61.3	78.8	28.9	100	0	130.6	71.5	1.49	4.55	Partially	Partially	3.26
Woodman 2003b	18	Hypertensive T2DM	61.2	73.5	30.3	100	0	130.6	70.5	1.65	4.55	Partially	Partially	4.5
Dyerberg 2004	50	Healthy males	38.4	100	24.5	0	21.6	127.5	77.6	1.18	4.95	None	None	3.75
Engler 2004	20	Dyslipidemic children	14	NR	21	0	0	115	58	1.57	7.33	NR	NR	5.9
Prabodh 2007	26	Healthy adults	31	65.4	23.4	0	0	113.2	64.3	1.24	NR	None	None	NR
Hill 2007a	65	Overweight adults	50.3	36.9	34.1	0	0	130.1	74.5	1.79	5.91	None	None	4.25
Hill 2007b	65	Overweight adults	50.3	36.9	34.1	0	0	130.1	74.5	1.79	5.91	None	None	4.25
Schiano 2008	32	PAD	66	90.6	26.9	46.9	94	NR	NR	1.77	5.1	Partially	Partially	6.9
Mindrescu 2008	30	Dyslipidemic South Asians	51	76.7	27.1	33.3	NR	123.3	73.1	1.56	4.87	All	Partially	−1.99
Wright 2008a	56	SLE	48.1	6.7	25.6	0	15	125.5	72.4	1.05	4.8	None	None	2.74
Wright 2008b	56	SLE	48.1	6.7	25.6	0	15	125.5	72.4	1.05	4.8	None	None	2.74
Rizza 2009	50	Normoglycemic offspring of T2DM	29.9	50	26.2	0	NR	115.1	76.9	1.33	4.99	None	None	7.9
Stirban 2010	32	T2DM	56.8	NR	31.2	100	NR	139	81	1.6	4.84	Partially	Partially	5.54
Wong 2010	97	T2DM	60.1	44.3	25.8	100	17.5	137.5	78	1.45	4.9	Partially	Partially	3
Hileman 2011	31	HIV-infected adults	51	100	25	0	74.3	118.5	79	1.4	4.64	None	None	3.13
Haberka 2011	40	AMI, successful PCI	60±9	80	28.9	20	52.5	NR	NR	1.59	5.05	All	All	9.5
Skulas-Ray 2011a	26	Hypertriglyceridemic	44.3	88.5	29	0	0	122.9	81.9	2.52	5.36	None	None	5
Skulas-Ray 2011b	26	Hypertriglyceridemic	44.3	88.5	29	0	0	122.9	81.9	2.52	5.36	None	None	5
Sanders 2011a	151	Nonsmokers	55	38.5	25.9	0	0	120.5	77	1.15	5.45	Partially	Partially	5.39
Sanders 2011b	150	Nonsmokers	55	38.7	26.4	0	0	121	77.5	1.14	5.45	Partially	Partially	4.85

Table 1. Cont.

Study	Participants	No. of subjects	Mean age years	Male %	BMI kg/m²	T2DM %	Smokers %	Mean SBP mmHg	Mean DBP mmHg	Mean TG mmol/L	Mean TC mmol/L	Statins used	ACEI/ARB used	Baseline FMD %
Sanders 2011c	Nonsmokers	151	55	38.9	25.9	0	0	120.5	76.5	1.13	5.45	Partially	Partially	5.01
Moertl 2011a	Nonischemic CHF	30	57.3	80	27.5	16.7	NR	110.5	72.9	NR	4.43	NR	All	8.41
Moertl 2011b	Nonischemic CHF	29	59.3	86.2	27.9	24.1	NR	110.3	74	NR	4.46	NR	All	8.46

The studies by woodman (2003), Skulas-Ray (2011) and Moertl (2011) include two intervention groups with different fish oil doses separately, and the study by Sanders (2011) includes three intervention groups with different fish oil doses. The studies by Hill (2007) and Wright (2008) each contain two comparisons with different fish oil treatment durations. BMI, body mass index; SBP, systolic blood pressure; DBP, diastolic blood pressure; TG, triglyceride; TC, total cholesterol; ACEI, angiotensin-converting enzyme inhibitor; ARB, angiotensin receptor blocker; FMD, flow-mediated dilation; T2DM, type 2 diabetes mellitus; PAD, peripheral artery disease; SLE, systemic lupus erythematosus; HIV, human immunodeficiency virus; AMI, acute myocardial infarction; PCI, percutaneous coronary intervention; CHF, chronic heart failure; NR, not reported.

Results

Search results

A total of 2663 records were identified, and 2626 were excluded because they were review articles or duplications, did not describe randomization or controlling, or because the objectives of these studies were irrelevant to the present meta-analysis. Of the 37 potentially relevant records screened, sixteen [13–28] met the selection criteria for the current meta-analysis (**Figure 1**). Twenty-one records were excluded because endothelial function was not measured by FMD in 13 records; only data of postprandial FMD were reported in 4 records; fish oil was supplemented for <3 days in 3 records; and because fasting FMD data were unavailable in one record.

Study characteristics

Overall, a total of 23 studies from 16 published articles were included in the current meta-analysis, which comprised a total of 1385 participants, 707 subjects in the fish oil group and 678 patients in the control group. The characteristics of these participants and the studies were shown in **Table 1** and **Table 2**. All of the included studies were prospective randomized controlled trials; of which nineteen studies [13,14,16–21,23–25,27,28] were of parallel-design and the other four [15,22,26] were crossover design. The sample size ranged from 18 to 151. Six studies [14,16,21,27] were made up of generally healthy populations; two studies [17] comprised of overweighed adults; the other fifteen studies included patients with ≥1 chronic conditions, such as dyslipidemia [15,19,26], T2DM [13,22,23], systemic lupus erythematosus (SLE) [20], peripheral artery disease (PAD) [18], human immunodeficiency virus (HIV) infection [24], acute myocardial infarction (AMI) [25] and chronic heart failure (CHF) [28]. The mean ages of the enrolled subjects ranged from 14 to 66 years old. The mean BMI ranged from 21 to 34.1 kg/m². The mean SBP and DBP varied from 110.3 to 139 mmHg and from 58 to 81.9 mmHg respectively. The mean baseline levels of TG and TC ranged form 1.05 to 2.52 mmol/L, and from 4.43 to 7.33 mmol/L. Ten studies included participants who were all [19,25] or partially [13,18,22,23,27] on statins, while ten studies only included participants who were not [14,16,17,20,21,24,26]. Twelve studies included participants who were all [25,28] or patially [13,18,19,22,23,27] on ACEIs/ARBs, while ten studies only included participants who were not [14,16,17,20,21,24,26]. The baseline FMD value varied from −1.99% to 9.5% (the FMD value below zero indicating constricted brachial artery in response to ischemia-induced hyperemia [9–11]). The dose of fish oil (defined as total dose of DHA and EPA) adopted in the included studies varied from 450 to 4530 mg/d, with the follow-up duration ranging from 2 to 52 weeks. Concerning the technical aspects of the FMD measurement, the occlusion position were forearms in 20 studies [13–20,22–27], and upper arms in the other 3 studies [21,28]; the occlusion duration was 5 min in 17 studies [13,15,17–19,21,23,24,26–28], 4.5 min in 4 studies [14,20,22] and 3 min in the rest 2 studies [16,25].

Data quality

The quality scores of the 23 studies ranged form 2 to 5. All of the included studies were randomized and controlled trials, with 19 studies in a double-blind design [13–15,17,20–24,26–28]. Only 6 studies reported the method of random sequence generation [23,27,28], and 11 reported allocation concealment [14,20,22–24,27,28]. Details of withdrawals were reported in all of the included studies.

Table 2. Characteristics of study design of included studies.

Study	Study design	Fish oil dose	DHA dose	EPA dose	Control	Duration	Occlusion position	Occlusion duration	Jadad Score
		mg/d	mg/d	mg/d		weeks		min	
Woodman 2003a	R, PC, DB	4000	0	4000	Olive oil	6	forearm	5	3
Woodman 2003b	R, PC, DB	4000	4000	0	Olive oil	6	forearm	5	3
Dyerberg 2004	R, PC, DB	3168	1320	1848	Palm oil	8	forearm	4.5	4
Engler 2004	R, PC, DB, CO	1200	1200	0	Corn/soy oil	6	forearm	5	3
Prabodh 2007	R, PC, SB	500	200	300	Corn oil	2	forearm	3	3
Hill 2007a	R, PC, DB	1920	1560	360	Sunflower oil	6	forearm	5	3
Hill 2007b	R, PC, DB	1920	1560	360	Sunflower oil	12	forearm	5	3
Schiano 2008	R, SB	1700	1063	637	No treatment	13	forearm	5	2
Mindrescu 2008	R, CO	4530	1950	2580	No treatment	4	forearm	5	2
Wright 2008a	R, PC, DB	3000	1200	1800	Olive oil	12	forearm	4.5	4
Wright 2008b	R, PC, DB	3000	1200	1800	Olive oil	24	forearm	4.5	4
Rizza 2009	R, PC, DB	1700	1020	680	Olive oil	12	upper arm	5	3
Stirban 2010	R, PC, DB, CO	1680	760	920	Olive oil	6	forearm	4.5	4
Wong 2010	R, PC, DB	2680	1000	1680	Olive oil	12	forearm	5	5
Hileman 2011	R, PC, DB	1660	730	930	Olive oil	24	forearm	5	4
Haberka 2011	R, SB	840	375	465	No treatment	4	forearm	3	2
Skulas-Ray 2011a	R, PC, DB, CO	840	375	465	Corn oil	8	forearm	5	3
Skulas-Ray 2011b	R, PC, DB, CO	3360	1500	1860	Corn oil	8	forearm	5	3
Sanders 2011a	R, PC, DB	450	180	270	Olive oil	52	forearm	5	5
Sanders 2011b	R, PC, DB	900	360	540	Olive oil	52	forearm	5	5
Sanders 2011c	R, PC, DB	1800	720	1080	Olive oil	52	forearm	5	5
Moertl 2011a	R, PC, DB	840	375	465	gelatin	13	upper arm	5	5
Moertl 2011b	R, PC, DB	3360	1500	1860	gelatin	13	upper arm	5	5

The studies by woodman (2003), Skulas-Ray (2011) and Moertl (2011) include two intervention groups with different fish oil doses separately, and the study by Sanders (2011) includes three intervention groups with different fish oil doses. The studies by Hill (2007) and Wright (2008) each contain two comparisons with different fish oil treatment durations.

DHA, docosahexaenoic acid; EPA, eicosapentaenoic acid; R, randomized; PC, placebo-controlled; DB, double-blinded; SB, single-blinded; CO, crossover.

Effects of fish oil supplementation on FMD

FMD was determined in all of the included studies by non-invasive ultrasound assessment of brachial artery endothelial responsiveness. The percentage change between baseline and endpoint induced by fish oil supplementation was used as the primary outcome. After data extraction and pooling, the meta-analysis was performed. The results of the pooled estimation revealed that fish oil significantly improved endothelial function in the included subjects, as demonstrated by increase of FMD (23 studies, 1385 subjects; WMD: 1.49%, 95% CI 0.48% to 2.50%, p = 0.004; **Figure 2**). However, significant heterogeneity existed in terms of fish oil supplementation-related improvements of FMD ($I^2 = 87\%$, p<0.001).

Meta-regression and subgroup analyses

Because differences in characteristics of participants and studies may contribute to the heterogeneity among the studies, we performed meta-regression analysis to explore the relationship between these study characteristics and the mean change in FMD after fish oil supplementation. The results of this meta-regression analysis revealed that among predefined variables (including number of the subjects in each study, mean age, BMI, SBP, DBP, TG, TC, mean baseline FMD of the participants, percentages of males, smokers and diabetic patients, dose of fish oil, EPA and DHA, follow-up duration and Jadad Score), Jadad Scores of the included studies were inversely associated with the FMD improvement after fish oil supplementation (regression coefficient $= -1.60$, p = 0.04; **Table 3**), indicating study quality may influence the overall effect of fish oil supplementation on FMD.

Subsequently, we performed subgroup analyses to evaluate how these predefined study characteristics may influence the effect of fish oil on endothelial function, as measured by FMD of brachial artery. The results of this analysis were similar to those in the meta-regression analysis described above. FMD improved significantly in studies of which the Jadad Score was 2 points (WMD: 8.99%, 95% CI 2.59% to 15.40%, p = 0.006; **Table 4**), but didn't in studies of which the Jadad Score were 3 points (WMD: 0.65%, 95% CI −0.63% to 1.94%, p = 0.32), 4 points (WMD: 1.07%, 95% CI −1.32% to 3.46%, p = 0.38) or 5 points (WMD: −0.22%, 95% CI −0.75% to 0.30%, p = 0. 40). Besides, pooled analysis of studies including only normoglycemic participants showed significant improvements in FMD compared with those including only diabetic patients (p = 0.04; **Table 4**); also, subjects with mean baseline DBP <75 mmHg seemed to experience more remarkable improvement of FMD than those with mean baseline DBP ≥75 mmHg (p = 0.01; **Table 4**). Sensitivity analysis by including only double-blind, placebo-controlled studies also showed that fish

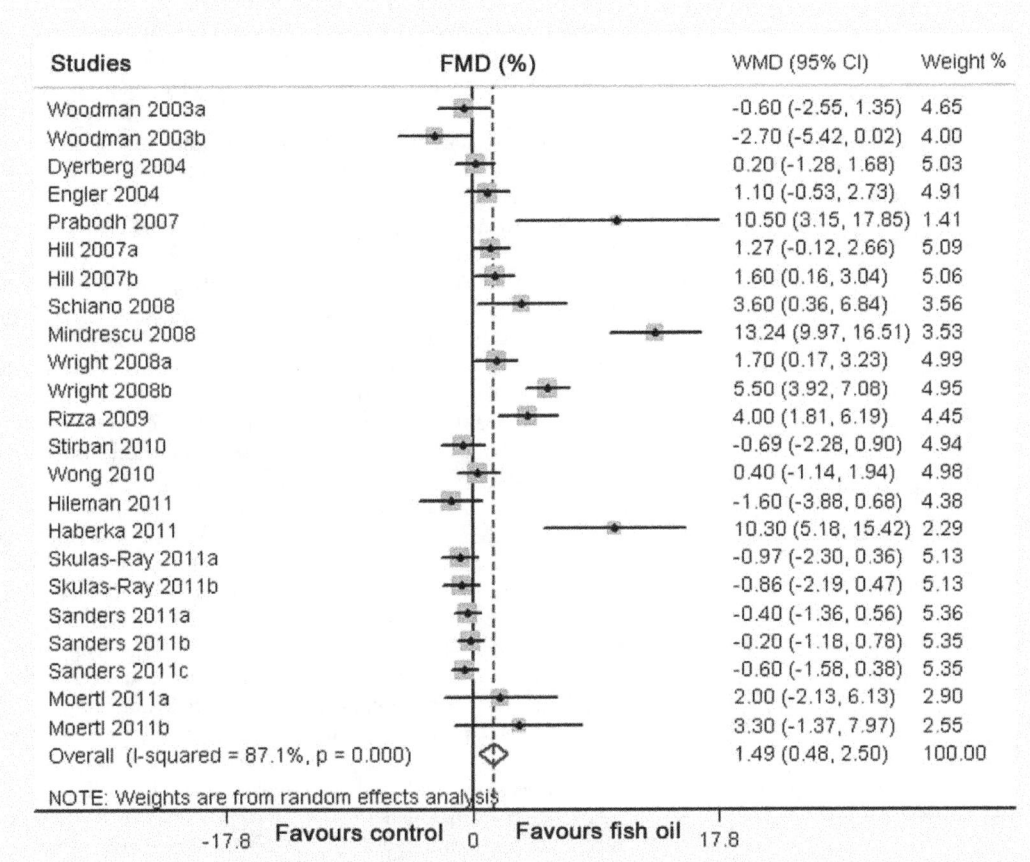

Figure 2. Forest plots from meta-analysis of weighed mean difference in flow-mediated dilation for subjects randomized to fish oil or control groups. The effect size of each study is proportional to the statistical weight. The diamond indicates the overall summary estimate for the analysis; the width of the diamond represents the 95% CI. FMD, flow-mediated dialtion; WMD, weighed mean difference; CI, confidence interval.

oil supplementation didn't significantly influence the value of FMD based on these high-quality studies (19 studies, 1227 subjects; WMD: 0.54%, 95% CI −0.25% to 1.33%, p = 0.18; **Table 4**), suggesting that pooled results derived from relatively high quality studies seemed not support a role of fish oil supplementation on improvement of FMD.

Publication bias

The funnel plot for the effect of fish oil supplementation on brachial FMD was asymmetrical, suggesting the presence of publication bias (**Figure 3**). Egger's significance test also indicated the existence of publication bias (p = 0.009). However, the result of the fail-safe N test indicated that it would take 266 unpublished null results (for all studies) to bring the combined p value to a nonsignificant level.

Discussion

In this study, by pooling the results of the available randomized controlled trials, we found that fish oil supplementation significantly improved endothelial function, as measured by FMD. However, the results of meta-regression and subgroup analyses suggested the quality of the included studies (evaluated by Jadad Score) may influence the effect of fish oil supplementation on FMD, and the significance of the results seemed to depend mainly on the contribution of the studies with relatively poor quality (Jadad Score <3). Furthermore, complication of T2DM and

baseline DBP of the included subjects may also influence the potential effects of fish oil on FMD. Normoglycemic participants and subjects with lower DBP (<75 mmHg) seemed to be associated with significant improvements of FMD after fish oil supplementation, while FMD of diabetic patients or the subjects with higher DBP didn't seem to benefit from fish oil supplementation.

Endothelial dysfunction has been recognized as an early pathophysiologic event during a variety of CVD, including hypertension and atherosclerosis [11]. More importantly, evidence from epidemiological studies and other clinical trials has suggested that impairment of endothelial function, as presented as a reduction of FMD, was an independent predictor for the incidence of cardiovascular events. Results of an early meta-regression analysis of 399 populations indicated that in populations at low CVD risk, endothelial function measured by FMD is related to principal cardiovascular risk factors, and to the estimated 10-year risk of coronary heart disease [5]. This is further supported by the evidence from a large population-based cohort study [6], which also found that brachial FMD is a predictor of incident cardiovascular events in multi-ethnic adults free of CVD, and may be of use to further improve the classification of subjects at low, intermediate and high CVD risk in addition to Framingham Risk Score. Besides, a recent meta-analysis [7], which mainly included studies in patients with established CVD, showed that impairment of FMD is significantly associated with future cardiovascular events and 1% reduction in brachial FMD was

Table 3. Characteristics associated with net change in flow-mediated dilation: univariate meta-regression analysis.

	FMD (%)		
	Coefficient	95% CI	p
Number of subjects	−0.017	−0.056 to 0.021	0.36
Mean age (years)	−0.03	−0.18 to 0.11	0.65
Males' percentage (%)	−0.006	−0.071 to 0.059	0.85
BMI (kg/m^2)	−0.15	−0.70 to 0.39	0.39
Smokers' percentage (%)	0.029	−0.027 to 0.084	0.29
SBP (mmHg)	−0.12	−0.33 to 0.09	0.26
DBP (mmHg)	−0.20	−0.49 to 0.09	0.14
Percentage of patients with T2DM (%)	−0.017	−0.036 to 0.002	0.10
Mean TG (mmol/L)	−1.46	−5.92 to 3.01	0.50
Mean TC (mmol/L)	−0.47	−3.03 to 2.08	0.70
Mean baseline FMD (%)	−0.14	−0.43 to 0.15	0.33
Fish oil dose (mg/d)	0.0003	−0.0011 to 0.0018	0.67
DHA dose (mg/d)	−0.0001	−0.0022 to 0.0020	0.92
EPA dose (mg/d)	0.0005	−0.0012 to 0.0023	0.54
Duration (weeks)	−0.06	−0.17 to 0.04	0.21
Jadad Score	−1.60	−3.13 to −0.08	0.04

FMD, flow-mediated dilation; CI, confidence interval; BMI, body mass index; SBP, systolic blood pressure; DBP, diastolic blood pressure; T2DM, type 2 diabetes; TG, triglyceride; TC, total cholesterol; DHA, docosahaexaenoic acid; EPA, eicosapentaenoic acid.

associated with 13% increase in risk of future cardiovascular events. In view of the above evidence, the result of our meta-analysis, which revealed that fish oil supplementation was associated with improvement of FMD by 1.49%, may be clinically relevant, indicating that the observed favorable effects of fish oil in primary and secondary prevention of CVD may be at least partially related to its effect on restoration of the endothelial function. Possible mechanisms underlying the beneficial effect of fish oil on endothelial function may include increase of membrane fluidity of endothelial cells, anti-inflammation, inhibition of platelets adhesion and aggregation, although the exact mechanisms involved are still unknown [3,34].

Recent published meta-analysis by Wang et al [35] concerning the similar topic found that supplementation of omega 3 fatty acids significantly improves FMD by 2.30% and this effect may be modified by the health status of the participants or the dose of supplementation. Their study [35] is different from ours because besides studies with EPA and DHA, studies with another type of omega-3 fatty acids, alpha-linolenic acid, were also included. On the other hand, our study mainly focused on the effect of fish oil supplementation on FMD, and included some studies which have not been included in the meta-analysis by Wang et al [17,24,28]. Furthermore, we collected more detailed information of the baseline characteristics (particularly of the CVD risk factors) and background medication, which enabled us to further explore whether the difference of these factors were potential source of heterogeneity.

However, the results of our pooled analysis concerning the effect of fish oil supplementation on FMD should be interpreted with caution, because according to the results of the meta-regression and subgroup analyses, quality of the included studies may

influence the overall effect, and the significance of fish oil's benefit on FMD was mainly driven by the studies of relatively poor quality. Sensitivity analysis by pooling only double-blind, placebo-controlled studies found that fish oil supplementation had no significant effect on FMD. This discrepancy may highlight the need of high-quality large scale trials in the future to evaluate the exact effect of fish oil supplementation on FMD and other markers of endothelial function.

Besides, we also found that complication of T2DM and levels of DBP of the included participants may influence the effects of fish oil on FMD. Specifically, FMD of participants with normal glucose metabolism or lower DBP seemed to be improved after fish oil supplementation, while those of diabetic participants or higher DBP didn't. The mechanisms underlying the above results were not known. In our opinions, it's possible that these results suggested that fish oil supplementation could only improve FMD in lower risk participants whose vascular function was largely preserved. This is because FMD is a biological process not only dependent on endothelial function (e.g. synthesis and release of NO and other endothelial derived vasodilators), but also dependent on the ability of vascular smooth muscle cells (SMC) to relax in response to the aforementioned vasodilators [9,11]. Many factors (such as hyperglycemia [36], hypertension [37] and insulin resistance [38]) may contribute to the injury of vascular SMCs and subsequently impair their relaxation ability. In these circumstances, even though the endothelial function can be restored, FMD may not be improved. However, this hypothesis needs to be further tested in future studies.

A few recently published large-scale clinical trials [39,40] and meta-analysis [41] failed to show a favorable effect of omega 3 fatty acids supplementation on cardiovascular events in patients who were of high risk or already with established CVD. It was suggested that perhaps the cardiovascular benefit of omega 3 fatty acids is limited with the improvements in cardioprotective drug treatment, such as use of statins and ACEIs/ARBs [42]. Because these two categories of drugs have been indicated to improve the endothelial function in former studies [43,44], we investigated whether background cardiovascular therapy with statins or ACEIs/ARBs could impact the effect of fish oil on FMD. However, since detailed information (such as percentage of participants used, types of medication, and the number of participants who were on medication of target dose)of statins or ACEIs/ARBs usage are generally lacking in the included studies, we had to categorize these studies according to whether all, partial or none of the participants were on statins or ACEIs/ARBs therapy. Results of the subgroup analyses didn't suggest different effect of fish oil on FMD according to whether participants on statins or ACEIs/ARBs were included or not. Obviously, we couldn't conclude that background statins or ACEIs/ARBs usage didn't influence the effect of fish oil on FMD based on the above analyses, and these results should be interpreted very cautiously.

Our meta-analysis has several limitations. First, great heterogeneity was found among the included studies. Although we tried our best to include potential variables of study characteristics into our meta-regression and subgroup analyses, the heterogeneity couldn't be completely explained by these factors. Because we included subjects with substantial clinical heterogeneity, some other factors, including the baseline status of omega-3 fatty acids and concurrent medicines (including statins, ACEIs/ARBs and many other medications which may affect endothelial function) may contribute to the heterogeneity among the studies. However, details of these variables were generally not accessible in the included studies and therefore couldn't be analyzed intensively. Second, the number of included studies and the total number of subjects in some

Table 4. Subgroup analyses for the effect of fish oil supplementation on flow-mediated dilation according to predefined study characteristics.

Study characteristics	FMD (%)				
	Studies (patients), n	I^2	WMD [95% CI]	p value for subgroup effects	p value for subgroup interaction
Number of subjects					
<40	12 (454)	88%	1.53 [−0.33, 3.39]	0.11	0.96
≥40	11 (931)	87%	1.58 [0.40, 2.77]	0.009	
Health status					
Generally healthy	6 (578)	78%	0.61 [−0.63, 1.86]	0.34	0.22
Chronic condition	17 (807)	88%	1.76 [0.39, 3.14]	0.01	
Mean age					
<51 years	10 (512)	86%	1.70 [0.32, 3.07]	0.02	0.72
≥51 years	13(873)	87%	1.33 [−0.11, 2.77]	0.07	
Percentage of males					
≤50%	9 (841)	87%	1.36 [0.14, 2.58]	0.03	0.44
>50%	12 (440)	89%	2.33 [0.21, 4.45]	0.03	
Mean BMI					
<27 kg/m^2	12 (890)	85%	1.32 [0.14, 2.50]	0.03	0.65
≥27 kg/m^2	11 (495)	89%	1.83 [−0.04, 3.70]	0.06	
Diabetic status					
Normoglycemic	14 (995)	85%	0.92 [−0.07, 1.91]	0.07	0.04
Diabetic	4 (199)	23%	−0.58 [−1.63, 0.47]	0.28	
Mean SBP					
<122 mmHg	9 (658)	73%	0.72 [−0.43, 1.87]	0.22	0.53
≥122 mmHg	12 (655)	91%	1.33 [−0.21, 2.86]	0.09	
Mean DBP					
<75 mmHg	11 (465)	89%	2.84 [0.90, 4.78]	0.004	0.01
≥75 mmHg	10 (848)	54%	−0.21 [−0.85, 0.42]	0.51	
Mean TG					
<1.5 mmol/L	11 (836)	87%	0.91 [−0.39, 2.20]	0.17	0.27
≥1.5 mmol/L	11 (490)	90%	2.17 [0.35, 4.00]	0.02	
Mean TC					
≤4.95 mmol/L	11 (511)	91%	1.73 [−0.29, 3.74]	0.09	0.44
>4.95 mmol/L	11 (848)	79%	0.85 [−0.11, 1.82]	0.08	
Statins used					
All or partially	10 (783)	90%	1.49 [−0.15, 3.13]	0.07	0.99
None	10 (503)	87%	1.48 [−0.00, 2.95]	0.05	
ACEIs/ARBs used					
All or partially	12 (842)	88%	1.61 [0.08, 3.14]	0.04	0.90
None	10 (503)	87%	1.48 [−0.00, 2.95]	0.05	
Baseline FMD					
<5%	11 (668)	91%	1.55 [−0.07, 3.17]	0.06	0.54
≥5%	11 (691)	77%	0.92 [−0.22, 2.06]	0.11	
Fish oil dose					
<1800 mg/d	11 (666)	80%	1.17 [−0.10, 2.45]	0.07	
≥1800 mg/d	12 (719)	91%	1.63 [0.08, 3.18]	0.04	0.65
DHA dose					
≤1100 mg/d	13 (894)	76%	0.62 [−0.38, 1.62]	0.22	0.14
>1100 mg/d	10 (491)	91%	2.22 [0.37, 4.07]	0.02	
EPA dose					
≤640 mg/d	11 (669)	77%	1.01 [−0.15, 2.18]	0.09	0.48

Table 4. Cont.

Study characteristics	FMD (%)				
	Studies (patients), n	I²	WMD [95% CI]	p value for subgroup effects	p value for subgroup interaction
>640 mg/d	12 (716)	91%	1.75 [0.09, 3.41]	0.04	
Duration					
<12 weeks	11 (487)	90%	1.83 [−0.00, 3.65]	0.05	0.70
≥12 weeks	12 (898)	85%	1.39 [0.20, 2.58]	0.02	
Jadad Score					
2 points	3 (132)	88%	8.99 [2.59, 15.40]	0.006	0.02
3 points	9 (388)	78%	0.65 [−0.63, 1.94]	0.32	
4 points	5 (257)	90%	1.07 [−1.32, 3.46]	0.38	
5 points	6 (608)	0%	−0.22 [−0.75, 0.30]	0.40	
Occlusion position					
Forearm	20 (1276)	88%	1.29 [0.24, 2.34]	0.02	–
Occlusion duration					
4.5~5 min	21 (1319)	87%	1.13 [0.15, 2.10]	0.02	–
Study design					
R, DB, PC	19 (1227)	79%	0.54 [−0.25, 1.33]	0.18	–

FMD, flow-mediated dilation; BMI, body mass index; SBP, systolic blood pressure; DBP, diastolic blood pressure; TG, triglyceride; TC, total cholesterol; ACEI, angiotensin-converting enzyme inhibitor; ARB, angiotensin receptor blocker; DHA, docosahexaenoic acid; EPA, eicosapentaenoic acid; R, randomized; DB, double-blinded; PC, placebo-controlled; WMD, weighed mean difference; CI, confidence interval.

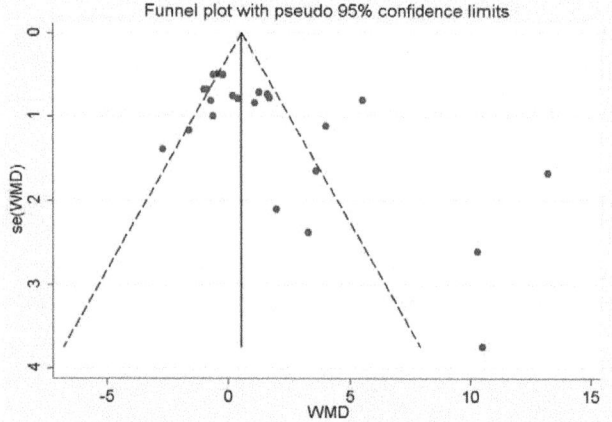

Figure 3. Funnel plots (with pseudo 95% CIs) of all individual studies in the meta-analyses of standard deviation of flow-mediated dilation. FMD, flow-mediated dialtion; WMD, weighed mean difference; CI, confidence interval.

subgroup analysis (e.g. the diabetic subgroup only included 4 studies with 199 patients) were relatively small. Therefore, interpretation the results of the subgroup analyses should be with caution. Third, publication bias was found for the current meta-analysis, although the fail-safe N test suggested that it would take over 200 unpublished null results to bring the combined p value to a nonsignificant level. Finally, FMD was determined in the included study using non-invasive methods, which may not fully represent endothelial function, especially in subjects with higher CVD risk. Studies with other reliable markers of endothelial function may be needed to evaluate the effect of fish oil supplementation.

In conclusion, results of our meta-analysis indicated that fish oil supplementation significantly improved endothelial function, as measured by FMD. However, these results seemed to be mainly driven by the studies with relatively poor quality. High quality large scale clinical trials with adequate statistical power are needed in the future to further evaluate the effect of fish oil supplementation on endothelial function in the context of optimized concurrent cardiovascular therapy before we can come to a definite conclusion, especially in patients with high risk of CVD.

Acknowledgments

The authors would like to thank Dr. Vittorio Schiano, Dr. Alison Hill, Dr. Peter Howe, Dr. Jorn Dyerberg, Dr. Maciej Haberka, Dr. Atman Shah, Dr. Deddo Moertl and Dr. Marguerite Engler for generous sharing of their unpublished baseline data of their studies.

Author Contributions

Conceived and designed the experiments: WX WW XL. Performed the experiments: WX WW. Analyzed the data: WX WW. Contributed reagents/materials/analysis tools: WX WW XL. Wrote the paper: WX WW XL.

References

1. Mozaffarian D, Rimm EB (2006) Fish intake, contaminants, and human health: evaluating the risks and the benefits. JAMA 296: 1885–1899.

2. Lavie CJ, Milani RV, Mehra MR, Ventura HO (2009) Omega-3 polyunsaturated fatty acids and cardiovascular diseases. J Am Coll Cardiol 54: 585–594.

3. Mozaffarian D, Wu JH (2011) Omega-3 fatty acids and cardiovascular disease: effects on risk factors, molecular pathways, and clinical events. J Am Coll Cardiol 58: 2047–2067.

4. De Caterina R (2011) n-3 fatty acids in cardiovascular disease. N Engl J Med 364: 2439–2450.

5. Witte DR, Westerink J, de Koning EJ, van der Graaf Y, Grobbee DE, et al. (2005) Is the association between flow-mediated dilation and cardiovascular risk limited to low-risk populations? J Am Coll Cardiol 45: 1987–1993.

6. Yeboah J, Folsom AR, Burke GL, Johnson C, Polak JF, et al. (2009) Predictive value of brachial flow-mediated dilation for incident cardiovascular events in a population-based study: the multi-ethnic study of atherosclerosis. Circulation 120: 502–509.

7. Inaba Y, Chen JA, Bergmann SR (2010) Prediction of future cardiovascular outcomes by flow-mediated vasodilatation of brachial artery: a meta-analysis. Int J Cardiovasc Imaging 26: 631–640.

8. Green DJ, Jones H, Thijssen D, Cable NT, Atkinson G (2011) Flow-mediated dilation and cardiovascular event prediction: does nitric oxide matter? Hypertension 57: 363–369.

9. Corretti MC, Anderson TJ, Benjamin EJ, Celermajer D, Charbonneau F, et al. (2002) Guidelines for the ultrasound assessment of endothelial-dependent flow-mediated vasodilation of the brachial artery: a report of the International Brachial Artery Reactivity Task Force. J Am Coll Cardiol 39: 257–265.

10. Bots ML, Westerink J, Rabelink TJ, de Koning EJ (2005) Assessment of flow-mediated vasodilatation (FMD) of the brachial artery: effects of technical aspects of the FMD measurement on the FMD response. Eur Heart J 26: 363–368.

11. Thijssen DH, Black MA, Pyke KE, Padilla J, Atkinson G, et al. (2011) Assessment of flow-mediated dilation in humans: a methodological and physiological guideline. Am J Physiol Heart Circ Physiol 300: H2–12.

12. Anderson JS, Nettleton JA, Herrington DM, Johnson WC, Tsai MY, et al. (2010) Relation of omega-3 fatty acid and dietary fish intake with brachial artery flow-mediated vasodilation in the Multi-Ethnic Study of Atherosclerosis. Am J Clin Nutr 92: 1204–1213.

13. Woodman RJ, Mori TA, Burke V, Puddey IB, Barden A, et al. (2003) Effects of purified eicosapentaenoic acid and docosahexaenoic acid on platelet, fibrinolytic and vascular function in hypertensive type 2 diabetic patients. Atherosclerosis 166: 85–93.

14. Dyerberg J, Eskesen DC, Andersen PW, Astrup A, Buemann B, et al. (2004) Effects of trans- and n-3 unsaturated fatty acids on cardiovascular risk markers in healthy males. An 8 weeks dietary intervention study. Eur J Clin Nutr 58: 1062–1070.

15. Engler MM (2004) Docosahexaenoic acid restores endothelial function in children with hyperlipidemia: Results from the early study. International Journal of Clinical Pharmacology and Therapeutics 42: 672679.

16. Prabodh Shah A (2007) Cardiovascular and endothelial effects of fish oil supplementation in healthy volunteers. Journal of Cardiovascular Pharmacology and Therapeutics 12: 213219.

17. Hill AM, Buckley JD, Murphy KJ, Howe PR (2007) Combining fish-oil supplements with regular aerobic exercise improves body composition and cardiovascular disease risk factors. Am J Clin Nutr 85: 1267–1274.

18. Schiano V, Laurenzano E, Brevetti G, De Maio JI, Lanero S, et al. (2008) Omega-3 polyunsaturated fatty acid in peripheral arterial disease: effect on lipid pattern, disease severity, inflammation profile, and endothelial function. Clin Nutr 27: 241–247.

19. Mindrescu C, Gupta RP, Hermance EV, DeVoe MC, Soma VR, et al. (2008) Omega-3 fatty acids plus rosuvastatin improves endothelial function in South Asians with dyslipidemia. Vasc Health Risk Manag 4: 1439–1447.

20. Wright SA, O'Prey FM, McHenry MT, Leahey WJ, Devine AB, et al. (2008) A randomised interventional trial of omega-3-polyunsaturated fatty acids on endothelial function and disease activity in systemic lupus erythematosus. Ann Rheum Dis 67: 841–848.

21. Rizza S, Tesauro M, Cardillo C, Galli A, Iantorno M, et al. (2009) Fish oil supplementation improves endothelial function in normoglycemic offspring of patients with type 2 diabetes. Atherosclerosis 206: 569–574.

22. Stirban A, Nandrean S, Gotting C, Tamler R, Pop A, et al. (2010) Effects of n-3 fatty acids on macro- and microvascular function in subjects with type 2 diabetes mellitus. Am J Clin Nutr 91: 808–813.

23. Wong CY, Yiu KH, Li SW, Lee S, Tam S, et al. (2010) Fish-oil supplement has neutral effects on vascular and metabolic function but improves renal function in patients with Type 2 diabetes mellitus. Diabet Med 27: 54–60.

24. Hileman CO, Carman TL, Storer N, Harrill Labbato D, White C, et al. (2011) Omega-3 fatty acids do not improve endothelial function in virologically-suppressed HIV-infected men: a randomized placebo-controlled trial. AIDS Res Hum Retroviruses.

25. Haberka M, Mizia-Stec K, Mizia M, Janowska J, Gieszczyk K, et al. (2011) N-3 polyunsaturated fatty acids early supplementation improves ultrasound indices of endothelial function, but not through NO inhibitors in patients with acute myocardial infarction: N-3 PUFA supplementation in acute myocardial infarction. Clin Nutr 30: 79–85.

26. Skulas-Ray AC, Kris-Etherton PM, Harris WS, Vanden Heuvel JP, Wagner PR, et al. (2011) Dose-response effects of omega-3 fatty acids on triglycerides, inflammation, and endothelial function in healthy persons with moderate hypertriglyceridemia. Am J Clin Nutr 93: 243–252.

27. Sanders TA, Hall WL, Maniou Z, Lewis F, Seed PT, et al. (2011) Effect of low doses of long-chain n-3 PUFAs on endothelial function and arterial stiffness: a randomized controlled trial. Am J Clin Nutr 94: 973–980.

28. Moertl D, Hammer A, Steiner S, Hutuleac R, Vonbank K, et al. (2011) Dose-dependent effects of omega-3-polyunsaturated fatty acids on systolic left ventricular function, endothelial function, and markers of inflammation in chronic heart failure of nonischemic origin A double-blind, placebo-controlled, 3-arm study. Am Heart J 161: 915.e911–919.

29. Moher D, Liberati A, Tetzlaff J, Altman DG (2009) Preferred reporting items for systematic reviews and meta-analyses: the PRISMA statement. BMJ 339: b2535.

30. Higgins J, Green S (2011) Cochrane Handbook for Systematic Reviews of Interventions Version 5.1.0. The Cochrane Collaboration. www.cochrane-handbook.org.

31. Moher D, Pham B, Jones A, Cook DJ, Jadad AR, et al. (1998) Does quality of reports of randomised trials affect estimates of intervention efficacy reported in meta-analyses? Lancet 352: 609–613.

32. Higgins JP, Thompson SG, Deeks JJ, Altman DG (2003) Measuring inconsistency in meta-analyses. BMJ 327: 557–560.

33. Egger M, Davey Smith G, Schneider M, Minder C (1997) Bias in meta-analysis detected by a simple, graphical test. BMJ 315: 629–634.

34. Di Minno MN, Tremoli E, Tufano A, Russolillo A, Lupoli R, et al. (2010) Exploring newer cardioprotective strategies: omega-3 fatty acids in perspective. Thromb Haemost 104: 664–680.

35. Wang Q, Liang X, Wang L, Lu X, Huang J, et al. (2012) Effect of omega-3 fatty acids supplementation on endothelial function: a meta-analysis of randomized controlled trials. Atherosclerosis 221: 536–543.

36. Madonna R, De Caterina R (2011) Cellular and molecular mechanisms of vascular injury in diabetes–part I: pathways of vascular disease in diabetes. Vascul Pharmacol 54: 68–74.

37. Ponnuchamy B, Khalil RA (2009) Cellular mediators of renal vascular dysfunction in hypertension. Am J Physiol Regul Integr Comp Physiol 296: R1001–1018.

38. Bornfeldt KE, Tabas I (2011) Insulin resistance, hyperglycemia, and athero-sclerosis. Cell Metab 14: 575–585.

39. Kromhout D, Giltay EJ, Geleijnse JM (2010) n-3 fatty acids and cardiovascular events after myocardial infarction. N Engl J Med 363: 2015–2026.

40. Bosch J, Gerstein HC, Dagenais GR, Diaz R, Dyal L, et al. (2012) n-3 fatty acids and cardiovascular outcomes in patients with dysglycemia. N Engl J Med 367: 309–318.

41. Kwak SM, Myung SK, Lee YJ, Seo HG (2012) Efficacy of Omega-3 Fatty Acid Supplements (Eicosapentaenoic Acid and Docosahexaenoic Acid) in the Secondary Prevention of Cardiovascular Disease: A Meta-analysis of Random-ized, Double-blind, Placebo-Controlled Trials. Arch Intern Med.

42. Hu FB, Manson JE (2012) Omega-3 Fatty Acids and Secondary Prevention of Cardiovascular Disease–Is It Just a Fish Tale?: Comment on "Efficacy of Omega-3 Fatty Acid Supplements (Eicosapentaenoic Acid and Docosahexaenoic Acid) in the Secondary Prevention of Cardiovascular Disease". Arch Intern Med.

43. Masoura C, Pitsavos C, Aznaouridis K, Skoumas I, Vlachopoulos C, et al. (2011) Arterial endothelial function and wall thickness in familial hypercholesterolemia and familial combined hyperlipidemia and the effect of statins. A systematic review and meta-analysis. Atherosclerosis 214: 129–138.

44. Shahin Y, Khan JA, Samuel N, Chetter I (2011) Angiotensin converting enzyme inhibitors effect on endothelial dysfunction: a meta-analysis of randomised controlled trials. Atherosclerosis 216: 7–16.

14

Whole Transcriptome Profiling of Successful Immune Response to *Vibrio* Infections in the Oyster *Crassostrea gigas* by Digital Gene Expression Analysis

Julien de Lorgeril[1], **Reda Zenagui**[1], **Rafael D. Rosa**[1,2], **David Piquemal**[3], **Evelyne Bachère**[1]*

1 Institut Français de Recherche pour l'Exploitation de la Mer, Centre National de la Recherche Scientifique, Montpellier, France, **2** Université Montpellier 2, and Institut de Recherche pour le Développement, UMR 5119 "Écologie des Systèmes Marins Côtiers", Montpellier, France, **3** Skuld-Tech, Cap Delta, ZAC Euromedecine II, Grabels, France

Abstract

The cultivated Pacific oyster *Crassostrea gigas* has suffered for decades large scale summer mortality phenomenon resulting from the interaction between the environment parameters, the oyster physiological and/or genetic status and the presence of pathogenic microorganisms including *Vibrio* species. To obtain a general picture of the molecular mechanisms implicated in *C. gigas* immune responsiveness to circumvent *Vibrio* infections, we have developed the first deep sequencing study of the transcriptome of hemocytes, the immunocompetent cells. Using Digital Gene Expression (DGE), we generated a transcript catalog of up-regulated genes from oysters surviving infection with virulent *Vibrio* strains (*Vibrio splendidus* LGP32 and *V. aestuarianus* LPi 02/41) compared to an avirulent one, *V. tasmaniensis* LMG 20012^T. For that an original experimental infection protocol was developed in which only animals that were able to survive infections were considered for the DGE approach. We report the identification of cellular and immune functions that characterize the oyster capability to survive pathogenic *Vibrio* infections. Functional annotations highlight genes related to signal transduction of immune response, cell adhesion and communication as well as cellular processes and defence mechanisms of phagocytosis, actin cytosqueleton reorganization, cell trafficking and autophagy, but also antioxidant and anti-apoptotic reactions. In addition, quantitative PCR analysis reveals the first identification of pathogen-specific signatures in oyster gene regulation, which opens the way for in depth molecular studies of oyster-pathogen interaction and pathogenesis. This work is a prerequisite for the identification of those physiological traits controlling oyster capacity to survive a *Vibrio* infection and, subsequently, for a better understanding of the phenomenon of summer mortality.

Editor: Jérôme Nigou, French National Centre for Scientific Research - Université de Toulouse, France

Funding: Funding from the Institut Français de Recherche et d'Exploitation de la Mer (Ifremer) and the Centre National de la Recherche Scientifique (Cnrs) grant from ANR (Agence Nationale de la Recherche) Genanimal project Cg-Physiogene (N°07/5210875/F). The funders had no role in study design, data collection and analysis, decision to publish, or preparation of the manuscript.

Competing Interests: The authors have declared that no competing interests exist.

* E-mail: ebachere@ifremer.fr

Introduction

Aquatic organisms and particularly marine invertebrates, such as the oyster *Crassostrea gigas*, harbour an abundant and diverse microflora on their surface (epibiosis) or inside their tissues (endobiosis) where *Vibrio splendidus* is found as a dominant culturable vibrio. With Evolution, the oysters have developed effective systems for maintaining their homeostasis and for controlling potentially harmful and pathogenic bacteria. However, for decades, the cultivated Pacific oyster *C. gigas* is suffering large scale summer mortalities that are reported in all areas of the world where this species is cultivated [1]. Mortalities result from the interaction between the environment, the oyster physiological and/or genetic status and the presence of pathogenic microorganisms [1]. Beside the recent identification of a virulent microvariant of an Herpes virus, the OsHV-1 [2], *Vibrio* strains of *V. splendidus* and *V. aestuarianus* groups have been repeatedly associated to summer mortality episodes [3] and the virulence of some strains has been demonstrated through *C. gigas* experimental infections [4,5].

Considerable effort has been invested in advanced genomic technologies to understand and characterize the major traits that govern the tolerance of oysters to stressful culture conditions or to pathogenic bacteria [6,7,8,9,10,11]. In particular, immune-related genes have been characterized from *C. gigas*. Briefly, a variety of antimicrobials have been fully characterized, namely a Bactericidal/Permeability-Increasing protein, *Cg*-BPI [12] and antimicrobial peptide families, *Cg*-Defensins (*Cg*-Defs) [13], *Cg*-Proline rich peptides (*Cg*-Prps) [14], and a great diversity has been demonstrated in terms of sequences and potential antimicrobial activities [15]. Tissue Inhibitor Metalloprotease (TIMP)-encoding gene has been shown to be induced upon microbial challenge, the expression of which would be controlled by a Rel/NF-κB pathway [16,17]. The oyster major plasma protein, *Cg*-EcSOD (extracellular Superoxide Dismutase) which appears to display anti-oxidant and LPS-binding properties [18] has recently been shown to be used as an opsonin for *V. splendidus* LGP32 phagocytosis through its RGD sequence [19]. Otherwise, oyster infection with *V. aestuarianus* results in a decrease in *Cg*-EcSOD transcripts in circulating hemocytes [20]. However, whereas most of these immune genes were shown to be modulated during infections, the molecular mechanisms by which the oyster can survive virulent *Vibrio* infections remained totally unknown.

Here, our objective was to develop a better understanding of the genetic-level responses of oysters to pathogenic vibrios and to identify genes that are involved in immune responsiveness to circumvent the infections. In this attempt, we have performed a comprehensive analysis of the transcriptome of oyster immunity (hemocytes), using Digital Gene Expression (DGE) [21], an improved version of the Serial Analysis of Gene Expression (SAGE) technique [22]. These methods generate genome-wide and high-throughput transcription profiles and provide qualitative and quantitative gene expression data that do not depend on prior identification of transcript information (sequence homologies or gene function). For the DGE approach, we have developed an original experimental infection protocol, considering individual monitoring of the oyster successful response in terms of survival to pathogenic *Vibrio* infections *versus* non pathogenic ones. Thus, two DGE libraries were constructed from hemocytes of oysters surviving infections by virulent *V. splendidus* LGP32 and *V. aestuarianus* LPi 02/41 on the one hand, and by *V. tasmaniensis* LMG 20012[T], an avirulent strain related to the *V. splendidus* group, on the other hand. The study aimed to compare the expression data of the two libraries and, beyond gene identification and functional annotation, to explore the putative functions related to the capability of oysters to circumvent and to survive vibrioses. This is the first report on genome-wide transcriptional analysis of oyster survival-responsiveness to virulent vibrios.

Results

Oyster survival to virulent *Vibrio* species

To strengthen the accuracy of our transcriptomic approach, two independent experimental infections have been performed for (i) the construction of hemocyte DGE libraries and for (ii) the exploration of the DGE data by qPCR, using respectively oysters from the Atlantic coast and Mediterranean Lagoon that differ in abiotic and biotic environmental conditions. After infections, the mortalities were individually monitored and Kaplan–Meier survival curves were generated for the different groups of oysters (**Figure 1**).

For the Atlantic oysters, the survival rates were 58.9% for the animals infected by virulent *V. splendidus* LGP32 (8.10[7] CFU/oyster), 33.3% for the oysters infected with the highly virulent *V.*

aestuarianus LPi 02/41 (2.10[7] CFU/oyster) and 96.9% for those injected with the avirulent *V. tasmaniensis* LMG 20012[T]. The peaks of mortalities were reached at 72 h and 48 h post-injection respectively for the *V. splendidus* and *V. aestuarianus* strains (**Figure 1a**). Kaplan–Meier survival curves revealed statistical differences between the oysters injected either with *V. splendidus* LGP32, *V. aestuarianus* LPi 02/41 or the avirulent strain ($p < 0.0001$, Log-Rank and Wilcoxon tests). For the Mediterranean Lagoon oysters, the two groups of individuals infected with *V. splendidus* LGP32 (4.10[8] CFU/animal) or *V. aestuarianus* LPi 02/41 (8.10[7] CFU/animal) displayed respectively 57% and 48% survival, with peaks of mortalities reached at 48 h post-infection for both bacteria strains. For the third group injected with *V. tasmaniensis* LMG 20012[T] (2.10[8] CFU/animal), 80% survival was observed at 24 h post-injection whereas 98% survival were recorded for the fourth non-injected oyster group used as control (**Figure 1b**). Generated Kaplan–Meier survival curves revealed significant differences ($p < 0.0001$, Log-Rank and Wilcoxon tests, DDL = 2) between the four experimental infections. The survival curves differed statistically between the two independent infections ($p < 0.0001$, Log-Rank and Wilcoxon tests, DDL = 5), but no differences were observed between Atlantic oysters injected with *V. tasmaniensis* and non injected Mediterranean oysters ($p > 0.05$, Log-Rank and Wilcoxon tests, DDL = 5). We showed that higher dose of *V. splendidus* LGP32 (4.10[8] CFU/oyster) was requested for Mediterranean oysters than for the Atlantic ones (8.10[7] CFU/oyster) to reach similar mortality rates. This would confirm variabilities previously observed in oyster susceptibility to infections according to their geographic origins (Duperthuy, pers. comm.).

To consider only animals that have been able to survive infections, hemolymph for hemocyte RNA extraction was individually collected at 96 h post-infection when mortalities were no more recorded.

General characteristics of hemocyte DGE libraries

To have access to the hemocyte transcriptome and to establish a complete quantitative and qualitative gene expression database for the oyster immune functions and survival to virulent vibrios, two DGE libraries were generated from pooled hemocyte RNAs of the Atlantic oysters (i) surviving virulent *V. splendidus* LGP32 and *V.*

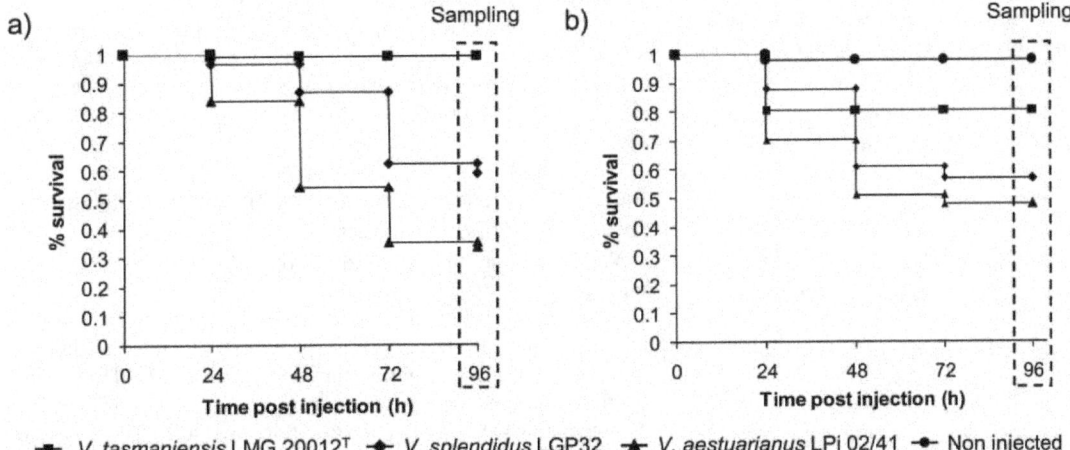

Figure 1. Kaplan-Meier survival curves of *C. gigas* **oysters during infections with virulent** *Vibrio* **strains,** *V. aestuarianus* **LPi 02/41 or** *V. splendidus* **LGP32, and with avirulent** *V. tasmaniensis* **LMG 20012**[T]**.** Infections for (a) construction of the hemocyte DGE libraries and (b) for gene expression exploration by qPCR analysis, with non injected oysters as control. For the two experiments, hemolymph was individually sampled from oysters that survived the infection at 96 h post-infection corresponding to the end of mortalities.

aestuarianus LPi 02/41 infections (SVir library) and (ii) those challenged with the avirulent strain, *V. tasmaniensis* LMG 20012^T (SaVir library). Sequencing of these two libraries resulted in a total of 6,983,680 DGE tags. Characteristics of the libraries are summarized in **Table 1**. According to P-value (0.001) and occurrence of tags (incidence of each tag in libraries) in the DGE libraries, 56,871 unique tags have been identified from both libraries (GEO accession numbers GSM667899 and GSM667900). Comparison of the tag occurrences between libraries revealed that 22,187 unique tags are differentially represented between libraries (fold change >2), composed of 9,815 tags more represented in SVir library and 12,372 tags more represented in SaVir library. By using Blast search on the *C. gigas* 29,745 unique ESTs (http://www.sigenae.org/aquafirst/), these DGE tags were found to be associated to 4,374 ESTs corresponding to 3,931 unique ESTs (1,610 for SVir and 2,321 for SaVir), which have been further considered for functional annotation of genes related to survival response to virulent *versus* non virulent vibrios. The relatively low percentage of homology found (19.7%) is mainly due to the fact that the ESTs assembled in the GigasDatabase were obtained from different developmental stages and oyster tissues that include only 13,898 ESTs from hemocytes [10].

Exploration of differential gene expressions according to *Vibrio* strains

The accuracy of our DGE approach have been validated by qPCR analysis. To reinforce this exploration, we used an independent experimental infection with oysters from Mediterranean Lagoon instead of Atlantic coast. For that, 18 genes more represented in SVir library and 17 genes more represented in SaVir library (2-fold change) were randomly chosen. Transcript levels were measured from hemocytes of oysters that survived infections with the virulent strains and the avirulent control. In addition, non-injected oysters were used as a control group (**Figure 1b**). Hierarchical clustering of the 18 SVir gene expression profiles distinguished, through clusters of conditions (CC), surviving oysters to virulent *Vibrio* strain infections (CC 3 and 4) to injected oysters with avirulent *Vibrio* strain (CC 2) and non injected oysters (CC 1) with a range of differential expression of 6.9-fold (**Figure 2a**). We can notice than up-regulated genes were found in all clusters of expression (CE) independently of the virulent *Vibrio* strains injected. Comparatively expression profiles of the 17 SaVir genes did not discriminate oyster response

according to the *Vibrio* strains or control in the clustering (**Figure S1**). Furthermore, for these genes, a 2.6-fold change of differential expression was observed, weaker than obtained for SVir genes.

Pathogen-specific signatures are evidenced among DGE genes

It is noteworthy that hierarchical clustering evidenced pathogen-specific and shared signatures upon bacterial challenges. Three major clusters of expressions (CE) were evidenced for SVir genes. Indeed, we observed genes up-regulated only in response to both virulent *Vibrio* compared to control or avirulent *Vibrio* injection (as examples: baculoviral IAP repeat containing protein 4, fascin or C-type lectin). Besides, genes were seen to be up-regulated in response to *V. splendidus* associated to cluster of expression 1 (as examples: Enhancer of kinase suppressor of ras 2, Metallothionein IV and Thioredoxin), while up-regulated genes in response to *V. aestuarianus* infection were found in clusters of expression 2 and 3 (as examples: Heat shock protein 22, Cystatin A2 and L-rhamnose-binding lectin) (**Figure 2b**). Various expression profiles were found for the SaVir genes tested, such as a down-regulation upon virulent *Vibrio* infections (as an example, the putative 26S proteasome non-ATPase regulatory subunit 7), up-regulation by the avirulent strain compared to virulent strains (Metaxin 3) or up-regulated indistinctly by the avirulent and virulent *Vibrio* compared to unchallenged oysters (no Blast hit sequence wy0aba12yi21fm1.1.a.cg.2).

Functional annotation

Altogether, the qPCR analyses which were performed on independent experimental infections and distinct biological material corroborated the DGE quantitative data. This supported further bioinformatic and Blast2GO and KEGG annotations [23] aiming to gain a general picture of the functional profiles of oyster survival- and infection-responsive genes identified by DGE approach. The 3,931 unique ESTs differentially represented between libraries (2-fold change) were annotated into 7 and 5 functional groups based on GO terms and KO terms respectively (**Figure S2**). SaVir up-regulated genes appeared to be assigned to "cell cycle and proliferation", "metabolism" and "genetic information processing" (**Figure 3 and Figure S2**). SVir up-regulated genes were assigned into groups related to "cellular processes" that included "cell motility and communication", "cell growth and death" and "cytosqueleton", to "signal transduction", and to "response to stimulus" (**Figure 3 and Figure S2**). Genes without GO terms were annotated manually (using Gene Cards informations) to enrich functional annotations in both DGE libraries.

Biological functions related to oyster successful response to virulent *Vibrio* infections

We focused on further identification of biological functions that may influence the successful immune response to virulent vibrios. Considering their occurrence in SVir *versus* SaVir DGE library, a list of genes has been established that highlights different pathogen-responsive immune functions (**Table S1**). This list has been generated from the analysis of full data provided in Supplementary file giving information on DGE tag sequences, *C. gigas* EST names, and on Blast2GO and KEGG functionnal annotation (**Table S2**). Several immune-related genes already characterized in oyster have been evidenced with no differential representation between the SVir and SaVir libraries (**Table S2**).

Immune response is characterized by genes related to signaling pathways notably involved in innate immunity (Toll-

Table 1. General characteristics of DGE libraries generated from oysters surviving virulent (SVir) and avirulent (SaVir) *Vibrio* infections.

DGE library	SVir	SaVir
Sequenced tags	3,349,884	3,633,796
Unique tags	52,224	52,528
Specific tags	3,647	4,343
Tags differentially represented (≥2 fold change)	9,812	12,372
Number of tags which match with EST	1,782	2,592
Unique EST	1,610	2,321
EST with B2GO annotation	393	766
EST with KEGG annotation	271	564

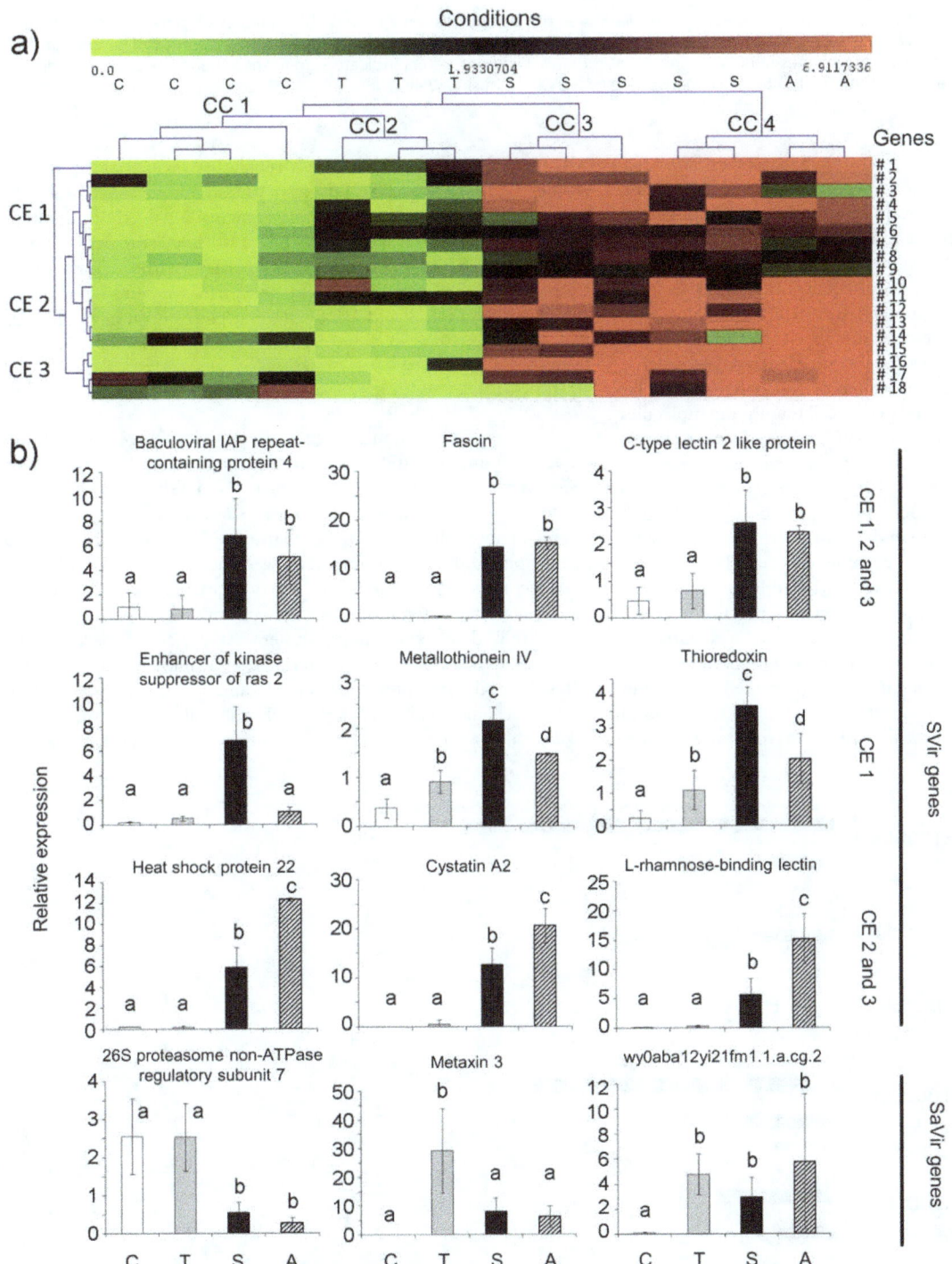

Figure 2. Gene expression analysis by qPCR. (a) *Hierarchical clustering* analysis and differential expression of 18 genes from SVir library. Hemocyte gene expression profiles were analysed in biological replicates from oysters: non injected as control (C), surviving infections with avirulent *V. tasmaniensis* LMG 20012[T] (T), with virulent *V. splendidus* LGP32 (S) and virulent *V. aestuarianus* LPi 02/41 (A). Each biological replicata was constituted by a pool of hemocyte RNA from ten oysters, and between 2 and 5 replicates were analysed for each experimental condition. Each cell in the matrix corresponds to the expression level of one gene in a sample. The intensity of the color from green to red indicates the magnitude of differential expression (see color scale at the bottom of the image). Relative expressions were calculated according the $2^{-(\Delta\Delta Ct)}$ method normalized with elongation factor-1α (EF-1α). Each value was calculated in reference to the mean of ΔCt of all conditions (relative expression = 1). The dendrograms at the top of the figures indicate relationship among experimental conditions which define clusters of conditions (CC). The dendrograms at the left of the figures indicate relationship among the profiles of the selected genes which define clusters of expression (CE), after clustering analysis using Multiple Array Viewer software. Gene #1: Inhibitor of apoptosis; #2: Baculoviral IAP repeat-containing protein 4; #3: Enhancer of kinase suppressor of Ras2; #4: Rac GTPase-activating protein1; #5: Proteasome 216S subunit, non-ATPase 11a; #6: Glyceraldhyde 3-phosphate dehydrogenase; #7: Thioredoxin; #8: C-type lectin 2 like protein; #9: Metallothionein IV; #10: F-box only protein37; #11: Cystatin B-like protein; #12: Heat shock protein 22 isoform 1; #13: L-rhamnose-binding lectin; #14: Microsomal glutathione S-tranferase; #15: Cystatin A; #16: Interferon-induced protein 44; #17: Cullin-associated and neddylation-dissociated 1. Hierarchical clustering was contructed with Multiple Array

Viewer software using average linkage clustering with Pearson correlation as the default distance metric. (b) *Examples of gene expression profiles defining pathogen- or challenge-specific signatures.* SVir genes; line 1: similar response to both virulent strains; line 2: response induced by *V. splendidus* LGP32; line 3: response induced by *V. aestuarianus* LPi 02/41; line 4: SaVir genes. Different letters indicate significant variation between conditions ($p < 0.05$) determined using the non-parametric multiple comparison test ANOVA of Kruskal-Wallis.

like/NF-κB and MAPK), from membrane receptors (Toll-like receptor) to regulating intermediaries (MAP kinases, G-protein, and NF-κB inhibitor) and transcription factors (LITAF-like protein). In addition, signaling and interaction molecules such as cytokine receptors were evidenced as well as recognition molecules such as lectins. Besides, proteases, protease inhibitors and stress proteins (such as heat shock proteins and metallothionein) were found with various immune effectors such as antimicrobial (lysozyme). ***Cell adhesion and communication*** category is dominated in SVir by regulating elements associated to cellular processes represented by genes related to cell membrane molecules like integrins, collagen or tetraspanins. ***Cytosqueleton reorganisation*** appears as a dominant group in terms of number of sequences associated to cellular and response processes related to actin (calcium dependent) and tubulin reorganization. Particularly, actin related genes associated to endocytosis were evidenced such as genes coding for membrane receptors associated to vesicle formation or to endosome and lysosome. In addition, trafficking and autophagy are highlighted by genes encoding several trafficking proteins and regulators of transport associated to cytosqueleton network. ***Respiratory chain***, dominated by regulating and response elements, is represented by genes associated to oxidative stress, several potent antioxidants and potential DNA repair elements associated to respiratory burst.

Apoptosis is noticed with both regulating pro-apoptotic related genes such as caspases but also anti-apoptotic ones, such as Baculoviral IAP repeat-containing proteins, involved in inhibition of tumor necrosis factor receptor-associated factors. Finally, ***Cellular differentiation and proliferation*** category is represented by regulating and response elements like genes involved in TGF-β and Wnt signaling pathways and positive or negative regulators of these pathways (such as small GTPase).

Discussion

Here, we report on the first deep sequencing study of the transcriptome of hemocytes, immunocompetent cells, of the oyster *Crassostrea gigas*. We have identified by DGE approach those defence mechanisms related to the oyster successful response and survival to virulent *Vibrios* comparatively to non virulent *Vibrio* infection. Only the animals that were able to survive infections were considered for the DGE library construction, *i.e.* hemocytes were collected after the peak of mortalities, 96 h post-infections. To fulfill this requirement, oysters have been experimentally infected by injection of the bacteria into the adductor muscle, since this method is the only reproducible and standardized procedure to induce mortality of oysters with controlled doses of virulent *Vibrio* [24]. Comparatively, the immersion in virulent *Vibrio-*

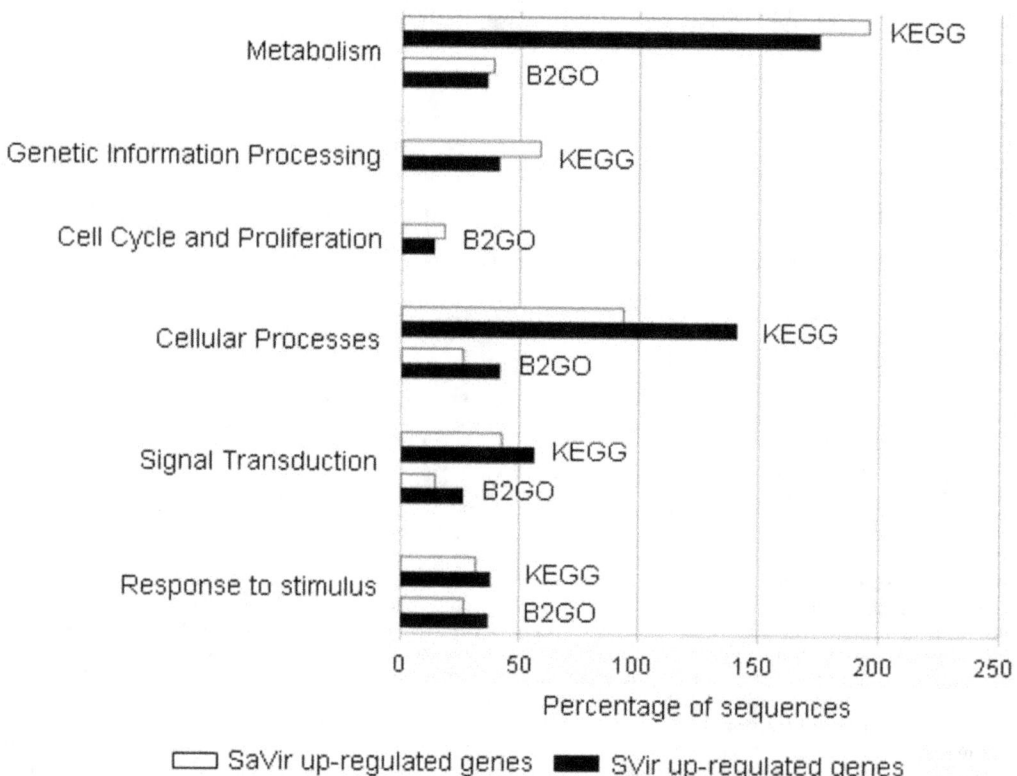

Figure 3. Functional annotation of unique ESTs differentially represented between SVir and SaVir libraries, respectively. Categorization was based on KO terms of the Kyoto Encyclopedia of Genes and Genomes (KEGG) using KEGG Automatic Annotation Server (KAAS), and on GO terms of Biological Process using Blast2GO software.

containing sea water does not consistently induce disease with well defined peaks of mortalities. It is noteworthy that, in our experimental conditions, bacteria virulence mechanisms that can breach the first line of oyster defences have been bypassed. Besides, potential recognition processes have not been highlighted in the analyses. Thus, the present study is not exhaustive. By considering surviving oysters after the peak of mortalities, we have focused on late processes of the immune responses relevant to bacteria elimination and infection resolution, instead of defence mechanisms occurring early post-infection.

Functional annotation of genes differentially represented between the SVir and SaVir DGE libraries has revealed biological processes that may characterize the successful immune response of oysters to pathogenic vibrios *versus* a non pathogenic one. However, because *C. gigas* is a non-model organism, it is noteworthy that such annotations may only suggest putative functions. The response to avirulent *Vibrio* was enriched in genes related to cell metabolism and cell cycle which may reflect hemocyte homeostasis recovery upon non-pathogenic challenge. Comparatively, in respect to oyster survival to infections, we found enrichment for genes related to signal transduction, signaling molecules and interactions, implicated in the regulation of immune response, as well as to cell adhesion and communication. In addition, the hemocyte transcript catalog we generated highlights cellular processes and defence mechanisms related to phagocytic events, actin cytosqueleton reorganization, cell trafficking and autophagy, but also to oxidative antioxidant and anti-apoptotic reactions.

The hemocyte gene repertoire of oysters able to circumvent pathogenic infections revealed the presence of **pathogen recognition molecules and elements of signaling pathways** involved in immune responses and inflammatory processes. Two lectins were seen up-regulated in response to virulent vibrios, a homologue to L-rhamnose-binding lectin (RBL) identified as constitutively expressed in *Hydractinia* [25] and the oyster C-type lectin. RBLs are pattern recognition receptors in vertebrates; they have opsonising properties and are activators of pro-inflammatory cytokines [26]. Interestingly, here, the putative RBL was seen to be exclusively expressed in response to the virulent vibrios according to a pathogen-specific expresssion signature for *V. aestuarianus*. C-type lectins are Ca$^+$-dependent carbohydrate binding proteins. Upon pathogen recognition, they are known as mediating various immune responses in invertebrates and vertebrates such as autophagy, phagosome maturation or apoptosis [27,28]. One can notice that in the DGE libraries, oyster galectin, a galactosidase-binding lectin known to be involved in infections [29], was not seen differentially expressed according to the bacterial virulence (**Table S2**). However, in vertebrates, galectins play dual functions as damage-associated molecular patterns (DAMPs) and as receptors for pathogen-associated molecular patterns (PAMPs) [30]. Among pattern recognition receptors (PRRs), a Toll-like receptor homologous to Toll-9 from *Anopheles* was evidenced as differentially expressed, whereas the Toll-like receptor 1, recently described in oyster [31], did not (**Table S2**). Our results enriched the Toll-like repertoire in oyster, but further investigations will determine the respective involvement of these molecules in innate immune pathways, particularly the Toll/NF-κB pathway. Among SVir up-regulated genes, we identified several components of *C. gigas* NF-κB pathway [17] as well as LITAF (LPS-Induced TNF-α Factor) transcription factor already identified in *C. gigas* [32]. LITAF signaling pathway plays a major role in regulating various mouse inflammatory cytokines in response to LPS stimulation [33]. Here, cytokine receptor and cytokine-induced protein categories related to inflammation

processes were also evidenced with up-regulated putative tumor necrosis factor ligand superfamily member 10, and Interferon-induced protein 44. Proinflammatory effects of virulent *Vibrio* infections were also revealed by over representation, in oyster SVir library, of components of MAPK signaling pathway. In abalones, pathogenic *V. harveyi* avoids both phagocytosis and ROS production and reduces p38 MAPK [34].

Besides cytokines and associated proteins, **protease inhibitors** were also found. Among them, *Cg*-TIMP, an inhibitor of metalloproteinases, which has NF-κB binding sites in promoting region, was shown to be implicated in oyster wound healing and defence mechanisms [16]. Several putative cystatins A and B and serpins were also identified whose putative role in invertebrate immune defence was suggested by previous genomic studies [35]. Vibrios secrete extracellular metalloproteases whose toxicity has been demonstrated both for *V. splendidus* LGP32 [36] and *V. aestuarianus* strain [37]. Thus, we can hypothesize that such protease inhibitors neutralize the *Vibrio* toxins. Moreover, members of cystatin superfamily have immunomodulatory properties and cytokine regulating properties thus preventing excessive inflammation (for review [38]). Cystatins may also interact with TIMPs and other proteases in patho-physiological processes that require tissue remodeling and that range from cell survival and proliferation, to differentiation and cell signaling [39]. High representation of sequences for stress proteins was shown, including various heat shock proteins (HSPs) and oyster metallo-thioneins (MTs). MTs are ubiquitous metal binding proteins though to be involved in detoxifying of heavy metals, in free radical scavenging but also in inflammatory responses [40].

Surprisingly, whereas we evidenced the potential involvement of immune response regulating pathways, few antimicrobials were seen to be significantly involved in the response to infections. Thus, no differential representation of antimicrobial transcripts was observed in the DGE data, namely for *Cg*-BPI, Bactericidal/Permeability-Increasing protein [12], *Cg*-Prps, proline-rich peptides [14], or for the new oyster big defensin family (Rosa *et al.* in prep). However, lysozyme transcripts were significantly increased in hemocytes from oyster surviving virulent *Vibrio* infection that motivates further characterization of this antimicrobial. Indeed, until now, three lysozymes have been characterized in oyster, mainly from mantle, gills or digestive gland with potential digestive functions [41]. Finally, one can mention the high representation of transcripts with homologies with Complement component 1q (C1q) in the SVir library. C1q domain containing (C1q-DC) proteins are ubiquitous in animal kingdom and many of them remain to be characterized. Nevertheless, C1q-DC proteins display many functions in immunity including clearance of pathogens in vertebrates [42] and invertebrates as well [43].

The DGE data presented here revealed the importance of genes involved in **cell adhesion and communication** that may contribute to oyster post-infection recovery. Among them, we showed several integrins including an oyster β-integrin that plays an important role in phagocytosis [44]. Integrins are also cell adhesion receptors for proteins of the extracellular proteins such as collagens evidenced here. Collagens maintain tissue integrity but also play significant role in regulating cell functions. Therefore, collagen production may indicate potential repairement of the lesions caused by virulent *Vibrio* infection. Besides, interestingly, members of tetraspanin superfamily were also significantly represented. Tetraspanins are small transmembrane proteins that, once interacting with immune receptors such as integrins, play important role in cellular processes such as migration, proliferation. They are immune modulators of signaling pathways (NF-κB and kinases) that induce cytokine production following pathogenic recognition [45]. Whereas tetraspanin associated to phagocytic receptors contribute to facilitate phagocytosis processes, microbial pathogens can also exploit tetraspanin to enter in host cell for further colonization and invasion [46]. Interestingly, *V. splendidus* LGP32 uses β-integrin through its outer membrane protein OmpU to invade oyster hemocyte and further impair defence fonctions [19]. Whether tetraspanins intervene also in *V. splendidus* LGP32 pathogenesis remains to be established. However, because tetraspanin transcript enrichment was shown here to be associated to the oyster successful response to circumvent *Vibrio* infection, it is likely that they display a role in modulating inflammatory responses. In several mollusk species, pathogenic vibrios (*i.e. V. aestuarianus*) use cytotoxins (metalloproteases and extracellular products - ECPs) as part of their virulence mechanisms by impairing cellular processes including adhesiveness, migration, morphogenesis and phagocytosis [47,48]. Thus, in our study, the oyster capability to have circumvented the infection is reflected by the enrichment of gene transcripts related to cell adhesion and motility, and consequently to cytosqueleton reorganization.

We evidenced numerous molecules involved in **cytoskeleton rearrangements** which are essential for various cellular processes dealing with morphogenesis, chemotaxis, migration but also phagocytosis and intracellular transports [49]. Thus, actin cytoskeleton changes are also involved in host-pathogen interactions. However, many pathogens subvert these cellular machineries for invading and surviving into host cells [50] as also reported for the oyster pathogen *V. splendidus* LGP32 [19]. Here, the DGE approach greatly contributed to characterize in *C. gigas* numerous genes, actors of cellular defence mechanisms, but consequently also potential targets for *Vibrio* virulence. Among those, one can cite overexpression of genes related to endocytosis, lysosome and active state of actin as well as to tubulin reorganization. All those cellular motor genes (myosin light chain, dynein arm light chain, kinesin-like protein) involved in intracellular trafficking or exocytosis of immune effectors may contribute to removal of invading pathogen and homeostasis recovery. This is particularly highlighted by the modulation of genes we classified in autophagy and microtubule transport. Authophagy is part of the microbicidal defence system. This conserved mechanism plays roles in degrading intracellular pathogens but also in regulator of innate immunity particularly the inflammatory or systemic immune response [51].

Apoptosis, another specialized form of **programmed cell death**, was also evidenced in our data with different pro- or anti-death effectors such as caspases, cytochrome c or IAPs. Baculoviral IAP repeat-containing 2/3/4 is described to inhibit apoptosis by binding to tumor necrosis factor receptor-associated factors [52]. Thus, hemocytes of surviving oysters up-regulate a variety of anti-apoptotic genes which could modulate inflammatory cytokine pathways and oxidative stress potentially toxic for the cell. Besides activation by inflammatory pathways, apoptosis as autophagy can be triggered by **reactive oxygen species** (ROS) [53]. Here, the successful response of oyster to virulent *Vibrio* infection revealed up-regulated genes related to respiratory chain and particularly in ROS production, a microbicidal defence reaction known in oyster [54], as well as putative genes coding for anti-oxidants. In *C. gigas*, hemocyte oxidative metabolism has been shown to be enhanced following virulent *V. aestuarianus* infection when superoxide dismutase *Cg*-EcSOD gene was down-regulated, likely being a pathogen adaption for impairing hemocyte functions and survival [20]. With respect to *V. splendidus* LGP32, we have shown that the *Vibrio* uses *Cg*-EcSOD as opsonin for hemocyte invasion and evades defence reaction by limiting ROS production [19]. Our DGE data provided several components of the oxidative stress that may contribute to pathogen elimination and to anti-inflammatory reactions, in the context of oyster survival to infections.

Our study highlighted signaling pathways and regulators we classified in **cell differentiation**, closely related to the different functions described above. Those are homologous to activins and TGF-β inducible early growth response proteins, members of the TGF-β signaling pathway which regulates a wide spectrum of cellular functions such as proliferation, apoptosis, differentiation and migration [55]. Additionally, the involvement of Wnt-signaling pathway and numerous members of the Rho protein family such as Cdc42 and small GTPases were evidenced. They are important actors for cellular processes such as migration, chemiotaxis or phagocytosis and they are involved in cellular functions related to actin cytoskeleton regulation. Thus, Rho GTPases are also known to be specific targets for bacterial cytotoxins [56].

Here, qPCR analyses aimed at validating the differential gene expression evidenced by DGE. So, independent experimental infections were performed separately with the different virulent and avirulent vibrios. Using oysters from the Mediterranean sea instead of the Atlantic ocean, we showed that the differential gene expression was independent of the oyster origin. Interestingly, pathogen-specific signatures were evidenced upon bacterial challenge. To our knowledge, this is shown for the first time in this invertebrate. The different oyster infections resulted in both responses shared by the three *Vibrio* strains whatever their virulence or avirulence, and specific responses to virulent vibrios. This is consistent with recent data on virulence mechanisms and pathogenesis that greatly differ between *V. aestuarianus* strain and *V. splendidus* LGP32. *V. aestuarianus* was shown to affect hemocyte phagocytosis and adhesiveness properties by the secretion of extracellular products and to enhance the production of ROS [48]. Regarding *V. splendidus* LGP32, a metalloprotease from ECPs (extracellular products) has been associated to toxicity and its outer membrane protein OmpU has been evidenced as a virulent factor [24,36]. Recently, *V. splendidus* LGP32 has been shown to be a facultative intracellular pathogen which invades the oyster hemocytes through OmpU adhesin/β-integrin recognition and survives by impairing phagosome acidification and ROS production [19]. In this work, we demonstrated that bacterial invasion induces hemocyte cytoskeleton reorganizations by analysing expression of genes evidenced in the present DGE study.

Concluding remarks

This genome-wide expression profiling aimed at a better understanding of the molecular mechanisms that control or contribute to the anti-infectious response of this marine bivalve mollusc with economical importance. The data we generated for characterizing the transcriptome of the oyster immune functions could be enriched by further progress on genomic resources of this non-model organism. Nevertheless, we provided here a catalog of genes and cellular or immune functions that are potential targets for mechanisms of pathogen resistance and escape to the immune response, that may concern not only *Vibrio* species but also viruses. Further exploitation of our DGE libraries opens the way to (i) the in-depth characterization of pathogen-specific gene expression signatures and (ii) the description of the effects of vibrio/virus co-infections that may impair oyster immune defences. Indeed, it is now considered that Herpes virus OsHV-1 together with *Vibrio* species, such as *V. splendidus* LGP32 present in the oyster microbiota, may contribute to *C. gigas* mortality outbreaks. Finally, this work is a prerequisite for the identification of those physiological traits controlling oyster survival capacity and subsequently for a better understanding of the phenomenon of summer mortality.

Materials and Methods

Bacterial strains

Two strains belonging to the *Vibrio splendidus* polyphyletic group were considered, namely the oyster pathogen *V. splendidus* LGP32 [4] and *V. tasmaniensis* LMG 20012^T used as an avirulent strain [57]. Additionally, virulent *V. aestuarianus* LPi 02/41 isolated during oyster mortality events was chosen as representative of this dominant *Vibrio* species [3]. The strains were grown under agitation at 20°C in artificial sea water (ASW) [58] supplemented with 4 g/l bactopeptone and 1 g/l yeast extract (referred to as Zobell medium) for 18 h. Bacterial concentrations were evaluated by optical density (OD) at 600 nm (UltraspecIII, Pharmacia Biotech), an OD value value of 1 corresponding to 10^9 colonies forming units (CFU)/ml for *V. aestuarianus* LPi 02/41 and to 2.10^9 for *V. splendidus* LGP32 and *V. tasmaniensis* LMG 20012^T. Bacteria were centrifuged (15 min, 3,000×g, 20°C) and suspended in autoclaved ASW at the concentration calculated for the experimental injections.

Oysters and experimental infections

For DGE library construction, adult (2 year-old) oysters, *Crassostrea gigas*, were purchased from an Atlantic oyster farm (La Tremblade, France) and acclimatized in the Ifremer laboratory (LGP, La Tremblade, France) over a 1-week period in aerated 0.45 μm-filtered seawater. The temperature was maintained at 20°C during the trial. A total of 390 oysters was individually tagged and distributed in three groups in separate tanks. For infection, oysters were first anesthetized for 3 h in aerated 50 g/l MgCl$_2$ bath (2/3 v/v sea water/freshwater) containing phytoplankton (*Chaetoceros gracilis* and *Isochrysis galbana*). Experimental infections were performed as previously described by injecting bacteria into the posterior adductor muscle [24]. One group of 180 oysters was injected with 8.10^7 CFU of *V. splendidus* LGP32 per animal under 100 μl, and a second group (180 osyters) with 2.10^7 CFU/animal of *V. aestuarianus* LPi 01/42. The third group (30 oysters) was injected with 8.10^7 CFU/animal of the avirulent *Vibrio tasmaniensis* LMG 20012^T.

A second experimental infection was performed for qPCR analyses with adult oysters (2 year-old) obtained from a Mediterranean commercial hatchery (Sodimer, Montpellier, France). Oysters were acclimatized at 20°C and maintained in tanks with UV-treated and biologically filtered sea water in the experimental aquaculture platform of Ifremer Palavas (France). To allow the intramuscular injection of bacteria suspensions, a small cut was made in the side of oyster shells, adjacent to the adductor muscle. A total of 300 oysters were divided into four groups. Two groups of 100 oysters were respectively infected by virulent *Vibrio* strains, 4.10^8 CFU/animal of *V. splendidus* LGP32 and 2.10^7 CFU/animal of *V. aestuarianus* LPi 02/41. The third group of 50 oysters was injected with 2.10^8 CFU/animal of the avirulent strain *V. tasmaniensis* LMG 20012^T. Finally, the fourth group of 50 non infected oysters was used as control for the experiment to assess mortality due to the handling of the animals.

For both experiments, mortalities were monitored daily and, at the end of mortalities, hemolymph was individually collected from surviving oysters by withdrawing 0.5 to 1 ml from the posterior muscle adductor using a precooled 2 ml syringe for further RNA extraction. The non-parametric method of Kaplan-Meier (XLSTAT 2008.7.02) test was used to estimate survival rates and the Log-Rank and Wilcoxon values for comparing differences between the groups. All experimental infections were performed according to the Ifremer animal care guideline and policy.

RNA extraction

Hemocyte samples from individual oysters were obtained by hemolymph centrifugation at 1,500×g for 15 min at 4°C. Each pellet was lysed in 1 ml of TRIzol® reagent (Invitrogen®) for total RNA extraction according to the manufacturer's instructions. Total RNA amount and purity were checked by using spectrophotometrer NanoDrop ND-1000 (Thermo Scientific, Les Ulis, France) and the integrity of total RNA was analyzed by agarose-electrophoresis.

DGE library construction and sequencing

Two DGE libraries were constructed from hemocyte total RNA of oysters surviving *Vibrio* infections: SVir from pooled RNA samples from surviving individuals of virulent infections with *V. splendidus* LGP32 and *V. aestuarianus* LPi 02/41; SaVir from individuals injected with the avirulent *V. tasmaniensis* LMG 20012^T.

Sequence tag preparation was done with Illumina's Digital Gene Expression Tag Profiling Kit according to the manufacturer's protocol (version 2.1B). For both libraries, 7 μg of total RNA (from 10 oysters, 0.7 μg per oyster) was incubated with oligo-dT beads. First- and second-strand cDNA syntheses were performed using superscript II reverse transcription kit according to the manufacturer's instructions (Invitrogen). The cDNAs were cleaved using the *NlaIII* anchoring enzyme. Subsequently, digested cDNAs were ligated with the GEX adapter 1 containing a restriction site of *MmEI*. The second digestion with *MmeI* was performed, which cuts 17 bp downstream of the CATG site. At this point, the fragments detach from the beads. The GEX adapter 2 was ligated to the 3' end of the tag. A PCR amplification with 15 cycles using Phusion polymerase (Finnzymes) was performed with primers complementary to the adapter sequences to enrich the samples for the desired fragments. The resulting fragments of 85 bp were purified by excision from a 6% polyacrylamide TBE gel. The DNA was eluted from the gel debris with 1× NEBuffer 2 by gentle rotation for 2 h at room temperature. Gel debris were removed using Spin-X Cellulose Acetate Filter (2 ml, 0.45 μm) and the DNA was precipitated by adding 10 μl of 3 M sodium acetate (pH 5.2) and 325 μl of cold ethanol, followed by centrifugation at 13,000×g for 20 min. After washing the pellet with 70% ethanol, the DNA was resuspended in 10 μl of 10 mM Tris-HCl (pH 8.5) and quantified by using Nanodrop 1000 spectrophotometer.

Cluster generation was performed after applying 4 pM of each sample to the individual lanes of the Illumina 1G flowcell. After hybridization of the sequencing primer to the single-stranded products, 18 cycles of base incorporation were carried out on the 1G analyzer according to the manufacturer's instructions. Image analysis and base calling were performed using the Illumina Pipeline, where sequence tags were obtained after purity filtering. This was followed by sorting and counting the unique tags.

DGE library characteristics and functional annotation of tags differentially represented between libraries

The sequence files of each DGE library were analyzed with BIOTAG software (Skuldtech, Montpellier, France). The statistical value of DGE data comparisons, as a function of tag counts, was calculated by assuming that each tag has an equal chance of being detected. For several highly expressed transcripts, we checked that tag frequencies in successive sequence batches were distributed in agreement with a binomial law [59]. Selected genes were chosen based on a comparison between the two libraries, combined with the significance threshold of the observed variations (p-value<0.01). Tag to gene mapping was performed using EST collection from *C. gigas* which contains 29,745 unique sequences (7,940 contigs and 21,805 singletons), generated from different oyster tissues including hemocytes, and stored on the platform of Sigenae-INRA Toulouse (http://www.sigenae.org/aquafirst/). For tag to gene mapping, the virtual tags were extracted from all contigs and singletons.

Sequence functional annotation analyses were performed using two ways of classification. First we used Blast2GO software v1.3.3. (http://www.blast2go.org/start_blast2go). Briefly, Blast2GO uses BlastX available through the National Center for Biotechnology Information (NCBI) with a user-defined threshold to find similar sequences from the NCBI (nr database). Sequences which found homology with annotated sequences were annotated according to the gene ontology (GO) terms. The hierarchical representation of the gene ontology is structured according to different levels, from the top (level 1) parents corresponding to the three main GO categories (cellular component, biological process and molecular function) to the lowest more specialized child terms level 2, 3, 4 etc. In the present research, GO annotations were represented at level 3 of biological process. Due to the redundancy of term attribution, we have manually condensed group functionally related gene and term into seven functional groups to clarify annotation. Second, we have used the KEGG Automatic Annotation Server (Kyoto encyclopedia of genes and genomes, http://www.genome.jp/tools/kaas/) with SBH method for EST annotation. This server provides functional annotation of genes by BlastX comparisons against the manually curated KEGG GENES database, for ortholog assignment and pathway mapping. In addition, we have used GenesCards version 3 to obtain functionnal description of indentified genes (http://www.genecards.org/).

Real-time quantitative PCR analysis

To validate the quantitative data of DGE libraries, we have quantified the expression levels of 35 selected genes from SVir or SaVir DGE libraries by quantitative PCR (qPCR) from pooled hemocyte total RNAs in biological replicates (10 oysters per replicata) from four experimental conditions: (i) uninfected oysters, (ii) oysters injected with avirulent *V. tasmaniensis* LMG 20012^T, (iii) oysters that survived infection with virulent *V. splendidus* LGP32 or (iv) *V. aestuarianus* LPi 02/41. These experimental conditions were analysed in biological replicates of ten oysters per replicata. The first cDNA strand was synthesized from 700 ng of purified total RNA from hemocytes, using MMLV Reverse Transcriptase kit, according to the manufacturer's instructions (Invitrogen®), in 20 µl of reaction volume. qPCR amplifications were performed in the LightCycler 480 (Roche) in a final volume of 5 µl containing 5 mM MgCl$_2$, 0.33 µM of each primer, 2.5 µl of reaction mix (LightCycler 480 SYBR Green I Master 2X) and 1 µl of each reverse transcribed RNA (diluted 1:9). The list of oligonucleotide primers used to amplify target genes is shown in (**Table S3**). Each qPCR reaction was performed with an initial denaturation step of 10 min at 95°C followed by an amplification of the target cDNA for 40 cycles, each cycle consisting of a denaturation at 95°C for 10 s, annealing at 57°C for 20 s and elongation at 72°C for 25 s. Specificity of the qPCR product was analyzed by melting curve analysis. To determine the qPCR efficiency of each primer pair used, standard curves were generated using six serial dilutions (1:1, 1:3, 1:7, 1:15, 1:31, 1:63) of a unique cDNA sample constituted from a pool of all cDNAs obtained from each condition; qPCR efficiencies of tested genes varied between 1.87 and 1.99. Results are shown as changes in relative expression normalized with the elongation factor 1-alpha reference gene (EF-1α, GenBank accession number **AB122066**) using the $2^{-(\Delta\Delta Ct)}$ method described by Pfaffl [60]. Global hierarchical clustering of qPCR data was performed with Multiple Array Viewer software (version

4.6.2, http://www.tm4.org/mev/) using avarage linkage clustering with Pearson correlation as the default distance metric. In addition, statitical analyses gene by gene were conducted using Statistica Statsoft software version 6 with the non-parametric multiple comparison test ANOVA of Kruskal-Wallis were considered significantly different at $p<0.05$.

Supporting Information

Table S1 List of functional groups and related hemocyte up-regulated genes from oysters surviving virulent *versus* avirulent *Vibrio* infections.

Table S2 List of annotated genes from hemocyte DGE libraries used for functional annotations.

Table S3 List of primers.

Figure S1 *Hierarchical clustering* analysis and differential expression of 17 genes from SaVir library. Hemocyte gene expression profiles were analysed in biological replicates from oysters: non injected as control (C), surviving infections with avirulent *V. tasmaniensis* LMG 20012^T (T), with virulent *V. splendidus* LGP32 (S) and virulent *V. aestuarianus* LPi 02/41 (A). Each cell in the matrix corresponds to the expression level of one gene in a sample. The intensity of the color from green to red indicates the magnitude of differential expression (see color scale at the bottom of the image). Relative expressions were calculated according the $2^{-(\Delta\Delta Ct)}$ method normalized with elongation factor-1α (EF-1α). Each value was calculated in reference to the mean of ΔCt of all conditions (relative expression = 1). The dendrograms at the top of the figures indicate relationship among experimental conditions which define clusters of conditions (CC). The dendrograms at the left of the figures indicate relationship among the profiles of the selected genes which define clusters of expression (CE), after clustering analysis using Multiple Array Viewer software.

Figure S2 Gene ontology assignment of 1,610 and 2,321 unique ESTs differentially represented between SVir and SaVir libraries, respectively. (a) Categorization based on GO terms assignment from 3nd level of Biological Process using Blast2GO software. (b) Categorization based on KO terms (level A and B) of the Kyoto Encyclopedia of Genes and Genomes (KEGG) using KEGG Automatic Annotation Server (KAAS).

Acknowledgments

We thank Denis Saulnier for support for experimental infections done at the Ifremer laboratory of La Tremblade, Julie Fievet and Marc Leroy for experiments at the Aquaculture Platform of Ifremer Palavas. RZ and RDR were supported by doctoral fellowships from Ifremer and Languedoc-Roussillon region and CNPq – Brazil, respectively. Data used in this work were partly produced through molecular genetic analysis technical facilities of the SFR "Montpellier Environnement Biodiversité" and the "Plate-forme qPHD UM2/Montpellier GenomiX".

Author Contributions

Conceived and designed the experiments: JdL DP EB. Performed the experiments: JdL RZ DP. Analyzed the data: JdL RDR DP EB. Contributed reagents/materials/analysis tools: JdL RZ RDR DP. Wrote the paper: JdL RDR DP EB.

References

1. Samain JF, McCombie H (2008) Summer mortality of Pacific oyster *Crassostrea gigas*. In: Quae, ed. 379 p.

2. Segarra A, Pepin JF, Arzul I, Morga B, Faury N, et al. (2010) Detection and description of a particular Ostreid herpesvirus 1 genotype associated with massive mortality outbreaks of Pacific oysters, *Crassostrea gigas*, in France in 2008. Virus Res 153: 92–99.

3. Garnier M, Labreuche Y, Garcia C, Robert M, Nicolas JL (2007) Evidence for the involvement of pathogenic bacteria in summer mortalities of the Pacific oyster *Crassostrea gigas*. Microb Ecol 53: 187–196.

4. Gay M, Renault T, Pons AM, Le Roux F (2004) Two *Vibrio splendidus* related strains collaborate to kill *Crassostrea gigas*: taxonomy and host alterations. Dis Aquat Organ 62: 65–74.

5. Garnier M, Labreuche Y, Nicolas JL (2008) Molecular and phenotypic characterization of *Vibrio aestuarianus* subsp. *francensis* subsp. nov., a pathogen of the oyster *Crassostrea gigas*. Syst Appl Microbiol 31: 358–365.

6. Gueguen Y, Cadoret JP, Flament D, Barreau-Roumiguiere C, Girardot AL, et al. (2003) Immune gene discovery by expressed sequence tags generated from hemocytes of the bacteria-challenged oyster, *Crassostrea gigas*. Gene 303: 139–145.

7. Tanguy A, Bierne N, Saavedra C, Pina B, Bachère E, et al. (2008) Increasing genomic information in bivalves through new EST collections in four species: development of new genetic markers for environmental studies and genome evolution. Gene 408: 27–36.

8. Lang RP, Bayne CJ, Camara MD, Cunningham C, Jenny MJ, et al. (2009) Transcriptome profiling of selectively bred Pacific oyster *Crassostrea gigas* families that differ in tolerance of heat shock. Mar Biotechnol 11: 650–668.

9. Taris N, Lang RP, Reno PW, Camara MD (2009) Transcriptome response of the Pacific oyster (*Crassostrea gigas*) to infection with *Vibrio tubiashii* using cDNA AFLP differential display. Anim Genet 40: 663–677.

10. Fleury E, Huvet A, Lelong C, De Lorgeril J, Boulo V, et al. (2009) Generation and analysis of a 29,745 unique Expressed Sequence Tags from the Pacific oyster (*Crassostrea gigas*) assembled into a publicly accessible database: the GigasDatabase. BMC Genomics 10: 341–356.

11. Fleury E, Moal J, Boulo V, Daniel JY, Mazurais D, et al. (2010) Microarray-based identification of gonad transcripts differentially expressed between lines of Pacific oyster selected to be resistant or susceptible to summer mortality. Mar Biotechnol (NY) 12: 326–339.

12. Gonzalez M, Gueguen Y, Destoumieux-Garzón D, Romestand B, Fievet J, et al. (2007) Evidence of a bactericidal permeability increasing protein in an invertebrate, the *Crassostrea gigas* Cg-BPI. Proc Natl Acad Sci U S A 104: 17759–17764.

13. Gueguen Y, Herpin A, Aumelas A, Garnier J, Fievet J, et al. (2006) Characterization of a defensin from the oyster *Crassostrea gigas*. Recombinant production, folding, solution structure, antimicrobial activities, and gene expression. J Biol Chem 281: 313–323.

14. Gueguen Y, Romestand B, Fievet J, Schmitt P, Destoumieux-Garzón D, et al. (2009) Oyster hemocytes express a proline-rich peptide displaying synergistic antimicrobial activity with a defensin. Mol Immunol 46: 516–522.

15. Schmitt P, Gueguen Y, Desmarais E, Bachère E, de Lorgeril J (2010) Molecular diversity of antimicrobial effectors in the oyster *Crassostrea gigas*. BMC Evol Biol 10: 23.

16. Montagnani C, Avarre JC, de Lorgeril J, Quiquand M, Boulo V, et al. (2007) First evidence of the activation of Cg-timp, an immune response component of Pacific oysters, through a damage-associated molecular pattern pathway. Dev Comp Immunol 31: 1–11.

17. Montagnani C, Labreuche Y, Escoubas JM (2008) Cg-IkappaB, a new member of the IkappaB protein family characterized in the pacific oyster *Crassostrea gigas*. Dev Comp Immunol 32: 182–190.

18. Gonzalez M, Romestand B, Fievet J, Huvet A, Lebart MC, et al. (2005) Evidence in oyster of a plasma extracellular superoxide dismutase which binds LPS. Biochem Biophys Res Commun 338: 1089–1097.

19. Duperthuy M, Schmitt P, Garzón E, Caro A, Rosa RD, et al. (2011) Use of OmpU porins for attachment and invasion of *Crassostrea gigas* immune cells by the oyster pathogen *Vibrio splendidus*. Proc Natl Acad Sci U S A 108: 2993–2998.

20. Labreuche Y, Lambert C, Soudant P, Boulo V, Huvet A, et al. (2006) Cellular and molecular hemocyte responses of the Pacific oyster, *Crassostrea gigas*, following bacterial infection with *Vibrio aestuarianus* strain 01/32. Microbes Infect 8: 2715–2724.

21. Morin R, Bainbridge M, Fejes A, Hirst M, Krzywinski M, et al. (2008) Profiling the HeLa S3 transcriptome using randomly primed cDNA and massively parallel short-read sequencing. Biotechniques 45: 81–94.

22. Velculescu VE, Zhang L, Vogelstein B, Kinzler KW (1995) Serial analysis of gene expression. Science 270: 484–487.

23. Conesa A, Gotz S, Garcia-Gomez JM, Terol J, Talon M, et al. (2005) Blast2GO: a universal tool for annotation, visualization and analysis in functional genomics research. Bioinformatics 21: 3674–3676.

24. Duperthuy M, Binesse J, Le Roux F, Romestand B, Caro A, et al. (2010) The major outer membrane protein OmpU of *Vibrio splendidus* contributes to host antimicrobial peptide resistance and is required for virulence in the oyster *Crassostrea gigas*. Environ Microbiol 12: 951–963.

25. Schwarz RS, Hodes-Villamar L, Fitzpatrick KA, Fain MG, Hughes AL, et al. (2007) A gene family of putative immune recognition molecules in the hydroid *Hydractinia*. Immunogenetics 59: 233–246.

26. Watanabe Y, Tateno H, Nakamura-Tsuruta S, Kominami J, Hirabayashi J, et al. (2009) The function of rhamnose-binding lectin in innate immunity by restricted binding to Gb3. Dev Comp Immunol 33: 187–197.

27. Zelensky AN, Gready JE (2005) The C-type lectin-like domain superfamily. FEBS J 272: 6179–6217.

28. Tanne A, Ma B, Boudou F, Tailleux L, Botella H, et al. (2009) A murine DC-SIGN homologue contributes to early host defense against *Mycobacterium tuberculosis*. J Exp Med 206: 2205–2220.

29. Vasta GR (2009) Roles of galectins in infection. Nat Rev Microbiol 7: 424–438.

30. Sato S, St-Pierre C, Bhaumik P, Nieminen J (2009) Galectins in innate immunity: dual functions of host soluble beta-galactoside-binding lectins as damage-associated molecular patterns (DAMPs) and as receptors for pathogen-associated molecular patterns (PAMPs). Immunol Rev 230: 172–187.

31. Zhang L, Li L, Zhang G (2011) A *Crassostrea gigas* Toll-like receptor and comparative analysis of TLR pathway in invertebrates. Fish Shellfish Immunol 30: 653–660.

32. Park EM, Kim YO, Nam BH, Kong HJ, Kim WJ, et al. (2008) Cloning, characterization and expression analysis of the gene for a putative lipopolysaccharide-induced TNF-alpha factor of the Pacific oyster, *Crassostrea gigas*. Fish Shellfish Immunol 24: 11–17.

33. Tang X, Metzger D, Leeman S, Amar S (2006) LPS-induced TNF-alpha factor (LITAF)-deficient mice express reduced LPS-induced cytokine: Evidence for LITAF-dependent LPS signaling pathways. Proc Natl Acad Sci U S A 103: 13777–13782.

34. Travers MA, Le Bouffant R, Friedman CS, Buzin F, Cougard B, et al. (2009) Pathogenic *Vibrio harveyi*, in contrast to non-pathogenic strains, intervenes with the p38 MAPK pathway to avoid an abalone haemocyte immune response. J Cell Biochem 106: 152–160.

35. Guillou F, Mitta G, Galinier R, Coustau C (2007) Identification and expression of gene transcripts generated during an anti-parasitic response in *Biomphalaria glabrata*. Dev Comp Immunol 31: 657–671.

36. Binesse J, Delsert C, Saulnier D, Champomier-Verges MC, Zagorec M, et al. (2008) Metalloprotease vsm is the major determinant of toxicity for extracellular products of *Vibrio splendidus*. Appl Environ Microbiol 74: 7108–7117.

37. Labreuche Y, Le Roux F, Henry J, Zatylny C, Huvet A, et al. (2010) *Vibrio aestuarianus* zinc metalloprotease causes lethality in the Pacific oyster *Crassostrea gigas* and impairs the host cellular immune defenses. Fish Shellfish Immunol 29: 753–758.

38. Ochieng J, Chaudhuri G (2010) Cystatin superfamily. J Health Care Poor Underserved 21: 51–70.

39. Kopitar-Jerala N (2006) The role of cystatins in cells of the immune system. FEBS Lett 580: 6295–6301.

40. Kanekiyo M, Itoh N, Kawasaki A, Matsuyama A, Matsuda K, et al. (2002) Metallothionein modulates lipopolysaccharide-stimulated tumour necrosis factor expression in mouse peritoneal macrophages. Biochem J 361: 363–369.

41. Xue Q, Hellberg ME, Schey KL, Itoh N, Eytan RI, et al. (2010) A new lysozyme from the eastern oyster, *Crassostrea virginica*, and a possible evolutionary pathway for i-type lysozymes in bivalves from host defense to digestion. BMC Evol Biol 10: 213.

42. Kishore U, Reid KB (2000) C1q: structure, function, and receptors. Immunopharmacology 49: 159–170.

43. Zhang H, Song L, Li C, Zhao J, Wang H, et al. (2008) A novel C1q-domain-containing protein from Zhikong scallop *Chlamys farreri* with lipopolysaccharide binding activity. Fish Shellfish Immunol 25: 281–289.

44. Terahara K, Takahashi KG, Nakamura A, Osada M, Yoda M, et al. (2006) Differences in integrin-dependent phagocytosis among three hemocyte subpopulations of the Pacific oyster "*Crassostrea gigas*". Dev Comp Immunol 30: 667–683.

45. Levy S, Shoham T (2005) The tetraspanin web modulates immune-signalling complexes. Nat Rev Immunol 5: 136–148.

46. van Spriel AB, Figdor CG (2010) The role of tetraspanins in the pathogenesis of infectious diseases. Microbes Infect 12: 106–112.

47. Allam B, Ford SE (2006) Effects of the pathogenic *Vibrio tapetis* on defence factors of susceptible and non-susceptible bivalve species: I. Haemocyte changes following in vitro challenge. Fish Shellfish Immunol 20: 374–383.

48. Labreuche Y, Soudant P, Gonçalves M, Lambert C, Nicolas JL (2006) Effects of extracellular products from the pathogenic *Vibrio aestuarianus* strain 01/32 on lethality and cellular immune responses of the oyster *Crassostrea gigas*. Dev Comp Immunol 30: 367–379.

49. Vicente-Manzanares M, Sanchez-Madrid F (2004) Role of the cytoskeleton during leukocyte responses. Nat Rev Immunol 4: 110–122.

50. Rottner K, Lommel S, Wehland J, Stradal TEB (2004) Pathogen-induced actin filament rearrangement in infectious diseases. J Pathol 204: 396–406.

51. Deretic V, Levine B (2009) Autophagy, immunity, and microbial adaptations. Cell Host Microbe 5: 527–549.

52. Zheng C, Kabaleeswaran V, Wang Y, Cheng G, Wu H (2010) Crystal structures of the TRAF2: cIAP2 and the TRAF1: TRAF2: cIAP2 complexes: affinity, specificity, and regulation. Mol Cell 38: 101–113.

53. Fulda S, Gorman AM, Hori O, Samali A (2010) Cellular stress responses: cell survival and cell death. Int J Cell Biol 2010: 214074.

54. Bachère E, Boulo V, Godin P, Goggin L, Hervio D, et al. (1991) *In vitro* chemiluminescence studies of marine bivalve defence mechanisms and responses against specific pathogens. Dev Comp Immunol 15: S102.

55. Subramaniam M, Hawse JR, Johnsen SA, Spelsberg TC (2007) Role of TIEG1 in biological processes and disease states. J Cell Biochem 102: 539–548.

56. Aktories K, Barbieri JT (2005) Bacterial cytotoxins: targeting eukaryotic switches. Nat Rev Microbiol 3: 397–410.

57. Thompson FL, Thompson CC, Swings J (2003) *Vibrio tasmaniensis* sp. nov., isolated from Atlantic salmon (Salmo salar L.). Syst Appl Microbiol 26: 65–69.

58. Saulnier D, Avarre JC, Le Moullac G, Ansquer D, Levy P, et al. (2000) Evidence that *Vibrio penaeicida* is the putative etiological agent of syndrome 93 in New Caledonia and development of a rapid and sensitive PCR assay for its detection in shrimp and sea water. Dis Aquatic Org 40: 109–115.

59. Piquemal D, Commes T, Manchon L, Lejeune M, Ferraz C, et al. (2002) Transcriptome analysis of monocytic leukemia cell differentiation. Genomics 80: 361–371.

60. Pfaffl MW (2001) A new mathematical model for relative quantification in real-time RT-PCR. Nucleic Acids Res 29: e45.

15

Using Temporal Sampling to Improve Attribution of Source Populations for Invasive Species

Sharyn J. Goldstien[1]*, Graeme J. Inglis[2], David R. Schiel[1], Neil J. Gemmell[1,3,4]

1 Marine Ecology Research Group, School of Biological Sciences, University of Canterbury, Christchurch, New Zealand, **2** National Institute of Water and Atmospheric Research, Aquatic Biodiversity and Biosecurity, Christchurch, New Zealand, **3** Centre for Reproduction and Genomics, Department of Anatomy, University of Otago, Dunedin, New Zealand, **4** Allan Wilson Centre for Molecular Ecology and Evolution, University of Otago, Dunedin, New Zealand

Abstract

Numerous studies have applied genetic tools to the identification of source populations and transport pathways for invasive species. However, there are many gaps in the knowledge obtained from such studies because comprehensive and meaningful spatial sampling to meet these goals is difficult to achieve. Sampling populations as they arrive at the border should fill the gaps in source population identification, but such an advance has not yet been achieved with genetic data. Here we use previously acquired genetic data to assign new incursions as they invade populations within New Zealand ports and marinas. We also investigated allelelic frequency change in these recently established populations over a two-year period, and assessed the effect of temporal genetic sampling on our ability to assign new incursions to their population of source. We observed shifts in the allele frequencies among populations, as well as the complete loss of some alleles and the addition of alleles novel to New Zealand, within these recently established populations. There was no significant level of genetic differentiation observed in our samples between years, and the use of these temporal data did alter the assignment probability of new incursions. Our study further suggests that new incursions can add genetic variation to the population in a single introduction event as the founders themselves are often more genetically diverse than theory initially predicted.

Editor: Simon Thrush, National Institute of Water & Atmospheric Research, New Zealand

Funding: A University of Canterbury Postdoctoral Fellowship supported this study, with additional funding from the Ministry of Agriculture and Forestry, Biosecurity New Zealand (contract B0202), the National Institute of Water and Atmospheric Research, and a subcontract to NJG from the Biodiversity and Biosecurity OBI (C01X0502). The funders had no role in study design, data collection and analysis, decision to publish, or preparation of the manuscript.

* E-mail: sharyn.goldstien@canterbury.ac.nz

Introduction

The theory of invasion genetics has been discussed in the literature for many decades [1], but advanced molecular tools have only been applied to invasion ecology within the last 25 years [1–3]. Allozyme markers were initially used to investigate genetic diversity between invasive and native populations within invaded regions [4–6]. Subsequently, the 21st Century saw increased use of PCR tools to assess the relationship between the genetic structure and the global geographic distribution of aquatic invasive species such as the Mediterranean fan worm *Sabella spallananzii* [7] and riverine invaders *Gammarus fossarum* and *Dreissena polymorpha* [8]. Numerous molecular studies have now been done on a wide variety of invasive species, the purpose of which was primarily to identify species and source populations, potential vectors and invasion pathways [3].

The primary and most conclusive outcome from the application of molecular tools has been the identification of cryptic species [3,9–12]. In contrast, of the many studies aiming to identify the source populations of invasions, very few have been able to do so due to admixture, multiple sources and secondary introductions [13–18]. For instance, the most extensively studied invasive marine species, the European green crab *Carcinus maenas* shows some clear genetic affinities for non-native populations in Tasmania and Nova Scotia but studies also highlight the veiling

of source populations and genetic affinities due to multiple incursions (in North American populations) and/or unsampled source populations (in Japanese populations) [16,19]. Molecular examination of another crustacean, *Caprella mutica*, also identified multiple pathways of introduction to coastal regions throughout the Atlantic, from Asia, Europe and America [20]. Exceptions to this pattern of multiple introductions include the invasive alga *Codium fragile* spp. *tomentosoides*, shown to have two distinct introductions into Europe [21], and the Pacific acorn barnacle *Balanus glandula*, for which independent incursions to Japan and Argentina were identified [22]. Part of the difficulty is that assignment of an incursion to a particular source requires good representation from the range of potential source populations, with the problem of missing populations well-described [23]. However, temporal sampling of genetic structure is equally important in populations at the invasion front, where high propagule pressure and rapid population turnover occur.

The contemporary evolution of invasive species has also been investigated over the past decade. Unfortunately, many of these studies have been initiated long after the first incursion and this 'snapshot' of genetic diversity has been used as a starting point to decipher the many possible mechanisms that may be driving the observed patterns. For example, *Undaria pinnatifida* [17], *Rapana venosa* [24], and *Styela plicata* [25] have all been transported around the world for more than 70 years with relatively few recent introductions recorded. In contrast, a study of *Carcinus maenas*, was conducted over an eight year period with large sample sizes and good historical data [26]. This study showed how temporal data can highlight the interplay between environmental conditions and natural dispersal in the regional spread of an introduced species along a coast from its point of introduction. *C. maenas* clearly showed asymmetric dispersal with new haplotypes working into existing populations and homogenising coastal populations through time [26].

New incursions, or recently established populations, may provide further insights into contemporary evolution. The initial period of invasion, which includes both founders and their progeny, is crucial for assessing the contribution of individuals to the persistence of the population and for predicting the evolutionary trajectory of the population. At this stage of the incursion, with relatively few colonisers, genetic changes are likely to occur as a result of stochastic processes in small populations [27] and the spatial dynamics of the newly formed and diverging population. The occurrence of rare alleles and genetic heterozygosity likely depend primarily on the effective population size at initial incursion and population growth, as alleles that persist through the genetic bottleneck become more common within the expanding population, and so large shifts in allelic frequency could be expected [1,27,28]. In addition, it is likely that, for invasive species, temporal variation in allele frequencies will result from the mixing of distant, genetically structured, populations from different sources; a Wahlund effect is known to occur when cohorts are mixed as a single population over time [29,30]. An increase in heterozygote deficiency and departure from Hardy-Weinberg equilibrium is expected under a temporal Wahlund effect as the number of pooled cohorts increases [31]. A similar effect should occur if invading populations were sourced from different locations over time. A recent study of an invasive ascidian *Perophora japonica* [32] in Europe, very nicely showed a reduction in genetic diversity over a 9-year period, beyond the initial bottleneck, with subsequent differentiation among European populations.

New Zealand is unique in that many relatively recent marine incursions have been well-documented as a result of regular monitoring of ports and marinas by the Ministry of Agriculture and Forestry, Biosecurity New Zealand (now the Ministry for Primary Industries - MPI). One recent and increasingly widespread ascidian invader, *Styela clava*, was first recorded in New Zealand in 2005 in widely separated populations of the North (Hauraki Gulf) and South Islands (Lyttelton, Fig. 1). It has subsequently spread to secondary locations throughout the country. This species is abundant in the northern Hauraki Gulf where it has been recorded at densities of up to 100 individuals per m^2 and occurs in much lower numbers of between one to 10 individuals per m^2 in the southern port of Lyttelton Harbour [33]. Previous studies on *S. clava* have provided a global and regional snapshot of the genetic diversity and population connectivity for this species, which shows that vessel activity is a major vector for the regional spread of *S. clava* in New Zealand and Britain [14,34,35]. Unlike *C. maenas*, which is known to disperse naturally over long distances, *S. clava* relies more heavily on anthropogenic transport for regional spread [35] and, therefore, exchange among populations should be more easily traced to internal movement among marinas or ongoing input from overseas ports. Here, we examined two common components in the application of genetics to the understanding of invasive species: 1) the utility of genotypic and haplotypic data when assigning new incursions, 2) The accuracy of genetic assignments with different levels of temporal and spatial sampling.

Materials and Methods

Sample Collections

To address the aims of the study we collected *Styela clava* individuals from hard surfaces within the top metre of the subtidal zone. Three levels of collections were made during 2007, and the data generated from this study were then compared with data collected from 2006 in two previous studies [14,35] (Fig. 1).

1) We obtained 46 specimens of *Styela clava* collected either from the hull of boats arriving to port (PAH), or from marinas where newly documented incursions had occurred (MAR, DUD, NEL, OPU), herein referred to as new incursions. These incursions were sampled during targeted surveillance and boat hull inspections in 2007, undertaken by the National Institute of Water and Atmospheric Research (NIWA) and MPI (Fig. 1) and were not found in previous surveillance inspections at these marinas.

2) In April 2007 we collected 171 *Styela clava* individuals from four locations in the Hauraki Gulf and one location in the Port of Lyttelton that were also sampled in a 2006 study [35].

To expand the spatial resolution of the dataset, in April 2007 we collected an additional 197 individuals from a further 10 locations throughout the Hauraki Gulf and Lyttelton Harbour. These locations were not covered in the 2006 study by Goldstien *et al.* [35] as the 2006 data set was focused on marinas and aquaculture farms, excluding the natural habitat studied here. These additional sites are not considered new incursions as *S. clava* was recorded from these locations preceding the 2006 study. No specific permits were required for the described field studies. No specific permissions were required for the locations or activities as collections were not on privately-owned land and did not involve endangered or protected species.

All specimens were preserved in 70% ethanol for storage. DNA extractions, mitochondrial DNA sequencing (new incursions only) and microsatellite genotyping (all samples) were done using the protocols of Goldstien *et al.* [35]. To confirm genotypes and avoid technical biases in the data, we also genotyped 30 individuals collected in 2006 [35] alongside these new samples. Six of the eleven microsatellites developed for *S. clava* [36] were used in this study (1A9, 2H9, 1D11, 2B12, 1H1, and 1C8). The remainder of the microsatellites did not achieve consistent amplification across the dataset and so were not used in this study

New incursion analyses – mitochondrial and microsatellite data

We used mtDNA and genotypic data to investigate the equality of these molecular markers in accurately assigning new incursions. Data from previous studies on the global distribution of mtDNA haplotypes and microsatellite genotypes [14,35] were compared using the assignment of new incursions sampled in this study. The relationship of haplotypes was assessed using statistical parsimony networks constructed in TCS [37], incorporating all haplotypes previously identified from New Zealand, Australia and North America [14]. Further statistical tests were not performed on these data due to the small and imbalanced sample sizes. Genalex6 [38] was used to assign microsatellite genotypes of the new incursions. New incursions were treated as an unknown population in the assignment analysis.

Microsatellite analyses

All individuals sampled in 2007 were genotyped for comparison to locations sampled in 2006 and to investigate the role of

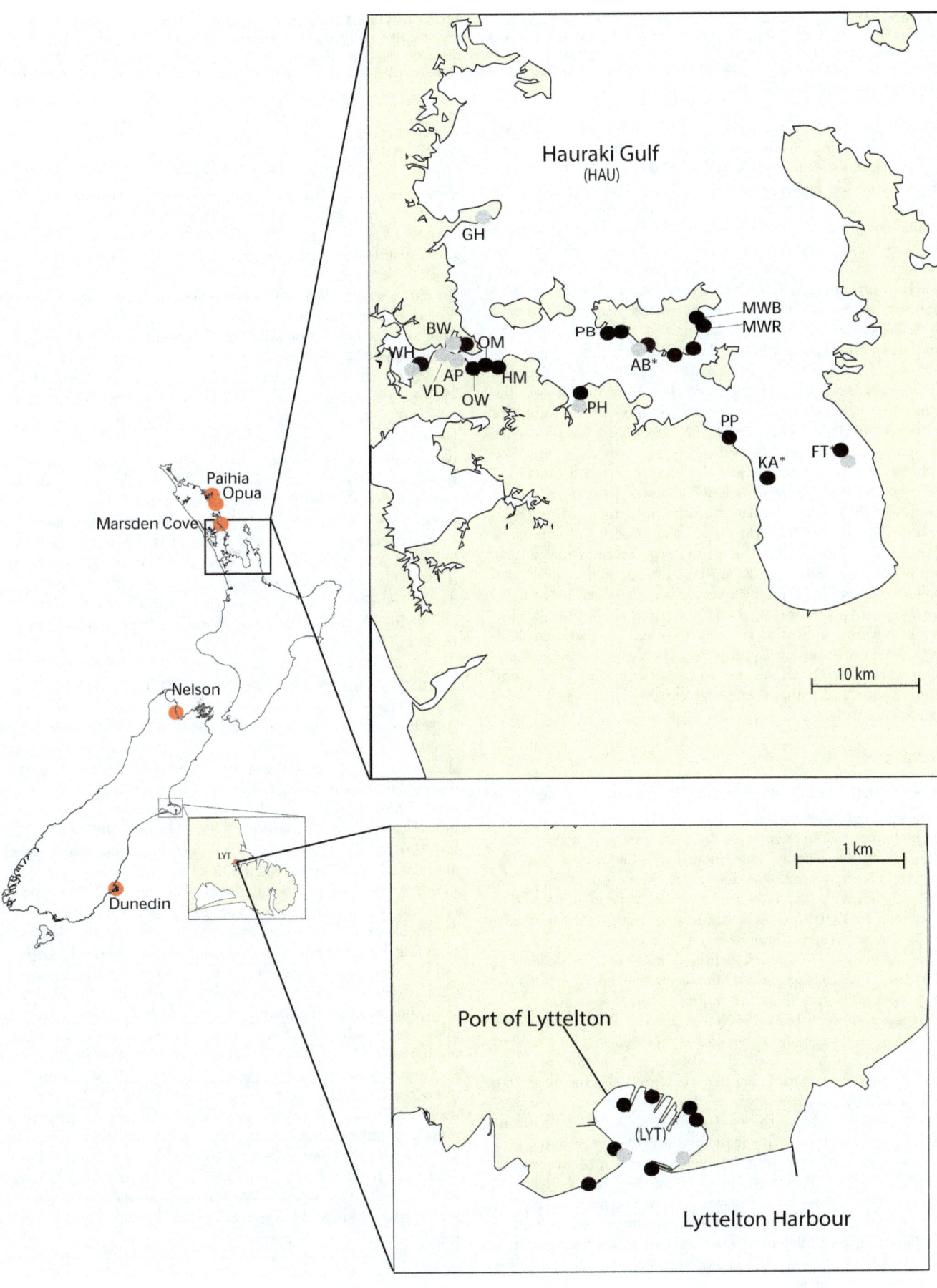

Figure 1. Sampling sites of _Styela clava_ throughout New Zealand in 2006 (grey) and 2007 (black). Abbreviations: AB, Awaawaroa Bay; AP, Auckland Port; BW, Bayswater marina; FT, Firth of Thames; GH, Gulf Harbour marina; HM, Half-Moon Bay marina; KA, Kaiaua; MW, Man-of-War Bay; O, Orapiu wharf; OM, Orakei marina; OW, Orakei Wharf; P, Puatiti Point; PB, Putiki Bay; PH, Pine Harbour marina; VD, Viaduct marina; WH, Westhaven marina; Port of Lyttelton (LYT). * Depicts populations sampled within aquaculture farms. Red circles represent locations of new incursions sampled in this study. Modified from Goldstien _et al._ [35].

temporal sampling in the assignment of the new incursions. Diversity indices such as expected heterozygosity (H_e) and allele frequency for loci and populations were estimated using Genalex6 [38], and a chi-squared test was run for each locus within each population, to assess Hardy-Weinberg equilibrium. To determine the allelic richness (Ar) of populations, which accounts for differences in sample sizes [39], we ran rarefaction in HP-RARE 1.0 [40], with rarefaction set to a sample size of 13. We then used t-statistics to test for differences in Ar among years. To compare H_e & F_{IS} statistics between 2006 and 2007 samples we used the randomisation procedure [41] in Fstat [41].

F_{ST} and pairwise distance statistics [42,43] were calculated using Arlequin v. 3.1 [44] and from these data we assessed the degree to which the five 2007 populations represented a sample of the five 2006 populations. To do this we applied statistics more common to the study of species diversity, whereby each individual was treated as a sample unit and the presence of alleles recorded for each individual across multiple loci was treated as species abundance. Species accumulation curves were done in Primer v.6 (Primer-E Limited, 2009) where the Chao2 index was compared against observations (S_{obs}). Due to the occurrence of rare alleles, Chao's Jaccard estimator (Chao-Jacc-Est) of abundance-based similarity index [45–47], which accounts for "unseen individuals' based on rare alleles was used and compared against the Jaccard and Bray-Curtis indices to estimate the similarity between 2006 and 2007 samples, using EstimateS v. 7.5 (Colwell R.K). Finally, using chi-squared analyses we assessed the efficacy of using datasets with and without temporal variation to assign the new incursions.

Results

New incursion analyses – similarities between genotypes and haplotypes

The haplotypic and genotypic data obtained were consistent in their ability to identify new incursions. Eight new and six previously assigned mtDNA haplotypes were observed in the new incursions (Fig. 2a). However, a haplotype previously found to be unique to Lyttelton, and occurring in high frequency there (H28), was not observed in any of the new incursions (Fig. 2b). In addition, unique haplotypes identified from new incursions at Marsden Cove, Opua and Dunedin were closely related to haplotypes previously found in multiple populations, including in the North and South Islands of New Zealand [14] (Fig. 2a). Alleles were consistently amplified from six microsatellite loci for a total of 46 individuals of _Styela clava_ sampled from four marinas and one boat hull in 2007. Considering the low sample size for these new incursions, the genotypic diversity (allelic richness and heterozygosity) was high relative to populations previously sampled from New Zealand (Table 1). Genotypic assignment of each of the new incursions consistently showed a high proportion of assignment (>60%) to North America (Fig. 2c) and only one population, Marsden Cove, showed a proportion of assignment to the Port of Lyttelton. Marsden Cove was also the most diverse site with six of the eight unique haplotypes. When all individuals from the new incursions were pooled, the assignment remained consistent with the separate groupings.

2007 microsatellite analyses

Alleles were consistently amplified from six loci for a total of 368 individuals of _Styela clava_ sampled from 15 populations in 2007 (Table 1). In addition, 30 individuals sampled in 2006 were re-genotyped to assess possible shifts in allele peaks due to changes in the instrument and running procedure. Four loci showed no shift in allele size, while two were adjusted for a shift of one base pair for data acquired across different years. All loci were polymorphic and significant genetic structure was observed between Hauraki Gulf and Lyttelton populations (F_{ST}, 0.106; P<0.01). The number of alleles per locus ranged from 8 (Sc2B12) to 19 (Sc1H1) (Table S1). Chi-square tests showed significant departures from Hardy-Weinberg equilibrium for 13 of the 16 sampled populations (with adjusted nominal level for multiple randomisations). Two populations from natural reef environments (PP and PB), comprising two small rock platforms, and Westhaven Marina showed no significant difference from H-W equilibrium for any of the loci sampled. For all other populations, two to four of the six loci were out of H-W equilibrium. There was no significant difference in the observed genetic diversity among sites. The greatest diversity was observed in populations from Bayswater Marina (A_r, 3.00; H_e, 0.54; Table 1) and Man-of-War beach (A_r, 3.04; H_e, 0.66; Table 1). The lowest genetic diversity was observed in a small intertidal reef population at Puatiti Point (A_r, 2.56; H_e, 0.56; Table 1), within Hauraki Gulf.

Temporal genetic change (2006/2007)

Five of the populations sampled in 2006 were re-sampled in 2007 to investigate the genotypic stability over this two-year period (Table 1). Comparisons between years for these five populations showed no significant difference for inbreeding index (F_{IS}), allelic diversity (Ar), or heterozygosity (H_E). Species (allele) accumulation curves show that 85–92% of the diversity estimated by the Chao2 index was captured in each year for the Hauraki Gulf (Fig. 3a) and 25–93% for Lyttelton (Fig. 3b). However, a plateau of allelic abundance was not observed in any of the data sets despite having sampled over 100 individuals in the Hauraki Gulf, indicating a relatively high allelic richness with a large proportion of singleton and doubleton alleles. The estimated similarity between populations was high when accounting for "unseen alleles" (Chao-Jacc-est) but was considerably lower for the classic Jaccard and Bray-Curtis estimates that do not account for "unseen alleles" (Table 2).

In the Lyttelton population, 15 'new' alleles were observed in 2007, of which only three were not observed in any other populations. Of the 29 alleles observed in Lyttelton in 2006, only five were not observed in the populations in 2007, but all of these did occur in the Hauraki Gulf populations (Table 2). Input of new alleles to populations within the Hauraki Gulf was high for one marina population but loss of alleles was more common in other populations. However, when all populations within the Hauraki Gulf were pooled in each year, the similarity estimates increased and the proportion of new and dropped alleles reduced, suggesting spatial shifts in allele frequencies (Table 2). There was no significant difference between years.

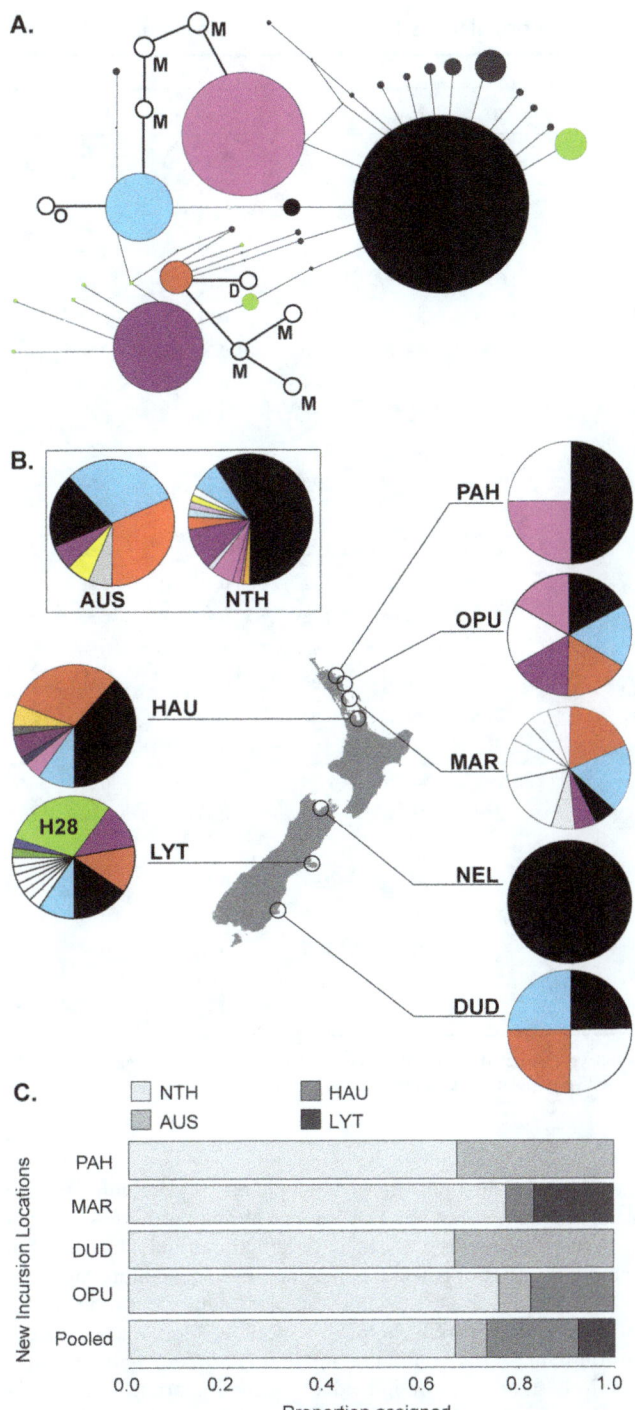

tions: PAH, Paihia; OPU, Opua; MAR, Marsden Cove; DUD, Dunedin; NEL, Nelson; LYT, Lyttelton; HAU, Hauraki Gulf; AUS, Australia; NTH, North America. 2A & B are modified from Goldstien *et al.* (2010 and Goldstien *et al.* (2011).

New incursion assignment efficacy with different data sets

Genotypic assignment of individuals from new incursions showed a low probability of assignment to New Zealand populations sampled in 2006 and 2007 (Fig. 4), although no single source could be assigned to these new incursions. The combined data from 2006 and 2007 showed a significant difference in the proportion of new incursions assigned to the northern Hauraki Gulf populations and the southern Lyttelton Port populations compared with the separate 2006 and 2007 data sets, suggesting their assignment probabilities depend on the temporal aspect of the data sets (Fig. 4; $X^2 = 49.95$, p<0.01). In all cases the new incursions were assigned with high probability to populations sampled in North America (0.67, 0.73, 0.61, 0.57 for the 2006, 2007 data, all individuals from 2006 and 2007 and the 2006/2007 shared populations data sets respectively). For all four of these data sets, the Hauraki Gulf had the highest number of individuals sampled, yet this does not appear to affect the results obtained. Temporal sampling increased the assignment to New Zealand, regardless of whether using only the shared populations or the whole data set, suggesting that this is not a function of increasing the sample size.

Discussion

The results obtained here highlight the importance of temporal genetic sampling and the application of baseline genetic data sets in determining the origin and spread of new incursions. To our knowledge this is the first study to use existing genetic data to assign individuals to sources as they arrive at new locations within a region. The key findings of this study are: 1) new incursions recorded in New Zealand exhibit high mitochondrial and nuclear genetic diversity and both markers were consistent in their assignment of individuals to populations; 2) genetic diversity has been maintained over a two-year period within approximately 10 years of the initial introduction; 3) significant spatial structure between the North and South Island populations is maintained over the two-year period, despite extensive regional admixture and population growth; 4) temporal sampling was important in the accurate assignment of new incursions.

Styela clava is transported around the globe in biofouling on vessels and relies heavily on anthropogenic transport for international and domestic spread [14]. Unlike the very clear picture of asymmetric dispersal displayed by the crab *Carcinus maenas* along the coast of Nova Scotia [26], the continuous translocation of *S. clava* makes for a very "chaotic" pattern of genetic admixture and allelic shifts. Models of maritime transportation networks have highlighted how complexity in the transport pathways, coupled with stochastic demographic events (e.g., recruitment to a vessel and establishment in a new location) can drive an unpredictable sequence of invasion from primary and subsequent incursions, with potential for considerable re-assortment of populations of frequently transported species [48–50]. Although it is expected that allelic change and genetic drift are directly related to population growth and demographic stochasticity [27,28], it is apparent that the stochasticity and temporal stability of transport pathways may be more influential in maintaining temporal stability of alleles in introduced species. For instance, the Hauraki

Figure 2. Haplotypic and genotypic data for the new incursions of *Styela clava* sampled in 2007: (A) The haplotypes unique to Marsden Cove (M), Opua (O) and Dunedin (D) are shown relative to their relationship with haplotypes recorded in 2006: The six haplotypes shared with data obtained in 2006 (colours match locations in figure 2B); the haplotypes present in locations outside of New Zealand (grey); and those present only in Lyttelton (green). Circles represent haplotypes and the size reflects the frequency of each haplotype within the data set. (B) The distribution and frequency of haplotypes present in the new incursion sampled in 2007 and in the populations sampled by Goldstien *et al.* (2010) in 2006. The colours are consistent with 2A. Pies represent the frequency of each haplotype within the populations. (C) The proportion of genotypes within the new incursion populations assigned to populations sampled by Goldstien et al. (2010) in 2006. Abbrevia-

Table 1. Descriptive statistics for genotypic data sampled from *Styela clava* populations throughout New Zealand.

Site		Year sampled	Habitat	n	N_{all}	A_r	F_{IS}	H_E
Lyttelton	LYT	*2006*	*Port*	*13*	*31*	*2.71*	*0.27*	*0.62*
		2007	Port	70	58	2.87	0.29	0.65
Hauraki Gulf (HAU)								
Awaawaroa Bay	AB	*2006*	*Oyster farm*	*49*	*45*	*2.85*	*0.27*	*0.63*
		2007	Oyster farm	27	41	2.87	0.26	0.62
Bayswater	BW	*2006*	*Marina*	*19*	*34*	*2.74*	*0.22*	*0.61*
		2007	Marina	32	49	3.00	0.22	0.66
Firth of Thames	FT*	*2006*	*Mussel farm*	*22*	*36*	*2.85*	*0.32*	*0.62*
		2007	Mussel farm	26	38	2.93	0.30	0.64
Westhaven	WH	*2006*	*Marina*	*29*	*40*	*2.90*	*0.26*	*0.65*
		2007	Marina	16	27	2.70	0.34	0.54
Half Moon Bay	HM	2007	Marina	26	35	2.87	0.30	0.64
Kaiaua	KA*	2007	Mussel Farm	27	36	2.70	0.28	0.66
Orakei	OM	2007	Marina	22	32	2.72	0.15	0.58
Orapiu Wharf	OW	2007	Wharf	6	22	2.78	0.29	0.57
Puatiti Point	PP	2007	Intertidal reef	5	17	2.56	−0.02	0.56
Man of War Bay	MWB	2007	beach	26	42	3.04	0.16	0.66
Man of War Bay	MWR	2007	Intertidal reef	24	39	2.92	0.07	0.66
Putiki Bay	PB	2007	Intertidal reef	11	27	2.74	0.10	0.56
Putiki Bay	PB*	2007	Oyster farm	29	40	2.97	0.29	0.65
TeMatuku Bay	TM*	2007	Oyster farm	21	36	2.70	0.22	0.58
New Incursions								
Pahia	PAH	2007	Hull	6	21	2.94	0.35	0.59
Nelson	NEL	2007	Marina	1	9	1.50	-	0.43
Marsdon Cove	MAR	2007	Marina	19	36	3.55	0.18	0.62
Dunedin	DUD	2007	Marina	11	32	3.56	0.08	0.62
Opua	OPU	2007	Marina	9	29	3.48	0.28	0.88

Genetic characteristics are represented across six polymorphic loci and include: Site location, year sampled, habitat type, sample size (n), number of alleles (N_{all}), allelic richness (A_r), fixation index (F_{IS}) and gene diversity (H_E). Note: Italicised text highlights the five populations sampled in 2006 and 2007. *highlights aquaculture farm populations.

Gulf experiences high levels of recreational vessel traffic. *Styela clava* is much more likely to be transported by recreational vessels than merchant vessels (because the former sit idle for longer, are less well maintained and travel at slower speeds), and this is reflected in the admixture observed among populations and between years in our data. In contrast, the Port of Lyttelton receives a high number of merchant vessels but much less recreational traffic, which is reflected in the significant genetic differentiation between the Lyttelton and Hauraki Gulf populations, maintained over the two-year period, or about four generations for *Styela clava* [51].

New incursions for *Styela clava* were not resolved to specific source locations in this study, but they were proportionally assigned to northern and southern New Zealand populations, or to overseas populations for which data were available. The pooling of data from populations sampled in 2006 and 2007 significantly changed the proportion of assignments to northern and southern New Zealand from 20% and 25% to 35% and 37%. The change observed with the temporal approach could simply be a function of increasing the sample size and capturing more alleles. However, using only the shared populations actually decreased the sample size for the Hauraki Gulf. In addition, the data collected from 2006 and 2007 showed no significant difference and all alleles

were captured in both datasets, albeit the distribution of these alleles did vary within years. It is more likely that there is a trade-off occurring here. For instance, the 2006 data for Hauraki Gulf includes the Port of Auckland, marinas and aquaculture farms; in 2007 Auckland Port was not sampled, but the marinas and aquaculture farms were sampled, as well as additional sites from natural habitats. By combining the 2006 and 2007 data, we have effectively sampled more individuals of the important populations around marinas and aquaculture farms, while the more minor sites have been eliminated. Analysis of this core data set, with a strong element of temporal resampling, indicates that temporal sampling may enhance the accuracy and likelihood of assignment to the true source region as unsampled (rare) alleles in the population may be important to assignments, particularly if these change in frequency from year-to-year.

The new incursions studied here highlight a key aspect of founding populations and the interplay of transport pathways and population dynamics in the genetic diversity of new incursions. Numerous genetic studies have shown that genetic diversity in populations of invasive species is higher than expected in founder populations under bottleneck conditions [52]. Several authors have definitively put this "paradox" to rest, highlighting that the

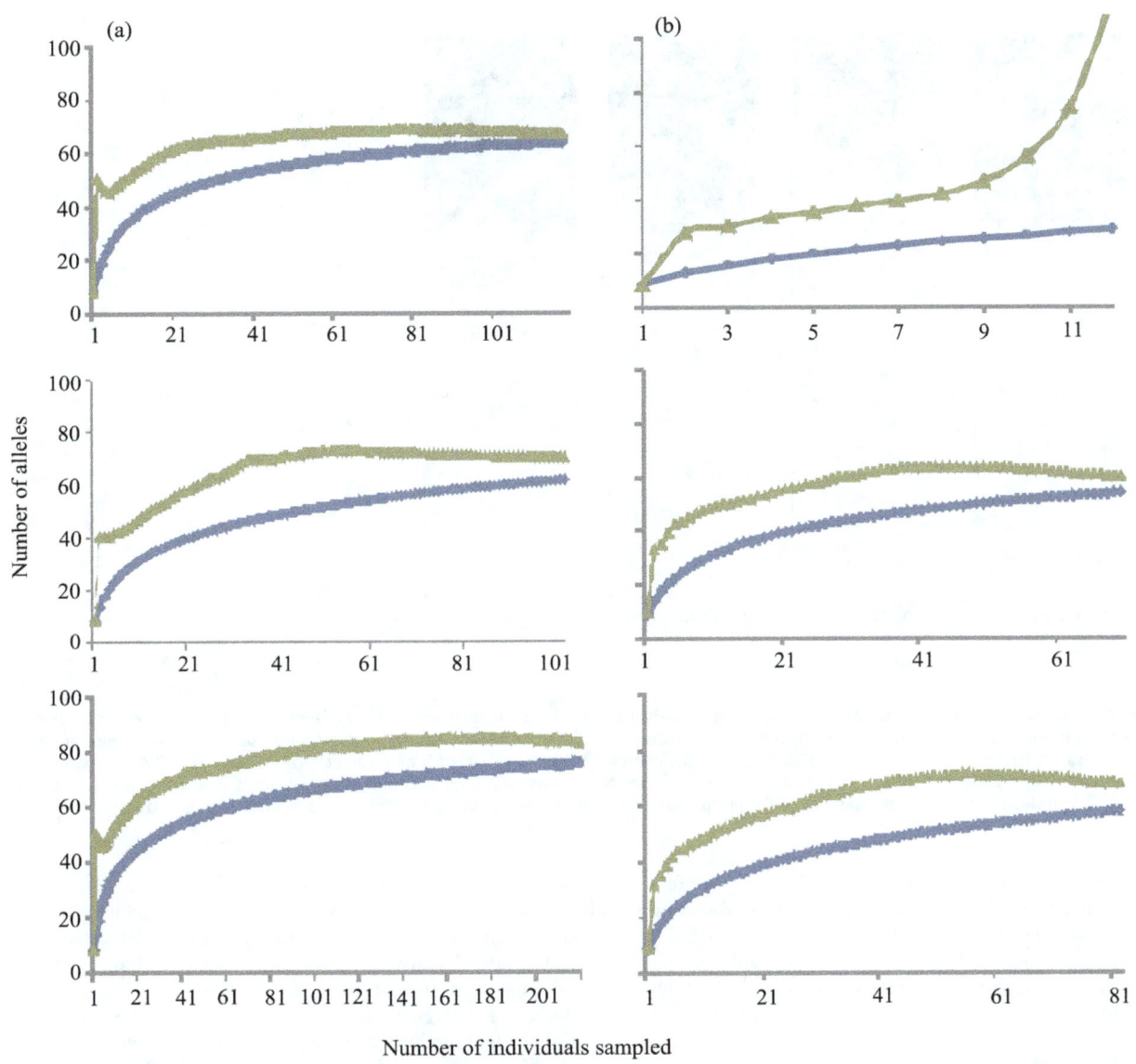

Figure 3. Allele accumulation curves for observed data (green) and the Chao2 index (blue). Populations represent Hauraki Gulf (a) and Lyttelton Port (b), from 2006 (top graph), 2007 (middle) and years combined (bottom graph).

gap between time since introduction and the genetic study, in most cases, would likely need no more than the temporal shifts and genetic divergences expected from growing populations to explain the diversity observed [53,54]. Our data are unique in that we have sampled populations within an estimated 20 generations from the initial incursion, as well as new incursions within the first

Table 2. Allele frequencies for *Styela clava* populations sampled in 2006 and 2007.

Population		N_{all}	N_{06}	N_{07}	n/d	CJ	J	BC
Bayswater	BW	53	34	49	19/5	0.6	0.5	0.1
Firth of Thames	FT*	43	36	38	7/7	0.9	0.5	0.6
Awaawaroa Bay	AB*	53	45	41	8/12	0.6	0.4	0.4
Westhaven	WH	42	40	27	2/15	0.6	0.4	0.4
Hauraki Gulf		**76**	**64**	**62**	**12/14**	**0.8**	**0.7**	**0.7**
Lyttelton Port	LP	**44**	**29**	**39**	**15/5**	**1.0**	**0.5**	**0.3**

Genetic characteristics are represented across six polymorphic loci and include: Site location with sample ID, total number of alleles (Nall), number of alleles for 2006 (N_{06}) and 2007 (N_{07}) and the ratio of alleles new in 2007 : alleles dropped in 2007 (n/d), Chao2-Jaccard-est index (CJ), Jaccard classic (J), Bray-Curtis Index (BC).

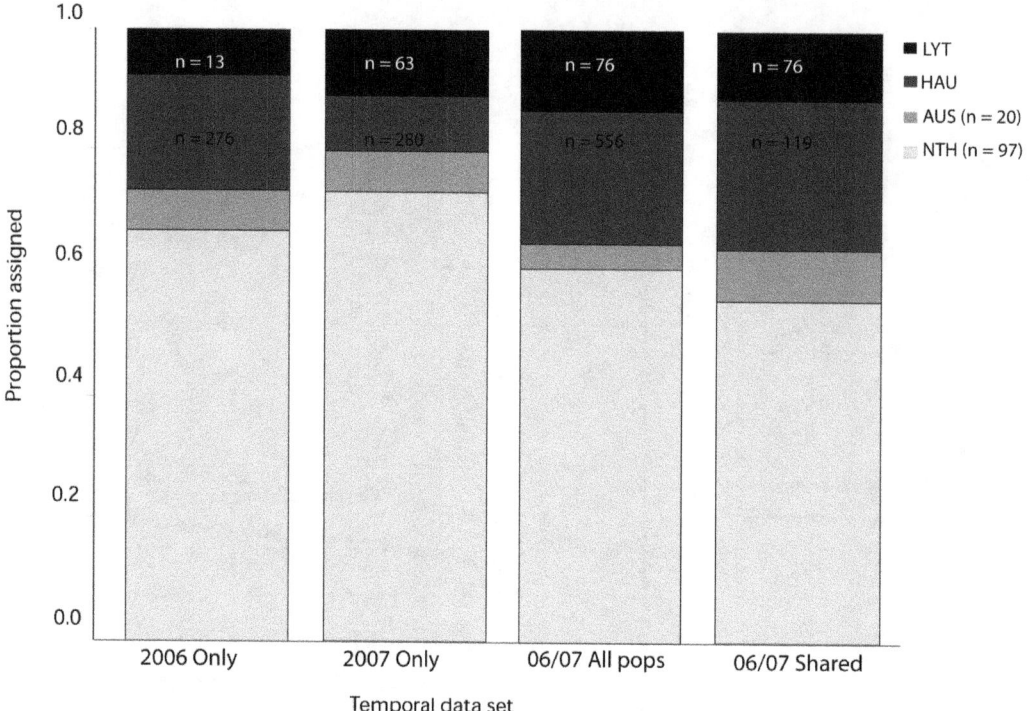

Figure 4. Proportional assignments of *Styela clava* **individuals sampled from new incursions.** Stacked bars show the proportion of individuals with the highest probability of genetic assignment to known populations in New Zealand (NZ) and overseas. Data sets were analysed separately for year-sampled combinations: 2006 and 2007 separately; combined 2006 and 2007 for all populations sampled (06/07 All pops) and the five populations sampled in 2006 and 2007 (06/07 Shared). Samples from North American and Australia were included in all assignment analyses. Abbreviations: LYT, Lyttelton Port; HAU, Hauraki Gulf; NTH, North America; AUS, Australia. Sample size (n) for each data set is shown.

generation of establishing within a location. From these first generation samples, it is clear that the diversity of the founding populations is not necessarily low; indeed some founders exhibit greater diversity than many of the established populations, and provide a diverse genetic pool for regional translocation, without invoking rapid population growth or multiple introduction processes.

These data also provide a powerful model for further monitoring the change in genetic diversity with population dynamic shifts over time. Several other studies have successfully used historical genetic data to investigate genetic change over time [55]. observed large genetic variance and temporal fluctuations in allele frequencies of the introduced fly *Rhagoletis completa* using samples taken from the initial founding population and again 30 years after its initial introduction. Similarly, hybrid zones of *Mytilus* spp. along the coast of California were sampled in 1994–95 and again in 2005–07 showing a large range shift toward the equator in the subtropical species *M. galloprovincialis* following a decade of climatic oscillation, and Peres-Portela [32] showed a reduction of diversity in a population established from a single introduction.

Perhaps genetic equilibrium is rarely met in introduced species, particularly at small initial population sizes, where an individual's contribution to the next generation is amplified and is critical to successful establishment and population growth. However, the importance of temporal variation in allele frequencies may differ between species with short- and long-generation times, with different per capita rates of growth (r) from high or low fecundity species, and between marine and terrestrial invaders. For example, iteroparous species may have more temporal buffering of reproductive success because they are spreading the risk of failure over multiple reproductive events and years, as seen in the sea

urchin *Paracentrotus lividus* [56]. Greater temporal variation may also be expected in populations of invaders where adults have life-histories adapted to broadcast spawning and high planktonic mortality and stochasticity, combined with disturbances experienced during the invasion process [57,58].

Increasing the genetic database for invasive species to include spatial and temporal variation of populations could prove to be an invaluable tool for pre-border management of NIS. Currently, genetics is predominantly used as a snapshot tool to identify species and their origin, yet there are very few instances where the pathway has been identified successfully from this one-off sampling approach. While much criticism has been focused on the lack of inclusion of source populations and spatial variation in genetic sampling, very little attention has been given to the change in genetic signature with time through the invasion process. Our work suggests that addressing the variability of source populations and the potential for genetic drift in small populations with the use of temporal sampling may be a critical element in the future use of genetic tools for invasive species management.

Supporting Information

Table S1   Allele frequencies for *Styela clava* populations sampled in 2006 and 2007.

Acknowledgments

The large sample size acquired for this study was made possible through the assistance of several agencies and many individuals. We also thank anonymous reviewers for constructive comments that improved this manuscript. Data used in this study are available on GenBank Accession numbers GU328006–GU328035. The unique sequences generated from this study are available on GenBank Accession numbers KC905096–KC905102.

Author Contributions

Conceived and designed the experiments: SJG GJI DRS NJG. Performed the experiments: SJG. Analyzed the data: SJG. Contributed reagents/materials/analysis tools: SJG NJG. Wrote the paper: SJG GJI DRS NJG.

References

1. Holland BS (2000) Genetics of marine bioinvasions. Hydrobiologia 420: 63–71.
2. Ruiz GM, Fofonoff PW (2000) Invasion of coastal marine communities in North America: Apparent patterns, processes, and biases. Annual Review of Ecology and Systematics 31: 481.
3. Darling J, Blum M (2007) DNA-based methods for monitoring invasive species: a review and prospectus. Biological Invasions 9: 751–765.
4. Dyer C, Richardson D (1992) Population genetics of the invasive Australian shrub *Hakea sericea* (Proteaceae) in South Africa. South African Journal of Botany 58: 117–124.
5. Andrew J, Ward R (1997) Allozyme variation in the marine fanworm *Sabella spallanzanii*: comparison of native European and introduced Australian populations. Marine Ecology Progress Series 152: 131–143.
6. Elderkin CL, Perkins EJ, Leberg PL, Klerks PL, Lance RF (2004) Amplified fragment length polymorphism (AFLP) analysis of the genetic structure of the zebra mussel, *Dreissena polymorpha*, in the Mississippi River. Freshwater Biology 49: 1487–1494.
7. Patti FP, Gambi MC (2001) Phylogeography of the invasive polychaete Sabella spallanzanii (Sabellidae) based on the nucleotide sequence of internal transcribed spacer 2 (ITS2) of nuclear rDNA. Marine Ecology-Progress Series 215: 169–177.
8. Muller J (2001) Invasion history and genetic population structure of riverine macro invertebrates. Zoology-Analysis of Complex Systems 104: 346–355.
9. Bucciarelli G, Golani D, Bernardi G (2002) Genetic cryptic species as biological invaders: the case of a Lessepsian fish migrant, the hardyhead silverside *Atherinomorus lacunosus*. Journal of Experimental Marine Biology and Ecology 273: 143–149.
10. Folino-Rorem NC, Darling JA, D'Ausilio CA (2009) Genetic analysis reveals multiple cryptic invasive species of the hydrozoan genus *Cordylophora*. Biological Invasions 11: 1869–1882.
11. Ryland JS, De Blauwe H, Lord R, Mackie JA (2009) Recent discoveries of alien Watersipora (Bryozoa) in Western Europe, with redescriptions of species. Zootaxa: 43–59.
12. Holland BS, Dawson MN, Crow GL, Hofmann DK (2004) Global phylogeography of Cassiopea (Scyphozoa : Rhizostomeae): molecular evidence for cryptic species and multiple invasions of the Hawaiian Islands. Marine Biology 145: 1119–1128.
13. Shefer S, Abelson A, Mokady O, Geffen E (2004) Red to Mediterranean Sea bioinvasion: natural drift through the Suez Canal, or anthropogenic transport? Molecular Ecology 13: 2333–2343.
14. Goldstien SJ, Dupont L, Schiel DR, Bishop J, Gemmell NJ (2011) Global phylogeography of the widely introduced NW. Pacific ascidian *Styela clava*. PLoS ONE 6: e16755.
15. Rius M, Pascual M, Turon X (2008) Phylogeography of the widespread marine invader *Microcosmus squamiger* (Ascidiacea) reveals high genetic diversity of introduced populations and non-independent colonizations. Diversity and Distributions 14: 818–828.
16. Darling J, Bagley M, Roman J, Tepolt C, Geller J (2008) Genetic patterns across multiple introductions of the globally invasive crab genus *Carcinus*. Molecular Ecology 17: 4992–5007.
17. Uwai S, Nelson W, Neill K, Wang W, Aguilar-Rosas L, et al. (2006) Genetic diversity in *Undaria pinnatifida* (Laminariales, Phaeophyceae) deduced from mitochondria genes - origins and succession of introduced populations. Phycologia 45: 687–695.
18. Audzijonyte A, Wittmann K, Vainola R (2008) Tracing recent invasions of the Pont-Caspian mysid shrimp *Hemimysis anomala* across Europe and to North America with mitochondrial DNA. Diversity and Distributions 14: 179–186.
19. Roman J, Palumbi SR (2004) A global invader at home: population structure of the green crab, *Carcinus maenas*, in Europe. Molecular Ecology 13: 2891–2898.
20. Ashton GV, Stevens MI, Hart MC, Green DH, Burrows MT, et al. (2008) Mitochondrial DNA reveals multiple Northern Hemisphere introductions of Caprella mutica (Crustacea, Amphipoda). Molecular Ecology 17: 1293–1303.
21. Provan J, Murphy S, Maggs CA (2005) Tracking the invasive history of the green alga *Codium fragile* ssp. *tomentosoides*. Molecular Ecology 14: 189–194.
22. Geller J, Sotka E, Kado R, Palumbi SR, Schwindt E (2008) Sources of invasions of a northeastern Pacific acorn barnacle, *Balanus glandula*, in Japan and Argentina. Marine Ecology Progress Series 358: 211–218.
23. Geller J, Darling JA, Carlton JT (2010) Genetic perspectives on marine biological invasions. Annual Review of Marine Science 2: 367–393.
24. Chandler E, McDowell J, Graves J (2008) Genetically monomorphic invasive populations of the rapa whelk, *Rapana venosa*. Molecular Ecology 17: 4079–4091.
25. Pineda MC, Lopez-Legentil S, Turon X (2011) The Whereabouts of an Ancient Wanderer: Global Phylogeography of the Solitary Ascidian *Styela plicata*. PloS ONE 6.
26. Pringle J, Blakeslee A, Byers JE, Roman J (2011) Asymmetric dispersal allows an upstream region to control population structure throughout a species' range. Proceedings of National Academy of Science USA 108: 15288–15923.
27. Grant P, Grant B, Petren K (2001) A population founded by a single pair of individuals: establishment, expansion, and evolution. Genetica 112–113: 359–382.
28. Dlugosch K, Hays C (2008) Genotypes on the move: some things old and some things new shape the genetics of colonization during species invasions. Molecular Ecology 17: 4583–4585.
29. Johnson MS, Black R (1984) The Wahlund effect and the geographical scale of variation in the intertidal limpet *Siphonaria sp*. Marine Biology 79: 295–302.
30. Sinnock P (1975) The Wahlund Effect for the two-locus model. American Naturalist 109: 565–570.
31. Ng W, Leung F, Chak S, Slingsby G, Williams G (2010) Temporal genetic variation in populations of the limpet *Cellana grata* from Hong Kong shores. Marine Biology 157: 325–337.
32. Perez-Portela R, Turon X, Bishop JDD (2012) Bottlenecks and loss of genetic diversity: spatio-temporal patterns of genetic structure in an ascidian recently introduced in Europe. Marine Ecology Progress Series 451: 93–105.
33. Gust N, Inglis GJ, Floerl O, Peacock L, Denny C, et al. (2009) Assessment of population management options for *Styela clava*. Wellington: MAFBNZ. 237 p.
34. Dupont L, Viard F, Dowell M, Wood S, Bishop J (2009) Fine-and regional-scale genetic structure of the exotic ascidian *Styela clava* (Tunicata) in southwest England, 50 years after its introduction. Molecular Ecology 18: 442–453.
35. Goldstien S, Schiel DR, Gemmell N (2010) Regional connectivity and coastal expansion: differentiating pre-border and post-border vectors for the invasive tunicate *Styela clava*. Molecular Ecology 19: 874–885.
36. Dupont L, Viard F, Bishop J (2006) Isolation and characterisation of twelve polymorphic microsatellite markers for the invasive ascidian *Styela clava* (Tunicata). Molecular Ecology Notes 6: 101–103.
37. Clement M, Posada D, Crandall KA (2000) TCS: a computer program to estimate gene genealogies. Molecular Ecology 9: 1657–1660.
38. Peakall R, Smouse P (2006) GenALEx6; genetic analysis in Excel. Population genetic software for teaching and research. Molecular Ecology Notes 6: 288–295.
39. Leberg P (2002) Estimating allelic richness: Effects of sample size and bottlenecks. Molecular Ecology 11: 2445–2449.
40. Kalinowski ST (2005) HP-RARE 1.0: a computer program for performing rarefaction on measures of allelic richness. Molecular Ecology Notes 5: 187–189.
41. Goudet J (1995) FSTAT (Version 1.2): A computer program to calculate F-statistics. Journal of Heredity 86: 485–486.
42. Reynolds J, Weir B, Cockerham C (1983) Estimation for the coancestry coefficient: basis for a short-term genetic distance. Genetics 105: 767–779.
43. Slatkin M (1995) A measure of population subdivision based on microsatellite allele frequencies. Genetics 139: 457–462.
44. Excoffier L, Laval L, Schneider S (2005) Arlequin ver 3.0: An integrated software package for population genetics data analysis Evolutionary Bioinformatics Online 1: 47–50.
45. Cabaço S, Santos R, Sprung M (2011) Population dynamics and production of the seagrass *Zostera noltii* in colonizing versus established meadows. Marine Ecology 33: 280–289.
46. Chao A, Chazdon R, Colwell R, Shen BC (2005) A new statistical approach for assessing similarity of species composition with incidence and abundance data. Ecology Letters 8: 148–159.
47. Chao A, Chazdon R, Colwell R, Shen T (2006) Abundance-based similarity indices and their estimation when there are unseen species in samples. Biometrics 62: 361–371.
48. Drake J, Lodge A (2004) Global hot spots of biological invasions: evaluating options for ballast-water management. Proceedings of the Royal Society B: Biological Sciences 271: 575–580.
49. Floerl O, Inglis G, Dey K, Smith A (2009) The importance of transport hubs in stepping-stone invasions. Journal of Applied Ecology 46: 37–45.
50. Kaluza P, Kolzsch A, Gastner MT, Blasius B (2010) The complex network of global cargo ship movements. Journal of The Royal Society Interface 7: 1093–1103.
51. Wong N, McClary D, Sewell MA (2011) The reproductive ecology of the invasive ascidian, *Styela clava*, in Auckland Harbour, New Zealand. Marine Biology 158: 2775–2785.
52. Roman J, Darling JA (2007) Paradox lost: genetic diversity and the success of aquatic invasions. Trends in Ecology & Evolution 22: 454–464.
53. Estoup A, Guillemaud T (2010) Reconstructing routes of invasion using genetic data: why, how and so what? Molecular Ecology 19: 4113–4130.
54. Hoos P, Miller A, Ruiz G, Vrijenhoek RC, Geller J (2010) Genetic and historical evidence disagree on likely sources of the Atlantic amethyst gem clam *Gemma gemma* (Totten, 1834) in California. Diversity and Distributions 16: 582–592.
55. Chen Y, Berlocher S, Opp S, Roderick G (2010) Post-colonization temporal genetic variation of an introduced fly, *Rhagoletis completa*. Genetica 138: 1059–1075.

56. Calderon I, Palacin C, Turon X (2009) Microsatellite markers reveal shallow genetic differentiation between cohorts of the common sea urchin *Paracentrotus lividus* (Lamarck) in northwest Mediterranean. Molecular Ecology 18: 3036–3049.

57. Clark GF, Johnston EL (2009) Propagule pressure and disturbance interact to overcome biotic resistance of marine invertebrate communities. Oikos 118: 1679–1686.

58. Siegel DA, Mitarai S, Costello CJ, Gaines SD, Kendall BE, et al. (2008) The stochastic nature of larval connectivity among nearshore marine populations. Proceedings of the National Academy of Sciences of the United States of America 105: 8974–8979.

16

A Nonluminescent and Highly Virulent *Vibrio harveyi* Strain Is Associated with "Bacterial White Tail Disease" of *Litopenaeus vannamei* Shrimp

Junfang Zhou[1]*, **Wenhong Fang**[1]*, **Xianle Yang**[2], **Shuai Zhou**[1], **Linlin Hu**[1], **Xincang Li**[1], **Xinyong Qi**[3], **Hang Su**[1], **Layue Xie**[1]

1 Key Laboratory of Marine and Estuarine Fisheries Resources and Ecology, East China Sea Fisheries Research Institute, Chinese Academy of Fisheries Science, Shanghai, China, **2** Aquatic Pathogen Collection Center of Ministry, Shanghai, China, **3** Shanghai Animal Disease Control Center, Shanghai, China

Abstract

Recurrent outbreaks of a disease in pond-cultured juvenile and subadult *Litopenaeus vannamei* shrimp in several districts in China remain an important problem in recent years. The disease was characterized by "white tail" and generally accompanied by mass mortalities. Based on data from the microscopical analyses, PCR detection and 16S rRNA sequencing, a new *Vibrio harveyi* strain (designated as strain HLB0905) was identified as the etiologic pathogen. The bacterial isolation and challenge tests demonstrated that the HLB0905 strain was nonluminescent but highly virulent. It could cause mass mortality in affected shrimp during a short time period with a low dose of infection. Meanwhile, the histopathological and electron microscopical analysis both showed that the HLB0905 strain could cause severe fiber cell damages and striated muscle necrosis by accumulating in the tail muscle of *L. vannamei* shrimp, which led the affected shrimp to exhibit white or opaque lesions in the tail. The typical sign was closely similar to that caused by infectious myonecrosis (IMN), white tail disease (WTD) or penaeid white tail disease (PWTD). To differentiate from such diseases as with a sign of "white tail" but of non-bacterial origin, the present disease was named as "bacterial white tail disease (BWTD)". Present study revealed that, just like IMN and WTD, BWTD could also cause mass mortalities in pond-cultured shrimp. These results suggested that some bacterial strains are changing themselves from secondary to primary pathogens by enhancing their virulence in current shrimp aquaculture system.

Editor: Ramy K. Aziz, Cairo University, Egypt

Funding: This work was supported by the Special Research Fund for the National Non-profit Institutes (East China Sea Fisheries Research Institute, grant# 2009 T06, 2011 T03 and 2007 M25), the Provincial Natural Science Foundation of Jiangsu (grant# BK2010269), the Three-project for Aquaculture in Jiangsu Province (grant# PJ2010-51) and China Postdoctoral Science Foundation (grant# 20080440654). The funders had no role in study design, data collection and analysis, decision to publish, or preparation of the manuscript.

Competing Interests: The authors have declared that no competing interests exist.

* E-mail: junfangzhou@yahoo.cn (JZ); whfang06@yahoo.com.cn (WF)

Introduction

Litopenaeus vannamei (*L. vannamei*) shrimp is the most extensively cultivated species worldwide for its high-yield and low-demand for concentration of salt. But with the continual expanding and intensifying of aquaculture, more and more serious viral diseases are emerging such as white spot syndrome virus (WSSV), infectious myonecrosis virus (IMNV) and *Penaeus vannamei* nodavirus (PvNV). Among these, IMNV and PvNV were documented to be causative pathogens of "white tail disease" (WTD)-like disease in marine shrimp following the *Macrobrachium rosenbergii* nodavirus (MrNV) identified in freshwater prawn [1–3]. The two viruses both primarily targeted the skeletal muscle and resulted in very similar gross signs (a white or opaque tail) and histopathological changes (focal to extensive areas of muscle necrosis and the formation of prominent lymphoid organ spheroids) in Penaeid shrimp. During or soon after stressful events, outbreaks of IMN are usually accompanied by high mortalities (PvNV is less virulent than IMNV) [3,4]. Since such viral diseases are significant in shrimp aquaculture, more research attention has been paid to the viral diseases and less to the

bacterial. However, with increasing global-warming and intensive aquaculture, bacterial diseases especially vibrioses are becoming another important threatening to the sustainable development of the Penaeid shrimp aquaculture industry [5].

In shrimp aquaculture system, vibrios are among the normal bacterial flora of cultural populations and the habitats [6,7], from which *Vibrio harveyi*, *V. alginolyticus* and *V. parahaemolyticus* are most frequently isolated [8–12]. As opportunistic pathogens, they may lead to mortality of affected aquatic animals due to stressful events, such as sudden changes in temperature [5,13] and salinity. Among these vibrios, *V. harveyi* is one of the most important pathogens, capable of causing devastation to diverse ranges of marine invertebrates including Penaeid shrimp [14–18]. In the past two decades, mass mortalities of Penaeid shrimp resulted from *V. harveyi* infections were frequently reported in hatcheries and grow-out ponds [10,11,19–22]. Notably, these pathogenic *V. harveyi* strains were generally luminescent.

In present study, a highly virulent *V. harveyi* strain HLB0905 was identified as the etiologic pathogen of bacterial white tail disease (BWTD) through microscopical examination, sequence analysis, bacterial isolation and challenge test. Interestingly, the strain not

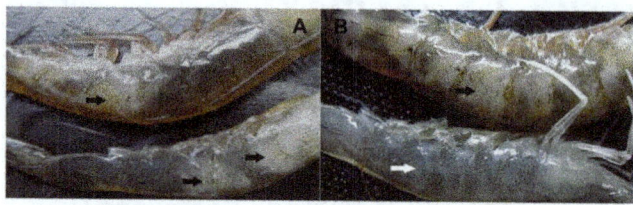

Figure 1. Gross signs of diseased *L. vannamei* shrimp. (A) Gross signs of *L. vannamei* shrimp naturally occurred in the farm. Focal to extensive whitish muscles in the tail (black arrows) with (top) or without (bottom) red discoloration in the body and appendages (B) Gross signs of *L. vannamei* shrimp laboratory-infected with the *V. harveyi* strain HLB0905. Compared to normal shrimp (bottom, white arrow), the infected shrimp exhibited an extensive whitish or opaque appearance in the tail (top, black arrow).

only mainly caused "white tail" in *L. vannamei* shrimp, but also was nonluminescent.

Results

Gross signs of affected shrimp

Soon after being reared in the indoor tank, many of the Penaeid shrimp, apparently healthy previously, started to show WTD-mimicking gross signs, which was characterized by focal to

extensive areas of whitish muscle, particularly in the distal abdominal segments, with or without a red discoloration in the body and appendages (Fig. 1A). One day later, perhaps due to a combination of the disease and sudden stresses such as collection by cast-netting and transportation, death occurred with a sudden high mortality. Six d later, the cumulative mortality of shrimp reached up to 76%.

Histopathological and ultrastructural analysis

Histopathological analysis showed that muscle fibers composing the whitish tail muscle were damaged in different degrees with focal to extensive fiber necrosis in both naturally- and artificially-infected *L. vannamei* shrimp (Fig. 2). The data indicated that the opaque or whitish appearance of diseased shrimp was due to muscle necrosis. Furthermore, an electron microscopical analysis showed that fiber cells composing the whitish muscle were damaged, including nuclear pyknosis, cell vacuolation, mitochondrial damage and myofibrils damaged in different degrees (Fig. 3). Notably, light and electron microscopical analysis both demonstrated that there were lots of rod-shaped bacteria (Fig. 2B, 3C and 3D) with a flagellum at one end (Fig. 3D) infiltrating in these necrotic muscles, and except for these, there were not any kinds of microorganisms such as viruses and parasites being observed over a large number of ultrathin sections cut consecutively. These analyses suggested that a *Vibrio*-like bacterium may be associated with the WTD-like disease.

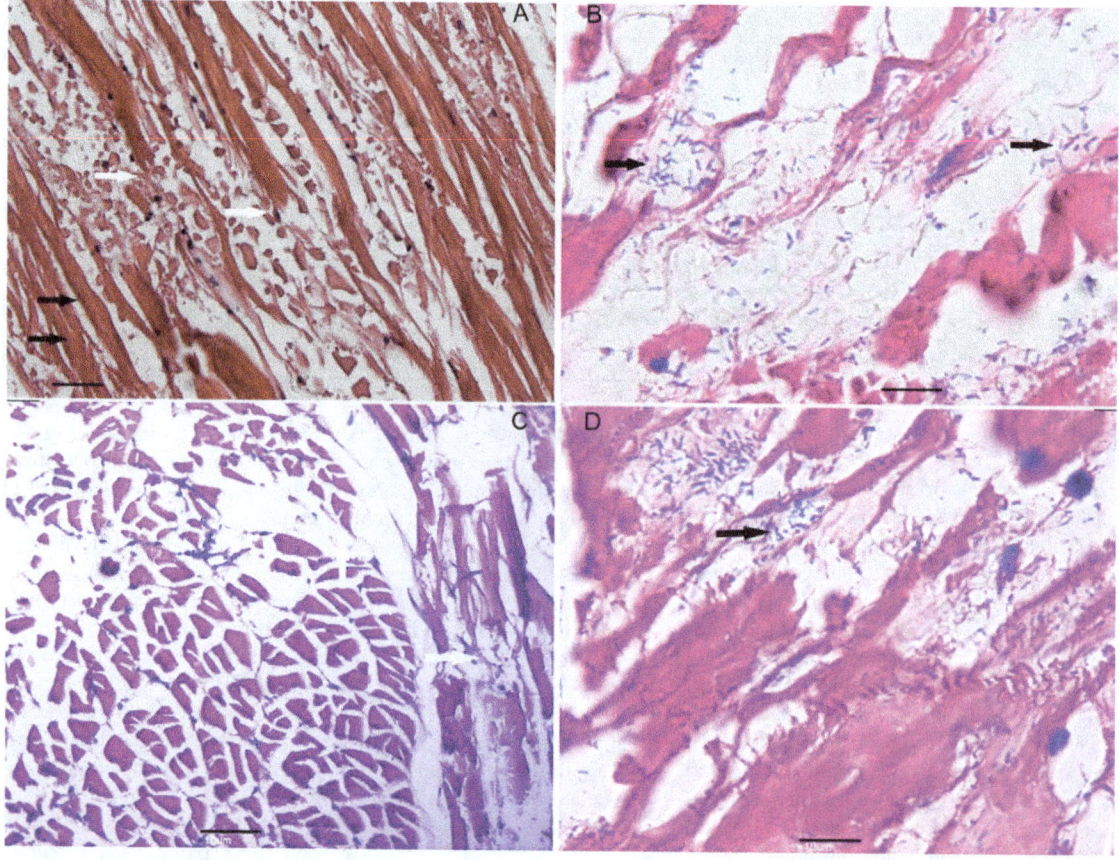

Figure 2. Histopathological changes in whitish muscles. (A and C) Normal (black arrows) and broken muscle fibers (white arrows). (B and D) Coagulative to liquefactive muscle necrosis and infiltration of a great number of rod-shaped bacteria (black arrows). Tissue A and B were sampled from *L. vannamei* shrimp with a WTD-like disease in the farm, while tissue C and D were sampled from *L. vannamei* shrimp laboratory-infected with the *V. harveyi* strain HLB0905(A: bar = 35 μm; B, C and D: bar = 10 μm).

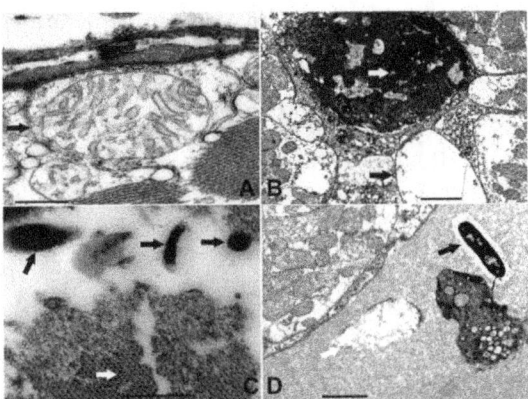

Figure 3. Ultrastructural changes in whitish muscles. (A) The mitochondrial membrane was broken (black arrow, bar = 0.5 μm). (B) The nuclear pyknosis (white arrow) and fiber cell vacuolation (black arrow, bar = 2 μm). (C) Damaged myofibrils (white arrow shows normal morphology of myofibrils) and invasion of bacteria (black arrow, bar = 0.5 μm). (D) Damaged fiber cells (white arrow shows damaged mitochondria) and infiltration of Vibrio-like bacteria (black arrow) (bar = 2 μm). All the tissues were sampled from L. vannamei shrimp with a WTD-like disease in the farm.

Bacteria isolation and identification

Twenty-four h post incubation of muscle smears, lots of colonies with the same phenotypes (i.e. round, milky and nonluminescent) appeared on the plates. All of the colonies analyzed were able to grow on thiosulfate citrate bile salts sucrose (TCBS) agar. They were Gram-negative, sensitive to the vibriostatic agent O/129 at 150 mg, and V. harveyi hemolysin-positive. All of these character-istics were in accordance with those owned by V. harveyi. Moreover, by negative staining electron microscopy, the V. harveyi isolate (designated as the strain HLB0905) was a short rod-shaped bacterium with a flagellum at one end (Fig. 4).

PCR detection and sequence analysis

Based on 16S rRNA gene sequence analysis, bacteria within the whitish muscle of diseased shrimp shared the highest (99%) identity

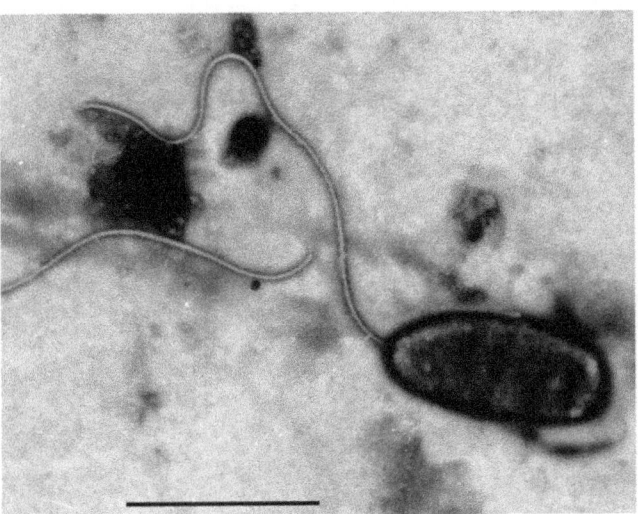

Figure 4. Morphology of isolated V. harveyi strain HLB0905 (bar = 2 μm).

with V. harveyi species (Genbank accession no. **HM590018**). Since the genus Vibrio contains a large number of closely related bacterial species with very small difference in the sequence of their 16S rRNA gene, total genomic DNAs respectively from the whitish muscle, hepatopancreas and hemolymph of diseased shrimp were all subsequently subjected to a PCR analysis for the V. harveyi-specific hemolysin gene. As shown in Fig. 5, all of the samples were V. harveyi hemolysin-positive. Furthermore, based on the BLAST searching and phylogenetic analysis (Fig. 6), the nucleotide sequence displayed the highest homology (99%) with the reported hemolysin gene of V. harveyi whereas no more than 80% with that of V. compbellii (which generally shows the closest relationship with V. harveyi [23]). These data further confirmed that the Vibrio-like bacterium infiltrating in the whitish muscle was V. harveyi.

Confirmation of the causative pathogen

In order to further identify whether the nonluminescent V. harveyi strain was the etiologic pathogen, bacterial challenge trials were conducted. The challenge tests showed that the HLB0905 strain was highly virulent. With a dose of 38 CFU of V. harveyi per shrimp, cumulative mortality of shrimp was 30% by 1 dpi, 60% by 2 dpi, and reached 100% by 4 dpi. In contrast, none of the shrimp in control groups died. Notably, most of those died in 2 dpi exhibited a focal to extensive whitish appearance in the tail (Fig. 1B). Further, the histopathological examination showed skeletal muscle necrosis and infiltration of a large number of rod-shaped bacteria (Fig. 2C and 2D). All of the observations were almost identical to those in clinical specimens. Finally, the PCR analysis of muscle DNA also indicated that the whitish muscles were strongly positive for V. harveyi hemolysin gene whereas muscles from shrimp in the control groups were negative (data not shown).

In conclusion, the present study demonstrated that the WTD-like disease, occurred in pond-cultured L. vannamei shrimp in Hainan Province, was caused by the nonluminescent but highly virulent V. harveyi strain HLB0905. Thus, to differentiate from other diseases with the similar sign of "white tail" but of non-bacterial origin such as IMN, WTD and PWTD, the present disease was named as "bacterial white tail disease (BWTD)".

Discussion

Vibrio species comprise the most frequently encountered bacterial pathogens of cultivated shrimp, and V. harveyi is amongst the most isolated [8–12]. Many Vibrio virulence factors including various enzymes (e.g., proteases and lipases), siderophores and proteinaceous toxins have been identified. They were considered to be the important determinants of iron binding and biofilm forming whereas to be regulated by quorum sensing [24–26]. However, since the composition and expression level of genes encoding these virulence factors generally vary with Vibrio spp. and strains, there is a large degree of virulence-variation and genetic diversity among different V. harveyi isolates. Some isolates couldn't cause shrimp mortality at high doses (10^5–10^7 cells per g shrimp body weight); while others were lethal at 10^3 per g shrimp body weight or less [27]. Notably, our isolate V. harveyi HLB0905 was able to cause 60% of the challenged shrimp to die during 48 h at a dose of 38 CFU per shrimp, which indicated that the isolate HLB0905 was highly virulent. It has been demonstrated that V. harveyi can change its virulence through horizontal gene transfer or quorum sensing [27–30]. For example, via transferring virulent genetic elements, bacteriophages may either promote the production of an existing toxin or somehow induce the production of a new one. As a result, the phenotype of a V. harveyi strain was

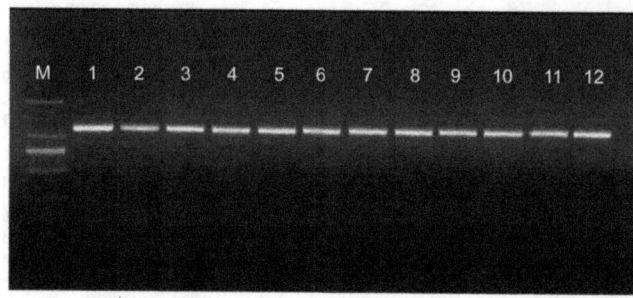

Figure 5. PCR products amplified from the *V. harveyi* hemolysin gene. These tissues were respectively sampled from four *L. vannamei* shrimp with a WTD-like disease in the farm. M: DL 2000 marker (Takara, Japan); 1–4, whitish muscle; 5–8, hemolymph; 9–12, hepatopancreas.

converted from non-virulent to virulent or was enhanced from less-virulent to highly virulent [27,31,32]. Furthermore, in some cases, the pore-forming activity of haemolysin is not restricted to erythrocytes, but extends to a wide range of other cell types, and enhances virulence by causing tissue damage [33]. In the present study, *V. harveyi* HLB0905 caused both severe muscle tissue necrosis and mass mortalities in the affected shrimp. Whether the *V. harveyi* HLB0905 has enhanced its virulence via one of the cases above or a combination of the several ones is being studied.

During the investigation, we found that occurrence of BWTD was closely climate- and aquaculture density-linked. For example, besides Qionghai district of Hainan Province, the *V. harveyi* HLB0905 was isolated and identified in Fuding district of Fujian Province (data not shown), and the average water temperature in these two provinces are generally greater than 29°C in summer. In contrast, in other provinces mainly with lower water temperature, the disease hasn't yet been detected. In fact, many researches [27–29,34] have showed that the cell density of *V. harveyi* can be promoted by sudden changes in environments and conditions in

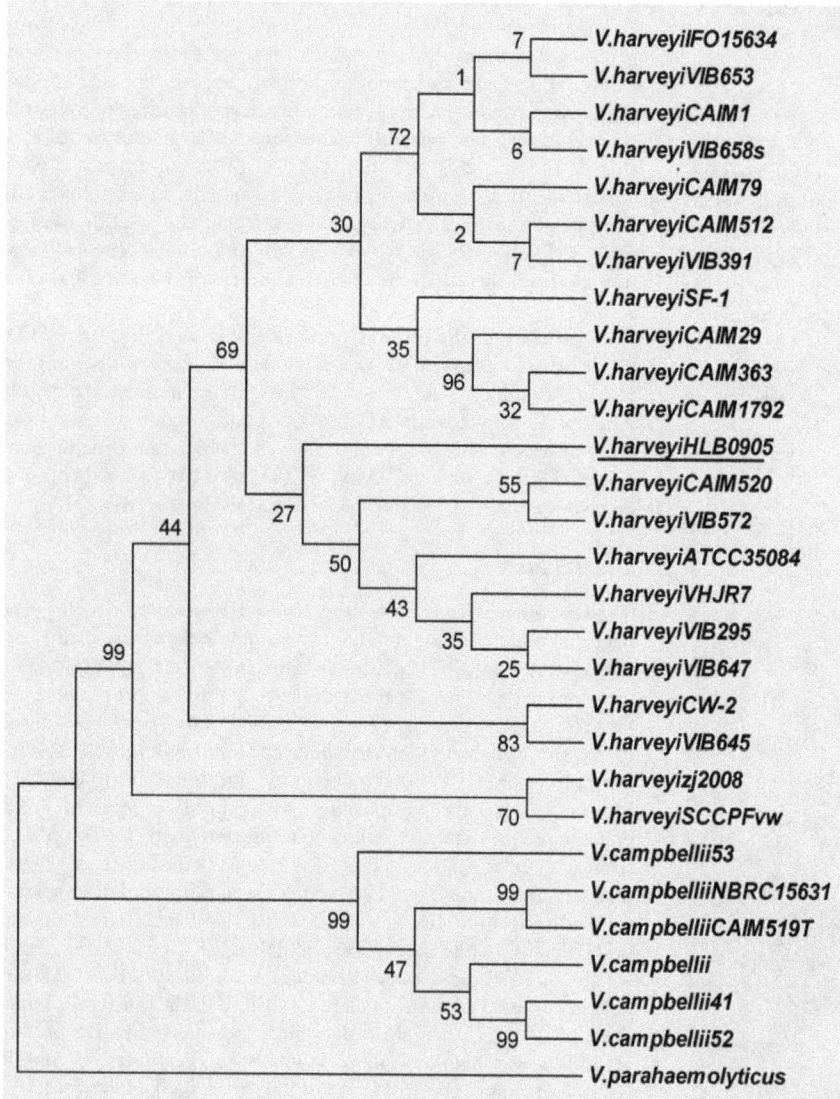

Figure 6. Phylogenetic tree based on the neighbor-joining method, using complete *Vibrio* hemolysin gene sequences. Bootstrap values are expressed as percentages of 1000 replications.

aquaculture such as temperature, salinity, nutrient concentration, niche switching and host animal density. The changed cell density then made horizontal gene transfer easier and/or induced quorum-sensing regulation. Moreover, selection pressure exerted by dramatic environmental changes generally perturbs. Many of the perturbed genes and regulatory networks were demonstrated to be preferentially modulating virulence mechanisms [35]. Obviously, virulence of *V. harveyi* strain was closely related to environmental factors, suggesting that environmental factors need to be deeply concerned in shrimp aquaculture.

Based on our observations, the typical sign of whitish appearance in *L. vannamei* shrimp caused by *V. harveyi* was very similar to that caused by some viruses such as IMNV [2] and PvNV [3] or some parasites such as microsporidium [36]. In order to distinguish the causative agent of the present BWTD in all these potential pathogens, all of the tissue sections from the whitish muscle were carefully examined for these pathogens besides vibrios by the light and electron microscopy and PCR method (IMNV- and PvNV-negative by RT-PCR, data not shown). The examinations showed that the nonluminescent *Vibrio* strain was the only pathogen in the whitish muscle, indicating that present BWTD was caused by *V. harveyi*. The results suggested that, under current management, some vibrios have changed themselves from secondary to primary pathogens by enhancing their virulence.

In conclusion, via isolation, subculture, reinfection and reisolation as well, the bacterium strain HLB0905 isolated from whitish muscle was identified as the causative agent of BWTD in the pond-cultured *L. vannamei* according to Koch's Postulates. The nonluminescent *V. harveyi* HLB0905 (its luminescent isogenic counterpart should be more virulent [37,38]) is highly pathogenic and generally causes epizootics with high mortalities in pond-cultivated Penaeid shrimp populations, especially following stressful events. Therefore, in practical terms for shrimp farmers, taking adequate prophylaxes such as the use of immunostimulants or probiotic bacteria to alleviate stresses and using pathogen-specific methods such as PCR to ensure the accurate diagnosis of WTD-like diseases would benefit to the prevention of severe disease outbreaks and economic losses in shrimp aquaculture industry [2,9,23,39].

Materials and Methods

Shrimp and tissue sections

L. vannamei shrimp (10 g mean weight), with or without an appearance of white tail, were collected from a commercial shrimp farm in Hainan Province in China, which were suffering from an outbreak of WTD-like disease. Fed with commercial diet at approximately 5% of body weight with a 50% water exchange and air supply daily at $29 \pm 1°C$, all the Penaeid shrimp were temporarily reared in an indoor tank. Meanwhile, moribund shrimp were picked out in time for analysis. Ultrathin sections were prepared from whitish muscles of diseased shrimp for light and electron microscopical examination. For light microscopy, the whitish muscle was fixed in 10% formaldehyde solution, embedded with paraffin and stained with hematoxylin and eosin [40]. For transmission electron microscopy, whitish muscle from the same shrimp was fixed in 2.5% glutaraldehyde in 0.1 M PBS (pH 7.4) for 2 h at 4°C, followed by in 1% osmium tetroxide for 2 h, embedded in Spurr's resin and stained with uranyl acetate and lead citrate.

PCR detection and DNA sequencing

For PCR detection, 10 moribund shrimp (6 for 16S rRNA analysis and 4 for hemolysin gene detection) with the WTD-

mimicking gross signs were randomly collected from the farm. The whitish muscle, hepatopancreas and hemolymph were respectively sampled, followed by preparation of total genomic DNA using commercially available TIANamp Marine Animals DNA Kit (Tiangen, China). Two pairs of primers, one for evolutionarily conserved 16S rRNA gene of bacteria [41] and the other for the whole *V. harveyi* hemolysin (*Vh*) gene (Table 1), were respectively synthesized. The amplification condition for both the two genes was pre-denaturation at 95°C for 30 s, 30 cycles of 95°C for 30 s, 54°C for 50 s, 72°C for 105 s, followed by elongation at 72°C for 8 min. The PCR products were subsequently sequenced.

Homology study was carried out using BLAST Searching (NCBI). The sequences were aligned using ClustalX, and phylogenetic trees were constructed by the neighbor-joining method [42]. The robustness of each topology was checked by 1000 bootstrap replications. Trees were drawn by using MEGA version 4.0.

Bacteria isolation and cultivation

Whitish muscles were aseptically sampled from freshly collected moribund *L. vannamei* shrimp suffering from a WTD-like disease in the farm. Muscle blocks from different shrimp were respectively touched and streaked on tryptic soy agar plates (TSA, supplemented with 2% NaCl [w/v]) followed by incubation at 29°C for 24 h. Five colonies were selected randomly from each plate respectively for examination of some biochemical characteristics, PCR detection of *V. harveyi* hemolysin gene (Table 1) and microscopical analysis. Identified colonies were subcultured and stored in deep tube TSA with 2% NaCl as stocks.

Bacterial challenge test

Apparently healthy subadult *L. vannamei* shrimp (10 g mean weight) were collected from another farm, where WTD-like disease had never occurred. Fed with commercial diet at approximately 5% of body weight with a 50% water exchange and air supply daily at $29 \pm 1°C$, these shrimp were cultured temporarily in indoor cement tanks for 7 days before they were used for bacterial challenge tests. For challenge trials, 45 shrimp, free from IMNV and PvNV by RT-PCR (Table 1), were randomly selected and put into three fiber tanks, each containing 15 shrimp and 90 L of seawater at $29 \pm 1°C$. The *V. harveyi* cells, pure cultured in tryptic soy broth (TSB) with 2% NaCl, were

Table 1. Primers used in this study.

Primer	Sequence (5′to 3′)	Amplicon Size	Reference
Vh F	ATGAATAAAACTATTACGTT	1254 bp	[26]
Vh R	GAAAGGATGGTTTGACAATT		
IMNV F (1-step)	CGACGCTGCTAACCATACAA	328 bp	[4]
IMNV R (1-step)	ACTCGGCTGTTCGATCAAGT		
IMNV F (2-step)	GGCACATGCTCAGAGACA	139 bp	[4]
IMNV R (2-step)	AGCGCTGAGTCCAGTCTTG		
PvNV F (1-step)	CTGTCTCACAGGCTGGTTCA	339 bp	[3]
PvNV R (1-step)	CCGTTTGAATTTCAGCAACA		
PvNV F (2-step)	CAAAACTGTGCCTTTGATCG	246 bp	[3]
PvNV R (2-step)	GCCTTATCCACACGAACGTC		
16S rRNA F	AGA GTT TGA TCC TGG CTC AG		[41]
16S rRNA R	AAG GAG GTG ATC AGC C		

collected and diluted with 0.01 M PBS (pH 7.4). For each treatment, each shrimp received 50 μl of an aliquot of *V. harveyi* (average 38 CFU/shrimp, decided by a pre-treatment test) by injected intramuscularly into the abdominal segment using a syringe with a 29-gauge needle. Shrimp in the control group were injected with an equal volume of 0.01 M PBS (pH 7.4). Animals were observed for the sign of "white tail" and the mortality after injection.

Acknowledgments

We would like to thank Dr. Haizhen Wu, Mr. Qinghua Yang and Qiang Song (Chinese Academy of Sciences) for their generous help to this study.

Author Contributions

Conceived and designed the experiments: JFZ WHF XLY. Performed the experiments: JFZ LLH SZ HS XYQ LYX. Analyzed the data: JFZ WHF XYQ. Contributed reagents/materials/analysis tools: HS XCL XYQ. Wrote the paper: JFZ.

References

1. Qian D, Shi Z, Zhang S, Cao Z, Liu W, et al. (2003) Extra small virus-like particles (XSV) and nodavirus associated with whitish muscle disease in the giant freshwater prawn, *Macrobrachium rosenbergii*. J Fish Dis 26: 521–527.
2. Lightner DV, Pantoja CR, Poulos BT, Tang KFJ, Redman RM, et al. (2004) Infectious myonecrosis: new disease in Pacific white shrimp. Glob Aquac Advocate 7: 85.
3. Tang KFJ, Pantoja CR, Redman RM, Lightner DV (2007) Development if in situ hybridization and RT-PCR assay for the detection of a nodavirus (PvNV) that causes muscle necrosis in *Penaeus vannamei*. Dis Aquat Org 75: 183–190.
4. Poulos BT, Lightner DV (2006) Detection of infectious myonecrosis virus (IMNV) of penaeid shrimp by reverse-transcriptase polymerase chain reaction (RT-PCR). Dis Aquat Organ 73(1): 69–72.
5. Lightner DV, Redman RM (1998) Shrimp diseases and current diagnostic methods. Aquaculture 164: 201–220.
6. Otta SK, Karunasagar I, Karunasagar I (1999) Bacterial flora associated with shrimp culture ponds growing *Penaeus monodon* in India. J Aquacult Trop 14: 309–318.
7. Shakibazadeh S, Saad CR, Christianus A, Kamarudin MS, Sijam K, et al. (2009) Bacteria flora associated with different body parts of hatchery reared juvenile *Penaeus monodon*, tanks water and sediment. Annals of Microbiology 59: 425–430.
8. Chitov T, Wongdao S, Thatum W, Puprae T, Sisuwan P (2009) Occurrence of potentially pathogenic *Vibrio* species in raw, processed, and ready-to-eat seafood and seafood products. Maejo Int J Sci Technol 3: 88–98.
9. Sung HH, Hsu SF, Chen CK, Ting YY, Chao WL (2001) Relationships between disease outbreak in cultured tiger shrimp (*Penaeus monodon*) and the composition of *Vibrio* communities in pond water and shrimp hepatopancreas during cultivation. Aquaculture 192: 101–110.
10. Uma A, Meena S, Saravanabava K, Muralimanohar B (2008) Identification of bacterial pathogens infecting *Penaeus monodon*, tiger shrimp by 16S rDNA amplification and sequencing. Tamilnadu J Veterinary and Animal Science 4: 188–192.
11. Vandenberghe J, Verdonck L, Robles-Arozarena R, Rivera G, Bolland A, et al. (1999) Vibrios associated with *Litopenaeus vannamei* larvae, postlarvae, broodstock, and hatchery probionts. Appl Environ Microbiol 65: 2592–2597.
12. Vezzulli L, Pezzati E, Moreno M, Fabiano M, Pane L, et al. (2009) Benthic ecology of *Vibrio* spp. and pathogenic *Vibrio* species in a coastal Mediterranean environment (La Spezia Gulf, Italy). Microb Ecol 58: 808–818.
13. Vaseeharan B, Ramasamy P (2003) Abundance of potentially pathogenic micro-organisms in *Penaeus monodon* larvae rearing systems in India. Microbiology Research 158: 299–308.
14. Abraham TJ, Palaniappan R (2004) Distribution of luminous bacteria in semi-intensive penaeid shrimp hatcheries of Tamil Nadu, India. Aquaculture 232: 81–90.
15. Haldar S, Maharajan A, Chatterjee S, Hunter SA, Chowdhury N, et al. (2010) Identification of *Vibrio harveyi* as a causative bacterium for a tail rot disease of sea bream Sparus aurata from research hatchery in Malta. Microbiol Res 165: 639–648.
16. Ransangan J, Mustafa S (2009) Identification of *Vibrio harveyi* isolated from diseased Asian Seabass *Lates calcarifer* by use of 16S ribosomal DNA sequencing. J Aquat Anim Health 21: 150–155.
17. Tendencia EA (2002) *Vibrio harveyi* isolated from cage-cultured seabass *Lates calcarifer* Bloch in the Philippines. Aquacult Res 33: 455–458.
18. Vezzulli L, Previati M, Pruzzo C, Marchese A, Bourne DG, et al. (2010) Vibrio infections triggering mass mortality events in a warming Mediterranean Sea. Environ Microbiol 12: 2007–2019.
19. Chrisolite B, Thiyagarajan S, Alavandi SV, Abhilash EC, Kalaimani N, et al. (2008) Distribution of luminescent *Vibrio harveyi* and their bacteriophages in a commercial shrimp hatchery in South India. Aquaculture 275: 13–19.
20. de la Pena LD, Lavilla-Pitogo CR, Paner MG (2001) Luminescent vibrios associated with mortality in pond-cultured shrimp *Penaeus monodon* in the Philippines: species composition. Fish Pathology 36: 133–138.

21. Lavilla-Pitogo CR, Baticados MCL, Cruz-Lacierda ER, de la Pena LD (1990) Occurrence of luminous bacterial disease of *Penaeus monodon* larvae in the Philippines. Aquaculture 91: 1–13.
22. Lavilla-Pitogo CR, Leano EM, Paner MG (1998) Mortalities of pond-cultured juvenile shrimp, *Penaeus monodon*, associated with dominance of luminescent vibrios in the rearing environment. Aquaculture 164: 337–349.
23. Thompson FL, Gomez-Gil B, Vasconcelos ATR, Sawabe T (2007) Multilocus sequence analysis reveals that Vibrio harveyi and V. campbellii form distinct species. Appl Environ Microbiol 73: 4279–4285.
24. Lee KK, Liu PC, Kou GH, Chen SN (1997) Investigation on the major exotoxin of *Vibrio harveyi* 770527 isolated from diseased *Penaeus monodon*. Rep Fish Dis Res 18: 33–42.
25. Liu PC, Lee KK (1999) Cysteine protease is a major exotoxin of pathogenic luminous *Vibrio harveyi* in the tiger prawn, *Penaeus monodon*. Lett Appl Microbiol 28: 428–430.
26. Zhong YB, Zhang XH, Chen JX, Chi ZH, Sun BG, et al. (2006) Overexpression, purification, characterization, and pathogenicity of *Vibrio harveyi* hemolysin VHH. Infect Immun 74: 6001–6005.
27. Flegel TT, Pasharawipas LO, Oakey HJ (2005) Evidence for phage-induced virulence in the shrimp pathogen *Vibrio harveyi*. In P. Walker, R. Lester, MG. Bondad-Reantaso, eds. Diseases in Asian Aquaculture V. pp 329–337.
28. Ahmed N, Dobrindt U, Hacker J, Hasnain SE (2008) Genomic fluidity and pathogenic bacteria: applications in diagnostics, epidemiology and intervention. Nat Rev Microbiol 6(5): 387–394.
29. Mok KC, Wingreen NS, Bassler BL (2003) *Vibrio harveyi* quorum sensing: a coincidence detector for two autoinducers controls gene expression. EMBO J 22: 870–881.
30. Zhu J, Miller MB, Vance RE, Dziejman M, Bassler BL, et al. (2002) Quorum-sensing regulators control virulence gene expression in *Vibrio cholerae*. Proc Natl Acad Sci USA 99: 3129–3134.
31. Munro J, Oakey J, Bromage E, Owens L (2003) Experimental bacteriophage-mediated virulence in strains of *Vibrio harveyi*. Dis Aquat Organ 54: 187–194.
32. Ruangpan L, Danayadol Y, Direkbusarakom S, Siurauratana S, Flegel TW (1999) Lethal toxicity of *Vibrio harveyi* to cultivated *Penaeus monodon* induced by a bacteriophage. Dis Aquat Org 35: 195–201.
33. Shinoda S (1999) Protein toxins produced by pathogenic vibrios. J Nat Toxins 8: 259–269.
34. Alavandi SV, Manoranjita V, Vijayan KK, Kalaimani N, Santiago TC (2006) Phenotypic and molecular typing of *Vibrio harveyi* isolates and their pathogenicity to tiger shrimp larvae. Lett Appl Microbiol 43: 566–570.
35. Aziz RK, Kansal R, Aronow BJ, Taylor WL, Rowe SL, et al. (2010) Microevolution of Group A Streptococci In Vivo: Capturing Regulatory Networks Engaged in Sociomicrobiology, Niche Adaptation, and Hypervirulence. PLoS ONE 5(4): e9798.
36. Ramasamy P, Jayakumar R, Brennan GP (2000) Muscle degeneration associated with cotton shrimp disease of *Penaeus indicus*. J Fish Dis 23: 77–81.
37. Lilley BN, Bassler BL (2000) Regulation of quorum sensing in *vibrio harveyi* by LuxO andσ-54. Mol Microbiol 36: 940–954.
38. Ruwandeepika HA, Defoirdt T, Bhowmick PP, Karunasagar I, Bossier P (2011) Expression of virulence genes in luminescent and nonluminescent isogenic *vibrios* and virulence towards gnotobiotic brine shrimp (*Artemia franciscana*). J Appl Microbiol 110: 399–406.
39. Selvin J, Lipton AP (2004) *Dendrilla nigra*, a marine sponge, as potential source of antibacterial substances for managing shrimp diseases. Aquaculture 236: 277–283.
40. Bell TA, Lightner DV (1988) A Handbook of Normal Penaeid Shrimp Histology. World Aquaculture Society, Baton Rouge, LA.
41. Weisburg WG, Barns SM, Pelletier DA, Lane DJ (1991) 16S ribosomal DNA amplification for phylogenetic study. J Bacteriol 173(2): 697–703.
42. Saitou N, Nei M (1987) The neighbor-joining method: a new method for reconstructing phylogenetic trees. Mol Biol Evol 4: 406–425.

17

Experts and Novices Use the Same Factors–But Differently–To Evaluate Pearl Quality

Yusuke Tani[1]*, **Takehiro Nagai**[2], **Kowa Koida**[3], **Michiteru Kitazaki**[1], **Shigeki Nakauchi**[1]

1 Department of Computer Science and Engineering, Toyohashi University of Technology, Toyohashi, Aichi, Japan, 2 Graduate school of Science and Engineering, Yamagata University, Yonezawa, Yamagata, Japan, 3 Electronics Inspired-Interdisciplinary Research Institute, Toyohashi University of Technology, Toyohashi, Aichi, Japan

Abstract

Well-trained experts in pearl grading have been thought to evaluate pearls according to their glossiness, interference color, and shape. However, the characteristics of their evaluations are not fully understood. Using pearl grading experiments, we investigate the consistency of novice (i.e., without knowledge of pearl grading) and expert participants' pearl grading skill and then compare the novices' grading with that of experts; furthermore, we discuss the relationship between grading, interference color, and glossiness. We found that novices' grading was significantly less concordant with experts average grading than was experts' grading; more than half of novices graded pearls the opposite of how experts graded those same pearls. However, while experts graded pearls more consistently than novices did, novices' consistency was relatively high. We also found differences between the groups in regression analyses that used interference color and glossiness as explanatory variables and were conducted for each trial. Although the regression coefficient was significant in 60% of novices' trials, there were fewer significant trials for the experts (20%). This indicates that novices can also make use of these two factors, but that their usage is simpler than that of the experts. These results suggest that experts and novices share some values about pearls but that the evaluation method is elaborated for experts.

Editor: Chris I. Baker, National Institute of Mental Health, United States of America

Funding: This work was supported by the Grant-in-Aid for Scientific Research on Innovative Areas (for SN, #22135005; http://www.jsps.go.jp/english/e-grants/grants01.html). The funders had no role in study design, data collection and analysis, decision to publish, or preparation of the manuscript.

Competing Interests: The authors have declared that no competing interests exist.

* E-mail: tani@vpac.cs.tut.ac.jp

Introduction

Pearls are known as jewels from the bottom of the sea. Their mystique from being produced by shellfish and their lustrous iridescence has attracted many people worldwide. The pearls produced by Akoya pearl oysters (*Pinctada fucata martensii*) have superior luster and impressive iridescence. In addition to these two features, their size, roundness, and the existence of scars or pocks are the key features inspected by farmers, traders, and craftsmen, who are collectively addressed as "experts" [1–3]. The quality or value of pearls is decided only by well-trained experts' visual inspection at north-facing windows on sunny mornings or afternoons. Further, consumers and novices accept these decisions. This situation suggests interesting questions: How do experts use visual information to evaluate pearls? What do they learn? What supports this tacit agreement between experts and novices?

About 100 years ago, pearl farming–a practice whereby Akoya pearl oysters are cultured and the spherical pearls are constantly harvested from them–began in Toba, Mie Prefecture, Japan. Even now, Toba is one of the principal areas of Akoya pearl farming and manufacturing.

A cultured pearl consists of a nucleus surrounded by hundreds to thousands of translucent layers of nacre. The nucleus is a spherical bead made of shell, and the nacre is a secretion of pearl oysters consisting of calcium carbonate ($CaCO_3$) and proteins like conchiolin. Calcium carbonate is an ingredient of both the nucleus and nacre; the former is calcite crystal, whereas the latter is aragonite crystal. The thicknesses of the aragonite crystal and the

protein membrane are approximately 300–500 nm and 10 nm, respectively. Thus, the thickness of a nacreous layer is in the range of the wavelength of visible light (Figure 1). These characteristics of the nacre are the origin of pearl's iridescence, one of the essences of pearliness. That is, the lustrous iridescence of pearl is due to the interference color, which is a kind of structural color caused by the multilayer thin film structure.

The strength and chromaticity of pearls' interference color depend on the thickness of the nacre layers and the length difference between the optical paths, respectively. Incident light travels through nacre in a complex way because of multiple reflections, refractions, and penetration in each nacreous layer. Therefore, the interference color is independent of the direction of the light source, and it depends on the viewing direction and the thickness of each nacreous layer [2,3]. As a result, nearly concentric chromatic patterns are seen on spherical pearls. In general, pearls regarded as good by experts have a typical concentric chromatic pattern, changing from greenish in the center to pinkish at the periphery [3].

Both the chromatic pattern and chromaticity of pearls' interference color correlate with the physical structure of the nacre; thus, experts evaluate the physical regularity of the pearls, in a sense. This leads to the following questions about novices, however: Can they evaluate pearls the same way as the experts do? If so, what differentiates the experts from novices?

In general, experts' senses seem superior to those of novices. For instance, most people believe that only experts can detect certain

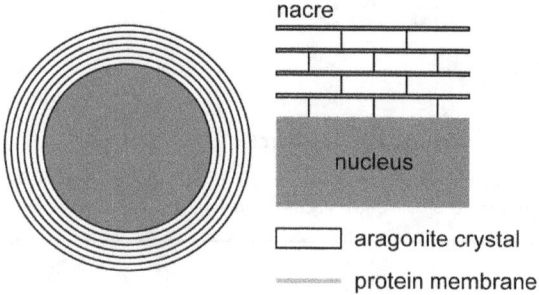

nacre

nucleus

☐ aragonite crystal

～～ protein membrane

Figure 1. Schematic diagram of internal structure of cultured pearl. The translucent nacre surrounds the nucleus. The nacre comprises hundreds to thousands of nacreous layers. The thickness of each layer is approximately 300–500 nm.

qualitative differences. Furthermore, judgments made by well-trained experts are often expected to be identical.

However, according to research on differences between experts and novices, such as that on athletes and artists, experts do not exceed novices in terms of lower-level abilities [4–6]. For example, the perception of both expert and novice painters was distorted by size constancy in a similar way [6]. The superiority of experts appears to manifest in a limited number of cases or situations [7]. That is, most distinctions can probably be attributed to differences in strategy or cognitive level [8]. In addition, top-level athletes have been found to have special learning abilities [9].

Whether concordance among experts is observed depends on the domain. For example, sommeliers and wine tasters outperform novices in olfactory discrimination and matching [10]. Well-trained tasters are able to rate the concentration of sodium chloride in mixtures of sodium chloride and sucrose solutions more correctly and consistently than novices can [11]. On the other hand, experienced violinists are divided in terms of preference for the tonal quality of violins [12].

In this paper, we compare experts and novices in terms of preference and within-individual consistency of pearl evaluation. If the tendency of rank ordering by novices resembles that of experts, then the experts' evaluation rules could be attributed to an innate sense of beauty, which should be shared by experts and novices; if not, then novices' rank-ordering tendency is likely artificial. Furthermore, if the consistency of rank ordering by the novices approaches chance levels, then the criterion they use is vague or unconscious; if it is high, the criterion that they use should be obvious. If the experts' consistency is higher than that of novices, experts indeed have elaborated the pearl evaluation method. Further, we investigate the functional relationship between pearl evaluation and optically measured glossiness and interference color that previous researchers [1–3] and experts have reported to be the factors considered in the evaluation of a pearl's quality.

Materials and Methods

Participants

Eight experts (mean = 43.63, SD = 4.56 years old) and eleven novices (mean = 41.73, SD = 4.96 years old) participated in the experiment. All of them were males and had normal or corrected-to-normal vision. The expert participants included five pearl oyster farmers with more than 20 years of experience in the field; two pearl marketers with more than 15 years of experience; and one scientist who had been studying improvements in the health of cultured pearls in order to raise the quality of the pearls generated in pearl oyster farming for five years. All experts worked in Toba,

Mie Prefecture, Japan. Although there are no national qualifications or licenses concerning pearls in Japan, in this study, experts are considered to be individuals who had learned about pearls from masters of the field, had been handling pearls in their work as professionals for a long period, and who were recognized as full-fledged experts by other experts.

The novice participants included ten scientists affiliated with Toyohashi University of Technology as assistant professors, associate professors, and professors and one university clerk. All of them were unfamiliar with pearls in both their research activities and daily life and were unaware of the purpose of the experiment. After the experimental procedure was explained, all participants gave written informed consent before the experiment began. This study was approved by the Committee for Human-Subject Studies at Toyohashi University of Technology.

Stimuli

We used 20 Akoya cultured pearls that had been labeled A-rank and B-rank. We got them from the trader in Kobe, a major center of pearl circulation (These pearls we used were labeled by some experts in Kobe other than our expert participants). All pearls used in the experiment were approximately 8 mm in diameter. The pearls were arbitrarily placed into two sets, with each set consisting of five A-rank ("good") pearls and five B-rank ("fair") pearls. Although Toyota and Nakauchi [3] used pearls ranging between A-rank to C-rank ("bad"), we avoided C-rank pearls because a preliminary examination revealed that C-rank pearls were easily distinguished from A- and B-rank pearls. To identify each pearl, a small (1 cm × 1 cm) piece of white paper, on the back of which an identification code was written, was attached to each pearl. The interference color and glossiness of each pearl were quantified using a device developed in our laboratory [3], which simultaneously provides both quantified interference color and glossiness.

In the device for qualifying pearls, a pearl is illuminated from the opposite side of the surface from where the camera is placed, and a transmission image is captured. The intensity map of the transmission image is calculated using both white and very narrow band light (around 520 nm). The device then calculates the intensity gradient with eccentricity using weighted coefficients; the resulting value is a quantitative measurement of interference color [3]. The skewness of the luminance histogram is used as the quantification value for the glossiness. Although Anderson et al. have emphasized that skewness is not a cue for perceptual glossiness [13–17], the correlation between skewness and glossiness–first discovered by Motoyoshi et al. [18]–was confirmed by the inventors using pearls ranging from A-rank to C-rank. The pearl was illuminated from the same side of the surface from where the camera is placed, and the image is captured. The skewness of the luminance histogram of the region where the pearl is in the image is calculated. We measured interference color and glossiness from fifteen points of view or directions and averaged them for each pearl (Figure 2).

Apparatus

The illumination we used had the same spectral pattern as sunny afternoon light, ranging from 370 nm to 780 nm (SERIC Ltd. SOLAX XC-100AF). We intended to imitate the lighting conditions under which experts look at pearls. We used two lamps and a diffuser to illuminate the desk; the distance between the lamps and the desktop was 100 cm. The illuminance on the desk was 109 lx.

The experiment was conducted in the cargo space of a truck that had been modified for use in psychological experiments, called Mobile-Labo (Figure 3). Mobile-Labo enabled us to quickly

A)

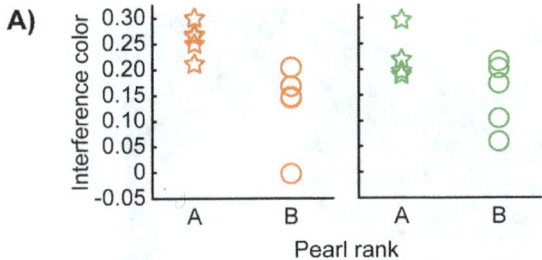

B)

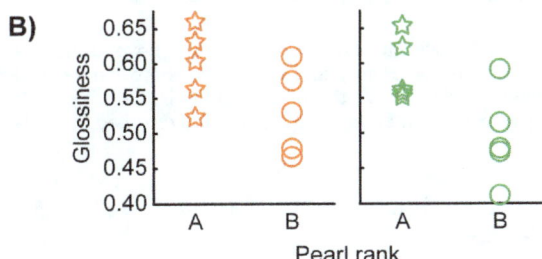

C)

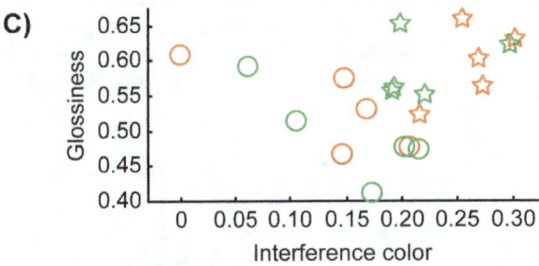

Figure 2. Measured values of interference color and glossiness. Each point shows the average values of 15 measures of a pearl. We used five pearls for each rank ("A" rank shown by star and "B" rank shown by circle) in both sets. Higher measured values seem to be associated with higher ranks ("A" rank) in both interference color and glossiness. (A) Relationship between interference color and pearl ranks in each stimulus set. (B) Relationship between glossiness and pearl ranks in each stimulus set. (C) Relationship between interference color and glossiness. Colors represent stimulus set. Pearson's r of the set in orange is 0.208, and that shown in green is 0.128.

perform the experiments, which prevented the pearls from degradation through aging. Because it was harvesting season for pearl farming and the experts were extremely busy, we visited their farms, offices, and institutes individually in this truck. The novice participants also performed the experiment in Mobile-Labo.

Procedure

The task was to evaluate ten pearls, which were randomly placed on a black desk mat, according to their goodness. The novices were not given any instruction about how experts grade pearls; they evaluated pearls according to their own subjective criteria or preference. On the other hand, the experts were asked to evaluate pearls according to their professional criteria. Although the experts group included participants with different profession and this could cause the diversity of criteria in this group, the economical relationships between their works were very close. For example, marketers selected and bought pearls selected by farmers. If there were great differences between their evaluations, the commercial transactions would never be established. That is,

we regarded the difference in their criteria derived from the difference in their profession as allowable. Although roundness and the existence of scars or pocks also affect the goodness of pearls, the participants were asked to ignore these aspects. They were allowed to look at the pearls while changing viewing positions or moving their heads, but were not allowed to pick up the pearls.

The participants observed and compared ten pearls, and then sorted the pearls by subjective rank order of goodness. After the participant declared that the ordering was complete, the experimenter (who sat next to the participant) checked the identification codes and recorded the rank order. The participants repeated this task five times for each set, alternating between sets. In total, the participants repeated this task ten times. They were not given any feedback about their ordering during the experiment.

Results

To compare the results between novices and experts, we first calculated the expert participants' average ranks for each pearl. Hereafter, these averaged ranks are referred to as the "reference rank" (or Ref-rank in Figure 4A). If a pearl ranked in the first half of the reference rank was ranked from first to fifth in a trial, or if a pearl ranked in the last half of the reference rank was ranked between sixth and tenth in a trial, the ordering was regarded as concordant with the reference rank. That is, we judged the concordance of each participant's rank ordering as whether it was categorically concordant with the reference rank; this was because experts' daily work is to categorize pearls according to their quality. Thus, we assessed the overall concordance of each trial by determining the rate of concordant orderings. Figure 4A shows individual experts' and novices' average concordance for each pearl set. Most experts, except one, showed high concordance; in contrast, around half of the novices (5 out of the 11) showed concordance rates of less than 0.5. According to two-way repeated analyses of variance (ANOVAs), there were statistical differences in experts' average concordances between pearl sets (M = 0.795 and 0.750, SD = 0.161 and 0.171, respectively; $F(1, 32) = 5.786$, $p = 0.022$); however, novices' concordances did not differ between pearl sets (M = 0.513 and 0.533, SD = 0.237 and 0.173, respectively; $F(1, 44) = 0.725$, $p = 0.399$). Although the differences among individuals were significant in both groups, the interaction was significant only in the novices ($F(1,10) = 4.933$, $p < 0.001$). That is, all experts evaluated both sets roughly equal from the point of view of concordance, on the other hand, the evaluations of some novices differed between pearl sets (shown by asterisks in Figure 4A). The group averages approached 0.773 (SD = 0.163) for experts and 0.522 (SD = 0.192) for novices, and there was a statistical difference between these averages (Figure 4B; an independent two-sample t-test, $t(17) = 2.832$, $p = 0.012$, $d = 1.39$).

Next, we compared the consistency of rank orderings between these two groups. Consistency referred to the similarity in rank orderings between sets of pearls within each participant. First, we calculated Kendall's coefficient of concordance (W) as an index of participants' consistency across five repeated rankings for each pearl set. Thus, we calculated two W values for each participant. Participants' within-subjects consistency was thus the average of the Ws (Figure 4C). These averaged consistencies in both groups were significantly higher than the chance W (red solid lines in Figure 4C). The significance was assessed by comparison between observed W and the 95% upper limit of a chance W (red dashed lines in Figure 4C). Although novices' average consistency was relatively high (M = 0.778, SD = 0.130), it was still significantly lower than that of experts (M = 0.903, SD = 0.048; an independent two-sample t-test, $t(17) = 2.449$, $p = 0.025$, $d = 1.22$). Secondly, we

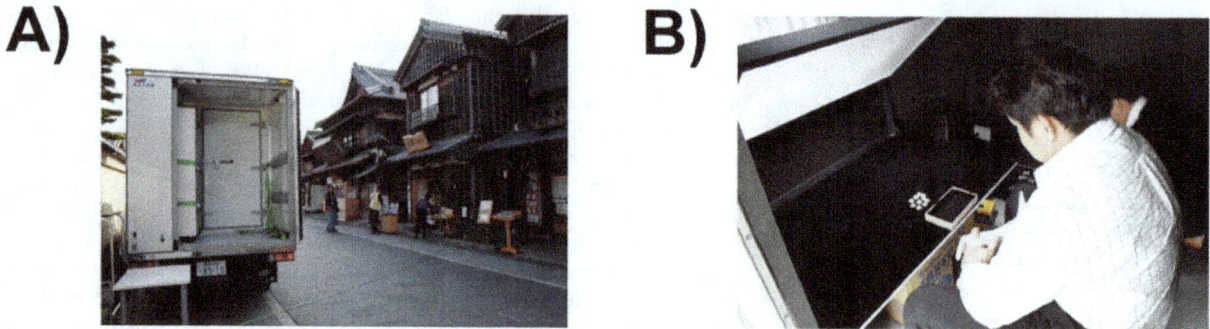

Figure 3. Photographs of the experimental conditions. (A) A snapshot of "Mobile-Labo" truck. The experimental space is beyond the inner door. For this photograph, we parked on a street in front of the office in the courtyard of the laboratory and conducted the experiment. (B) Experimental setting for the experts. The person sitting in front is the participant. The white board above the black desktop is the diffuser. The participant is asked to sort the pearls according to a subjective criterion in the small box. The experimenter received the box and recorded the order.

calculated Spearman's rank correlation coefficients (ρ) between the pairs of trials for each pearl set. That is, 20 separate ρ coefficients were calculated for each participant. The averages of the ρ coefficients are shown in Figure 4D. A Mann-Whitney U test again revealed that the experts' consistency was statistically higher than that of novices ($M_{experts} = 0.878$, SD $= 0.060$, $M_{novice} = 0.721$, SD $= 0.163$, respectively; $U(8, 11) = 19$, $p<0.05$). The significance was confirmed after a z-transformed comparison of the correlation

coefficients (an independent two-sample t-test; $t(17) = 2.519$, $p = 0.045$, $d = 1.24$).

Finally, comparisons were made between subjective grading and pearls' optical properties. To examine whether the rank orderings were explained by the physical properties of the pearls, we carried out four regression analyses for each trial; three simple linear regressions and one multiple regression, in which optically measured interference color and glossiness were used as explan-

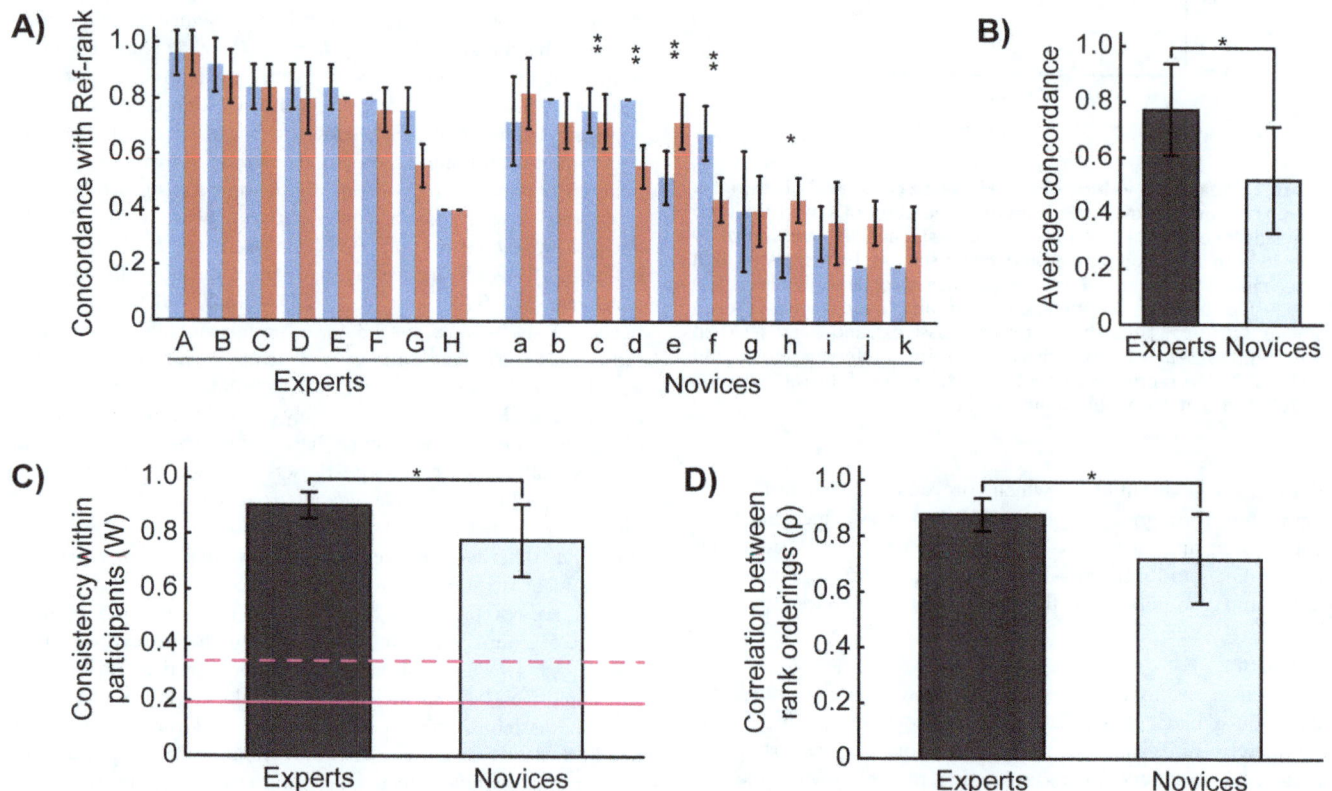

Figure 4. Comparisons between experts and novices. (A, B) Individual averages of concordance rates for each pearl set (blue and red). In A, the participants are sorted into high to low concordance. The asterisks signify that the difference between concordances for each pearl set was significant (*: $p<0.05$; **: $p<0.01$). (C) The average concordances of both groups are shown. As shown, there is a significant difference between them. (D) For consistency, the values of experts and novices were significantly higher than chance and the 95% upper limit. The red solid and dashed lines indicate the expected chance value of W and the values corresponding to the 95% upper limit of chance value of W, respectively, which was calculated from a randomized resampling (n $= 1,000,000$). The difference between them is statistically significant (*: $p<0.05$). (E) The averages of all ρ coefficients in both groups. The difference between them is statistically significant (*: $p<0.05$). The error bars in graphs refer to the standard deviations.

atory variables. Although a single multiple regression with two variables would have sufficed, we wanted to demonstrate to what extent each parameter (interference color, glossiness, and the product of these two variables) could explain each trial. Thus, we conducted three simple regressions as well. First, we performed separate simple linear regressions with interference color (Figure 5A) and glossiness as explanatory variables (Figure 5B). Both simple regressions showed that the rank orderings could not be fully explained by a single explanatory variable. Only 6 of 80 (7.5%) trials for the experts and 41 of the 110 (37.3%) trials for the novices showed significant correlations between the rank ordering and the interference color (Figure 5A), and 8 of 80 (10.0%) trials for the experts and 43 of 160 (39.1%) trials for the novices showed significant correlations between the rank ordering and the glossiness (Figure 5B). Next, we carried out a multiple regression analysis for each trial. That is, we tested the hypothesis that a pearl's rank was determined by a linear combination of its interference color and glossiness. We found that the regression was significant in only 20.0% (16 of 80) and 60.0% (66 of 110) of trials among experts and novices, respectively (Figure 5C). Finally, we conducted a simple linear regression analysis in which the product of the values of interference color and glossiness was used as an explanatory variable, following the notion that interference color and glossiness cannot be segregated perceptually. The regression was significant for 20.0% (16 of 80) and 60.0% (66 of 110) of trials among experts and novices, respectively (Figure 5D). The trials in which the regression analyses were significant were found in the experts who showed relatively low concordance and in all novices with the exception of the participant "f" who showed moderate concordance. Then, we calculated the correlation coefficient between participants' concordance level and the number of trials where the regression was significant. For the experts, the correlations were not found in any regressions. On the other hand, for the novices, positive correlations (0.44 ~ 0.54) were found in all regressions, however, all correlations were not statistically significant ($p = 0.08 \sim 0.18$).

Discussion

In this study, we found that the average rank ordering of novices was significantly less concordant with the standard ranks than that of experts. Indeed, more than half of novices showed a concordance rate lower than 0.5 (i.e., only 50% of rankings were concordant). Thus, their rank orderings were apparently the opposite of the standard ranks for experts in the field, suggesting that the experts' method of evaluating pearls or pearliness is not used by novices who had not been given any instructions about how to evaluate pearls.

Thus, novices could be divided into two groups; those whose rank orderings showed the same tendency as the standard ranks, and those whose rank orderings showed the opposite tendency. In other words, novices may use the same criteria as experts for evaluating pearls, but they utilize these criteria differently.

Interestingly, experts' average concordance was not very high. This suggests that experts' evaluations were not identical. One possible reason for this was that they were unaccustomed to experimental settings such as ours. Although our experimental procedure and environment were designed to imitate the conditions in which experts usually engaged in their daily work, some residual differences might have affected their performance. For example, their work task may be to categorize large numbers of pearls in A-rank, B-rank, or C-rank, while in our experimental task, experts had to rank individual pearls. Thus, the difference between categorization and rank ordering could have been larger

than expected. The other candidate should be the variety or complexity of the pearl evaluation. We took five factors into consideration; size, roundness or shape, the existence of scars or pocks, interference color, and glossiness. We controlled the former three factors in choosing pearls and instructed participants to ignore them, and we tried to explain the rank orderings by optically measured interference color, and glossiness (the latter two variables). However, experts might have examined other factors that we had not considered. For example, one expert had a rank-ordering tendency that was opposite that of the other experts. In addition to the factors we had considered, he might have strongly depended on other factors that we did not consider, which would have caused his results to differ from the reference rank.

We should note that there seems to be no absolute scale of pearl beauty, as is the case for music [12]; this can be attributed to learning methods, especially reward or positive feedback. The effects of reward and positive feedback would be broad, not restricted [19,20]. If the conditions in which the reward is given can be defined strictly, like in discrimination [10,11], then the effects of reward would become concentrated, and some specific responses would be enhanced. On the other hand, if the condition in which the reward is given cannot be defined strictly, as in judgments of goodness or beauty [12], then the effects of reward would be diffuse. The evaluation of pearl and our results would correspond to the latter case.

Both experts and novices showed sufficient consistency in their rank orderings of pearls. However, their consistencies were not equal; the experts' consistency was significantly higher than that of the novices. In addition, during introspection after the experiment, most of the experts noticed that only two sets of pearls were presented repeatedly. On the other hand, none of the novices said that they noticed this. This indicates that the experts were superior in terms of consistency. This could also indicate a better recognition of the pearls by experts. This could be explained by a better attention to specific pearls' physical features, that novices did not notice, and that experts would have used in their judgment.

A series of regression analyses did not fully support the previous findings that experts use both interference color and glossiness in their judgments [1–3]. However, usage of these two qualities may occur in a complex, nonlinear way, because experts' rank orderings could not be fully explained by a linear combination of the two variables. As such, experts likely employ other variables or nonlinear processing for which we did not account. In the context of perceptual learning, Watanabe et al. [21,22] revealed that conscious effort is not essential in processing counterintuitive facts and that implicit processing has a more significant role in perceptual learning. Experts are likely to have received intensive training in their field, with or without conscious effort. Therefore, there could be other variables that the experts themselves are not aware of. On the other hand, more trials of the novices could be explained by two variables in a linear fashion. That is, the novices might evaluate pearls in a simpler manner than the experts do.

In this paper, we investigated how experts and novices evaluate pearls, although this problem had already been investigated in different ways [1,2]. Using a semantic differential method and a multivariate analysis, Nagata et al. extracted the deciding factors used by experts in judging pearls [1]. They also examined pearl-like quality as assessed by novices in a paired comparison experiment [2]. Their aim was to model, visualize, and synthesize computer graphics of pearls; thus, their work focused on the visual features that distinguish pearls from other materials. On the other hand, this paper focused on the visual features that decide the goodness of pearls. We introduced optically measured and

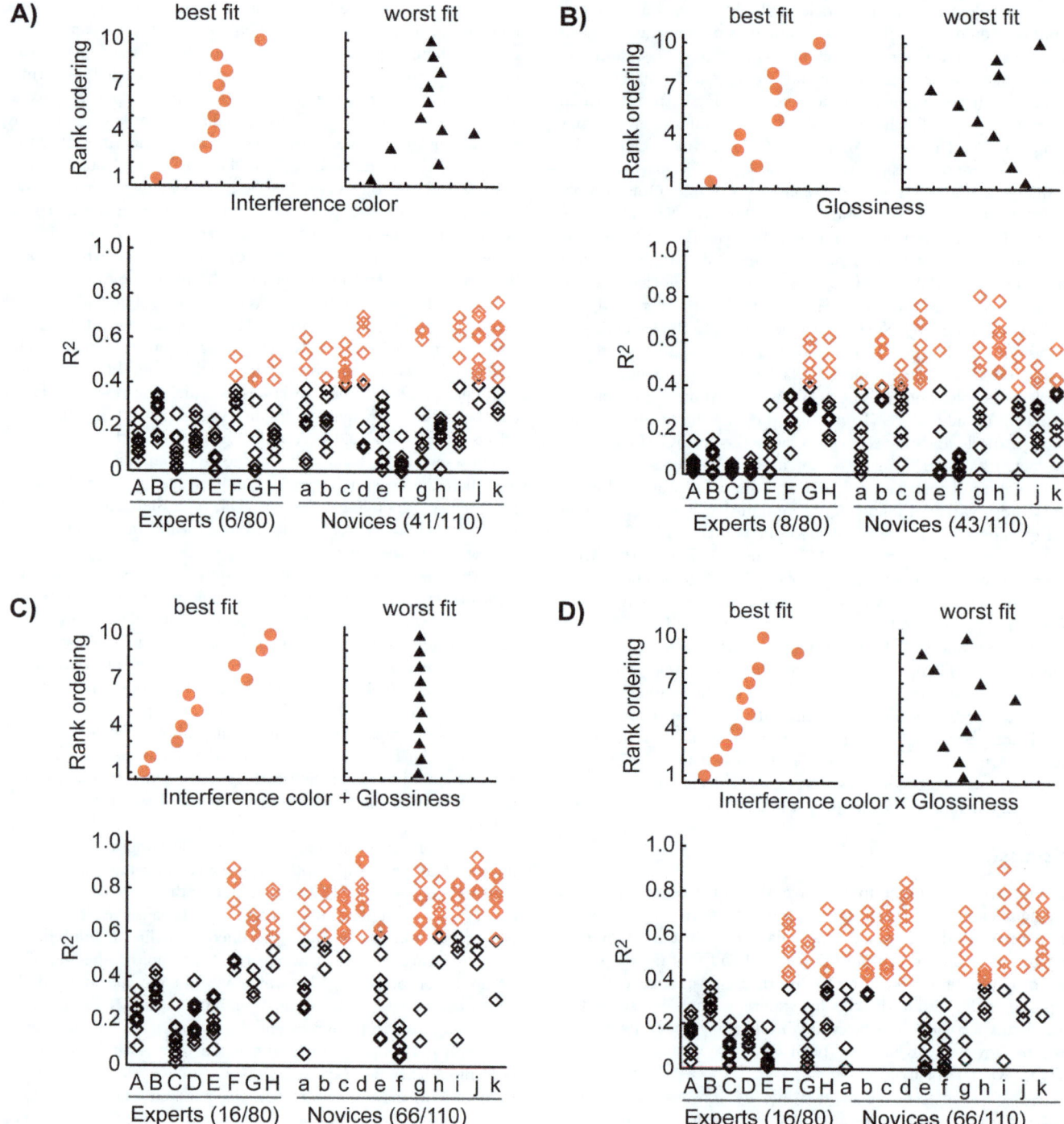

Figure 5. Regression analyses by measured variables. In all panels, the upper insets show the best (left) and worst (right) regression results. The vertical axes show the rank orderings, and the horizontal axes show either the explanatory variable (in A, B, D) or the prediction from the regression equation (in C). The lower graphs show the results of all the regressions. In all graphs, each point represents one trial, and each row represents one participant. The numbers written in parentheses are the numbers of trials for which the regression is significant (shown by red points) and the total number of trials. The vertical axes show the coefficients of determination (R^2). (A and B) Results of simple linear regression analyses on interference color and glossiness, respectively. (C) Results of multiple regression analyses on both variables separately. (D) Results of simple linear regression analyses on the product of both variables.

quantified variables to explain pearl grading. A new technique [3] enabled us to do this, and this paper is the first attempt to explain pearl evaluation quantitatively. Our data suggested that the experts' conception of pearl goodness was somewhat accepted by the naive participants (at least half of them), who were the same as most consumers. This means that consumers accept expert-decided pearl values not only for economic reasons but also because of individual aesthetic preferences. Our data also suggested that naive participants use pearls' interference color and glossiness to evaluate them, but probably in a simpler way than experts do. However, we do not yet have an explicit explanation for the experts' usage of interference color and glossiness. This is the target of our future work.

References

1. Nagata N, Kamei M, Akane M, Nakajima H (1992) Development of a pearl quality evaluation system based on an instrumentation of 'Kansei'. Trans IEE Japan 112-C: 111–116.
2. Nagata N, Dobashi T, Manabe Y, Usami T (1997) Modeling and visualization for a pearl-quality evaluation simulator. IEEE Trans Vis Comput Graph 3: 307–315.
3. Toyota T, Nakauchi S (2013) Optical measurement of interference color of pearls and its relation to subjective quality. Opt Rev 20: 50–58.
4. McLeod P (1987) Visual reaction time and high-speed ball games. Perception 16: 49–59.
5. Gombrich EH (1987) Art and illusion: A study in the psychology of pictorial representation. Oxford: Phaidon. 512 p.
6. Perderau F, Cavanagh P (2013) Is artists' perception more veridical? Front Neurosci 7: 6. doi: 10.3389/fnins.2013.00006.
7. Voss M, Kramer AF, Prakash RS, Roberts B, Basak C (2009) Are expert athletes "expert" in the cognitive laboratory? A meta-analytic review of cognition and sport expertise. Appl Cogn Psychol 24: 812–826.
8. Williams AM, Davids K (1997) Visual search strategy, selective attention, and expertise in soccer. Res Q Exerc Sport 69: 111–128.
9. Faubert J (2013) Professional athletes have extraordinary skills for rapidly learning complex and neutral dynamic visual scenes. Sci Rep 3: 1154. doi: 10.1038/srep01154.
10. Bende M, Nordin S (1997) Perceptual learning in olfaction: Professional wine tasters versus controls. Physiol Behav 62: 1065–1070.
11. Masuda T, Wada Y, Okamoto M, Kyutoku Y, Yamaguchi, etal. (2013) Superiority of experts over novices in trueness and precision of concentration estimation of sodium chloride solutions. Chem Senses 38: 251–258. doi: 10.1093/chemse/bjs137.
12. Fritz C, Curtin J, Poitevineau J, Morrel-Samuels P, Tao FC (2012) Player preferences among new and old violins. Proc Natl Acad Sci USA 109: 760–763.
13. Anderson BL, Kim J (2009) Image statistics do not explain the perception of gloss and lightness. J Vis 9(11), 10: 1–17. doi: 10.1167/9.11.10.
14. Kim J, Anderson BL (2010) Image statistics and the perception of surface gloss and lightness. J Vis 10(9), 3: 1–17. doi: 10.1167/10.9.3.
15. Kim J, Marlow P, Anderson BL (2011) The perception of gloss depends on highlight congruence with surface shading. J Vis 11(9), 4: 1–19. doi: 10.1167/11.9.4.
16. Marlow P, Kim J, Anderson BL (2011) The role of brightness and orientation congruence in the perception of surface gloss. J Vis 11(9), 16: 1–12. doi: 10.1167/11.9.16.
17. Marlow PJ, Kim J, Anderson BL (2012) The perception and misperception of specular surface reflectance. Curr Biol 22: 1909–1913.
18. Motoyoshi I, Nishida S, Sharan L, Adelson EH (2007) Image statistics and the perception of surface qualities. Nature 447: 206–209.
19. Seitz A, Kim D, Watanabe T (2009) Rewards evoke learning of unconsciously processed visual stimuli in adult humans. Neuron 61: 700–707.
20. Murayama K, Kitagami S (2013) Consolidation power of extrinsic rewards: Reward cues enhance long-term memory for irrelevant past events. J Exp Psychol Gen. Feb 18. doi: 10.1037/a0031992.
21. Watanabe T, Náñez JE, Sasaki Y (2001) Perceptual learning without perception. Nature 413: 844–848.
22. Watanabe T, Náñez JE, Koyama S, Mukai I, Liederman J, et al. (2002) Greater plasticity in lower-level than higher-level visual motion processing in a passive perceptual learning task. Nat Neurosci 5: 1003–1009.

Author Contributions

Conceived and designed the experiments: YT TN KK MK SN. Performed the experiments: YT. Analyzed the data: YT TN KK. Contributed reagents/materials/analysis tools: SN. Wrote the paper: YT KK MK.

18

Larval Dispersal Modeling of Pearl Oyster *Pinctada margaritifera* following Realistic Environmental and Biological Forcing in Ahe Atoll Lagoon

Yoann Thomas[1]*, **Franck Dumas**[2], **Serge Andréfouët**[1]

1 Institut de Recherche pour le Développement, unité de recherche CoRéUs, Nouméa, New Caledonia, **2** Institut Français de Recherche pour l'Exploitation de la Mer, unité DYNECO, Plouzané, France

Abstract

Studying the larval dispersal of bottom-dwelling species is necessary to understand their population dynamics and optimize their management. The black-lip pearl oyster (*Pinctada margaritifera*) is cultured extensively to produce black pearls, especially in French Polynesia's atoll lagoons. This aquaculture relies on spat collection, a process that can be optimized by understanding which factors influence larval dispersal. Here, we investigate the sensitivity of *P. margaritifera* larval dispersal kernel to both physical and biological factors in the lagoon of Ahe atoll. Specifically, using a validated 3D larval dispersal model, the variability of lagoon-scale connectivity is investigated against wind forcing, depth and location of larval release, destination location, vertical swimming behavior and pelagic larval duration (PLD) factors. The potential connectivity was spatially weighted according to both the natural and cultivated broodstock densities to provide a realistic view of connectivity. We found that the mean pattern of potential connectivity was driven by the southwest and northeast main barotropic circulation structures, with high retention levels in both. Destination locations, spawning sites and PLD were the main drivers of potential connectivity, explaining respectively 26%, 59% and 5% of the variance. Differences between potential and realistic connectivity showed the significant contribution of the pearl oyster broodstock location to its own dynamics. Realistic connectivity showed larger larval supply in the western destination locations, which are preferentially used by farmers for spat collection. In addition, larval supply in the same sectors was enhanced during summer wind conditions. These results provide new cues to understanding the dynamics of bottom-dwelling populations in atoll lagoons, and show how to take advantage of numerical models for pearl oyster management.

Editor: Arga Chandrashekar Anil, CSIR- National institute of oceanography, India

Funding: This study was funded by the Agence Nationale de la Recherche (ANR, www.agence-nationale-recherche.fr), as part of the POLYPERL project (www.polyperl.org). The authors' research in Ahe atoll has also been supported since several years by the Direction des Ressources Marines de Polynésie française. The funders had no role in study design, data collection and analysis, decision to publish, or preparation of the manuscript.

Competing Interests: The authors have declared that no competing interests exist.

* E-mail: yoann.thomas1@gmail.com

Introduction

Understanding the population connectivity of marine species is necessary to fully comprehend population dynamics, community dynamics and structure, genetic diversity, and the resilience of natural populations to human exploitation [1]. Population connectivity is driven by interactions between organism physiology, morphology and behavior, and the biological and physical environments [2–4]. The main factors explaining spatial and temporal variability in recruitment are: (1) the larval supply, dependent on broodstock density, fecundity and spawning seasonality [5], (2) the larval dispersal, driven by the currents and organism behavior [6], (3) the larval development, primarily controlled by temperature and trophic resources availability [7], (4) the larval mortality, as a consequence of predation, physiological stress (e.g., temperatures, salinity or oxygen stress), disease and parasites [8], (5) the habitat suitability [9] and finally (6) post-settlement survival, which depends on delay of metamorphosis, biological and physical disturbances, hydrodynamics, physiological stress, predation, and competition [10]. Given the complexity of the processes controlling recruitment, and given the inherent

difficulties in measuring larval dispersal and development in the field, biophysical models are increasingly used to explore how biological and physical factors influence larval dispersal and settlement [2,3,11–13]. Connectivity modeling studies typically aim to identify sources (i.e., where larvae are spawned), destinations (i.e., where larvae settle), and fluxes. If *in situ* recruitment data are available, a sensitivity analysis can identify the most important factors explaining the observed variability [14,15]. However, very few studies have used good validation and forcing data.

The main objective of the present study was to identify the factors most affecting the larval dispersal kernel (defined as the probability function of the dispersal distance) of the black-lip pearl oyster (*Pinctada margaritifera* (L.) var. *cumingii* (J.)). The study is conducted at the scale of a specific atoll lagoon (Ahe atoll), in the context of *P. margaritifera* culture, but it also contributes to the general understanding of the dynamics of bottom-dwelling populations in atoll lagoons. The black-lip pearl oyster is found throughout the Indo-Pacific region, from the Red Sea to Central America. This species is particularly abundant in Polynesian archipelagoes and is the foundation species of the black pearl industry, one of the main economical resources of French Polynesia and Cook Islands since the 1990s. Intensive research has taken place in Polynesian lagoons in the past decades, enhancing knowledge on atoll lagoon hydrobiology, lagoon planktonic food resources, and P. margaritifera ecophysiology (see review in [16]). Recently, larval ecology was studied in priority to better understand spat collecting variability [7,17]. Indeed, the black pearl industry production requires steady rates of spat collection. This process yields the necessary oysters that will be subsequently grafted to provide the valued black pearls. The most recent findings on P. margaritifera larval ecology have been acquired in Ahe atoll in the Tuamotu Archipelago, French Polynesia [16].

To model larval dispersal, the first step was to implement, calibrate and validate for the Ahe lagoon, a 3D Eulerian transport model coupled with an empirical vertical swimming sub-model [18,19]. This biophysical model was validated against a series of larval census campaigns [19]. These first studies showed the good agreement between observed and simulated dispersal kernels. They also showed the significant influence of the wind on larval dispersal, thus a major driver of the potential connectivity of P. margaritifera in Ahe lagoon. The present study aims to refine these results by looking at how various forcing variables affect the pearl oyster larvae dispersal kernels both spatially and temporally. For this, the influence of climatological wind forcing, depth and location of larval release, destination location, vertical swimming behavior, pelagic larval duration factors and broodstock location was investigated, to draw a realistic picture of Ahe atoll lagoon-scale connectivity. The consequences of our results for pearl oyster culture management are also discussed.

Materials and Methods

Ethics statement

This work did not involve the manipulation of animals, thus it did not involve endangered or protected species. Modeling methods did not require specific permission from any relevant body as they are harmless and meet all applicable standards for the ethics of experimentation and research integrity.

Pinctada margaritifera: ecology, aquaculture rearing stock and natural stock

We present hereafter, useful background information on P. margaritifera biology, ecology and culture for the connectivity modeling parameterization.

P. margaritifera is a protandrous hermaphrodite, male in early life and female later on, with sex ratio being balanced by age. P. margaritifera maturity is reached during the first year, followed by an important gonad development during the second year [20–22]. Gametogenesis is quick (1 month), and is observed throughout the year with a significant asynchronism. However, austral summer is the more favorable period [22]. At each spawning event, females can propagate up to 40–50 million eggs (50 µm) and males 10–100 time more spermatozoa (5 µm). Fecundation occurs in the water. The first larval stage (D-larva, 80 µm), is reached after 24 h. A ciliate organ (the velum) allows swimming and feeding activities. The pelagic larval duration (PLD) may vary from 15 to more than 30 days [7]. After metamorphosis, young spats fix themselves to the substratum with their byssus, between 0 and 50 m depth. Adult life duration may be more than 12 years, with a theoretical maximum length of 18 cm, but larger shells are commonly seen on several atolls [23,24]. Preferred substrates are along the flanks of coral pinnacles and on deep coral patches, as well as debris from coral and mollusk shells on lagoon sandy bottom. Oysters were naturally abundant in deep atoll lagoons such as Manihi, Scilly,

Takapoto atolls and Mangareva island [25,26]. Historical stock assessments revealed a population peak around 20–40 m depth. In Takapoto atoll, the bulk of the natural stock, estimated at 4.3 million oysters in 1995, was found around 30 m depth [27].

In French Polynesia, pearl culture spread across 27 islands, among which 15 developed a spat collecting activity. Spat collection supports the entire Polynesian production. Indeed, the pearl culture relies entirely on the supply of wild juveniles collected on artificial substrates. Typically, collectors are placed along 200 m longlines submerged at 5 to 10 m depth. Spat collection devices can be set anywhere depending on the farmer's experience (or estimate). After 12-to 24 months, the oysters on the collectors have grown by 5–10 cm and are afterwards set on oyster-rearing chains suspended on lines at 6 to 10 m depth. These 200 m oyster-rearing chains support between 4 000 to 10 000 oysters. After grafting 18 month-old oysters, the last stage is the pearl harvest. This occurs after another rearing period of about 18 months. Unlike spat collection areas, rearing sectors are legally defined by marine concessions boundaries, which are periodically controlled and mapped. Rearing concession capacities are legally limited to a maximum of 12.000 oysters per hectare. Therefore, the rearing stock and its location can be estimated precisely for any given atoll.

Study site: Ahe atoll and lagoon

Ahe atoll is located in the northwestern part of the Tuamotu Archipelago (14.48S–146.30W), about 500 km northeast of Tahiti (Figure 1a). The lagoon is a 142-km^2 deep-water body with an average depth of 41 m, reaching up to 70 m depth. Numerous pinnacles rise to the surface. The deeper areas are made of honeycomb-like cellular structures. The volume of the inner water body is 5.9×10^9 m^3.

The lagoon is an almost closed water body connected to the open ocean through a 11-meter deep, 200-meter wide, pass located on the north-west side of the atoll rim (Figure 1b). Several reef-flat spillways connect the ocean to the lagoon. These spillways, named hoa (a Tahitian designation now part of the formal geomorphology vocabulary), are about 30 cm deep, between 10 and 300-meter wide, with a total cumulated width of 4 km. Thus, they represent about 5% of the rim perimeter (77 km). Hoa are only present on the southern side and on the northwestern side of the rim. Other rim sections are completely closed to water exchanges. All year round, Ahe receives wind waves generated locally by dominant easterly trade winds, which are typically stronger from April to October [28]. In contrast, Ahe is subjected to an attenuated swell regime due to its northward position, in the lee of the large Tuamotu atolls that block the predominant south swells all year round [28].

Ahe atoll's lagoon water circulation is driven by wind [18]. The primary circulation pattern is a downwind surface flow and a returning upwind deep flow. Dumas et al. (2012) [18] described two main barotropic circulation structures under climatological tradewinds. First, a north large anticlockwise circulation pattern occupied two-thirds of the water body with residual currents of around 5 cm.s^{-1}. Second, a weaker clockwise circulation pattern occurred in the south. The renewal time of the water body has been estimated at 252 days. The average e-flushing time, which corresponds to the time needed to decrease a concentration of tracers (e.g. larvae) by a factor e = 2.718, was 80 days. It ranged from 50 days in the vicinity of the pass to 140 in the center of the northern circulation cell [18].

Larval production of Ahe lagoon

Larval production (W_s) was assumed to be directly linked to the broodstock. We did not consider a potential spatial heterogeneity

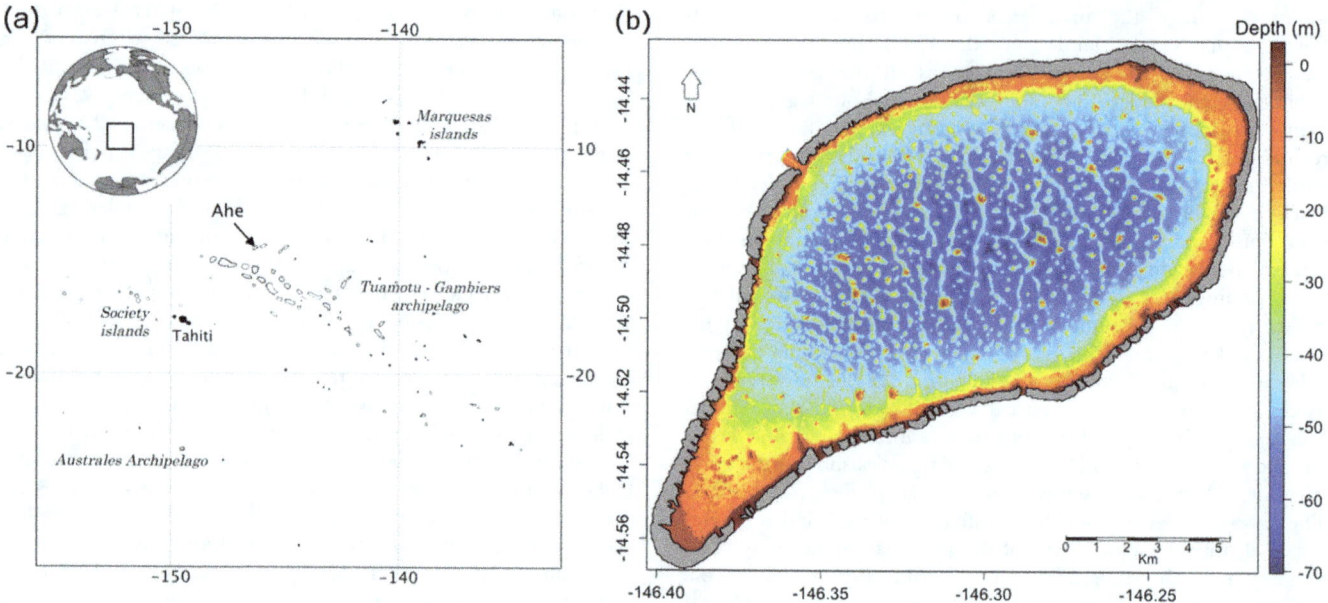

Figure 1. The Ahe atoll. (a) Ahe atoll location and (b) morphology and bathymetry of the Ahe lagoon.

of fecundity. Two stocks were considered: natural and reared. The cultivated stock was mapped using the official concession boundary information (Figure 2a, Marine Resource Direction, pers. comm.) and considering 12.000 oysters per hectare of concession. Conversely, the distribution of the natural stock was still unknown in Ahe lagoon at the time of this study. Therefore, we applied the density by depth level found by Zanini and Salvat (2000) [27] in Takapoto atoll (Table 1, Figure 2b). Table 2 provides the reared and natural stocks for each of the 12 spawning sectors considered here (defined below).

Two different release scenarios were further considered. First, the reared population emitted larvae between 5 and 10 m depth, where longlines are typically immerged. Second, the natural stock emitted larvae from the bottom layer, where the natural stock is located.

Spawning and destination sites

Twelve sectors of similar surface areas, representing both spawning and destination sites, were defined for the connectivity analysis (Figure 3). The extent of each sector was obtained by clustering the longitudes and latitudes from the model grid. The

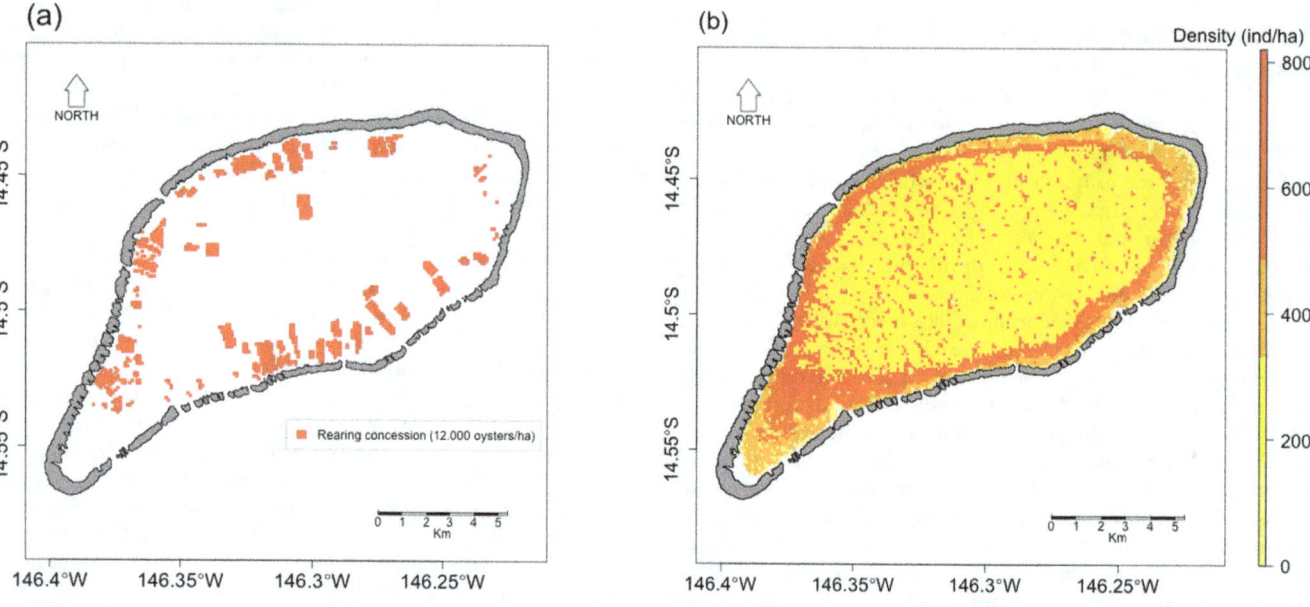

Figure 2. The pearl oyster broodstock. Maps of (a) the pearl oyster rearing concessions, (b) the natural pearl oyster density. The pearl oyster concessions correspond to the situation in 2012, with 12.000 oysters/ha (source: Direction des Ressources Marines). The natural pearl oyster density is an estimation after [27] (scale: oysters per hectare).

Table 1. Density of wild pearl oyster according to the bathymetric level (after [27]).

Bathymetric level	Density (oyster/100 m²)
0–10 m	1
10–20 m	3.6
20–30 m	5.2
30–40 m	8.2
>40 m	2.5

number of sites was a compromise; a high enough number chosen to realistically represent the various parts of the lagoon, described in previous studies [19] and also low enough to optimize the model computing time. Only the near shore sectors, with less than 5 m depth and thus unsuitable for farming activity, were not included. These shallow sectors and the open ocean were nonetheless considered in the connectivity study as destination locations only. Shallow sectors and open ocean were used to estimate the loss rate and export rate respectively.

The pelagic larval duration (PLD)

There is currently very little *in situ* information on PLD for *P. margaritifera*. Most of our knowledge comes from hatchery experiments. They provided a mean PLD of 21 days [29]. However, depending on environmental conditions (e.g., temperature, food concentration), PLD may vary from 15 to more than 30 days [7]. Connectivity patterns were thus computed after 15, 20, 25 and 30 days of simulation. This indirectly provided the potential effects of food and/or temperature on the larvae and *in fine* on connectivity.

Bio-physical transport model

Larval transport and dispersal were simulated with the 3-D hydrodynamic model MARS3D, which resolves the ocean dynamics equations [30]. A full description of the bio-physical larval transport model can be found in Thomas et al. (2012) [19]

Table 2. Estimation of the reared and natural oyster stock, cumulated by spawning site (see Figure 3).

Spawning site	Reared stock	Natural stock
1	1.092.000	379.280
2	972.000	633.580
3	2.364.000	450.540
4	1.848.000	518.370
5	1.212.000	452.690
6	0	347.360
7	2.616.000	449.960
8	0	297.540
9	2.184.000	413.620
10	864.000	381.790
11	948.000	431.570
12	264.000	399.560
Total	14.364.000	5.155.860

and model implementation and validation are explained by Dumas et al. (2012) [18]. In short, the model is constrained by a horizontal cell size of 100 m by 100 m. The vertical resolution of the model includes 23 sigma-vertical layers. These sigma-vertical layers are tightened close to the bottom and to the surface in order to better represent velocity gradients at the interface layers.

The hydrodynamic model was coupled with an advection/dispersal module, itself integrating a model reproducing the larvae dial vertical swimming behavior. The larval transport followed an Eulerian scheme since the state variables were calculated at fixed locations. Larval distribution was thus described as a grid of larval concentrations, transported by the water flow network and by the vertical swimming displacement. The larvae exported to sectors unsuitable for farming activity, into the ocean, the pass, and shallow waters (less than 5 m-deep), can be remobilized into the global larval pool via inbound exchanges through the pass and the hoa, or from the shallow waters to the deeper one.

The swimming model simulated the vertical displacement velocity following a sinusoid centered on 0, with a positive velocity (going up) during the night and a negative velocity (going down) during the day [19].

Biophysical models are able to simulate dispersal accurately, but they require assumptions on relevant physical processes, boundary conditions, atmospheric forcing, and larval behavior [2]. The present biophysical model was validated against numerous field records [18], and good agreement between observations and simulations of larval concentrations was achieved, both at the vertical and lagoon scales [19].

Potential and realistic connectivity

Following Watson et al., (2010) [13], the potential connectivity is defined as the probability of larval transport from a spawning site (*i*) to a destination location (*j*). Conversely, realistic connectivity is defined as the actual (or modeled) number of larvae that travel from *i* to *j*. We intentionally modified the term "realized" used by Watson et al. (2010) [13], by "realistic", considering that the realized connectivity depends on effective recruitment, after settlement and post-settlement [4,31]. In contrast, the realistic connectivity only depends on dispersal processes until the stage of competent larvae. The reason to compare the potential and the realistic connectivity is to assess the relative importance of the spatial distribution of the broodstock (i.e., natural and reared) on the patterns of larval connectivity. In other words, the potential connectivity was calculated following a homogeneous distribution of spawners, while realistic connectivity accounts for the spatial distribution of the stocks.

The potential connectivity P_{sd} was calculated according to the physical (*i.e.*, hydrodynamics) and behavioral (*i.e.*, vertical migration) factors, following:

$$P_{sd}(t) = \frac{Q_{sd}(t)}{Q_s}$$

Where Q_{sd} is the number of larvae found in the destination location *d* at time *t* and coming from the spawning site *s*, and Q_s is the total number of larvae released in the spawning site *s*. For the simulations, the initial larvae concentration in the spawning sites was set at 100 individuals per hectare.

In a second step, the realistic connectivity L_{sd} at time *t* was defined as the product of potential connectivity P_{sd}, and larval production W_s, which corresponds to the number of spawners in the considered spawning site:

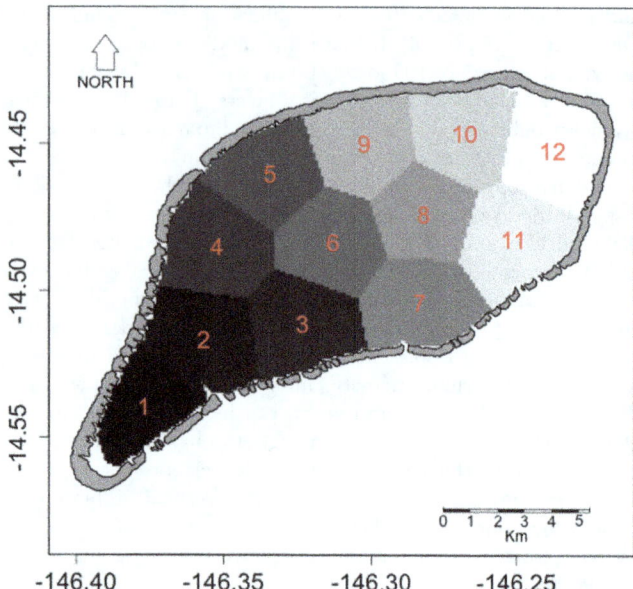

Figure 3. Map of the 12 spawning sites and destination locations, of the connectivity study. The extent of each sector was obtained through a clustering method performed on the longitudes and latitudes of the model grid. Only the near shore sectors, with less than 5 m depth, were not included.

$$L_{sd}(t) = P_{sd}(t) \cdot W_s$$

Fecundity and mortality rates were not explicitly included in this study. The realistic connectivity thus gives a theoretic number of larvae, only dependent on the number of adults in the different spawning sites.

Weather regimes

To evaluate the effects of wind on larval dispersal and lagoon-scale connectivity, we identified a number of typical 30-day realistic wind sequences for the simulations using a long-term meteorological reanalysis. Meteorological dataset from ERA-Interim analysis (http://www.ecmwf.int/research/era/do/get/era-interim) were extracted daily at the closest grid point to Ahe from 1st January 1979 to 31th December 2011. The 30-day sequences were selected with an overlap of 20 days between each, giving 1203 sequences for the entire 32-year period.

Principal component analysis (PCA) and clustering techniques are widely used to identify climatological regimes in observations and model results [32]. PCA was performed on the wind stress components since it is linearly related to the transport within Ekman layer. It could be expressed as:

$$\tau_x = \rho_a \cdot Cd \cdot W_s \cdot u$$

$$\tau_x = \rho_a \cdot Cd \cdot W_s \cdot v$$

where ρ_a is the air density, Cd the drag coefficient, W_s the wind speed and (u, v) its zonal and meridian components. To a certain extent (i.e., under moderate wind conditions and in a medium

where sea roughness remains moderate, like in an enclosed area such as a lagoon), Cd can be considered constant (e.g., 0.0016 following Deacon and Webb 1962 [33]). Under this hypothesis, our classification could be performed using:

$$\frac{\tau_x}{\rho_a \cdot Cd} = W_s \cdot u$$

$$\frac{\tau_x}{\rho_a \cdot Cd} = W_s \cdot v$$

which scale homothetically with the wind stress itself and thus does not influence the classification. W_s, u, and v come directly from the ERA interim re-analysis data set.

In order to optimize the PCA calculation and still preserve the intra-sequence variability (e.g., distinguish a permanent null sequence from "loop shaped" sequence), each sequence was summarized into 15 sub-sequences of 2-day averages. This was found to be the best trade off. PCA was performed on non-standardized dataset, in order to maintain the absolute variability.

A k-means clustering was performed on the PCA scores using the *pam* function of the *cluster* package in R software [34]. No significant gain in cluster dissimilarity was found beyond 12 clusters, and 12 was kept as the optimal number of clusters, corresponding to a 10% threshold in dissimilarity. The real wind sequence the closest to the barycenter of each cluster was selected as reference for the cluster. This finally yielded the 12 wind regimes used for the different connectivity scenarios.

Finally, the occurrence probability was calculated for each cluster as the ratio between the number of sequences included in each cluster and the total number of sequences (i.e., 1203).

Connectivity scenarios, statistical analysis and connectivity matrices

Table 3 summarizes the range of biophysical factors used in the different connectivity scenarios and the number of treatment per factor.

To test the effect of these factors on connectivity results, we applied ANOVA using *aov* function of the R software [35]. Since ANOVA requires normal data, the connectivity output data were transformed. The potential connectivity probabilities (P_{sd}) were normalized following a $\mathrm{Arcsine}(\sqrt{P_{sd}})$ transformation. The realistic connectivity results (L_{sd}) were subjected to a BoxCox transformation: $\frac{L_{sd}^{\lambda} - 1}{\lambda}$ with lambda estimated at 0.3.

A transition probability matrix formalized the potential connectivity. For the realistic connectivity, the theoretic number of larvae replaced the probability in the transition matrix. To rank the different sites in terms of 'source' and 'sink' potential, the cumulated connectivity of each spawning site (i.e., column of the matrices) and destination location (i.e., row of the matrices) are computed for both the potential and realistic connectivity matrices. To represent the spatial heterogeneity of cumulated connectivity, the summed data were standardized using the standard score z calculation: $z = (V–MV)/SD$, where V, MV and SD are respectively the data value, the mean value and the standard deviation.

Results

Wind regimes

Twelve wind regimes (WR) of 30 days each were identified with the clustering (Table 4). The overall frequencies of each WR are between 4.9% (WR 12) and 13.3% (WR 8). The highest mean wind speed is achieved for WR 12, with 7.6 ± 1.1 m.s^{-1} and the lowest for WR 3 with 3.7 ± 0.9 m.s^{-1}. The wind directions are mainly eastern, with the most southern one at $100.0\pm12.8°$ (WR 12) and the most northern one at $42.5\pm91.4°$ (WR 3) (Figure 4).

Analysis of the connectivity variance

The variance of the potential connectivity (P_{sd}) (Table 5) is mainly related to the "Destination" (D) and "Spawning site" (S) factors that explain 59.4% and 25.9% of the total variance respectively. The interaction of (S : D) explains an additional 7.9% of the variance. Finally the PLD factor explains another 5% of the variance. Some other interactions have a significant effect on the connectivity variability. Namely (WR : D), (PLD : D), (WR : S : D) and (PLD : S : D), but with contributions all under 1%. The factors "Release level", "Swim" and "Wind Regime" do not contribute significantly to the connectivity variance.

For the realistic connectivity (L_{sd}), the factor "Release Level" (RL) alone explains 41% of L_{sd} variance. More than the depth level of larvae release, this factor reflects the effect of broodstock density on connectivity, since the release level is not significant on the potential connectivity. The "Destination", "Spawning Site" and "PLD" factors have a significant contribution to L_{sd} variance, with 39.8%, 2.9% and 1.3%, respectively. The interactions (RL : S) and (S : D) contribute 8.4% and 5.2% of the variance, respectively. Finally, the interactions (PLD : S), (RL : D), (RL : S : D) and (PLD : S : D) contribute significantly to the realistic connectivity variance, but at less than 1%. The "Swim" and "Wind Regime" factors do not contribute significantly to the realistic connectivity variance.

Spatial patterns of connectivity

The potential and realistic connectivity matrices computed after 20 days of simulation are presented in Figure 5 and Figure 6 respectively.

The intra-lagoonal potential connectivity (P_{sd}) reaches $5.9\pm2.7\%$ on average, giving a mean coefficient of variation (CV = SD/MV) of 45%. Patterns of P_{sd} show a symmetry close to the diagonal, with two main sectors showing high retention: the northeastern part of the lagoon (i.e., sites 8, 9, 10 and 12), with highest values between sites 8, 9 and 10 (reaching 18%), and the southwestern part (i.e., sites 2, 3, 4, 6 and 7), with lower connectivity levels. The lowest connectivity is measured in the destination location (D) 1 and 5. The loss through export in the

shallow waters appears very low, with 0.5% in average. However, the export through the pass to the open ocean averaged 5.2%, with higher level coming from spawning sites 7 and 11 (7.9% in average). Site 11 shows an asymmetric pattern since it exports a high number towards the southwestern destination locations and also receives larvae from the northeastern spawning sites.

The intra-lagoonal realistic connectivity (L_{sd}) is respectively $25.9\pm11.3\times10^{3}$ and $70.3\pm63.7\times10^{3}$ in average for the natural and reared scenarios, giving a coefficient of variations of 44% and 91%, respectively. Patterns of realistic connectivity due to the wild oysters show an increase of the retention in the western sector mainly due to an increase in the contribution of spawning sites (S) 2 to 4. Conversely, connectivity level in the northeastern sector decreases. The realistic connectivity due to reared oysters shows a more asymmetric pattern. Spawning sites 6 and 8, without broodstocks, do not contribute to the global connectivity. Conversely, sites S 3, 7 and 9, with the highest broodstock densities, have their contributions reinforced in the reared scenario.

The potential cumulated connectivity shows an increasing west to east gradient in spawning site contributions. Higher levels are recorded eastward, which is consistent with the higher water renewal rate of the western sectors (Figure 7a). However, this pattern is highly modified in the realistic cumulated connectivity, with an increase of the western sites contribution and a decrease of the eastern sites contribution for both the natural and reared scenarios. Only spawning site S1 shows a decreased contribution in the natural scenario. In the reared scenario, the three spawning sites previously identified (3, 7 and 9, with the highest broodstock densities), have an increased contribution while spawning sites S 6, 8 and 12 contribution is decreasing, due to the heterogeneity of their broodstocks.

The cumulated realistic connectivity in the destination locations (Figure 7b) suggests an increase of connectivity levels in the western destinations (i.e., from D 1 to D 7) and a decrease in the eastern destinations (from D 8 to D 12). These tendencies are strengthened in the reared scenario compared to the natural one. In potential connectivity, D 8 receives the highest number of larvae while this position is achieved by D 4 in the realistic scenarios, Conversely, D 1, 5 and 12 collect the lowest number of larvae in all scenarios.

Influence of the PLD factor

The connectivity variation (CV) explained by the PLD factor is 23% on average (Figure 8). Figure 8 represents the spatial variability of this CV, in mean wind conditions, and the CV variability depending on the wind direction. Southwestern destination locations (DL 1 to 6) are thus the most variable (Figure 8a). In this sector, the DL 1 has the highest variation level

Table 3. Summary of the factors tested in the connectivity analysis.

Factor	Treatments	Nb. of treatments
Larval release	5–10 m depth/bottom	2
Swimming	With/without	2
Wind regime	12 sequences of 30 days	12
Spawning sites	12 spawning sites	12
Destination location	12 destination locations + intra-lagoon shallow water + open ocean	14
PLD	d15, d20, d25, d30	4

Table 4. Summary of the 12 wind regime characteristics.

ID cluster	Start date	Nb. of sequences per cluster	Wind regime probability (%)	Wind speed (m.s⁻¹) Moy.	Wind speed (m.s⁻¹) SD	Wind direction (°) Moy.	Wind direction (°) SD
1	28/08/81	135	11.2	7.1	0.9	96.1	11.3
2	16/03/82	98	8.1	4.6	1.9	87.6	66.2
3	20/12/84	87	7.2	3.7	0.9	42.5	91.4
4	14/05/86	83	6.9	4.9	1.8	94.3	51.1
5	24/05/86	94	7.8	5.9	1.5	93.2	18.8
6	12/01/94	83	6.9	4.2	1.4	75.8	63.7
7	22/01/94	88	7.3	4.8	1.4	81.3	61.6
8	02/05/94	160	13.3	4.4	1.3	80.9	37.7
9	06/01/99	108	9.0	5.7	1.2	89.0	17.6
10	07/03/99	86	7.1	6.2	1.2	94.6	20.0
11	06/04/99	122	10.1	5.4	1.3	87.8	14.1
12	25/09/08	59	4.9	7.6	1.1	100.0	12.8

according to the PLD, with a mean CV of 34%. The export rate toward the ocean is also highly dependent on the PLD factor, with in average CV = 31%.

Figure 8b provides the relation between the averaged wind direction of each wind regime (WR) and the mean coefficients of variation (CV) of the connectivity according to the PLD factor. A negative linear relation emerges following variations of wind directions from north to south. WR 3, the most northern wind, is an obvious outlier. The other northern winds have the highest connectivity variation, compared to east and southeast wind regimes.

Seasonal patterns of connectivity

Despite little contribution of the wind regime (WR) to the connectivity variance, a significant effect of the interaction between the WR factor and destination location was measured (Table 5). Therefore, this interaction was further explored to identify a potential seasonal pattern.

Firstly, we investigated the spatial pattern of overall connectivity variation associated to the WR factor (Figure 9a). The overall average of CV is 24%. CV appears inversely related to the potential connectivity (Figure 9b). Eastern spawning sites, from S 7 to S 12, display the highest levels of variation, with an extreme for the S 7 and D 10 cell. This cell reaches 89% and is explained by a high level of connectivity only during northeastern winds (wind regime 3, data not shown). The destination location D 1 shows the highest level of variation (mean 40%). Conversely, D 5 and 6 have the lowest levels of variation with 16% and 17% on average, respectively.

Secondly, we explored the seasonality in connectivity patterns. The probability of occurrence of each wind regime in each month was calculated and the wind regimes were ranked to reveal their seasonal patterns (Figure 10a). Wind regimes: 3, 8, 9 and 11 are mostly observed in the austral winter (i.e., June to October) while regimes 2, 4, 5 and 6 are observed in the austral summer (i.e., November to May). Other wind regimes are present during transition periods, with regular occurrences all year long.

To evaluate how the destination locations may be favored during a specific season, realistic connectivity (L_{sd}) was cumulated by destination location and averaged for each of the three periods identified (i.e., winter, transition, summer) (Figure 10b). The austral summer winds generate the highest connectivity levels, especially in the southwestern part of the lagoon (i.e., destination locations: 2, 3, 4, 6, 7 and 8). The austral winter and transition conditions create the lowest overall connectivity levels in the lagoon.

Discussion

Understanding recruitment patterns across a broad range of scale is necessary to improve the management and conservation of marine communities [4,36]. This is especially true for species used as a resource in a fishing or aquaculture context. Their population dynamics can lead to overfishing and quick decline in case of low recruitment. Very few studies have addressed recruitment patterns of mollusks in atoll lagoons [37] and this study gives new insights on the biological and physical factors affecting the dispersal and *in fine* the recruitment of one species: the pearl oyster *P. margaritifera*, which has a high commercial and social value in French Polynesia. In the present case, the entire pearl oyster production remains dependent on the natural collection of juveniles on artificial substrates [16]. In the last decade, spatial and temporal variations in spat collection, including poor production years, led to professional and institutional concerns and demand for knowledge

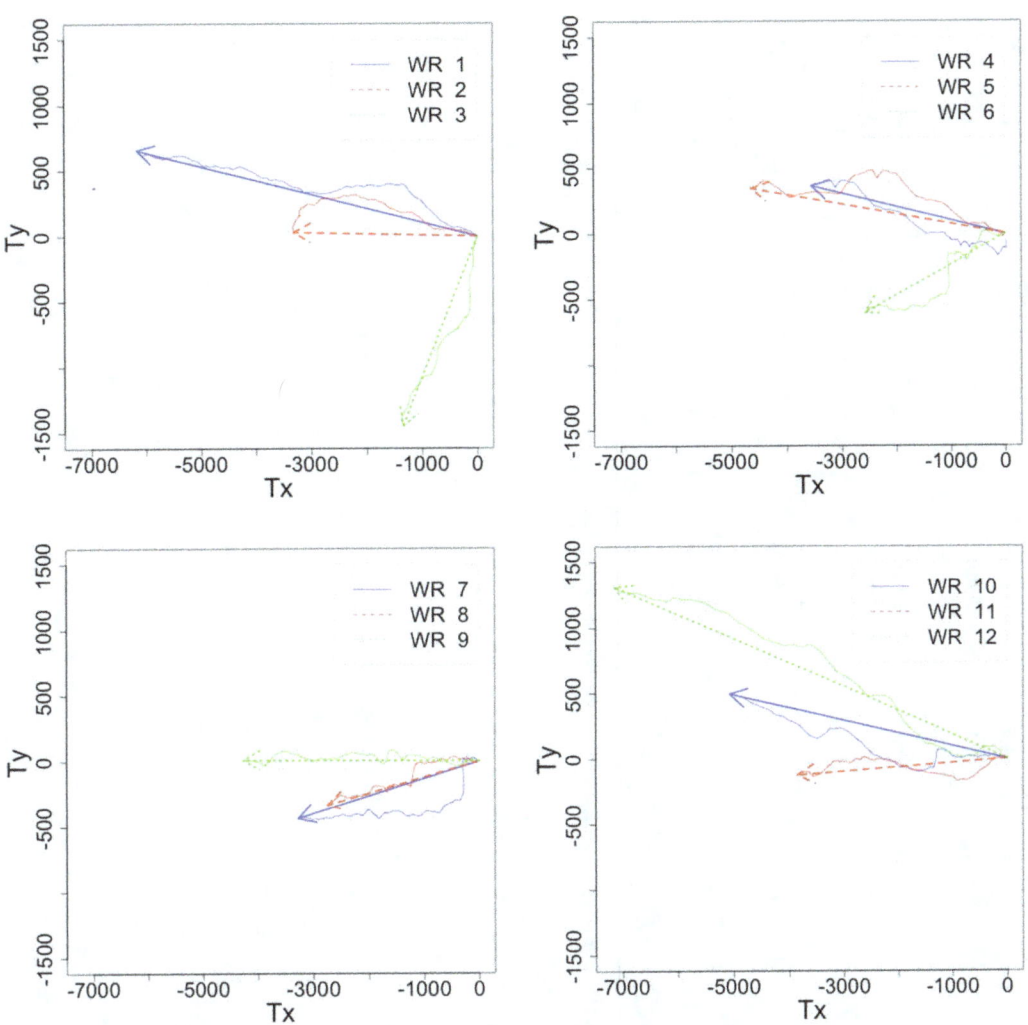

Figure 4. Hodographs of the 12 wind regimes. These graphs are performed with the wind stress components Tx and Ty. Arrows give the mean pattern for the 30 days. The numbers correspond to the wind regime indices (see Table 4).

and tools to improve practice [38]. After the first step of implementing, calibrating and validating with actual larvae counts a transport model [18,19], this study refined the results by looking at how forcing variables act spatially and temporally on larval dispersal, and connectivity. These results have practical and management implications. In particular, the calculation of a realistic connectivity has shown the sensitivity of the larval supply to the location and density of the reared broodstock.

Not surprisingly, the spawning site, destination location and pelagic larval duration (PLD) are the main drivers of the connectivity patterns in Ahe atoll lagoon. In contrast, vertical swimming, depth of release and wind regime taken alone did not significantly contribute to the connectivity variability. Almost 60% of the potential connectivity variability was explained by the destination location factor only. This emphasizes the careful choice of collection areas for successful spat collection. Despite the simple saucer-shaped lagoon morphology, the larvae were far from being homogeneously distributed throughout the lagoon. In particular, the two northeastern and southwestern sectors had high retention levels and a low degree of connection between them. This is due to the two main barotropic circulation structures described by Dumas et al. (2012) [18]. The connection between the pass sector (i.e., site 4 in the west) and the eastern sector 11

(Figure 5) is itself explained by the downwind surface flow and upwind mid-depth return flow described in Dumas et al. (2012) [18].

The pelagic larval duration (PLD) of any species affects its broad-scale connectivity [39]. The PLD is also related to the growth rate of larvae, which itself depends on the environmental conditions (i.e., trophic state and temperature) [7]. In Ahe lagoon, because the currents are weak [18], larval dispersal is strongly dependent on the PLD even if the lagoon is small (i.e., 142 km^2). Our results (Figure 8) show that PLD mostly modulates connectivity between the northeast and southwest parts of the lagoon. These connections are promoted by longer PLD and by northeastern winds. A linear relation between the wind orientation and the connectivity variation in time (Figure 8b) showed that the most southern wind quickly homogenizes larval concentrations, which rapidly leads to steady connectivity patterns. In contrast, a northern wind leads to slower homogenization and thus to more variable connectivity in time. Longer PLD also increase export rate, with thus a decrease in overall larval supply. Here, we considered PLD homogeneous in time and throughout the lagoon. However, *P. margaritifera* larvae growth rate can be spatially and temporally heterogeneous according to temperature, available trophic resources and their spatial distribution [7]. Depending on

Table 5. Results of the variance analysis on potential and realistic connectivity.

Factor	DF	Potential connectivity			Realistic connectivity		
		p value	Signif.	%Var	p value	Signif.	%Var
Release Level (RL)	1	0.476		0.0	$<2.10^{-16}$	***	41.0
Swim (Sw)	1	0.883		0.0	0.935		0.0
Wind Regime (WR)	11	0.374		0.0	0.493		0.0
PLD	3	$<2.10^{-16}$	***	5.0	$2.6.10^{-10}$	***	1.3
Spawning Site (S)	11	$<2.10^{-16}$	***	25.9	$<2.10^{-16}$	***	2.9
Destination (D)	11	$<2.10^{-16}$	***	59.4	$<2.10^{-16}$	***	39.8
RL : S	11	0.389		0.0	$<2.10^{-16}$	***	8.4
PLD : S	33	$6.13.10^{-6}$		0.5	0.001	***	0.3
RL : D	11	0.706		0.0	$4.10-4$	***	0.4
WR : D	121	0.003	**	0.2	0.127		0.1
PLD : D	33	0.019	*	0.1	0.282		0.0
S : D	121	$<2.10^{-16}$	***	7.9	$<2.10^{-16}$	***	5.2
RL : S : D	121	0.871		0.0	0.013	*	0.0
WR : S : D	1331	0.040	*	0.1	0.430		0.0
PLD : S : D	363	$2.94.10^{-8}$	***	0.7	0.005	**	0.3
			sum	99.96		sum	99.93

Only significant interactions are listed. DF corresponds to the degree of freedom and %Var gives the percentage of the total variance explained by each factor.
Signif. codes : '***' 0.001, '**' 0.01, '*' 0.05, ' ' 1.

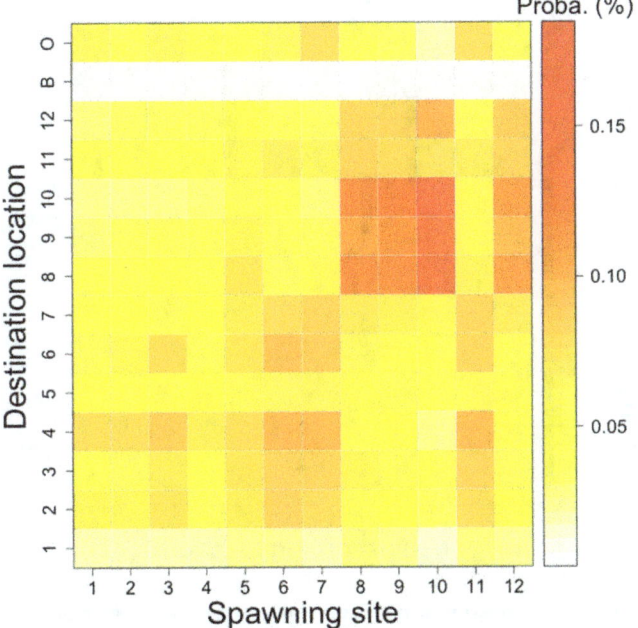

Figure 5. Potential connectivity matrix (*i.e.,* probability). This potential connectivity is an average of the 12 wind regimes scenarios. The potential connectivity is calculated as the ratio between the number of larvae in the destination location *j* after 20 days of dispersal, coming from the spawning site *i*, and the total number of larvae emitted in the spawning site *i*. In the destination locations, "B" represents the shallow waters and "O" the open ocean.

the dispersal pathways, the larvae may experience different trophic conditions leading to different growth patterns. This might reinforce the contribution of some of the spawning site locations and thus might change the connectivity patterns. Further work

could seek to refine the dispersal model using a dynamic PLD in time and space.

Swimming behavior of larvae could also significantly control larval dispersal outcomes [6]. However, our results showed that the larvae dial vertical migration does not have any effect on the connectivity patterns. Here, larval behavior was not modulated spatially and temporally while several factors may play a role in the vertical migration of planktonic larvae (e.g., light, food, salinity discontinuity, temperature, predators, larval size/stage) [6,40,41]. We also did not change the behavior with larval size or development stage. Indeed, there is currently no evidence that *P. margaritifera* larvae change their behavior at each developmental stage. We are aware that *P. margaritifera* spat recruitment on collectors is at maximum 5 m depth and collapse deeper [17]. This may suggest a change of behavior at some late development stage but this remains poorly understood. A complementary sensitivity analysis performed on the maximum swimming speed (i.e., with a 3 times higher parameter α) did not provide significant effect (i.e., mean connectivity variation coefficient = 0.2%, data not shown). This result reinforces our first observations and seems consistent with Kim et al. (2010) [41], who showed no significant effect of the swimming behavior on the *Crassostrea virginica* larvae dispersal in Mobile Bay, Alabama. These authors explained their results by frequent destratifications of the water column. In Ahe lagoon, vertical stratification, in temperature and salinity, is weak and transient [18]. The lagoon waters are well mixed [18] and homogenize larval concentrations in the water column. Vertical migration and turbulent mixing could also explain the similarity of connectivity patterns between bottom and surface releases, since larval concentrations can be quickly mixed.

This study suggests a small effect of the wind regime taken alone on connectivity patterns. This can be related to the relative constancy of the wind direction, coming mainly from the east combined with the simple geometric configuration of Ahe lagoon. Indeed, a consistent wind direction can not really generate complex patterns, in contrast to more complex embayment

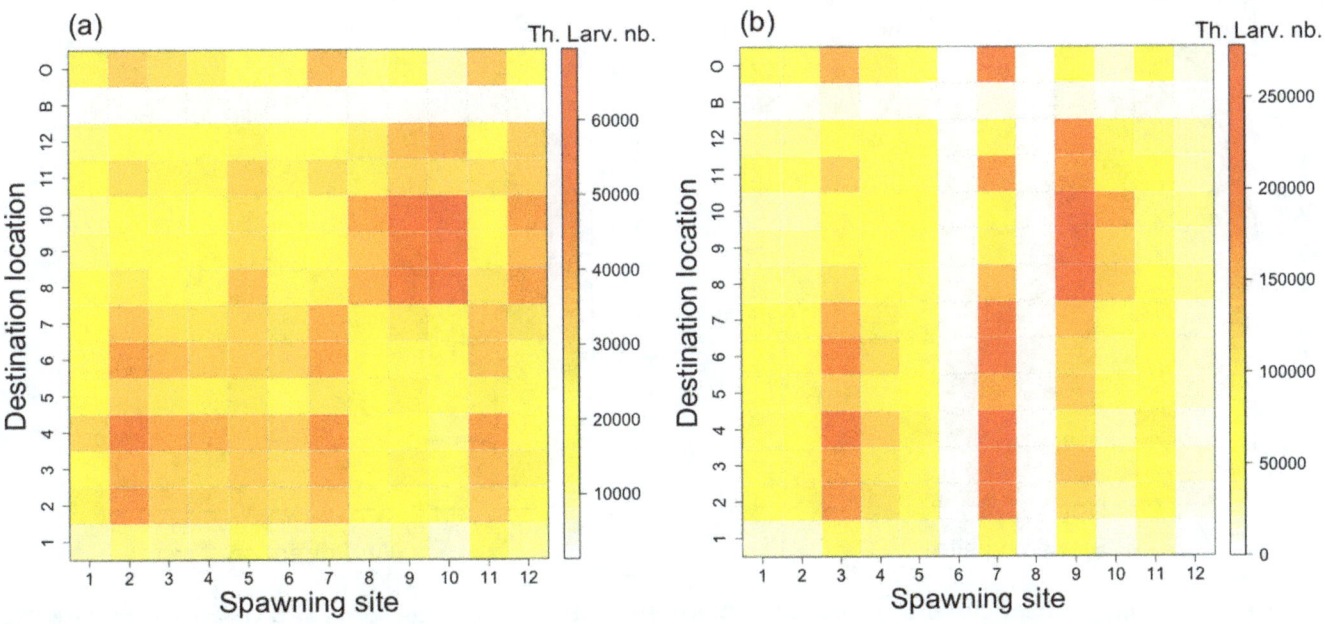

Figure 6. Realistic connectivity matrix (*i.e.,* theoretic larval number). The realistic connectivity corresponds to the mean potential connectivity (see Figure 5), spatially weighted in spawning sites by (a) the natural broodstock density and (b) the reared broodstock density, respectively. In the destination locations, "B" represents the shallow waters and "O" the open ocean.

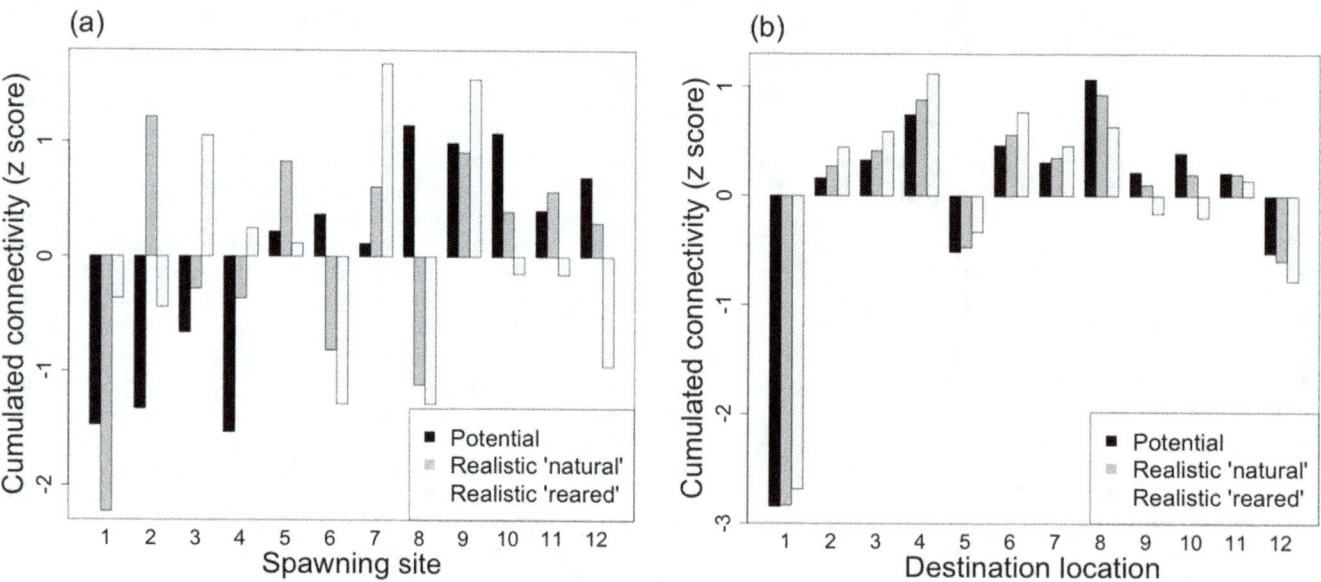

Figure 7. Cumulated connectivity according to (a) the spawning site and (b) the destination location. To compare results with different units, the cumulated connectivity for each scenario (potential and realistic: 'natural' and 'reared') is standardized using the standard score z.

configurations like Poole Harbour where manila clam dispersal has been studied [42]. However, wind significantly contributes to the connectivity variability when coupled with destination location. The destination locations with the lowest connectivity levels are those showing higher variation coefficient when the wind regimes change. Transient unusual connections can occur, like, for example, between spawning site 7 and destination location 10 (i.e., south north connection). These locations are usually weakly connected but the infrequent northeastern winds strengthen the

connections. Our study takes into account only the typical wind regimes occurring around Ahe. Regimes were defined statistically by clustering and from the barycenter of each cluster. Thus, we did not consider extreme events such as storms with wind coming from the west during 2 to 3 days [28]. Specific connectivity patterns related to extreme weather might exist, albeit with a low probability of occurrence during any given year.

The time scales over which larval connectivity varies is extremely important to understand the demographics of marine

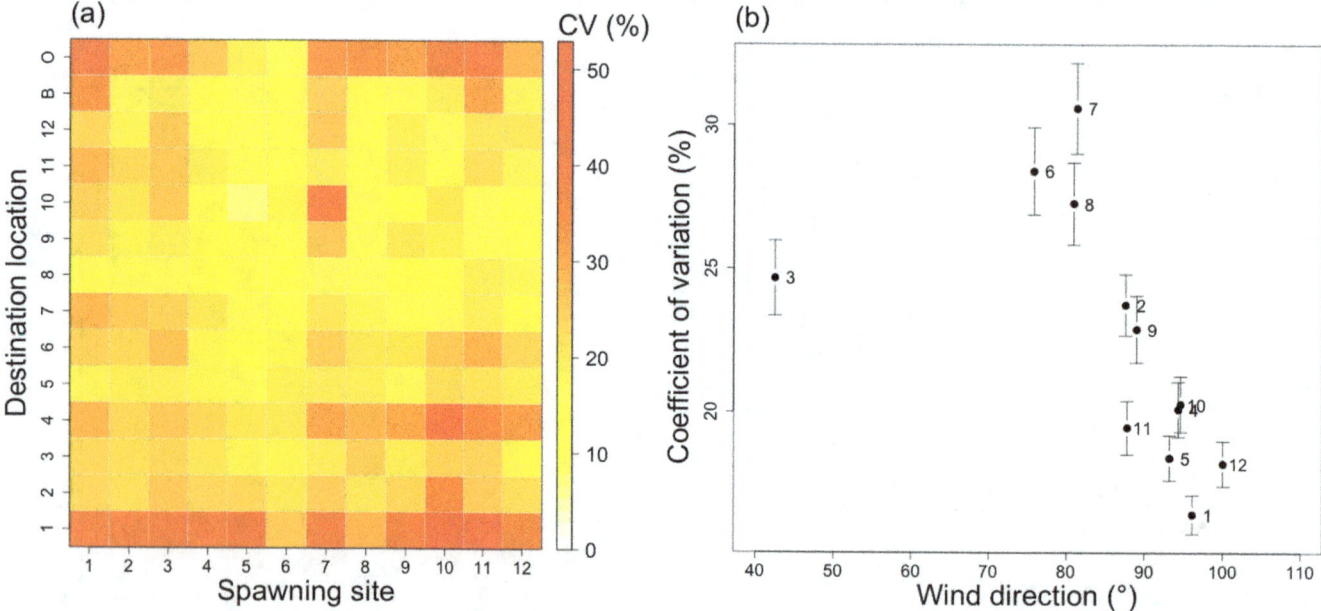

Figure 8. Connectivity variation (CV, %), according to PLD. The mean connectivity variations coefficients are calculated according to the PLD factor (*i.e.*, considering 15, 20, 25 and 30 days), by dividing the connectivity standard deviation calculated according to the PLD factor by the average. The matrix (a) is used to represent the spatial variability of CV, during mean wind conditions (average of the 12 wind scenarios). The graph (b) represents the CV variability according to wind direction. In the latter case, the error bars represent the spatial standard deviation. The numbers on graph (b) correspond to the wind regime indices (see Table 4).

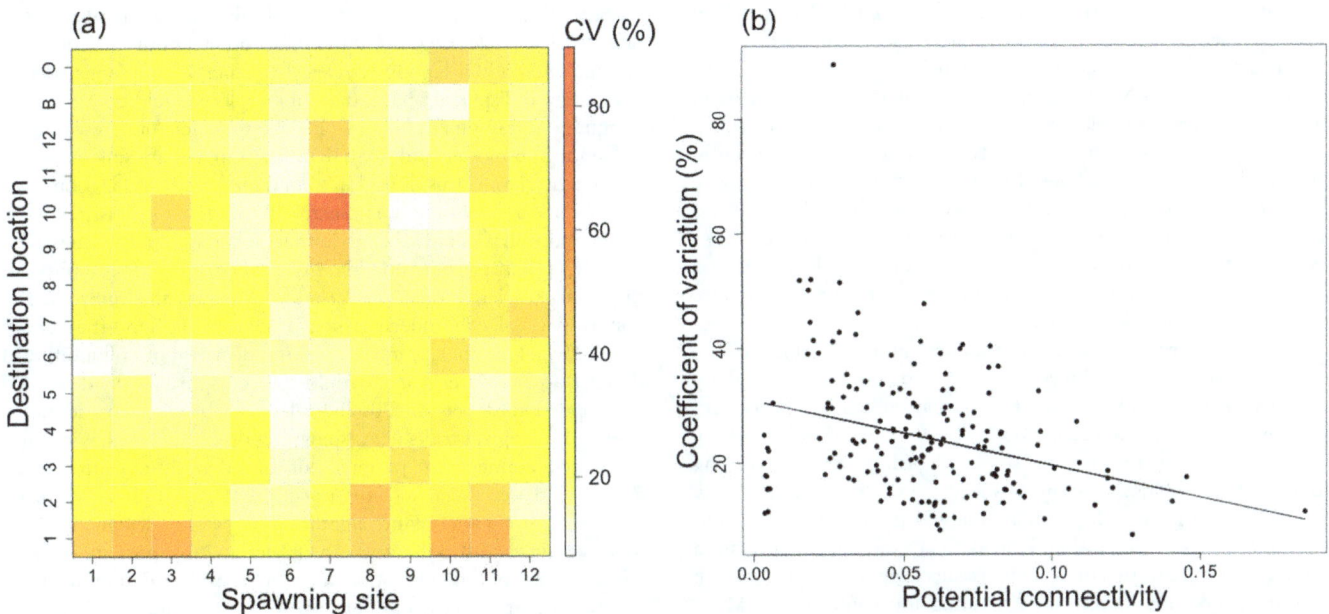

Figure 9. Connectivity variation (CV, %), according to wind. (a) Matrix of the connectivity variation and (b) relation between the potential connectivity and CVs. The coefficients of variation (CV, %) of connectivity were calculated after 20 days of simulations, by dividing the connectivity standard deviation calculated according to the wind factor (12 wind scenarios) by the average. In the destination locations, "B" represents the shallow waters and "O" the open ocean. In (b), the line is the linear regression.

species [13]. In our study, the summer conditions gave the highest levels of larval supply in destination locations, mainly in the west of the Ahe lagoon. Winter and transition periods gave the lowest levels. The optimal dispersal period, giving the highest larval supplies is thus concomitant to the seasonal cycle of observed reproduction patterns. Indeed, reproduction occurs all year long, but with a maximum in austral summer [22], which is also the

season with maximum observed larval concentrations and spat collection [17,43]. However, inter-annual variability in the occurrence of each wind scenario, calculated over 32 years, indicates higher occurrences of winter winds with 9.9% in average, compared to summer and transition winds, with 7.4% and 7.6% in average, respectively (Table 4). These occurrences might be linked to large-scale climatological events, like the El Niño Southern

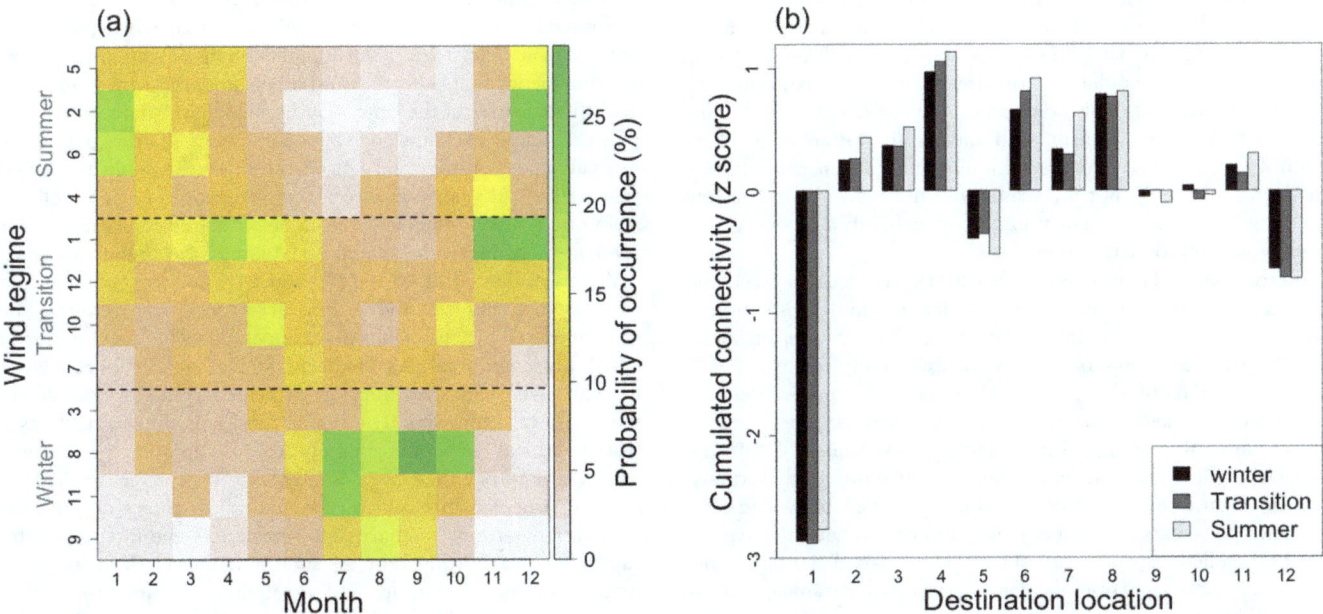

Figure 10. Seasonal pattern of connectivity. (a) Monthly probability of occurrence (in %) of each wind regime and (b) cumulated realistic connectivity, by season. The data used for realistic connectivity are the sum of the reared and natural scenarios. The connectivity is standardized using the standard score z.

Oscillation and La Niña conditions, which may affect physical processes (e.g., increased temperature, upwelling, circulation modification) and thus recruitment intensity [13,44]. In the north Tuamotu sector, El Niño periods are associated with a lower wind directed further north, close to our summer conditions. Conversely, La Niña periods strengthen winds that shift more to the south, close to our winter conditions. El Niño conditions may thus give better dispersal conditions and increase the potential of recruitment throughout Ahe lagoon. However, other cross-effects (e.g., increased temperature) may increase predation or decrease food availability, thus negatively affecting the recruitment potential, and should have to be taken into account [44].

Our aim here was to gain a better understanding of the influence of the broodstock location on larval connectivity patterns. Taking into account the wild broodstock density and location as observed in Takapoto atoll by Zanini and Salvat (2000) [27], as well as the reared stock according to the rearing concessions, provided a more realistic view of the connectivity patterns. Even if the wild stock in Ahe was not exactly structured by depth range and as abundant as in Takapoto, the potential connectivity *vs* realistic connectivity results suggest that broodstock location play a major role in structuring the connectivity. More importantly, the reared broodstock contribution is three times higher than the natural one due to the high densities located on breeding concessions. The spatial distribution of reared stock, more heterogeneous, also induces a strong spatial variability of connectivity. However, both natural and reared scenarios resulted in an increase of the larval supply in western sectors. This result is consistent with professional practices that heavily rely on this sector for spat collecting.

Given the importance of the broodstock location on larval dispersal patterns, the influence of the age/size population structures between the different populations warrants further investigation. Indeed, farmers tend to favor young oysters to harvest pearls of superior quality [45]. This, in fact, promotes a cultured male stock since *P. margaritifera* is protandric hermaphrodite. This gives even more importance to the age and size distribution of the wild population. If the wild stock is declining, or has declined, a cultured stock massively dominated by male individuals may quickly lead to a collapse in spat collecting. Along with the sex/age population structure, the reproductive potential of the populations will also depend on food availability [46]. The introduction of mechanisms and factors that spatially explicit control reproductive effort is an interesting challenge for future modeling work [13]. For this reason, our next efforts will focus on the coupling of the current transport model with a bioenergetics growth and reproduction model.

According to Thomas et al. (2012) [19], the good agreement between larval concentrations simulations and observations suggests that most mortality occurs in the first 2 days of larval life. Hence, an estimation of this early mortality might allow a proper estimate of the real larval supply. The post-settlement mortality also needs to be better quantified. In their study, Friedman et al. (1998) [47] measured a spat mortality of 42% on collector, before harvest, mainly due to predation. This mortality has significant consequences on actual spat availability and also modifies the rearing practices (i.e., timing of harvest, rearing location, collector protection). These points, larval mortality and recruitment are also essential for future model parameterization and validation.

The various points discussed above call for additional research to fine tune the outputs according to a complex series of interacting factors. This will likely be a long path requiring additional field and experimental work. However, results already seem realistic and robust enough to think about generalizing the dedicated *Pinctada margaritifera* model to other important commercial and functional lagoon species. In principle, for other species with a larval dispersal phase, the same toolbox could be used, with adequate parameterization of PLD, stock location, swimming behavior sub-model and spatial units (source and sink sites). In particular, giant clams, found in abundance in several lagoons, are the focus of ongoing management models for fishery and aquaculture that could take advantage of lagoon-scale connectivity models [48,49]. Beyond mollusks, lagoon-scale biophysical dispersal models could be used to assess the intra-lagoon connectivity of echinoderms such as sea cucumbers, the spread of invasive algae species in lagoons, the dispersal of pollutants and derelict aquaculture gear. A multi-species approach as described by Lòpez-Duarte et al. (2012) [50] may also lead to identify a range of biological traits (e.g., spawning time, PLD, behavior) and associated connectivity patterns for broad multi-objective management plans dedicated or not to bottom-dwelling cultured species in atolls and island lagoons. For passive drifters, such as invasive algae and pollutants, the Eulerian transport model and wind regime parameterization alone could help define accumulation risk zones very accurately. However, working in other atolls and lagoons than Ahe imply the acquisition of proper baseline data such as bathymetry, rim geomorphology, wave regime, and current measured in strategic places to set and validate the hydrodynamic model. Then other hydrobiological measurements may be needed according to the gradient of atoll morphology [51]. These are not trivial, simple and cheap tasks in logistically challenging remote places.

Conclusion: Implications for Management

Understanding of the distribution of suitable spawning zones, larval transport processes and connectivity between spawning and nursery grounds are useful information for coastal spatial planning, management and fishery regulation. Our results are in agreement with professional farmers' empirical knowledge, who, for instance, favor the western sectors as nursery grounds (Marine Resource Direction, pers. comm.). What the model brings that was not available before is a better understanding of why some sectors are suitable for spat collecting and others not, and why sectors that are efficient most of the time may be ineffective some years later. Farmers' empirical knowledge after sometimes 30 years of practice is invaluable, yet, they still ignore the reasons leading to a good year and a bad year, and forecasting has proved to be unreliable from one year to another, with major economic consequences for some farmers. Here, the spatial variability of the connectivity (i.e., CV) was evaluated at 45% 44% and 91% for the potential and realistic (i.e., natural and reared scenarios), respectively. The current reared broodstock location appeared as the main source of variability in larval dispersal patterns, a fact that was not conceived by many who witnessed the start of pearl farming at time where the stock was only wild. The use of models is thus valuable as it brings critical new information to sustain the professional practices, improve their plans and collection strategies in the long run. In particular, broodstock location and the choice of collecting areas will determine the harvest performances and then the oyster supply for pearl production. As such, ongoing work includes the transfer of model outputs and possibility to query the model according to specific wind conditions to the French Polynesia Marine Resource Division.

The differences between the potential and realistic connectivity showed the significant contribution of the pearl oyster broodstock location to its own dynamics. As such, spawning sanctuaries are cited as an effective strategy to sustain oyster stock replenishment [52]. The selection of sanctuary areas, with an effective connectivity potential toward sectors traditionally used by professionals, and from a high reproductive potential broodstock (i.e., sex-ratio, maturation efficiency), may optimize spat collection. In Ahe atoll lagoon, the potential connectivity showed the high level of retention in two hydrodynamic cells. An increased broodstock in these two sectors should enhance spat collection potential. As fecundity is highly related to the trophic resource [46], sanctuaries should be located in the richest sectors. In Ahe, the highest levels of plankton concentration are found on the southwestern sector [53], and contribute to the larval supply in the collection area promoted by the farmers. The southwest area might be a suitable sanctuary, allowing an increase in the collecting yield by optimizing the population structure in this sector. In general, this work points to the need to rethink how the adult, non-reared stock, is managed. Instead of routinely discarding shells no longer used for pearl production, it may be wise to promote these sanctuaries in suitable areas. The idea is widely accepted as shown during the restitution of the first modeling work to farmers (in November 2010). The implementation is now in local managers' hands.

Beyond Ahe, the present work offers modeling tools and connectivity concepts applied to a specific aquaculture context. Similar approaches could be applied to lagoons with different or more contrasted weather, different configurations, and higher or lower degrees of exchange with the ocean. The conceptual and technical bases can now be used for further applications in other lagoons, for other scientific and management issues.

Acknowledgments

Authors are very grateful to the Laboratoire des Sciences de l'Environnement Marin (LEMAR), unité Ifremer PFOM/PI, for hosting of YT. We acknowledge I. Bernard, for his conceptual and technical support in R code development. Three anonymous reviewers helped to clarify several aspects of this manuscript.

Author Contributions

Conceived and designed the experiments: YT FD SA. Performed the experiments: YT FD SA. Analyzed the data: YT FD SA. Contributed reagents/materials/analysis tools: YT FD SA. Wrote the paper: YT FD SA.

References

1. Cowen RK, Gawarkiewicz G, Pineda J, Thorrold SR, Werner FE (2007) Population connectivity in marine systems: An overview. Oceanography 20: 14–21.

2. Botsford L, White J, Coffroth M, Paris C, Planes S, et al. (2009) Connectivity and resilience of coral reef metapopulations in marine protected areas: matching empirical efforts to predictive needs. Coral Reefs 28: 327–337. doi:10.1007/s00338-009-0466-z.

3. Lett C, Ayata S-D, Huret M, Irisson J-O (2010) Biophysical modelling to investigate the effects of climate change on marine population dispersal and connectivity. Progress in Oceanography 87: 106–113. doi:10.1016/j.pocean.2010.09.005.

4. Pineda J, Reyns N, Starczak V (2009) Complexity and simplification in understanding recruitment in benthic populations. Population Ecology 51: 17–32. doi:10.1007/s10144-008-0118-0.

5. Hughes TP, Baird AH, Dinsdale EA, Moltschaniwskyj NA, Pratchett MS, et al. (2000) Supply-side ecology works both ways: the link between benthic adults, fecundity, and larval recruits. Ecology 81: 2241–2249.

6. North E, Schlag Z, Hood RR, Li M, Zhong L, et al. (2008) Vertical swimming behavior influences the dispersal of simulated oyster larvae in a coupled particle-tracking and hydrodynamic model of Chesapeake Bay. Mar Ecol Prog Ser 359: 99–115.

7. Thomas Y, Garen P, Pouvreau S (2011) Application of a bioenergetic growth model to larvae of the pearl oyster Pinctada margaritifera L. Journal of Sea Research 66: 331–339. doi:10.1016/j.seares.2011.04.005.

8. Eckman JE (1996) Closing the larval loop: linking larval ecology to the population dynamics of marine benthic invertebrates. Journal of Experimental Marine Biology and Ecology 200: 207–237. doi:10.1016/S0022-0981(96)02644-5.

9. Pante E, Adjeroud M, Dustan P, Penin L, Schrimm M (2006) Spatial patterns of benthic invertebrate assemblages within atoll lagoons: importance of habitat heterogeneity and considerations for marine protected area design in French Polynesia. Aquatic Living Resources 19: 207–217. doi:10.1051/alr:2006021.

10. Hunt H, Scheibling R (1997) Role of early post-settlement mortality in recruitment of benthic marine invertebrates. Mar Ecol Prog Ser 155: 269–301.

11. Metaxas A, Saunders M (2009) Quantifying the "bio–" components in biophysical models of larval transport in marine benthic invertebrates: advances and pitfalls. Biological Bulletin 216: 257–272.

12. Treml E, Halpin P, Urban D, Pratson L (2008) Modeling population connectivity by ocean currents, a graph-theoretic approach for marine conservation. Landscape Ecology 23: 19–36. doi:10.1007/s10980-007-9138-y.

13. Watson JR, Mitarai S, Siegel DA, Caselle JE, Dong C, et al. (2010) Realized and potential larval connectivity in the Southern California Bight. Marine Ecology Progress Series 401: 31–48.

14. Edwards KP, Hare JA, Werner FE, Seim H (2007) Using 2-dimensional dispersal kernels to identify the dominant influences on larval dispersal on continental shelves. Mar Ecol Prog Ser 352: 77–87.

15. Levin LA (2006) Recent progress in understanding larval dispersal: new directions and digressions. Integr Comp Biol 46: 282–297.

16. Andréfouët S, Charpy L, Lo-Yat A, Lo C (2012) Recent research for pearl oyster aquaculture management in French Polynesia. Marine Pollution Bulletin 65: 407–414. doi:10.1016/j.marpolbul.2012.06.021.

17. Thomas Y, Garen P, Bennett A, Le Pennec M, Clavier J (2012) Multi-scale distribution and dynamics of bivalve larvae in a deep atoll lagoon (Ahe, French Polynesia). Marine Pollution Bulletin 65: 453–462. doi:10.1016/j.marpolbul.2011.12.028.

18. Dumas F, Le Gendre R, Thomas Y, Andréfouët S (2012) Tidal flushing and wind driven circulation of Ahe atoll lagoon (Tuamotu Archipelago, French Polynesia) from in situ observations and numerical modelling. Marine Pollution Bulletin 65: 425–440. doi:10.1016/j.marpolbul.2012.05.041.

19. Thomas Y, Le Gendre R, Garen P, Dumas F, Andréfouët S (2012) Bivalve larvae transport and connectivity within the Ahe atoll lagoon (Tuamotu Archipelago), with application to pearl oyster aquaculture management. Marine Pollution Bulletin 65: 441–452. doi:10.1016/j.marpolbul.2011.12.027.

20. Tranter D (1958) Reproduction in Australian Pearl Oysters (Lamellibranchia). IV. Pinctada margaritifera (Linnaeus). Mar Freshwater Res 9: 509–525.

21. Thielley M (1993) Etude cytologique de la gamétogenèse, du la sex-ratio et du cycle de reproduction chez l'huître perlière Pinctada margaritifera (L) var. cumingii (Jameson) Université de Polynésie Française.

22. Pouvreau S, Gangnery A, Tiapari J, Lagarde F, Garnier M, et al. (2000) Gametogenic cycle and reproductive effort of the tropical blacklip pearl oyster, Pinctada margaritifera (Bivalvia: Pteriidae), cultivated in Takapoto atoll (French Polynesia). Aquatic Living Resources 13: 37–48. doi:10.1016/S0990-7440(00)00135-2.

23. Sims NA (1994) Growth of wild and cultured black-lip pearl oysters, Pinctada margaritifera (L.) (Pteriidae; Bivalvia), in the Cook Islands. Aquaculture 122: 181–191. doi:10.1016/0044-8486(94)90509-6.

24. Pouvreau S, Prasil V (2001) Growth of the black-lip pearl oyster, Pinctada margaritifera, at nine culture sites of French Polynesia: synthesis of several sampling designs conducted between 1994 and 1999. Aquatic Living Resources 14: 155–163. doi:10.1016/S0990-7440(01)01120-2.

25. Intes A (1982) Programme nacre. Etude des stocks naturels. Rapport de mission, atoll de Manihi. ORSTOM et Service de la Pêche de Polynesie francaise.

26. Intes A, Laboute P, Coeroli M (1985) Le stock naturel de nacre (Pinctada margaritifera L.) dans l'atoll de Scilly (Archipel de la Société, Polynésie française). Notes et Documents Oceanographiques. ORSTOM Tahiti.

27. Zanini JM, Salvat B (2000) Assessment of deep water stocks of pearl oysters at Takapoto Atoll (Tuamotu Archipelago, French Polynesia). Coral Reefs 19: 83–87. doi:10.1007/s003380050231.

28. Andréfouët S, Ardhuin F, Queffeulou P, Le Gendre R (2012) Island shadow effects and the wave climate of the Western Tuamotu Archipelago (French Polynesia) inferred from altimetry and numerical model data. Marine Pollution Bulletin 65: 415–424. doi:10.1016/j.marpolbul.2012.05.042.

29. Doroudi MS, Southgate PC (2003) Embryonic and larval development of Pinctada margaritifera (Linnaeus, 1758). Molluscan Res 23: 101–107.

30. Lazure P, Dumas F (2008) An external-internal mode coupling for a 3D hydrodynamical model for applications at regional scale (MARS). Advances in Water Resources 31: 233–250. doi:10.1016/j.advwatres.2007.06.010.

31. Burgess SC, Treml EA, Marshall DJ (2012) How do dispersal costs and habitat selection influence realized population connectivity? Ecology 93: 1378–1387.

32. Lefèvre J, Marchesiello P, Jourdain NC, Menkes C, Leroy A (2010) Weather regimes and orographic circulation around New Caledonia. Marine Pollution Bulletin 61: 413–431. doi:10.1016/j.marpolbul.2010.06.012.

33. Deacon EL, Webb EK (n.d.) Small scale interactions. The Sea 1: 43–87.

34. Maechler M, Rousseeuw P, Struyf A, Hubert M, Hornik K (2012) cluster: Cluster Analysis Basics and Extensions. R package version 1142.

35. R Development Core Team (2012) R: A language and environment for statistical computing. R Foundation for Statistical Computing, Vienna, Austria http://wwwR-project.org/. Available: http://www.R-project.org/.

36. Cowen RK, Paris CB, Srinivasan A (2006) Scaling of Connectivity in Marine Populations. Science 311: 522–527.

37. Adjeroud M, Andréfouët S, Payri C, Orempüller J (2000) Physical factors of differentiation in macrobenthic communities between atoll lagoons in the Central Tuamotu Archipelago (French Polynesia). Mar Ecol Prog Ser 196: 25–38.

38. Andréfouët S, Ouillon S, Brinkman R, Falter J, Douillet P, et al. (2006) Review of solutions for 3D hydrodynamic modeling applied to aquaculture in South Pacific atoll lagoons. Marine Pollution Bulletin 52: 1138–1155. doi:10.1016/j.marpolbul.2006.07.014.

39. Treml EA, Roberts JJ, Chao Y, Halpin PN, Possingham HP, et al. (2012) Reproductive Output and Duration of the Pelagic Larval Stage Determine Seascape-Wide Connectivity of Marine Populations. Integrative and Comparative Biology 52: 525–537.

40. Dekshenieks MM, Hofmann EE, Klinck JM, Powell EN (1996) Modeling the vertical distribution of oyster larvae in response to environmental conditions. Marine Ecology Progress Series 136: 97–110.

41. Kim C-K, Park K, Powers SP, Graham WM, Bayha KM (2010) Oyster larval transport in coastal Alabama: Dominance of physical transport over biological behavior in a shallow estuary. J Geophys Res 115: n/a–n/a. doi:10.1029/2010JC006115.

42. Herbert RJH, Willis J, Jones E, Ross K, Hübner R, et al. (2012) Invasion in tidal zones on complex coastlines: modelling larvae of the non-native Manila clam, *Ruditapes philippinarum*, in the UK. Journal of Biogeography 39: 585–599. doi:10.1111/j.1365-2699.2011.02626.x.

43. Brié C (1999) Étude expérimentale du collectage de naissain de *Pinctada margaritifera* (Linné, 1758) à Takapoto, atoll des Tuamotu, en Polynésie Française [Mémoire]. EPHE.

44. Gaymer CF, Palma AT, Vega JMA, Monaco CJ, Henriquez LA (2010) Effects of La Niña on recruitment and abundance of juveniles and adults of benthic community-structuring species in northern Chile. Marine Freshwater Ressources 61: 1185–1196.

45. Chávez-Villalba J, Soyez C, Huvet A, Gueguen Y, Lo C, et al. (2011) Determination of Gender in the Pearl Oyster *Pinctada margaritifera*. Journal of Shellfish Research 30: 231–240. doi:10.2983/035.030.0206.

46. Fournier J, Levesque E, Pouvreau S, Pennec ML, Moullac GL (2012) Influence of plankton concentration on gametogenesis and spawning of the black lip pearl oyster *Pinctada margaritifera* in Ahe atoll lagoon (Tuamotu archipelago, French polynesia). Marine Pollution Bulletin 65: 463–470. doi:10.1016/j.marpolbul.2012.03.027.

47. Friedman KJ, Bell JD, Tiroba G (1998) Availability of wild spat of the blacklip pearl oyster, *Pinctada margaritifera*, from 'open' reef systems in Solomon Islands. Aquaculture 167: 283–299. doi:10.1016/S0044-8486(98)00286-5.

48. Andréfouët S, Van Wynsberge S, Gaertner-Mazouni N, Menkes C, Gilbert A, et al. (2013) Climate variability and massive mortalities challenge giant clam conservation and management efforts in French Polynesia atolls. Biological Conservation 160: 190–199. doi:10.1016/j.biocon.2013.01.017.

49. Van Wynsberge S, Andréfouët S, Gilbert A, Stein A, Remoissenet G (2013) Best Management Strategies for Sustainable Giant Clam Fishery in French Polynesia Islands: Answers from a Spatial Modeling Approach. PLoS ONE 8: e64641. doi:10.1371/journal.pone.0064641.

50. López-Duarte PC, Carson HS, Cook GS, Fodrie FJ, Becker BJ, et al. (2012) What Controls Connectivity? An Empirical, Multi-Species Approach. Integrative and Comparative Biology 52: 511–524.

51. Dufour P, Andréfouët S, Charpy L, Garcia N (2001) Atoll morphometry controls lagoon nutrient regime. Limnology and Oceanography 46: 456–461.

52. Schulte DM, Burke RP, Lipcius RN (2009) Unprecedented restoration of a native oyster metapopulation. Science 325: 1124–1128.

53. Charpy L, Rodier M, Fournier J, Langlade M-J, Gaertner-Mazouni N (2012) Physical and chemical control of the phytoplankton of Ahe lagoon, French Polynesia. Marine Pollution Bulletin 65: 471–477. doi:10.1016/j.marpolbul.2011.12.026.

A Regional-Scale Ocean Health Index for Brazil

Cristiane T. Elfes[1]*, **Catherine Longo**[2], **Benjamin S. Halpern**[2,3,4], **Darren Hardy**[2,5], **Courtney Scarborough**[2], **Benjamin D. Best**[2], **Tiago Pinheiro**[6], **Guilherme F. Dutra**[7]

1 Department of Ecology, Evolution and Marine Biology, University of California Santa Barbara, Santa Barbara, California, United States of America, 2 National Center for Ecological Analysis and Synthesis, Santa Barbara, California, United States of America, 3 Bren School of Environmental Science and Management, University of California Santa Barbara, Santa Barbara, California, United States of America, 4 Imperial College London, Silwood Park Campus, Berkshire, London, United Kingdom, 5 Digital Library Systems, Stanford University, Stanford, California, United States of America, 6 Atlantic Forest Program, Conservation International Brazil, Belo Horizonte, Minas Gerais, Brazil, 7 Marine Program, Conservation International Brazil, Rio de Janeiro, Rio de Janeiro, Brazil

Abstract

Brazil has one of the largest and fastest growing economies and one of the largest coastlines in the world, making human use and enjoyment of coastal and marine resources of fundamental importance to the country. Integrated assessments of ocean health are needed to understand the condition of a range of benefits that humans derive from marine systems and to evaluate where attention should be focused to improve the health of these systems. Here we describe the first such assessment for Brazil at both national and state levels. We applied the Ocean Health Index framework, which evaluates ten public goals for healthy oceans. Despite refinements of input data and model formulations, the national score of 60 (out of 100) was highly congruent with the previous global assessment for Brazil of 62. Variability in scores among coastal states was most striking for goals related to mariculture, protected areas, tourism, and clean waters. Extractive goals, including Food Provision, received low scores relative to habitat-related goals, such as Biodiversity. This study demonstrates the applicability of the Ocean Health Index at a regional scale, and its usefulness in highlighting existing data and knowledge gaps and identifying key policy and management recommendations. To improve Brazil's ocean health, this study suggests that future actions should focus on: enhancing fisheries management, expanding marine protected areas, and monitoring coastal habitats.

Editor: Athanassios C. Tsikliras, Aristotle University of Thessaloniki, Greece

Funding: Beau and Heather Wrigley provided the founding grant for the original Ocean Health Index work and the Pacific Life Foundation is the Founding Presenting Sponsor. Additional financial and in-kind support was provided by the Thomas W. Haas Fund of the New Hampshire Charitable Foundation, the Oak Foundation, Akiko Shiraki Dynner Fund for Ocean Exploration and Conservation, Darden Restaurants Inc. Foundation, Conservation International, New England Aquarium, National Geographic, and the National Center for Ecological Analysis and Synthesis, which supported the Ecosystem Health Working Group as part of the Science of Ecosystem-Based Management project funded by the David and Lucile Packard Foundation. The funders had no role in study design, data collection and analysis, decision to publish, or preparation of the manuscript.

Competing Interests: The authors have declared that no competing interests exist.

* E-mail: cristiane.elfes@lifesci.ucsb.edu

Introduction

Brazil's coastline spans more than 7,000 km with a vast diversity of ecosystems, including extensive mangrove areas in the Amazon basin, coral reefs in the Northeast, and lagoons, estuaries and saltmarshes in the south. These systems play a fundamental role in the economy and identity of the country. As Brazil's economy continues to grow – for 2012 it was listed as the seventh largest economy in the world [1] – interest in using and benefiting from coastal and marine systems is also growing. Often the various activities tied to these benefits, such as fisheries, coastal development and tourism, come into conflict, and resource managers and policy makers are faced with decisions about how and where to allow and regulate each of them, with the ultimate goal of maintaining and ideally improving the overall health of the ocean and the communities that use it.

Given this context, there is a great need in Brazil for tools to assess and monitor the overall health of coastal ecosystems, as well as the status of components of the system. A framework was recently developed to do just that, and was applied to every coastal country in the world [2]. This index to assess the health and benefits of the ocean (Ocean Health Index) evaluates the condition of coupled human-ocean systems by tracking the current status and likely future state of ten publicly held goals, ranging from food provision to jobs, tourism, and coastal protection (Table 1).

The Index is based on the understanding that humans are part of ecosystems and that the health of natural and human systems are tightly coupled [3,4]. From this coupled human-natural systems perspective, a healthy ocean is defined as one that provides a range of benefits to people now and in the future [2]. As such, the Index measures the amount of benefits relative to a sustainable optimum. The Index is not intended to be a measure of how pristine an area of ocean or coastline is.

The novelty of the Index is that it provides an integrated framework in which to quantitatively assess and compare the condition of these benefits, thus providing a portfolio perspective useful for informing management decisions. The Index can also be used to track progress in achieving specific management goals, because it establishes a target or reference point to which current status and likely future condition are compared ([2,5], Table S1 in Text S1).

Here we present a case study, applying the Ocean Health Index framework to Brazil at the national and sub-national levels. The

Table 1. Ten public goals and sub-goals showing benefit measured under each.

Goal	Subgoal	Benefit measured
Food Provision (FP)	Fisheries (FIS)	Seafood sustainably harvested for human consumption from wild, or cultured stocks
	Mariculture (MAR)	
Artisanal fishing opportunity (AO)		Opportunity to engage in artisanal fishing as a social, cultural and livelihood activity
Natural products (NP)		Amount of sustainably harvested natural products (other than for food provision)
Carbon storage (CS)		Conservation of coastal habitats affording carbon storage and sequestration
Coastal protection (CP)		Conservation of coastal habitats affording protection from inundation and erosion
Tourism and recreation (TR)		Opportunity to enjoy coastal areas for recreation for locals and tourists
Coastal livelihoods and economies (LE)	Livelihoods (LIV)	Employment (livelihoods) and revenues (economies) from marine-related sectors
	Economies (ECO)	
Sense of place (SP)	Iconic species (ICO)	Sense of place and cultural connectedness to the ocean afforded by lasting special places and iconic species
	Lasting special places (LSP)	
Clean waters (CW)		Clean waters that are free from pollution, debris and safe to swim in
Biodiversity (BD)	Habitats (HAB)	Conservation of biodiversity of species and habitats for their existence value
	Species (SPP)	

global analysis [2] precluded use of higher resolution datasets that are available for individual countries, data that can provide a more accurate assessment of a country's ocean health as well as sub-national assessments. As such, the global analysis is too coarse to guide specific interventions at national and regional levels, particularly for a country as large and heterogeneous as Brazil.

Applying the Ocean Health Index to Brazil provides an important opportunity to test the scalability and flexibility of the Index to be adapted to country-specific concerns by including higher resolution information, place-specific targets and regional proxies for calculating goals. The case study also highlights a number of challenges related to data quality and quantity for assessing the range of benefits evaluated under the Index framework. Here, we show how the Index can be adapted to the Brazilian context, and discuss the main patterns and policy implications emerging from our analysis. Our intent is that the lessons learned from this case study can be used to guide future assessments and management strategies in Brazil, and help to inform other current and future regional applications of the Ocean Health Index.

Methods

Details on calculation of the Index are provided in Halpern *et al.* [2]. Here we give a brief summary, and elaborate on goal-specific methods and data layers used in this case study in Text S1.

The Index is comprised of ten widely-held public goals: Food Provision, Artisanal Opportunities, Natural Products, Carbon Storage, Coastal Protection, Coastal Livelihoods and Economies, Tourism and Recreation, Sense of Place, Clean Waters and Biodiversity (Table 1). As the Index is focused on the sustainable provision of benefits, we do not include activities such as oil and gas exploration. The location of oil and gas deposits and productivity of such reserves are not indications of a healthy, or sustainably managed ocean. It is worth noting, however, that oil is incorporated as a pollutant in the pressure calculations (see below) and the status of the Clean Waters goal.

For each goal, a score is calculated from four dimensions – current status, recent trend, existing pressures and expected resilience in the near-term based on current management actions. The Index value (I) is determined as a linear weighted sum of the scores for each of the public goal indices ($I_1, I_2, ..., I_{10}$) and the appropriate weights for each of the goals ($\alpha_1, \alpha_2, ... \alpha_{10}$), such that:

$$I = \alpha_1 I_1 + \alpha_2 I_2 + ... \alpha_{10} I_{10} \tag{1}$$

The weights determine the relative importance of each goal in the overall Index score and ideally reflect people's values within the region. Here we used equal weighting, as an in-depth interview process with stakeholders from all Brazilian coastal states was outside the scope of this case study (for an example, see [6]).

Each goal score, I_i, is calculated as the average of its present status x_i, and an estimate of its likely near-term future status $\hat{x}_{i,F}$, such that:

$$I_i = \frac{x_i + \hat{x}_{i,F}}{2} \tag{2}$$

The present status of goal i, x_i, is its present status value, (X_i), relative to a reference point, $X_{i, R}$ uniquely chosen for each goal and scaled 0 to 100.

$$x_i = \frac{X_i}{X_{i,R}} \tag{3}$$

The reference point, $X_{i, R}$, is determined a number of ways depending on the purpose (management objective) and data constraints of each goal. The main ways of establishing a reference point are: through a known functional relationship (e.g. a target value of extracting the maximum sustainable yield of a given fish stock), a time series approach (e.g. historical habitat extent), a spatial comparison (e.g. the country with highest wages in marine-related sectors), or through a known or established target value (e.g. no species at risk of extinction, or 30% of marine waters designated in protected areas). A more detailed discussion of the considerations and process for selecting reference points is found in Samhouri *et al.* [5], and in Text S1 for Halpern *et al.* [2]. Our case study also used a spatio-temporal comparison, in which the present status of a goal for all Brazilian coastal states was compared to the best performing state over the analysis period. For example, the Tourism and Recreation goal uses as its reference value the highest score achieved across all states and all years of data available (i.e. Rio de Janeiro in 2011). A full description of the goal-specific reference points is provided in the Supporting Information and listed in Table S1 in Text S1.

The likely near-term future status of a goal, $\hat{x}_{i,F}$, is given as:

$$\hat{x}_{i,F} = (1+\delta)^{-1}[1+\beta T_i + (1-\beta)(r_i-p_i)]x_i \qquad (4)$$

where r_i is Resilience, p_i is Pressures, and T_i is the Trend. A discount rate (δ) was included in the equation, but was approximated to 0, because the likely future state is an assessment in the very near future [2]. Beta (β) represents the relative importance of the Trend versus the Resilience and Pressure terms in determining the likely trajectory of the goal status into the future. We assume $\beta = 0.67$ based on the idea that the direct measure of Trend is a better indicator of the near future than the indirect measures of Pressure and Resilience, and therefore carries twice the weight [2].

Trend is calculated as the change in Status (slope) over the previous five years. The annual rate of change was multiplied by five to give an estimation of the Status in the near-term future [2]. To calculate Pressures for each goal (p_x) we evaluate both ecological (p_E) and social pressures (p_S). Ecological pressures are comprised of five broad categories: fishing pressure, habitat destruction, climate change, water pollution and species introductions. The contribution of individual pressures to the overall ecological pressure score is based on a weighting scheme having pressures ranked as 'high' (weight = 3), 'medium' (weight = 2), and 'low' (weight = 1) impacts on the goal, sub-goal, or component (see Table S7 in Text S1). Ecological pressures (p_E) are calculated as the weighted-average of the Pressure categories relevant to each goal. Rankings were determined by literature review [2], modified slightly through expert judgment on Brazilian systems. Social pressures were based on a metric developed by The Economist Intelligence Unit (UIE; see Text S1), which ranks management effectiveness in Brazilian states. The UIE index is comprised of eight categories: Political Environment, Economic Environment, Tributary and Regulatory Environment, Policies for International Investment, Human Resources, Infrastructure, Innovation, and Sustainability determined at the state level. We used the aggregate score of all components of the UIE index for each coastal state.

The Pressure for each goal (p_x) is therefore calculated as:

$$P_x = \gamma * (P_E) + (1-\gamma) * (P_S) \qquad (5)$$

where γ is the relative weight for ecological vs. social pressures and is set equal to 0.5. Total Pressure scores range between 0 and 100 with 100 being the highest threat.

To calculate Resilience (Table S8 in Text S1) for each goal (r_x) we assess three types of measures: ecological integrity (Y_E), goal specific regulations aimed at addressing ecological pressures (G),

and social integrity (Y_S). When all three aspects are relevant to a goal, Resilience is calculated as:

$$r_x = \gamma * \left(\frac{Y_E + G}{2}\right) + (1-\gamma) * Y_S \qquad (6)$$

where the three measures are scaled 0–100, and gamma is assumed to be 0.5 (such that ecological and social Resilience components are equivalent). Ecological integrity (Y_E) is measured as the relative condition of assessed marine species (see Text S1), regulations (G) are laws and institutional measures that support that goal and is calculated as the weighted average of those measures, and social integrity measures (Y_S) is simply the UIE Index.

The Index was calculated for each Brazilian coastal state (Figure 1) and for the entire country as an area-weighted average of coastal state scores. The scale of goal or sub-goal calculations was dependent on data resolution. Based on available input data and goal formulation, the spatial scale and geographic domain of analysis differed between goals (Table 2). When possible, we took advantage of state-level statistics such as tourism data, population census counts, and habitat data with direct relevance to the state's terrestrial coastline and coastal waters (0–12 nmi offshore). National level data included variables such as national level statistics, and data pertaining to the entire Brazilian EEZ (0–200 nmi). When only national data were available, the values for the goal's status and trends across states were identical and any variation in final score (Table 3) was due to the influence of pressure and resilience, which differed between states.

To see how the Ocean Health Index compares with another across-sector index, we compared current versus likely future status scores for each state with an independent metric used in Brazil to track development status (FIRJAN Development Index score, IFDM). IFDM is an index of human development, measured in three areas: jobs and income, education, and health, providing a useful comparison to our evaluation of ocean health.

Results

The overall Index score for Brazil was 60 out of 100, with state-level scores ranging from 47 to 71 (Table 3). Highest scoring goals were those relating to habitat condition, including Coastal Protection (score: 92) and Carbon Storage (89). The Biodiversity score for Brazil was 85, averaged across Habitats (95) and Species (74) sub-goals.

Mariculture (6) scored lowest in the Index at the national level. Other extractive goals or sub-goals, such as Natural Products (29) and Fisheries (42) were also low. Iconic species (47), scored lower than the Species sub-goal, indicating that a high proportion of culturally and aesthetically valued species are threatened (Table 3).

Goals and sub-goals for which state-level data were used showed high variability among regions (Table 3, Figure 2). The most variable scores among states were for Tourism and Recreation, which ranged from a low of 1 in Pará to 100 in Rio de Janeiro. Similarly, Lasting Special Places ranged from 10 (Piauí) to 98 (Amapá), and the Mariculture sub-goal ranged from 0 (Maranhão, Pará, Rio de Janeiro and Rio Grande do Sul) to 66 (Santa Catarina). Clean Waters scored highest in Amapá (90) and São Paulo (95), and lowest in Piauí (31) and Sergipe (47).

Scores for the Artisanal Opportunity goal (62) and Livelihoods (56) and Economies (48) sub-goals were low, but were evaluated at the country-level, likely masking important regional differences. Similarly, Carbon Storage (89), Coastal Protection (92) and the

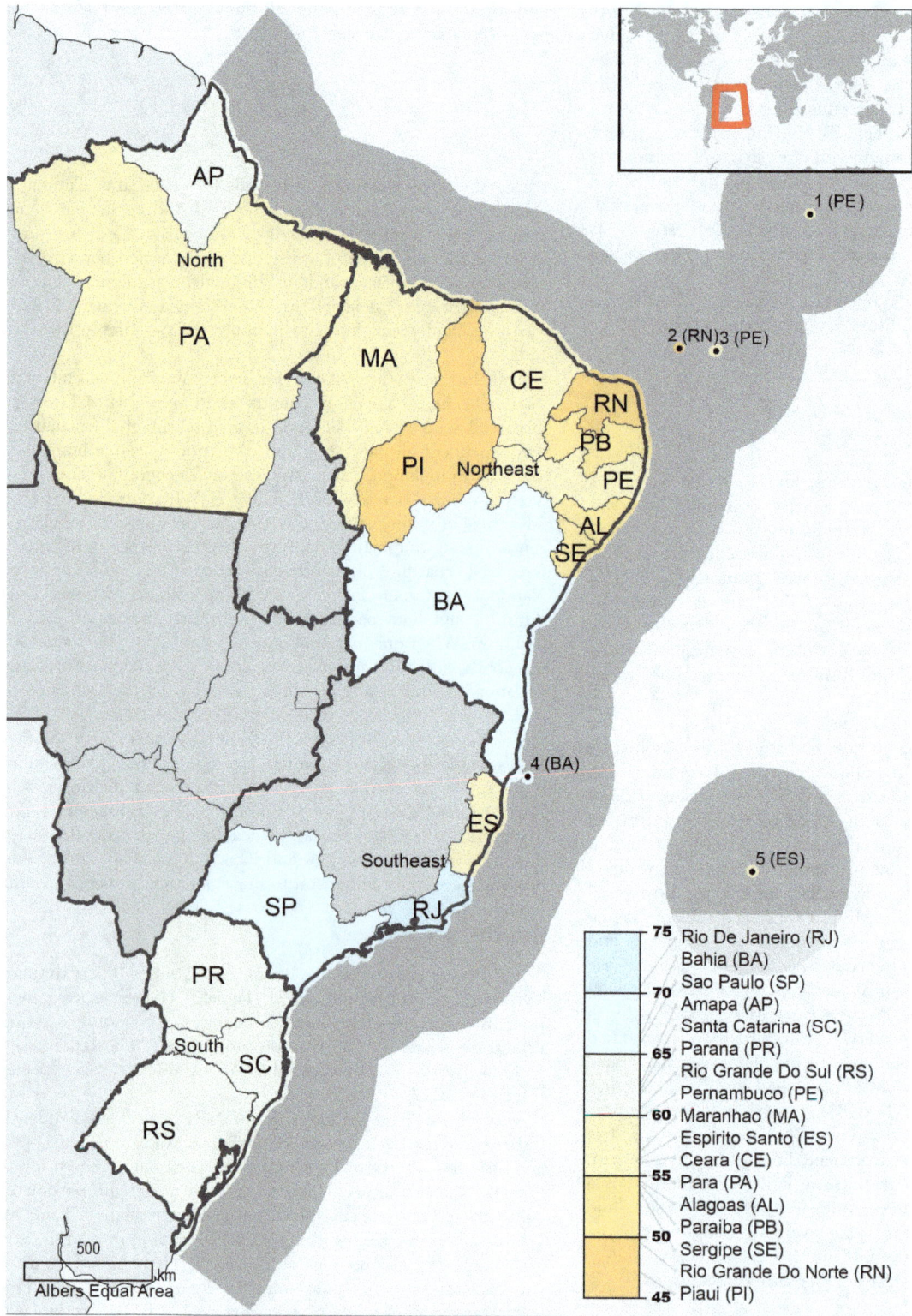

Figure 1. Brazil study region showing coastal states, colored by final OHI score and listed in legend by ranked score. The following islands were considered within the jurisdiction of states specified in parenthesis: 1. São Pedro & São Paulo Archipelago (PE), 2. Rocas Atoll (RN), 3. Fernando de Noronha (PE), 4. Abrolhos Archipelago (BA), 5. Trindade & Martim Vaz (ES).

Table 2. Spatial resolution and geographic domain of goal calculations based on available input data and goal formulation.

Spatial resolution	Goal or sub-goal	Geographic Domain		
		Terrestrial coastline	Coastal waters (0–12 nmi)*	Federal waters (0–200 nmi)
Coastal State	Clean Waters	x	x	
	Tourism and Recreation	x		
	Mariculture (FP)	x	x	
	Lasting Special Places (SP)	x	x	
Mixed State and National	Carbon Storage	x	x	
	Coastal Protection	x	x	
	Habitats (BD)	x	x	
National	Fisheries (FP)			x
	Artisanal Opportunities			x
	Natural Products			x
	Livelihoods (LE)	x		x
	Economies (LE)	x		x
	Iconic Species (SP)			x
	Species (BD)			x

Where sub-goals are shown, the respective goal is indicated in brackets (for acronyms see Table 1).

Habitats sub-goal of Biodiversity (95) showed high scores, with little variation among regions (Figure 3). Habitat data varied greatly in quality and quantity. Effects of these data constraints on the patterns we observed are discussed below.

Comparisons of current and likely future status scores for each state's combined Index score revealed that the level of development of a state (assessed using the independent measure of development status, IFDM) was influential in determining the state's likely future score (Figure 4). The most developed states, São Paulo and Rio de Janeiro, are likely to improve their scores into the future. Paraná, Rio Grande do Sul, Santa Catarina and Bahia would likely maintain similar scores, while the remaining states are expected to have lower future scores.

Discussion

This study is the first integrated assessment of the health of Brazil's ocean, based on the Ocean Health Index framework [2], and incorporating regional datasets. In the following sections we discuss sub-national patterns, lessons learned from this case study analysis, and key policy implications for Brazil.

Spatial patterns

Differences in Food Provision scores between coastal states were driven by the Mariculture sub-goal (Table 3), as the Fisheries sub-goal was evaluated only at the national level. Despite the importance of wild-capture fisheries to coastal communities in Brazil, fisheries monitoring data is historically deficient, and it was not possible to determine state-specific landings. Mariculture scores were generally low due to low production (i.e. opportunity lost relative to potential production from mariculture) or production of unsustainable species. For example, whiteleg shrimp (*Litopenaeus vannamei*) was the most commonly cultivated species, with high production levels in the Northeast region, in particular Ceará and Rio Grande do Norte. However, these states did not achieve high Mariculture scores (Figure 2) due to the low sustainability of production for this species. Indeed, environmental

and social problems associated with shrimp farming in Brazil are numerous, including severe mangrove loss, coastal erosion, pollution, land-use conflicts and loss of traditional livelihoods [7,8]. Highest scores were achieved by Santa Catarina (score: 66) and Paraná (27), the two states with highest landings of bivalves relative to coastline length.

Habitat-based goals, including Carbon Storage, Coastal Protection and the Habitat sub-goal of Biodiversity, scored high across most states, with the exception of Rio Grande do Norte (Table 3, Figure 3) which has seen high rates of mangrove loss due to rapid expansion of shrimp farms (see Text S1). The high and relatively homogenous habitat-related scores are likely related to two reasons. The first is that for marine habitats such as coral reefs and seagrasses, there were challenges in obtaining data at the state-level, and historic reference points were not available within Brazil, such that Caribbean or South Atlantic averages were used. Such averages likely masked important localized declines, the result being remarkably similar scores between states (Figure 3). The second reason is related to Federal Law providing protection to mangrove and saltmarsh habitats. This situation may soon drastically change, as a recent revision of the Brazilian Forest Code legislation opens the possibility of using the salt flat portions of mangroves (locally known as "apicuns") for mariculture, up to 10% in the Amazon Biome and 35% in the remaining coastal regions of the country (Brazilian Federal Law 12,651 of 2012).

The Lasting Special Places sub-goal was assessed using a national database of protected areas (including fully-protected and sustainable use designations at federal, state and municipal levels) and Indigenous lands. The remote state of Amapá achieved a score of 98, almost reaching the target value of 30% protection of the coastal zone (Table 3, Figure S1 in Text S1). Amapá contains the largest continuous extent of protected areas within the country in what is called the Biodiversity Corridor of Amapá. In the coastal zone, large areas have been set aside as fully protected and contain representative ecosystems of the Amazonian region, including the greatest extent of preserved mangroves in the Americas. Relatively high scores were also achieved by São Paulo (75), Bahia (69) and

Table 3. Overall Index, goal and sub-goal scores for Brazil (country) and each Brazilian coastal state.

Region		FP			AO	NP	CS	CP	LE			TR	SP			CW	BD	
	Index	FIS	FP	MAR					LIV	LE	ECO		ICO	SP	LSP		HAB	BD	SPP
Brazil	60	42	36	6	62	29	89	92	56	52	48	31	47	48	48	77	95	85	74
Alagoas (AL)	55	40	33	1	59	28	90	89	55	51	46	22	46	33	20	60	94	82	70
Amapá (AP)	62	42	42		62	28	93	94	54	50	46	3	47	73	98	90	96	85	74
Bahia (BA)	66	41	34	1	61	29	93	93	56	52	48	88	47	58	69	71	97	85	73
Ceará (CA)	56	41	36	12	60	29	75	76	55	51	47	34	47	35	24	85	90	81	73
Espírito Santo (ES)	57	42	35	3	61	29	95	94	56	52	48	15	47	38	28	62	97	85	74
Maranhão (MA)	57	40	34	0	60	28	87	88	55	50	46	9	46	53	60	79	93	82	72
Pará (PA)	55	41	34	0	60	28	92	93	55	50	46	1	46	37	29	74	96	84	72
Paraíba (PB)	55	40	33	1	59	28	87	89	55	51	46	11	46	44	43	62	93	82	71
Pernambuco (PE)	60	41	34	2	60	29	85	88	56	52	48	58	47	41	35	70	94	83	73
Piauí (PI)	47	40	33	1	59	27	81	82	54	50	45	2	45	27	10	31	91	80	69
Paraná (PR)	60	42	40	27	63	29	95	96	56	53	49	3	48	53	59	85	99	87	76
Rio De Janeiro (RJ)	71	44	36	0	65	30	99	99	57	54	50	100	50	57	65	77	99	88	78
Rio Grande Do Norte (RN)	50	40	34	5	59	28	33	74	55	50	46	33	46	32	17	79	77	74	71
Rio Grande Do Sul (RS)	60	43	36	0	63	30	100	100	57	53	49	5	49	42	35	84	100	88	77
Santa Catarina (SC)	62	42	46	66	62	29	93	94	56	52	49	37	48	39	31	77	99	87	75
Sergipe (SE)	54	40	34	2	60	28	89	90	55	51	47	11	46	45	45	47	95	83	71
São Paulo (SP)	66	45	37	1	66	30	97	97	58	54	51	29	51	63	75	95	99	89	80

Empty cells are goals not relevant to that region. Goals (two-letter codes) and sub-goals (three-letter codes) are reported separately; LE, SP and BD goals are the average of sub-goal scores; FP scores are the weighted average of sub-goal scores. Acronyms are the same as in Table 1.

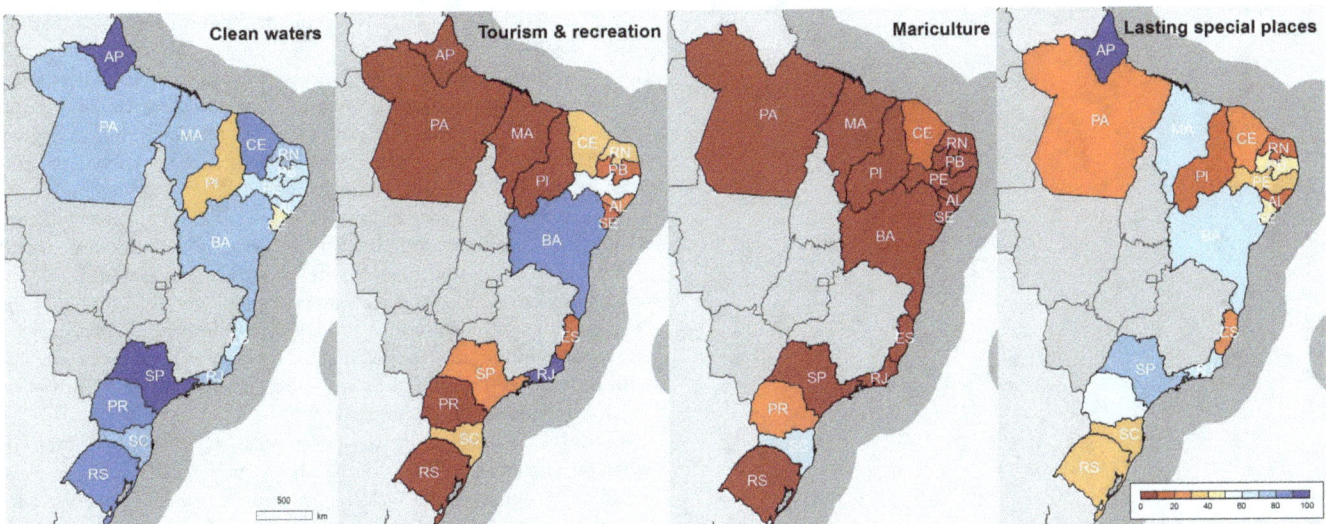

Figure 2. Goal and sub-goals calculated using state-level data.

Rio de Janeiro (65) states, which contain a mosaic of areas, with a larger contribution of sustainable use areas and indigenous lands. States in the Northeast region of Brazil had the lowest scores, in particular those with small coastal areas (Table 3, Figure S1 in Text S1). Here, the majority of the population resides in urban areas along the coast, and even sustainable use areas are few. In our analysis, we chose not to include the category "Área de Proteção Ambiental" (APA), which typically comprise vast areas used for zoning multiple uses, not necessarily reflecting areas with specific protection (see Text S1). We note that a high Lasting Special Places score does not necessarily imply good biodiversity conservation, as this goal is driven by the cultural values people place on coastal areas (Table 1, Text S1).

The Tourism and Recreation goal had large variation in scores (Table 3, Figure 2), reflecting the variable importance of coastal tourism among regions. Rio de Janeiro (100), Bahia (88) and Pernambuco (58) had the highest scores as these states have a combination of high numbers and density of tourists (estimated by the density of coastal hotel employees per state; see Text S1), and are well known for the touristic attraction of their beaches and coastal cities. States in the North and South of Brazil had the lowest scores (with the exception of Santa Catarina).

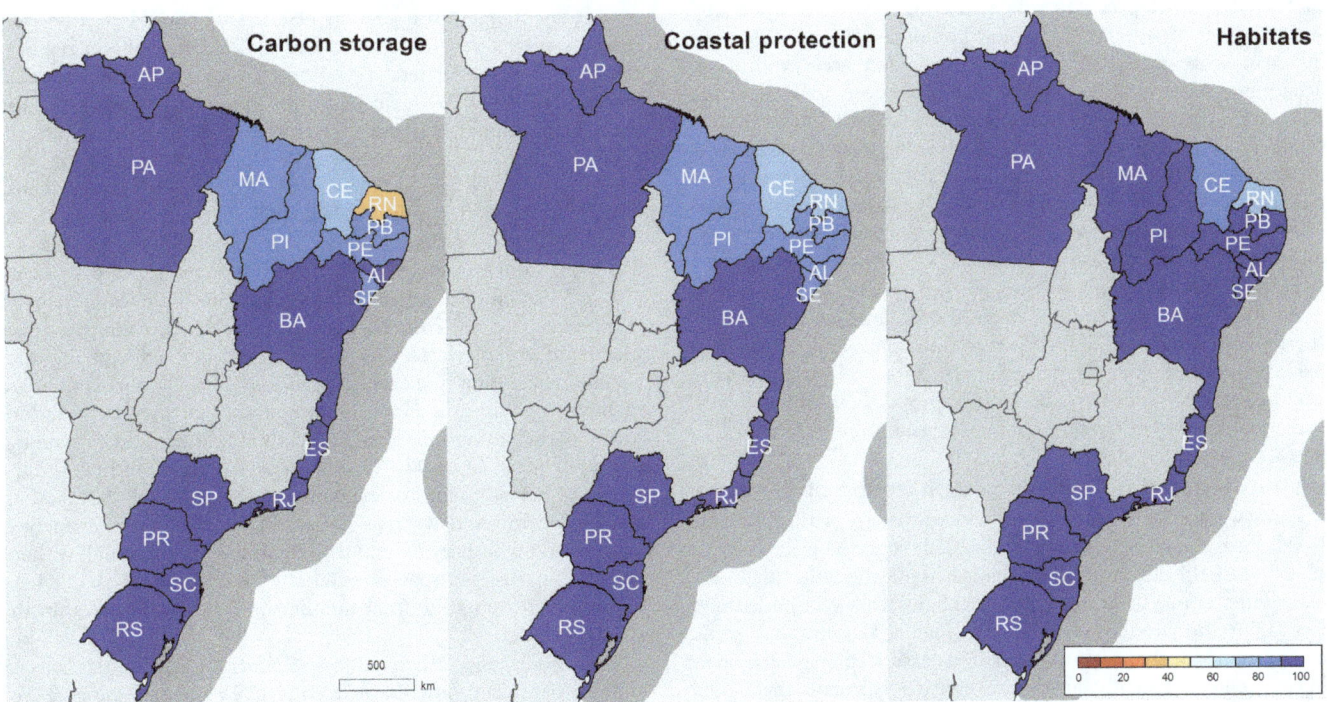

Figure 3. Goal and sub-goals using data from mixed national and state-level scales.

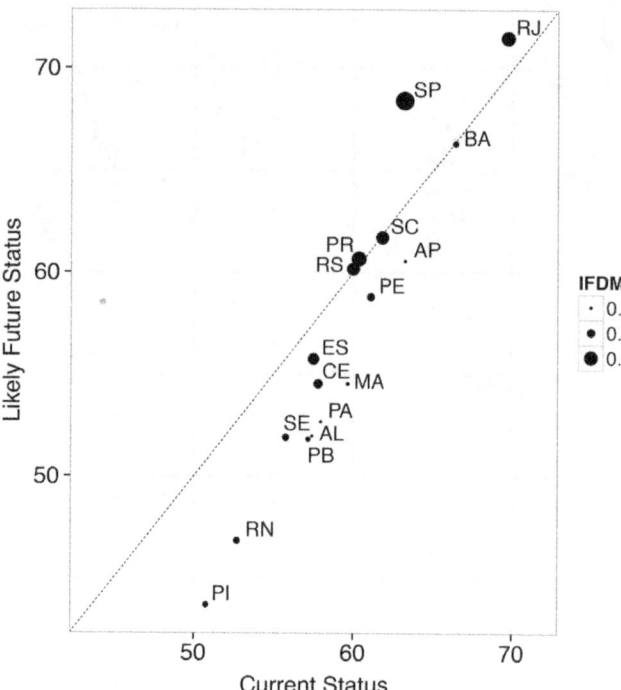

Figure 4. Current and likely future status for each state's overall Index score (axis values) and the value from an independent measure of development status (IFDM) used in Brazil (size of data point). Points below the dashed line are trending negatively into the future, and above are trending positively. IFDM scores range from 0 to 1 (low development = 0–0.4, average development = 0.4–0.6, moderate development = 0.6–0.8, and high development = 0.8–1).

Scores for the Clean Waters goal (Table 3, Figure 2) revealed that poorly developed states with low access to sanitation and waste management services, but low population densities (Amapá: 90), can score similarly to densely populated areas with good access to services (São Paulo: 95). However, only resident population density of coastal municipalities was used in estimating the Trash and Pathogen components of this goal. Many coastal urban areas receive a large influx of tourists in summer months, putting pressure on local infrastructure, including sanitation and waste management services [9], and likely polluting coastal waters, but these data are not available.

Perhaps unsurprisingly, a state's level of development influences its current and likely future status score. States with stronger economies, and better infrastructure and management, such as São Paulo and Rio de Janeiro, are more likely to pursue sustainable development paths and improve their Index scores, while less developed states show the opposite trend (Figure 4).

Lessons Learned from Regional Application of the Index

Here we compare results from this case-study with scores for Brazil from the global analysis (year 2012; reported in Halpern *et al.* [2]) to help illustrate how regional applications of the Index may differ from global ones, and highlight what can be learned from efforts to conduct regional assessments. We note that some methodological changes and data updates affecting goal models have occurred recently and were used to calculate the global Ocean Health Index for year 2013; they also have been used to recalculate the Index for all countries for year 2012 (see oceanhealthindex.org/about/methods for details). However, we

have chosen to focus on how results from the present case study relate to those from the initial, published assessment in Halpern *et al.* [2] as both share more methodological approaches. This allows a more direct comparison between global scores for Brazil and the results from this regional study, as differences are due to higher quality regional data, more direct measures of ocean health, and case-study specific model changes.

For some goals, the national scores remained similar to those from the global assessment [2], for example Mariculture and Clean Waters (Figure 5), even when significant model and/or data changes were made. As a consequence, the overall Index score for Brazil (60) was remarkably similar to the country score derived from the global analysis (62; Figure 5). These similarities suggest the Index may be able to broadly characterize ocean health even with poorer-quality, global-scale data.

The Index framework was designed to be flexible to different societal values and data contexts. Important model changes were made to some goals to reflect local conditions or to incorporate higher quality data. Below we discuss some changes to goal models and data layers and how they impact resulting scores.

The Food Production goal was analyzed using the same conceptual framework as in Halpern *et al.* [2], with some model adjustments (see Text S1). For the Mariculture sub-goal, we used national harvest statistics reported by each state, rather than country-level FAO statistics. Such data were not available for assessing the wild-capture component, the Fisheries sub-goal, of Food Production. For Mariculture, we were able to improve the estimation of the reference yield for each cultured species using historic time series of production; we also refined the total area of production available within each state (Text S1).

The Artisanal Opportunity goal scored lower than the global analysis (62 versus 88). The global model considered several aspects potentially related to the need and opportunity for people to fish artisanally, including regulations targeted to artisanal, subsistence and recreational fishing [2]. Until 2004, most fisheries in Brazil were considered "uncontrolled species" for which no management measures were in place [10]. For this reason, our model assumes that the opportunity to fish artisanally in Brazil is not limited by regulated access, but mainly by the condition of stocks. The lower Artisanal Opportunity score is indicative of the high proportion of stocks that fall within the overexploited, collapsed and rebuilding categories (for the most recent year (2006): 13%, 25% and 15%, respectively). The goal was calculated nationally, though significant regional differences likely exist. At present, standardized time series of small-scale fisheries landings are scarce, but would greatly increase the accuracy of this goal.

For Tourism and Recreation we used fine-scale data on hotel employment at the coastal municipality level, providing a better picture of ocean-related tourism than that derived from international arrivals data (used in Halpern *et al.* [2]). Data used here are of better quality and more comprehensive, as they also capture local participation in recreation. This altered the national aggregate score from 0 to 31 (Figure 5). The case-study approach has since helped to inform the global analysis, such that evaluation of the Tourism and Recreation goal in the updated assessment is now based on employment in the tourism sector, although without the benefit of higher spatial resolution data, as were available here (global data report the total number of jobs, not jobs within the coastal region).

For Biodiversity, regional assessments of threatened species showed significant differences from IUCN global assessments. For species assessed both globally and nationally, 58% held the same threat category, 33% had a higher risk of extinction, and 9% had a lower risk of extinction in national assessments. This difference

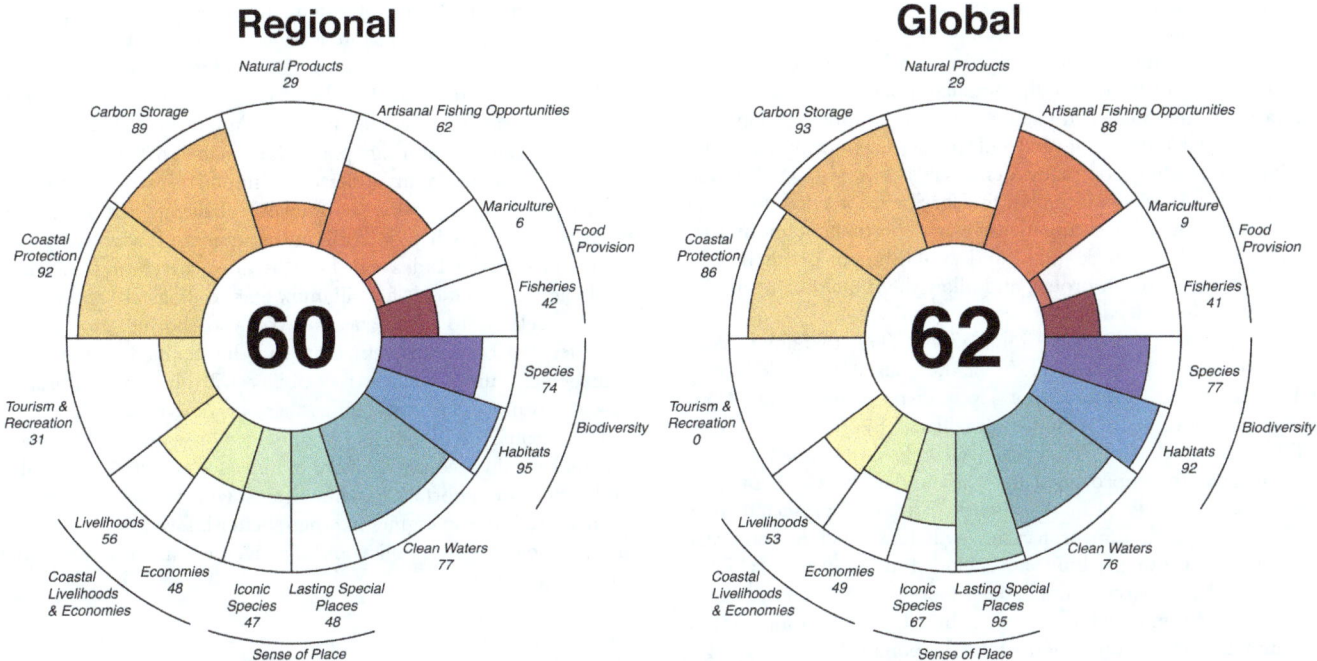

Figure 5. Goal and sub-goal scores for Brazil regional analysis (left), and Brazil global analysis (right). Key differences are found in Artisanal Fishing Opportunities, Tourism and Recreation, Lasting Special Places and Iconic Species. Overall Index scores (center) for the regional study are remarkably similar to global results for Brazil.

was particularly notable for sharks and rays where 39% are considered threatened in Brazilian waters based on regional assessments [11] compared to only 17% based on global assessments [12]. Regional assessments, when available, were also used for the Iconic Species sub-goal of Sense of Place. The list of Iconic species was expanded (Table S6 in Text S1) to include a number of seabirds, in particular eight threatened species of albatross, and the Critically Endangered Atlantic goliath grouper (*Epinephelus itajara*). These changes and the inclusion of regional assessments led to a decrease in the Iconic Species score from 67 to 47 (Figure 5).

Clean Waters scores were nearly identical for both regional and global studies (77 and 76). The regional analysis used the same models for chemical and nutrient pollution [13], but modified the pathogens and trash pollution components to include datasets at the coastal municipality level, including: urban population densities, presence or absence of sewage treatment service (pathogens), and four types of waste management services (marine debris).

Finally, we adapted measures of pressures and resilience to address those that are important to the local context. We incorporated state-level data on ecological resilience (protected areas) and social resilience (UIE management index of Brazilian states). Pressure layers with new regional data sources included: Human Pathogens, Trash, Intertidal Habitat Destruction, and Shrimp farming in mangroves (See Text S1). The latter in particular was identified as an important aspect at the regional scale, with impacts on several goals (Table S7 in Text S1).

Our assessment shows that the Index can utilize data of varied quantity and quality. Our aim was to adapt to the regional context, while recognizing gaps and understanding potential limitations of the available information. We sought to gather the best currently available data that met minimum requirements for Index score calculation. For this, data needed to be collected with

similar protocols across regions, available for all 17 coastal states (or sampled across all states, or ocean areas, but aggregated to the national level), and have enough spatial and/or temporal resolution for a reference point to be determined. When such requirements were not met, we used global data with country-level resolution. Better quality data sets were available for localized regions, which have been the focus of more intense research efforts in Brazil (e.g. Lagoa dos Patos region in Rio Grande do Sul state). Although such data are valuable for analyzing issues of a specific region, they are less suited for the integrated, comparative look used in the Index.

Key Policy Implications

Our analysis reveals some important trends across states and at the country level. Here we focus on two key policy implications related to fisheries management and habitat protection.

First, Brazil has substantial room for improvement in sustainable food production. Landings from wild capture fisheries far exceeded sustainable target levels in the main portion of the Brazilian coast and Trindade and Martim Vaz islands (Figure S2 in Text S1). Similar results were found from fisheries assessments from a multi-year Brazilian research program called REVIZEE, with the majority of stocks either fully (23%) or over-exploited (33%) and little room for expansion into new fisheries [10]. Yet despite these clear indications of overexploitation, policy initiatives from the Ministry of Fisheries and Aquaculture have focused on increasing harvests. A new governmental initiative (Plano Safra) will invest 4.2 billion Reais (~1.8 billion USD) in the fishing and aquaculture sectors with the goal of increasing total production to 2 million tons per year by 2014. Although the Mariculture scores suggest room for increased production in this sector, a greater focus on sustainable mariculture practices is needed if better ocean health is to be achieved.

Fisheries management in Brazil has been characterized by decades of open access to most fisheries and consequently to high fishing exploitation levels impacting both sustainability and profitability of its fisheries [10]. Brazilian policies for foreign fleets operating in the outer continental shelf and continental slope lack even minimal monitoring and enforcement, contributing to the decline of landings of many demersal stocks [14]. Substantial changes in current management of fisheries are needed, including implementation of a comprehensive and permanent monitoring system to evaluate stock status, and establishment of catch limits and other measures to protect and allow the rebuilding of marine resources, where needed.

The second set of policy implications relate to habitat-based goals. In our analysis, only 12% of the coastal zone (defined as 1 km inland and 3 nmi offshore) was in protected areas. These areas only cover 0.35% of the Brazilian EEZ (as noted above, APAs were excluded). This falls well below the target of 10% marine area to be protected by 2020 under the Convention of Biological Diversity [15]. A network of marine protected areas (MPAs) is important for protecting key habitats such as coral reefs [16], as well as other habitats not included in our analysis [17,18]. For example, the largest contiguous rhodolith bed in the world has recently been mapped off eastern Brazil, and is estimated to account for 5% of the world's total carbonate banks, playing a significant role in carbon storage [19], but remains totally unprotected.

With an immense coastline and diverse coastal habitats, Brazil still lacks systematic mapping and monitoring data for its marine habitats. Although initiatives for broad-scale mapping and monitoring of marine and coastal habitats are emerging in Brazil (e.g. SISBiota: www.sisbiota.ufsc.br, and Rebentos: http://rebentos.org/), data from these projects were not yet available at the time of this study. For our analysis, we found that seagrass beds monitored with similar protocols currently exist for only 3 sites and coral reef data from repeated surveys were available for only 11 sites. Continuous, rather than sporadic or one-time monitoring of key sites for coastal and marine habitats is a priority. In the future it will be important to include habitats not yet incorporated in this study, such as seamounts, mesophotic and deep corals, and algae banks. More comprehensive and systematic mapping of marine habitats provides benchmarks that are useful to understand the effects of new developing activities and enables to detect in a timely manner the effects of potentially competing interests, such as oil development and offshore leasing occurring in Brazil, so as to regulate these activities based on management priorities.

While data are available for some economic sectors (e.g. tourism, mariculture, waste disposal, and protected areas), the country lacks monitoring plans for many of the types of information required to understand human uses of the marine environment, thus posing a practical challenge for long-term management of the health of marine ecosystems. Notwithstanding, we found that the Index can be a useful metric, using currently available information for illuminating ecological and social patterns related to ocean health. We also showed how it is a scalable, flexible approach that can be applied at different management units, and this flexibility will allow incorporating newer and more relevant data as these become available.

The results presented here represent a first attempt to assess ocean health in a comprehensive manner for Brazil. As such, this study offers an important baseline against which future change can be measured. It also highlights where better information is needed, and can help to guide policy and management actions at national and sub-national scales.

Acknowledgments

We are grateful to the Brazilian Ministry of the Environment, and the Chico Mendes Institute for Biodiversity Conservation for help with obtaining regional data. We would like to thank Monica Brick Peres, Estevão Carino Fernandes de Souza, Ugo Eichler Vercillo and Steve Katona for their valuable input. A number of additional scientists contributed advice on regional datasets. We would also like to thank two anonymous referees for their useful feedback.

Author Contributions

Conceived and designed the experiments: CTE CL BSH CS TP GD. Performed the experiments: CTE CL TP. Analyzed the data: CTE DH BDB BSH CS. Contributed reagents/materials/analysis tools: CTE BSH DH BDB GD TP. Wrote the paper: CTE CL BSH GD.

References

1. IMF (2013) Report for selected countries and subjects. World Economic Outlook Database, International Monetary Fund. Available at: http://www.imf.org/external/pubs/ft/weo/2012/02/weodata/index.aspx Accessed 3 July 2013.

2. Halpern BS, Longo C, Hardy D, McLeod KL, Samhouri JF, et al. (2012) An index to assess the health and benefits of the global ocean. Nature 488: 615–622.

3. Rapport DJ, Böhm G, Buckingham D, Cairns J Jr., Costanza R, et al. (1999) Ecosystem health: the concept, the ISEH, and the important tasks ahead. Ecosystem Health 5:82–90.

4. Millennium Ecosystem Assessment (2005) Ecosystems and Human Well-Being: Synthesis Report. Island Press, Washington, DC.

5. Samhouri JF, Lester SE, Selig ER, Halpern BS, Fogarty MJ, et al. (2012) Sea sick? Setting targets to assess ocean health and ecosystem services. Ecosphere 3: art41.

6. Halpern BS, Longo C, McLeod KL, Cooke R, Frischhoff B, et al. (2013) Elicited preferences for components of ocean health in the California Current. Marine Policy 42:68–73.

7. Tobey J, Clay J, Vergne P (1998) Maintaining a balance: the economic, environmental and social impacts of shrimp farming in Latin America. Coastal Resources Center, University of Rhode Island, Narragansett, RI, USA.

8. Guimarães AS (2005) Carcinicultura marinha brasileira: sustentabilidade, reflexões históricas e situação atual. Monograph, Department of Oceanography, Federal University of Pernambuco.

9. Polette M, Raucci GD (2003) Methodological proposal for carrying capacity analysis in sandy beaches: a case study at the central beach of Balneário Camboriú (Santa Catarina, Brazil). Journal of Coastal Research SI 35: 94–106.

10. MMA (2006) Relatório Executivo Programa REVIZEE: Avaliação do potencial sustentável de recursos vivos na Zona Econômica Exclusiva. Ministério do Meio Ambiente (MMA), Brasília.

11. Peres MB, Barreto R, Lessa R, Vooren C, Charvet P, et al. (2012) Heavy fishing puts Brazilian sharks and rays in great trouble. Abstract. 6th World Fisheries Congress, 7–11 May 2012, Edinburgh, Scotland. p.21.

12. Polidoro BA, Livingstone SR, Carpenter KE, Hutchinson B, Mast RB, et al. (2008) Status of the world's marine species. In: Vié JC, Hilton-Taylor C, Stuart SN, editors. The 2008 Review of the IUCN Red List of Threatened Species. IUCN, Gland, Switzerland.

13. Halpern BS, Walbridge S, Selkoe KA, Kappel CV, Micheli F, et al. (2008) A global map of human impact on marine ecosystems. Science 319: 948–952.

14. Perez JAA, Pezzuto PR, Lucato SHB, Vale WG (2003) Frota de arrasto de Santa Catarina. In: Cergole MC, Rossi-Wongtscowski CLDB, editors. Dinâmica das frotas pesqueiras – Análise das principais pescarias comerciais do Sudeste-Sul do Brasil. Avaliação do potencial sustentável de recursos vivos na Zona Econômica Exclusiva, Programa REVIZEE, Score Sul. Evoluir, São Paulo, pp. 117–183.

15. CBD (2010) Strategic plan for biodiversity 2011–2020, including Aichi biodiversity targets. Available from: http://www.cbd.int/sp/targets. Accessed 5 November 2012.

16. Selig ER, Bruno JF (2010) A Global Analysis of the Effectiveness of Marine Protected Areas in Preventing Coral Loss. PLoS ONE 5: e9278. DOI:10.1371/journal.pone.0009278

17. Koslow JA, Gowlett-Holmes K, Lowry JK, O'Hara T, Poore GCB, et al. (2001) Seamount benthic macrofauna off southern Tasmania: community structure and impacts of trawling. Marine Ecology Progress Series 213:111–125.

18. Marone E, Dutra GF, Prates APL, Soares MLG, Gonçalves MA (2012) Biomas Costeiro e Marinho. In: Scarano FR. Biomas Brasileiros: retratos de um país plural. Rio de Janeiro: Casa da Palavra e Conservação Internacional.

19. Amado-Filho GM, Moura RL, Bastos AC, Salgado LT, Sumida PY, et al. (2012) Rhodolith beds are a major $CaCO_3$ bio-factories in the tropical south west Atlantic. PLoSONE 7: e35171.

Survival, Growth and Reproduction of Cryopreserved Larvae from a Marine Invertebrate, the Pacific Oyster (*Crassostrea gigas*)

Marc Suquet[1]*, Catherine Labbé[2], Sophie Puyo[3], Christian Mingant[1], Benjamin Quittet[3], Myrina Boulais[1], Isabelle Queau[1], Dominique Ratiskol[1], Blandine Diss[4], Pierrick Haffray[3]

1 Ifremer, UMR 6539, PFOM Department, Station Expérimentale d'Argenton, Argenton, France, 2 INRA, UR 1037, LPGP, Rennes, France, 3 SYSAAF, LPGP, Rennes, France, 4 Satmar, Barfleur, France

Abstract

This study is the first demonstration of successful post-thawing development to reproduction stage of diploid cryopreserved larvae in an aquatic invertebrate. Survival, growth and reproductive performances were studied in juvenile and adult Pacific oysters grown from cryopreserved embryos. Cryopreservation was performed at three early stages: trochophore (13±2 hours post fertilization: hpf), early D-larvae (24±2 hpf) and late D-larvae (43±2 hpf). From the beginning (88 days) at the end of the ongrowing phase (195 days), no mortality was recorded and mean body weights did not differ between the thawed oysters and the control. At the end of the growing-out phase (982 days), survival of the oysters cryopreserved at 13±2 hpf and at 43±2 hpf was significantly higher (P<0.001) than those of the control (non cryopreserved larvae). Only the batches cryopreserved at 24±2 hpf showed lower survival than the control. Reproductive integrity of the mature oysters, formely cryopreserved at 13±2 hpf and 24±2 hpf, was estimated by the sperm movement and the larval development of their offspring in 13 crosses gamete pools (five males and five females in each pool). In all but two crosses out of 13 tested (P<0.001), development rates of the offspring were not significantly different between frozen and unfrozen parents. In all, the growth and reproductive performances of oysters formerly cryopreserved at larval stages are close to those of controls. Furthermore, these performances did not differ between the three initial larval stages of cryopreservation. The utility of larvae cryopreservation is discussed and compared with the cryopreservation of gametes as a technique for selection programs and shellfish cryobanking.

Editor: Tilmann Harder, University of New South Wales, Australia

Funding: The present research was supported by the national project CRECHE (Ofimer 136/08/C) and CRYOAQUA (GIS IBISA) and by the European Union (FEP 30906-2009). The funders had no role in study design, data collection and analysis, decision to publish, or preparation of the manuscript.

Competing Interests: The private hatchery, Satmar, declares its affiliation in the competing interests section of the online manuscript form, along with any other relevant declarations relating to employment, consultancy, patents, products in development or marketed products.

* E-mail: marc.suquet@ifremer.fr

Introduction

Shellfish farming worldwide is mostly based on wild spat. The development of hatcheries combined with the recent adoption of genetic technologies such as selection or polyploidization (see for review [1], [2]) has created new genotypes with higher genetic value than wild stocks. The mid- and long-term cryopreservation of these new genetic resources has become of prime interest for research, genetic improvement (selection, hybridization or polyploidization) and for restoration programs for endangered species or populations. Obtaining optimized rearing potential for thawed embryos or larvae would result in an extension of the type of material that can be preserved in cryobanks [3].

Since the pioneering work carried out by Renard [4], most publications on the cryopreservation of shellfish embryos or larvae have been focused on the definition of protocols (cooling rate, type and concentration of cryoprotectant, embryo concentration in straws, etc...) and they have rarely reported good survival of thawed individuals or good performances after their settlement. In blue mussel (*Mytilus galloprovincialis*), the survival of 21-day old

larvae after thawing was 12.5% of the control group [5]. Only 2.8% of the frozen/thawed Greenshell mussel (*Perna canaliculus*) trochophores survived up to competent pediveligers and their size was lower than that observed on unfrozen larvae [6].

In Pacific oysters, only one thawed larva succeeded in settling after 29 days of rearing [7]. During the growing-out phase, the survival of 850 four month old Eastern oyster (*Crassostrea virginica*), produced from thawed trochophore larvae, was not different from the control group [8]. However, this work did not provide information on later stages of development that could have validated larval cryopreservation as an effective tool for the maintenance of genetic resources. The recent improvement in survival reported from Pacific oyster larvae frozen after the trochophore stage [9] provided an opportunity to compare performances of one full generation in order to assess the feasibility of larval cryopreservation for genetic resources management purposes.

The aim of the present work was to describe the long-term effects of cryopreservation on the subsequent performances of Pacific oysters. The survival and growth rate of oysters which were cryopreserved at larval stages were assessed during the ongrowing and growing-out phases. Furthermore, the reproductive performances of these oysters were estimated up to the D-larval stage of their progenies.

Materials and Methods

Obtaining thawed larvae

In June 2009, oyster embryos were obtained from crosses made at the Ifremer experimental hatchery in Argenton (Northern Brittany, France; no specific permissions were required): three females were stripped during the natural spawning period. For each female 500 000 oocytes were fertilized in 300 ml seawater with a pool of sperm collected from three males (1.3 10^6 spermatozoa ml^{-1}) [10]. For each female, the embryos were incubated in 5L beakers (30 000 embryos L^{-1}, 20°C). They were then pooled and batches were cryopreserved at 13±2 hours post fertilization: hpf (trochophore stage), 24±2 hpf (early D-larval stage) or 43±2 hpf (late D-larval stage), according to a protocol previously published [11] modified as follows: briefly, the larvae at the required stage were filtered on 20 μm mesh and diluted at a 1:1 volume ratio in cryoprotectant (10% ethylene glycol with 1% PVP and 200 mM trehalose in bi-distilled water, final concentration). The larvae were frozen in 0.5 ml straws (23 000 larvae per straw, 18 straws per larval stage), using a Kryo 10 (Planer, Sunbury, U.K.) The freezing curve was: −1°C min^{-1} from 0 to −10°C, hold for 5 min at −10°C and then, −0.3°C min^{-1} from −10 to −35°C. Finally, the straws were plunged into liquid nitrogen. In September 2009, three months after cryopreservation, the straws were thawed in a water bath (37°C, 10 s) and the larvae were reared in a 5 L flow through culture system for 19 days (24°C), according to a method previously developed by Rico-Villa et al. [12]. A control group was created using a pool of oocytes and spermatozoa collected from three females and three males. At the end of the larval rearing phase which lasted 21 days, the survival was: 28.9±25.8% in the fresh control, 0.1±0.0% in the 13±2 hpf batch, 0.9±0.7% in the 24±2 hpf batch and 0.4±0.1% in the 43±2 hpf batch. Spat (juvenile oysters) were then filtered on 180 μm mesh, transferred to mesh bottomed sieve tanks (45×20×6 cm, mesh size: 150 μm) and maintained in a race way (flow rate: 12 L h^{-1} tank^{-1}, seawater 24°C). They were fed daily with a mixture of two micro-algae (*Isochrysis galbana* and *Chaetoceros calcitrans*: at 65% and 35% respectively, maintaining a volume of 1 500 μm^3 algae μl^{-1} seawater) for a 2 month period (November-December 2009).

Ongrowing phase (this phase refers to 1 to 4 g oysters and aged of 88 to 195 days: the age is given without taking into account the period during which larvae were maintained in liquid nitrogen)

After filtration on a 4 mm mesh, spat were maintained in the same mesh bottom sieve tanks and reared according to the conditions described above. Spat number in each tank and the number of tanks were adapted to the number of spat available in each group, this number being related to the initial spat survival for each treatment (control: 280 spat/3 tanks; oysters produced from larvae cryopreserved at 13±2 hpf: 110 spat/1 tank, 24±2 hpf: 290 spat/3 tanks and 43±2 hpf: 245 spat/1 tank). Spat were counted and weighed individually (n = 30 per batch) at the end of the 4-month period (January to April 2010). The survival was calculated from the number of spat at the beginning of this ongrowing phase.

Growing-out phase (4–80 g, 196–982 days)

The spat were then transferred to the natural site of Aber Benoit (North Brittany, France) where water temperature ranges from 9 to 17°C (The owner of the site gave permissions to conduct the field study and this study did not involve endangered or protected species). Spat were maintained in oyster bags (90×45×6 cm, mesh size 1.5 cm), spat number and number of bags were the same as those of the tanks in the ongrowing-phase. Oyster survival and mean weight (n = 30 per batch) were assessed at the end of the 26 month rearing period (April 2010-June 2012).

Reproduction phase (>80 g, >982 days)

In June 2012, adult oysters were transferred from the Aber Benoit rearing site to the Ifremer experimental hatchery in Argenton, during the natural spawning period and their reproductive performances were assessed. Five groups of parents were created using five males and five females in each case: one group was the control made with the oysters which had never been frozen, two groups were composed of oysters formerly cryopreserved at 13±2 hpf (13.1 and 13.2) and the two last groups contained oysters produced from larvae cryopreserved at 24±2 hpf (24.1 and 24.2).

Gametes from each group were individually collected and pooled according to previously published techniques [10]. Therefore, each group produced one pool of sperm (from the five males in the group), and one pool of eggs (from the five females in the group). Sperm movement characteristics were assessed for each of the five pools of five males (13.1, 13.2, 24.1, 24.2 and control). Sperm motility was estimated using a two-step dilution procedure: firstly, 10 μL sperm (sperm concentration: 10^9 to 10^{10} sperm ml^{-1}) were diluted in 90 μL SW (seawater, salinity 35‰ +5 g L^{-1} BSA, 19°C), secondly, 2 μL of this first suspension was diluted again in 100 μL SW (final dilution 1:500). Sperm samples of 7 μL were transferred to a Thomas cell and the movement characteristics were observed under a phase contrast microscope (Olympus BX51, ×20 magnification), connected to a camera (Qicam Fast 1394). The percentage of motile spermatozoa and their VAP (Velocity of the Average Path) were assessed using a CASA plug-in developed for the Image J software [13]. Calibration settings of the software were as follows, frame rate: 25 frame sec^{-1}; sperm size range: 0.9 to 7.5 μm; minimum VAP for motile sperm: 10 μm sec^{-1}; minimum track length: 6 frames; minimum number of sperm observed: 30.

The fertilization capacity of the four sperm pools (13.1, 13.2, 24.1, 24.2) was then estimated in triplicate (50 000 oocytes for each pool; fertilization volume 100 ml) [10]. Firstly, each sperm pool was used to fertilize a pool of oocytes collected from the control group (four crosses). Secondly, the sperm was used to fertilize the oocytes of its respective female groups (four crosses). Because sperm quality was being studied, a limiting sperm to oocyte ratio of 300, corresponding to 150 000 sperm ml^{-1}, was used. Furthermore, oocyte quality was estimated using the four oocyte pools (13.1, 13.2, 24.1, 24.2) and a pool of fresh sperm collected from the control group (four crosses). The pools of oocytes were fertilized using the method described above (except that a non limiting sperm to oocyte ratio of 800 was used to study oocyte quality, corresponding to 400 000 spermatoza ml^{-1}). Then, a control (male control x female control) was added. For both sperm and oocyte quality studies, the number of D-larvae was counted 48 hours post fertilization (3×100 μL samples) for the 13 crosses carried out. The D-larval survival was calculated as a percentage of the oocyte number used for fertilization.

Statistical analysis

Data are presented as mean ± standard deviation. Percentages were arcsin square root transformed before analysis. Means were compared using one way ANOVA. When differences were significant (P<0.05), a Tukey a posteriori test was used for mean comparison.

Results

At the end of the ongrowing phase, no mortality was recorded in the four different oyster batches (control, oysters produced from larvae cryopreserved at 13±2 hpf, 24±2 hpf and 43±2 hpf). Furthermore, the mean weights of the four oyster batches were not significantly different (Figure 1).

At the end of the growing-out phase, the survival of two cryopreserved batches (13±2 hpf and 43±2 hpf) was significantly higher than the control (Figure 2). Only the 24±2 hpf had survival lower than the fresh control (P<0.001). The mean weight of the cryopreserved batches was either higher (13±2 hpf and 24±2 hpf; P<0.05) or similar (43±2 hpf) to that of the control.

During the reproductive phase, the percentage of motile spermatozoa observed for pool 13.2 (cryopreservation 13±2 hpf of the gamete pool n°2) was significantly higher (P<0.001) than that observed for the four other pools (control, 13.1, 24.1 and 24.2; Figure 3). A percentage of motile spermatozoa higher than 70% was recorded for four of the pools, but not for pool 24.2. This observation highlights the good quality of the sperm produced in four pools out of five. The VAP observed for the pool 24.2 was significantly higher (P<0.001) than that of the control. Furthermore, the VAP observed for the pools 13.2 and 24.1 was lower than that recorded for the control. In all, the VAP recorded for three (24.2, 13.1 and 13.2) out of the four sperm pools from thawed oysters was never below 70% of the VAP value recorded for the control.

Larges variations in the D-larval survival (1.5 to 49.6%; Figure 4) were observed depending on the crossing but they were not significantly different from the control (male control x female control), except for the progenies issued from control x 13.2 and 13.2×13.2 (P<0.001).

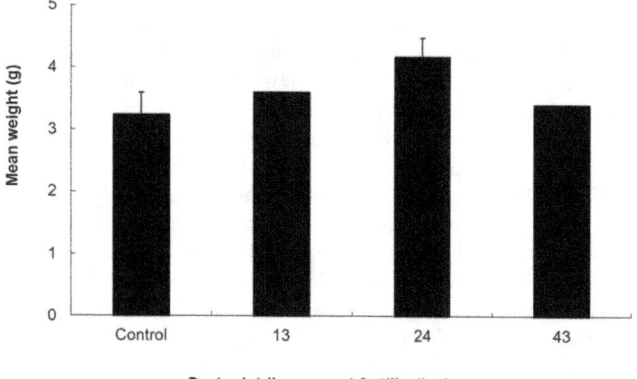

Figure 1. Effect of larvae cryopreservation at different stages post fertilization on the subsequent growth rate of oysters at the end of the ongrowing phase (195 days; spat number in each tank/number of tanks, control: 280/3, 13±2 hpf batch: 110/1, 24±2 hpf batch: 290/3, 43±2 hpf batch: 245/1).

Discussion

This study reports survival and growth to reproduction at 2.7 years of age of thawed Pacific oyster larvae previously cryopreserved at three different larval stages. This is the first report of the successful development of cryopreserved individuals to reproduction in molluscs and the second report of viable progenies produced from cryopreserved parents in aquatic invertebrates. Indeed a study was already published on the rotifer, *Brachionus plicatilis* [14]. In this species, the thawed larvae started to lay eggs 2 to 3 days after thawing. Therefore, any damage caused by cryopreservation may not have had time to be expressed compared with oysters in the present work, as these reproduced more than 2 years after. Moreover, gamete quality was not evaluated in *B. plicatilis* because it reproduces by parthenogenesis, thereby producing amictic eggs. Similar positive results had already been reported in other invertebrates such as in insects (see review in [15]). However, major differences exist between such studies and the present model: permeabilization of the insect egg shell, deshydratation and vitrification are not required in Pacific oyster. The simpler cryopreservation procedure used in oyster is associated with the eco-physiology of marine mollusc larvae which do not have a thick external cuticle that could limit osmotic exchanges and cryoprotectant efficacy.

In the present study, no marked differences were observed in the ongrowing, growing-out and reproductive capacities of oysters produced from cryopreserved larvae compared with those of oysters that had been grown directly from fresh larvae with no cryopreservation step. These results confirm a pioneering study carried out by Paniagua-Chavez and co-workers [8], assessing the effects of cryopreservation at the trochophore stage on the subsequent development of Eastern oysters until the age of four months after thawing: the survival of larvae produced from thawed embryos was not different from the control. Furthermore, our work compares the effect of cryopreservation at three different stages of development. No marked differences were found in the rearing performances between these treatments and the control group. This indicates that after the initial dramatic mortalities observed during the larval development, the surviving larvae yielded similar post-settlement rearing performances to the ones that have never been frozen. The protocol of the present study differs from that used by Paniagua-Chavez et al. [8] in many respects (equilibration phase, cooling rate, cryoprotectant and thawing temperature). Because of similar conclusions at least for the immature stage, it can be suggested that there is only a limited interaction between the freezing protocol and the performances of the thawed larvae after their settlement. However, this hypothesis must be confirmed after the improvement of initial larvae survival observed just after thawing.

Some limitations of the study include the long-term protocol used. Firstly, oyster density maintained in tanks during the ongrowing phase and in bags during the growing-out phase differed between batches. Oyster density of each batch depended on the survival observed during the previous rearing phase. However, these densities are considered as non limiting for oyster growth [16]. Furthermore at the end of the ongrowing phase, the low survival observed for the 24±2 hpf batch which had the highest oyster density was moderated by its high growth rate assessed during the same rearing phase and its high subsequent reproductive performances. Secondly, parental effects on rearing performances of spat produced from thawed larvae were not studied in the present work. Such effects could influence a number of different rearing parameters: the survival of D-larvae produced from thawed oocytes collected from eight Pacific oyster females

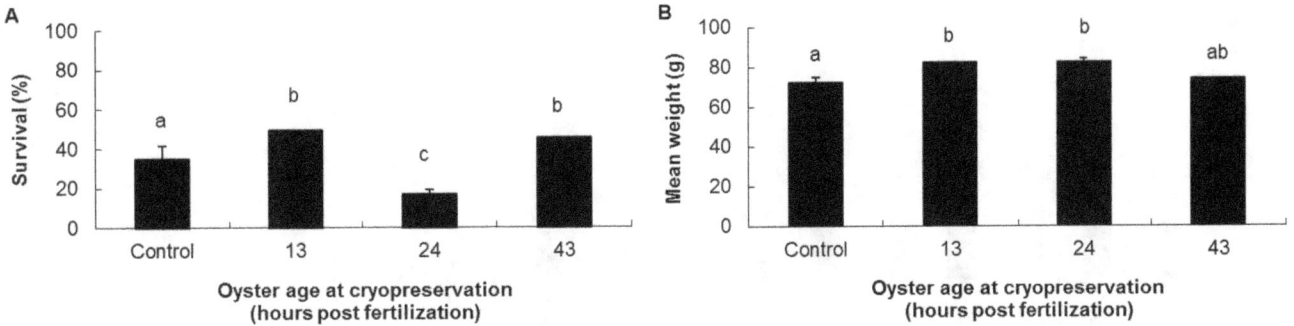

Figure 2. Effect of larvae cryopreservation at different stages post fertilization on the subsequent performances of oysters at the end of the growing-out phase (982 days). A: survival, B: mean weight observed at the end of the phase (spat number in each bag and number of bags are the same as those reported for Figure 1; different letters indicate significantly different results).

ranged from 0.1 to 30.1% [11]. However, the present 32 month protocol would be too difficult to conduct because it would require keeping separated progenies from different families. As a consequence, a pool of females and a pool of males were used to obtain the larvae used for cryopreservation. This choice is similar to previous studies aiming to describe the long-term rearing performances of molluscs produced from thawed larvae [5,6].

The high rearing performances of surviving thawed larvae may suggest an absence of genome alterations on the individuals that succeeded in settling, allowing subsequent development of these larvae. DNA alteration of thawed embryos has mostly been investigated in mammalian species [17] but rarely in invertebrates. Several instances of DNA damage were identified in the ragworm (*Nereis virens*) [18] and in insects [15], the most life-threatening of these is thought to be damages to the gut epithelium in ragworm. DNA integrity has been more greatly studied in cryopreserved fish gametes and several techniques to assess DNA integrity have been developed, including Tunel (Terminal deoxynucleotidyl transfer-ase-nick-end-labelling), SCSA (Sperm Chromatin Structure Assay) and Comet assays (single gell electrophoresis) [19]. Inter-species differences have been found: cryopreservation caused damage to rainbow trout (*Onchorhynchus mykiss*) sperm compared with fresh sperm, but not to sperm of gilthead sea bream (*Sparus aurata*) [20]. In molluscs, the percentage of Pacific oyster spermatozoa with damaged DNA significantly increased after cryopreservation [21] but the consequences of such damage for the F_1 progenies were not estimated. Because a principal application of cryopreservation is to maintain the genetic variability of domesticated and wild

animal species, the long-term development capacities of thawed larvae must be studied.

The discussion sections of scientific papers on embryo cryopreservation are often generally limited to scientific or technical questions, without addressing the relative benefits offered by larvae cryopreservation compared to gamete cryopreservation. It was shown that the development capacities of surviving D-larvae up to spat stage was not altered by oocyte cryopreservation [11]. Together with the success of sperm cryopreservation, and with the results showed in our study on larvae, it can be inferred that both haploid and diploid genomes of Pacific oyster can now be cryopreserved. The freezing of haploid spermatozoa or oocytes or of diploid spermatozoa collected from tetraploid males may be useful for controlled crosses using parents presenting desired characteristics, thus supporting the production of original geno-types for genetic selection and of triploids. The rationale of larvae cryopreservation is different since it cannot create new genetic variation *per se* but preserves diploid family genomes already created by controlled crosses. The cryopreservation of larvae has several other interests for shellfish aquaculture. First, it secures investments in selection programs as the families created can be conserved frozen. This could allow breeder to return to the best families of earlier-made crosses for selection purposes (with a loss of a one generation interval for sib selection). Second, it could allow a rapid restoration of genetic variation from previous generations if genetic drift or inbreeding depression are observed. Third, it could allow inter-generation estimation of genetic progress, reducing costs and risks associated with maintenance of

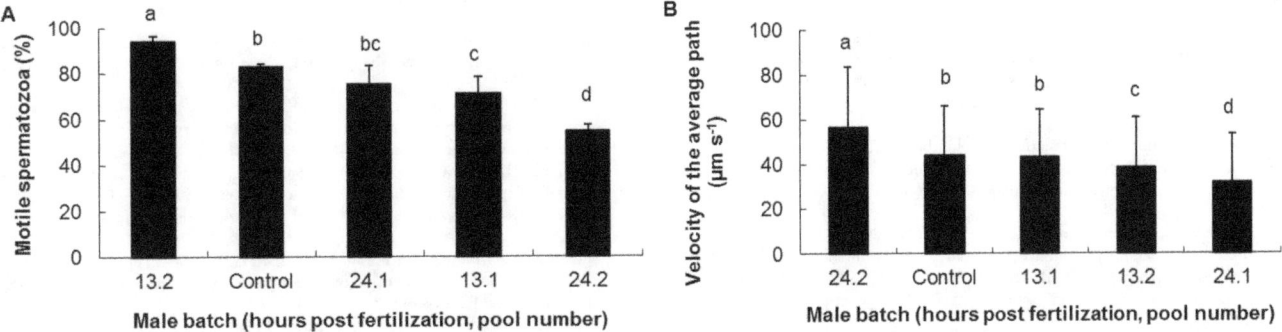

Figure 3. Effect of larvae cryopreservation at different stages post fertilization on the subsequent performances of oysters during the reproductive phase (>982 days). A: percentage of motile spermatozoa, B: Velocity of the Average Path (n>30 spermatozoa scored for each male gamete pool; n = 5 males contributed to each pool; different letters indicate significantly different results).

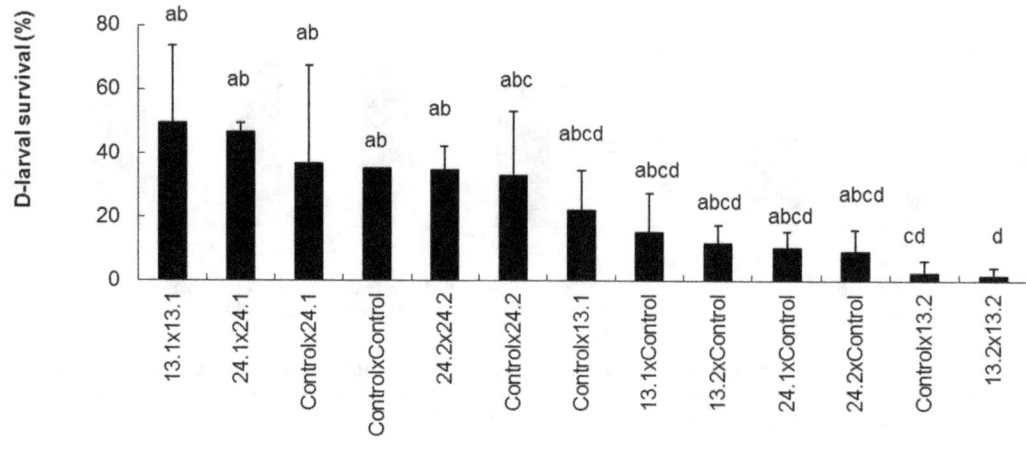

Figure 4. Effect of larvae cryopreservation at different stages post fertilization on the subsequent D-larval survival in progenies. (n = 5 oysters contributed to each gamete pool; different letters indicate to significantly different results).

live control lines that are never really equivalent to a true control because of genetic drift across generations, non-intentional selection or domestication. More generally, the cryopreservation of diploid progenies may also provide suitable means for diffusing genetic progress and genetic exchange of material worldwide in using only sanitary approved genotypes ready to be used as future broodstock. The application of a cryopreservation procedure developed for diploid larvae may also facilitate maintenance and genetic improvement of tetraploid oyster stocks by cryopreservation of tetraploid larvae [2].

Conclusion

In the present work, We demonstrated that the subsequent growing-out and reproductive performances of surviving thawed oysters are similar to those observed for unfrozen ones. From thawing up to the end of the growing out phase (982 days), oyster survival observed was respectively 0.05% (13±2 hpf), 0.15% (24±2 hpf), 0.18% (43±2 hpf) and 10.06% (control). Because of the very high fecundity observed in Pacific oyster (10 to 50×10^6 oocytes per female), larvae cryopreservation allows the production of 11,000 to 42,000 good quality adult oysters for each female,

depending on oyster development stage at cryopreservation. These yields can be largely increased by the further optimisation of the cryopreservation protocols designed for oyster larvae.

Pacific oyster is the first aquatic farmed species for which sperm, oocyte and larval genomes can be stored in this way. Freezing larvae will help to preserve the genomes of wild oyster populations and of selected oyster families. Furthermore, thawed oysters may be used to evaluate the genetic progress recorded between successive generations. Such advances in embryo cryopreservation will contribute to the establishment of oyster cryobanks and maintenance of the genetic diversity of domestic and wild stocks of other mollusc species.

Acknowledgments

Many thanks to H. McCombie-Boudry for corrections of the English.

Author Contributions

Conceived and designed the experiments: MS CL SP PH. Performed the experiments: MS CL SP CM BQ MB IQ DR BD PH. Analyzed the data: MS CL SP PH. Contributed reagents/materials/analysis tools: CM BQ MB IQ DR BD. Wrote the paper: MS CL PH.

References

1. Boudry P, Dégremont L, Haffray P (2008) The genetic basis of summer mortality in Pacific oyster spat and potential for improving survival by selective breeding in France. In: Samain JF, McCombie H, editors. Summer mortality of Pacific oyster - The Morest Project. Versailles, France: Quae Editions. pp.153–196.
2. Piferrer F, Beaumont A, Falguière JC, Colombo L, Flajshans M, et al. (2009) The use of induced polyploidy in the aquaculture of fish and shellfish for performance improvement and genetic containment. Aquaculture 293: 125–156.
3. Hiemstra SJ, van der Lende T, Woelders H (2006) The potential of cryopreservation and reproductive technologies for animal genetic resources conservation strategies. In: Ruane J, Sonnino A, editors. The role of biotechnology in exploring and protecting agricultural genetic resources. Rome: : FAO. pp. 45–59.
4. Renard P (1991) Cooling and freezing tolerances in embryos of the Pacific oyster, Crassostrea gigas: methanol and sucrose effects. Aquaculture 92: 43–57.
5. Wang H, Li X, Wang M, Clarke S, Gluis M, et al. (2011) Effects of larval cryopreservation on subsequent development of the blue mussels, Mytilus galloprovincialis Lamarck. Aquaculture Res 42: 1816–1823.

6. Paredes E, Adams SL, Tervit HR, Smith J, McGowan LT, et al. (2012) Cryopreservation of Greenshell mussel (Perna canaliculus) trochophore larvae. Cryobiol 65: 256–262.
7. Usuki H, Hamaguchi M, Ishioka H (2002) Effects of developmental stage, seawater concentration and rearing temperature on cryopreservation of Pacific oyster Crassostrea gigas larvae. Fish Sci 68: 757–762.
8. Paniagua-Chavez CG, Buchanan JT, Supan JE, Tiersch TR (2000) Cryopreservation of sperm and larvae of the Eastern oyster. In: Tiersch TR, Mazik PM, editors. Cryopreservation in aquatic species. Baton rouge, Louisiana: WAS. pp. 230–239.
9. Suquet M, Le Mercier A, Rimond F, Mingant C, Haffray P, et al. (2012) Setting tools for the early assessment of the quality of thawed Pacific oyster (Crassostrea gigas) D-larvae. Theriogenol 78: 462–467.
10. Song YP, Suquet M, Quéau I, Lebrun L (2009) Setting of a procedure for experimental fertilisation of Pacific oyster (Crassostrea gigas) oocytes. Aquaculture 287: 311–314.
11. Tervit HR, Adams SL, Roberts RD, McGowan LT, Pugh PA, et al. (2005) Successful cryopreservation of Pacific oyster (Crassostrea gigas) oocytes. Cryobiol 51: 142–151.

12. Rico-Villa B, Woerther P, Mingant C, Lepiver D, Pouvreau S, et al. (2008) A flow- through rearing system for ecophysiological studies of Pacific oyster, *Crassostrea gigas*, larvae. Aquaculture 282: 54–60

13. Wilson-Leedy JG, Ingermann RL (2007) Development of a novel CASA system based on open source software for characterization of zebrafish sperm motility parameters. Theriogenol 67: 661–672.

14. Toledo JD, Kurokura H (1990) Cryopreservation of the euryhaline rotifer *Brachionus plicatilis* embryos. Aquaculture 91: 385–394.

15. Leopold RA, Rinehart JP (2010) A template for insect cryopreservation. In: Denlinger DL, Lee RE, editors. Low temperature biology of insects. Cambridge: Cambridge University Press. pp. 325–341.

16. Honkoop PJC, Bayne BL (2002) Stocking density and growth of the Pacific oyster (*Crassostrea gigas*) and the Sydney rock oyster (*Saccostrea glomerata*) in Port Stephens, Australia. Aquaculture 213: 171–186

17. Smith GD, Silva CASE (2004) Developmental consequences of cryopreservation of mammalian oocytes and embryos. Reprod Bioled Online 9, 171–178.

Available: http://ac.els-cdn.com/S1472648310621268/1-s2.0-S14726483106 21268-main.pdf?_tid = 92971c4c-7f41-11e2-874f-00000aab0f6b&acdnat = 13617 93218_93bf401c2c411ee833e5454b804763fe. Accessed 25 January 2013.

18. Wang WB, Olive JW (2000) Effects of cryopreservation on the ultrastructural anatomy of 3-segment *Nereis virens* larvae (Polycheta). Inv Biol 119: 309–318.

19. Cabrita E, Sarasquete C, Martinez-Paramo S, Robles V, Beirao J, et al. (2010) Cryopreservation of fish sperm: applications and perspectives. J Appl Ichthyol 26: 623–635.

20. Cabrita E, Robles V, Rebordinos L, Sarasquete C, Herraez MP (2005) Evaluation of DNA damage in rainbow trout (*Onchorhynchus mykiss*) and giltead sea bream (*Sparus aurata*) cryopreserved sperm. Cryobiol 50: 144–153.

21. Gwo JC, Wu CY, Chang WS, Cheng HY (2003) Evaluation of damage in Pacific oyster (*Crassostrea gigas*) spermatozoa before and after cryopreservation using comet assay. Cryoletters 24: 171–180.

Spatial and Temporal Dynamics of Mass Mortalities in Oysters Is Influenced by Energetic Reserves and Food Quality

Fabrice Pernet[1,2]*, Franck Lagarde[1], Nicolas Jeannée[3], Gaetan Daigle[4], Jean Barret[1], Patrik Le Gall[1], Claudie Quere[2], Emmanuelle Roque D'orbcastel[1]

1 Ifremer, Laboratoire Environnement Ressource du Languedoc Roussillon, Bd Jean Monnet, Sète, France, 2 UMR LEMAR Ifremer/CNRS/UBO/IRD, Technopole de Brest-Iroise, Plouzané, France, 3 Geovariances, Avon, France, 4 Université Laval, Département de mathématiques et de statistique, Pavillon Alexandre-Vachon, Québec, Québec, Canada

Abstract

Although spatial studies of diseases on land have a long history, far fewer have been made on aquatic diseases. Here, we present the first large-scale, high-resolution spatial and temporal representation of a mass mortality phenomenon cause by the Ostreid herpesvirus (OsHV-1) that has affected oysters (*Crassostrea gigas*) every year since 2008, in relation to their energetic reserves and the quality of their food. Disease mortality was investigated in healthy oysters deployed at 106 locations in the Thau Mediterranean lagoon before the start of the epizootic in spring 2011. We found that disease mortality of oysters showed strong spatial dependence clearly reflecting the epizootic process of local transmission. Disease initiated inside oyster farms spread rapidly beyond these areas. Local differences in energetic condition of oysters, partly driven by variation in food quality, played a significant role in the spatial and temporal dynamics of disease mortality. In particular, the relative contribution of diatoms to the diet of oysters was positively correlated with their energetic reserves, which in turn decreased the risk of disease mortality.

Editor: Niko Speybroeck, Université Catholique de Louvain, Belgium

Funding: This work was supported by grants from France Agrimer, Région Languedoc-Roussillon and Conseil Général de l'Hérault and is part of the GIGASSAT project funded by ANR-AGROBIOSPHERE No. ANR-12-AGRO-0001-01. The funders had no role in study design, data collection and analysis, decision to publish, or preparation of the manuscript.

Competing Interests: The authors have the following interests. Nicolas Jeannée is an employee of Géovariances. There are no patents, products in development or marketed products to declare.
online in the guide for authors.

* E-mail: fabrice.pernet@ifremer.fr

Introduction

Since the mid-1970s, large-scale episodic events such as disease epidemics, mass mortalities, harmful algal blooms and other population explosions have been occurring in marine environments at a historically unprecedented rate [1–3]. Bivalve farming, because it relies directly upon natural marine environments and feeding resources, is presently faced with a number of severe risks and limiting factors. Historically, infectious diseases have seriously affected the marine bivalve industry. In the early 1970s, the French oyster industry suffered a serious crisis when irido-like virus infections decimated the Portuguese oyster, *Crassostrea angulata*, in European Atlantic waters [4]. The Pacific oyster *C. gigas* has since been introduced for culture along the coasts of France. Episodes of Pacific oyster mortality have been occurring for more than five decades in the major oyster-producing countries [5]. Most of these episodes have been classified as "summer mortality", typically affecting animals during the warmer months of the year. Since 2008, massive mortality events in *C. gigas* oysters have been reported in almost all farming areas in France when seawater temperature exceeds 16°C [6–8]. These mortality events have been attributed to a combination of adverse environmental factors combined with the presence of ostreid herpes virus 1 (OsHV-1) [7–

10]. Evidence suggests that OsHV-1 infection is a necessary cause, and a particular genotype, named OsHV-1 µvar [11], appears to have been the dominant viral genotype in the most serious mortality events of 2008–2011. Oyster mortalities were considerable in this period, particularly among seed stocks [6–8]. The devastation of oyster beds in France observed every year since 2008 has now resulted in a shortfall in shellfish supplies corresponding to approximately 40% of commercial production.

Although mass mortalities of oysters are clearly related to infectious diseases, they can also reflect an unfavourable energetic balance. In 2008, oysters in Mediterranean Thau lagoon (France) showed a sharp decrease in carbohydrate concentration and arrested accumulation of lipid reserves before mortalities occurred [10]. It was suggested that young oysters go through a phase of energetic weakness during the spring that renders them more susceptible to pathogen infections. Correspondingly, oysters with high energy reserves are probably less affected by pathogen infections than those with low levels [7], as reported for vertebrate species [12–15]. It is therefore likely that differences in energetic condition of oysters have a significant influence on the spatial and temporal dynamics of disease mortality. Given that the energetic condition of marine bivalves varies depending on food quality and availability [16–18], these parameters may also be related to the

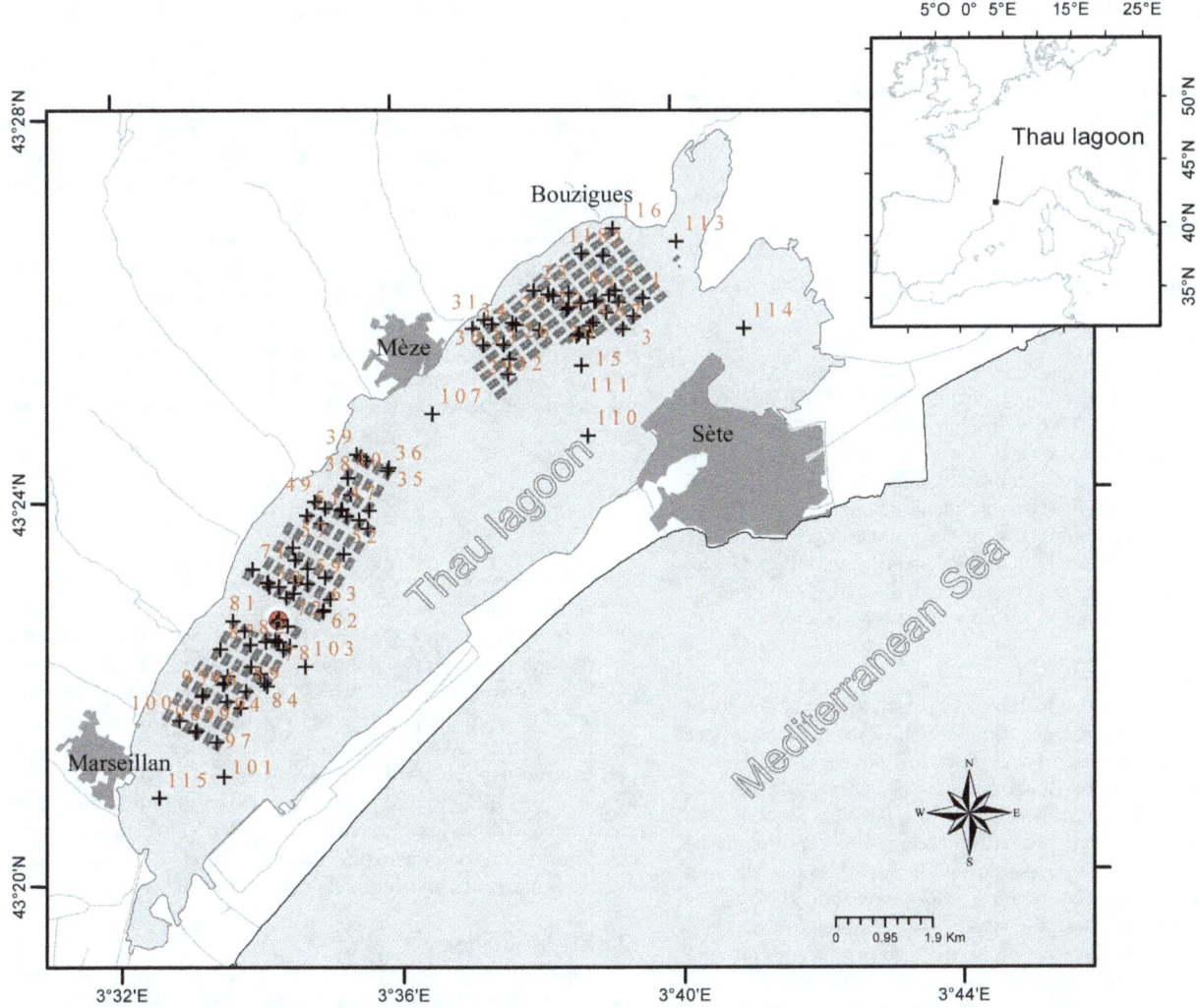

Figure 1. Sampling sites (black crosses along with non-continuous numbering) located in the Thau lagoon. Areas with grey boxes: bivalve farms; red circle: temperature probe.

disease mortality of oysters. Although the quality and availability of food consumed by bivalves have drawn little attention in terms of host–parasite interactions [19], they have been demonstrated to modulate defense related mechanisms [18,20–22].

The first objective of this paper is to explore the spatial pattern of disease mortality in oysters *C. gigas*. For instance, describing the distribution and dynamics of infectious disease is important for management [23,24]. The spatial distribution of a disease is often examined by applying statistical methods to data collected during disease surveillance and then generating a map that describes spatial variation in risk. However, this task is often complicated by heterogeneities in the population of interest [14]. For example, in oysters, susceptibility to the disease varies with age, origin of oysters, farming practices, and life history traits [7]. These heterogeneities, if not accounted for, may obscure observation of important disease trends that may be of interest to resource managers. In order to circumvent this issue, disease mortality of oysters was investigated in susceptible animals deployed at 106 locations in the Thau lagoon, on the French Mediterranean coast, before the start of the annual epizootic event. Therefore, variance in disease mortality in these "sentinel oysters" reflects the location itself, while minimizing confounding effects. Disease mortality is

expected to show strong spatial and temporal dependence due to the epidemic process of local transmission. Although terrestrial studies on the spatial aspects of diseases have a long history, driven by a necessity to understand diseases of humans, crops, farm animals and wildlife [25], spatial studies of aquatic diseases are much less frequent and generally focused on distribution of macro-parasites [26–33].

The second objective is to examine the relationships between energetic condition, food quality and the probability of disease mortality in oysters. Our hypothesis is that the spatial and temporal dynamics of disease mortality in oysters are partly driven by local differences in energetic reserves and food quality. To test this hypothesis, the probability of disease mortality of oysters was analysed in relation to their energetic reserves and the fatty acid composition of their neutral lipids. Marine bacteria, diatoms, dinoflagellates, terrestrial inputs and vascular plants show different combinations of specific fatty acids. Given that fatty acids are incorporated largely unaltered into the reserve lipids of primary consumers like oysters, they generally reflect the fatty acid profiles of the food consumed and therefore reveal useful information about trophic sources [34,35].

Materials and Methods

Animals

Diploid oysters were produced at the Ifremer hatchery in La Tremblade in July 2010 and then maintained in the nursery at the Ifremer marine station in Bouin from August 2010 onwards. These oysters were maintained free of disease mortality from hatching until deployment in Thau lagoon on 15 March 2011. Mean shell length was 10 mm, mass was 0.7 g and age of oysters was 8 mo. These oysters had shown no disease mortality prior to deployment and OsHV-1 DNA was not detected in their tissues.

Study Area

The Thau lagoon on the French Mediterranean coast is an oyster farming area that accounts for about 10% of Pacific oyster production in France. Oyster growth rates in the lagoon are among the highest in France for this species. Thau lagoon is 19 km long, 4.5 km wide and 5 m deep on average (Figure 1). Shellfish are cultured in 3 areas of the lagoon, namely Bouzigues, Mèze and Marseillan, covering about 20% of the total surface area. The lagoon is almost completely closed off from the Mediterranean Sea, with only narrow connections through the Sète channel; other connections being negligible in terms of water exchange.

Experimental Design

The oysters were placed in 106 pearl nets at an initial density of 150 individuals per pearl net, as commonly practised by oyster farmers. These pearl nets were deployed within the Thau lagoon in the farming areas of Bouzigues (n = 35), Mèze (n = 34) and Marseillan (n = 29) and also outside of the farming areas (n = 8). The sampling design was primarily developed to capture spatial and temporal trends within the farming areas. Fifteen additional sampling stations outside farming areas were added to allow capturing spatial trends at the lagoon scale, although only eight were maintained the duration of the study to produce useful data (Figure 1). The distribution of sampling stations corresponds to a random coverage of shellfish farms. Permission for deploying oysters outside of farming areas was issued by the French Ministry of Ecology and Sustainable Development, dept. of maritime affairs, in February 2011. For locations within farming areas, the owner of the farm gave permission to conduct the study on this site. The present field studies did not involve endangered or protected species.

Live and dead oysters were counted in each pearl net on 31 March; 6, 9, 13, 16, 19, 21, 24, 26 and 29 April; 3, 6, 10, 13 and 20 May 2011. Ten live oysters were randomly sampled from each pearl net on 6 and 16 April for laboratory analyses. Soft tissues of oysters were carefully removed from the shells, then pooled together and dipped into liquid nitrogen immediately after sampling before being stored at −80°C. These two sampling dates corresponded (1) to the time of the year when seawater temperature attains 16°C, a threshold temperature above which disease transmission is optimal and mortalities occur [6–8,36], and (2) to the onset of the mass mortality, respectively (Figure 2).

Temperature and salinity (data not shown) were recorded every 10 min during the entire period of study, using autonomous CTD multiparameter recorders (NKE Instrumentation) deployed at Marseillan, within the bivalve farming area (43°22′44.86″ N, 3°34′15.88″ W, Figure 1). Temperature and salinity were also measured at each location for each sampling date using a WTW ProfiLab LF597-5 salinometer (Sigma-Aldrich, Lyon, France) to obtain information for the entire lagoon. Periodic temperature measurements taken at each location fit with the continuous temperature profile recorded at Marseillan (Figure 2). Overall, spatial variation in seawater temperature was ±0.4°C in Thau lagoon.

Laboratory Analyses

Pooled oysters (n = 10 individuals) were ground with a MM400 homogeniser (Retsch) under liquid nitrogen, and the resulting powder was stored at −80°C and sub-sampled for pathogen detection and biochemical analyses.

Pathogen and energetic analyses were conducted as previously described ([7,10] and see File S1). Briefly, the detection and quantification of OsHV-1 DNA was carried out using a previously published real-time PCR protocol [37]. The energetic reserves triacylglycerol and carbohydrate were expressed in mg g^{-1} dry mass of tissues. The relative variation of these energetic reserves was calculated between 6 and 16 April according to the following formula:

$$\Delta_{reserve} = \frac{(X_{06\ April} - X_{16\ April})}{X_{06\ April}}$$

where X represents triacylglycerol or carbohydrate.

The fatty acid markers investigated in this study were the ratio of 16:1n−7/16:0 and 20:5n−3, both of which indicate the contribution of diatoms to the diet of oysters and mussels; the sum of 18:2n−6 and 18:3n−3, which is generally considered as a marker of terrestrial inputs; the ratio of 18:1n-9/18-1n-7, which is generally used as an indicator of carnivory; the ratio of polyunsaturated/saturated fatty acid (PUFA/SFA), which is an indicator of freshness; and the sum of iso- and anteiso-branched chain fatty acids and unbranched 15:0 and 17:0, which reflects the contribution of bacteria to the organic matter. These fatty acid food web markers are extensively used in trophic ecology [34,35].

Statistical Analyses

Nonparametric estimates of the survivor function were computed by the Kaplan–Meier method [38]. Survival time was measured as days from 6 April (time origin), when seawater temperature reached 16°C. The data were read as the number of dead animals within each experimental unit at each time interval for 13 intervals. Sampling areas were used as strata and the resulting survival estimates were compared by using the log-rank test of homogeneity of strata.

Two-way mixed model ANOVAs were conducted to determine potential differences in energetic reserves of oysters according to sampling area (Bouzigues, Mèze, Marseillan, outside) and time (6 and 16 April 2011). The term 'Time' was a random factor with two levels of repeated measurements. One-way ANOVAs were conducted to determine potential differences in fatty acid food web markers of oysters according to sampling area. Where differences were detected, Fisher's protected LSD multiple comparison tests were used to determine which means were significantly different. This procedure controls the familywise error rate since those tests are only applied following a significant effect in the ANOVA table.

Detection frequencies of OsHV-1 DNA in oysters sampled on 6 April were analysed by chi-square tests of independence according to area (4 areas×2 outcomes [not detected/detected]).

The survival time curves of oysters in different sampling areas were compared using the Cox proportional hazards regression model [39], after adjustment for the effect of some static covariates like whether or not OsHV-1 DNA is detected, the levels of triacylglycerol and carbohydrate measured on 6 April, and the relative variation of these energetic reserves between 6 and 16 April. The tests on regression parameters were made using the

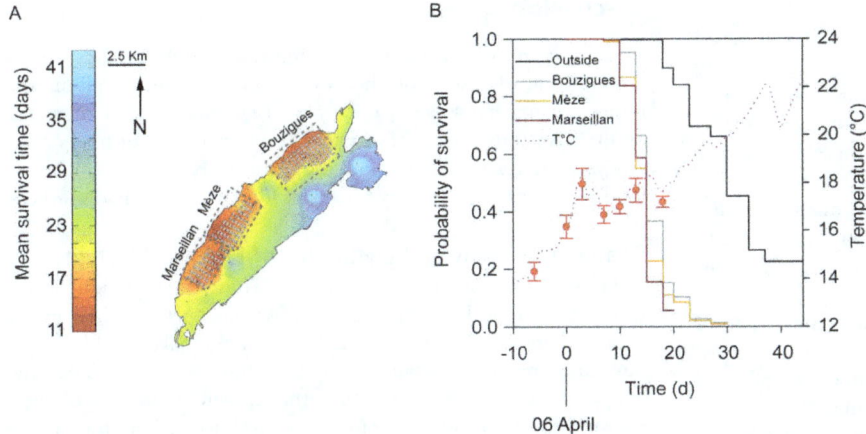

Figure 2. Survival of oysters in the Thau lagoon. (A) Kriged map of the mean survival time of oysters in the Thau Mediterranean lagoon. Mean survival time was measured as days from 6 April, when seawater temperature reached 16°C. Black points represent sampling sites, areas with grey boxes represent bivalve farms and dashed rectangles correspond to the three farming areas. (B) Left axis: survival functions of oysters for each area of the lagoon. Right axis: evolution of seawater temperature during the period of study. Temperature was recorded continuously at one location within the bivalve farming area of Marseillan (dotted line), and punctually at all locations (red circles, data are means ± SD).

robust sandwich method [40]. The proportionality of hazards was checked by testing the interaction between time and levels of treatment (areas).

Multiple regression models were used to examine the relationship between energetic reserves and food web fatty acid markers of oysters measured on 6 April. Analyses were done on the best transformation of response variables among the Box-Cox family [41]. Triacylglycerol and carbohydrate were square root transformed while Δ triacylglycerol and Δ carbohydrate were log transformed to eliminate skewness and to reach normality. For each model, explanatory variables (fatty acid food web markers), including cross-product terms and interactions, were selected based on the Akaike information criteria (AIC) in a stepwise method. This means that explanatory variables were added into the model one by one, by selecting at each step the one that minimizes the AIC criteria. This was done until the stopping rule was met, which consists of the smallest residual sum of squares using the leave-one-out cross-validation technique.

These statistical analyses were conducted using LIFETEST, FREQ, MIXED, PHREG, CORR, REG and GLMSELECT procedures of the SAS software package (SAS 9.3, SAS institute, Carry, USA).

Spatial structures of the mean survival time, energetic reserves and trophic markers of oysters were described through variograms, allowing quantification of the spatial dependency and its partitioning among distance classes [42]. Statistical models (linear, exponential and spherical) were fitted to the variograms to produce interpolated maps by kriging for each variable using ISATIS [43].

Results

Disease Mortality of Oysters

The sentinel oysters deployed in Thau lagoon were severely hit by the mass mortality phenomenon. Final survival was lower than 15% in 104 out of 106 experimental units (Figure S2.1 in File S2). Survival was almost 100% in the two remaining experimental units located in the south-eastern extremity of the lagoon (Figure S2.1 in File S2).

Survival time was computed in days from 6 April, when seawater temperature reached 16°C, and used for spatial analyses

(Figure 2A). Survival time of oysters varied from 8.9 d to 45.0 d, the latter corresponding to the entire duration of the experiment. Survival time of oysters showed strong spatial dependence (Figure S2.2 in File S2). Survival functions of oysters differed among areas (Figure 2B, log-rank test $p<0.001$). Mean survival time of oysters was particularly low within the bivalve farming areas (16.0 d ±0.06, 14.9 d ±0.06, and 14.1 d ±0.04 at Bouzigues, Mèze and Marseillan, respectively) compared to that of animals deployed outside the farming areas where it reached 29.7 d ±0.29.

On 6 April, when seawater temperature attained 16°C, OsHV-1 DNA was detected in 33 experimental units out of 104 (Table 1, Figure S3.1 in File S3). There was no significant effect of sampling areas on the detection of OsHV-1 DNA in oysters (χ^2 test, p = 0.095, Figure S3.1 in File S3). On 16 April, when the mass mortality event started (Figure 2B), OsHV-1 DNA was detected in all experimental units (Table 1). Therefore, the mass mortality of oysters coincided with the general spread of OsHV-1 in oysters of the Thau lagoon.

Table 1. Detection of OsHV-1 DNA according to areas and time in oysters deployed in the Thau lagoon.

Areas	OsHV-1 DNA					
	6 April		Σ	16 April		Σ
	Not detected	detected		not detected	detected	
Within farming areas	66	30	96	0	92	92
Bouzigues	23	12	35	0	30	30
Mèze	27	5	32	0	33	33
Marseillan	16	13	29	0	29	29
Outside farming areas	5	3	8	0	8	8
Σ	71	33	104	0	100	100

Data are number of pearl nets containing the sampled oysters (n = 10 pooled animals for each analysis).

Energetic Reserves

Level of triacylglycerols in oysters sampled on 6 April varied markedly, from 2.0 mg g^{-1} wet tissues to 53.5 mg g^{-1}, depending on location in the lagoon (Figure 3A). On average, the level of triacylglycerol in oysters maintained outside the farming areas was 1.5, 1.8 and 2.7 times higher than the levels of those held in the farming areas of Mèze, Bouzigues and Marseillan, respectively (Figure 3B). Ten days later, at the onset of the mass mortality event, levels of triacylglycerol dropped by 49%, 36% and 29% in oysters held at Bouzigues, Mèze and outside of the farming areas, respectively (Figures 3A, B). The level of triacylglycerol in oysters held outside the farming areas remained higher than that of oysters maintained within the farming areas. The level of triacylglycerol in oysters held at Marseillan remained low and did not change significantly over the sampling time.

Level of carbohydrates in oysters sampled on 6 April varied depending on location in the lagoon (Figure 3C) and had dropped by 51% and 19% at Bouzigues and Mèze, respectively, by 16 April (Figures 3C, D). Also, the level of carbohydrates in oysters held outside the farming areas remained higher than that of oysters maintained within the farming areas on 16 April. As observed for triacylglycerol, the level of carbohydrate in oysters held at Marseillan remained low and did not change significantly between sampling times (Figures 3C, D).

Levels of triacylglycerols and carbohydrates in OsHV-1 positive oysters on 6 April were slightly lower or similar to levels in OsHV-1 negative animals (Triacylglycerol: 14.2 $vs.$ 17.4 mg g^{-1} wet tissues, p $=0.047$, Carbohydrate: 149.6 vs. 139.6 mg g^{-1} wet tissues, p $=0.337$).

Proportional Hazards Model

Table 2 shows the final model describing the effects of sampling areas, detection of OsHV-1 DNA and energetic reserves on the risk of mortality in oysters (hazard ratio). The risk of oyster mortality was 9.6 to 14.9 times higher in the farming areas of the Thau lagoon than outside these areas (Table 2). The risk of mortality of oysters was similar among the three farming areas (p> 0.05). Risk of mortality was 1.4 times higher in oysters when OsHV-1 DNA was detected on April 6, than when it was not. Although a relation between the risk of mortality and energetic reserves of oysters measured on April 6 was not significant, survival time increased with difference of energetic reserves between 6 and 16 April (Table 2). Therefore, the capacity of oysters to use their energetic reserves during the infection period was associated with a lower risk of mortality. It is, however, noteworthy that energetic reserves of oysters measured on April 6 were correlated with their utilisation between 6 and 16 April (r values were 0.377 and 0.511 for triacylglycerols and carbohydrates, respectively, p<0.001).

Fatty Acid Food Web Markers

Kriged maps of fatty acid food web markers show high spatial variation for diatom (16:1n-7/16:0 and 20:5n-3), terrestrial (18:2n-6+18:3n-3), animal (18:1n-9/18:1n-7) and bacterial contributions to the diet of oysters (Figure S4.1 in File S4).

Multiple regression models showed that the two diatom markers, 16:1n-7/16:0 and 20:5n-3, explained 28.3% of the spatial variance in triacylglycerol levels of oysters, whereas 20:5n-3 and PUFA/SFA explained 41.3% of the spatial variance in carbohydrate (Table 3). The diatom marker 20:5n-3 explained the largest part of the variance in Δ triacylglycerol and Δ carbohydrate (15.8% and 18.4% respectively, Table 3).

Discussion

Here we provide a large-scale high-resolution spatial representation of a disease mortality event affecting an economically important marine species, the oyster *Crassostrea gigas*. The spatial and temporal dynamics of disease-related mortality of oysters was analysed in relation with host energetic reserves and quality of food resources estimated by means of food web fatty acid markers.

Spatial and Temporal Dynamics of Disease Mortality

Transmission of OsHV-1 within an oyster population occurs when susceptible hosts encounter infectious particles shed in the environment by neighboring infected individuals [36,44,45]. In our study, disease mortality of oysters showed strong spatial and temporal dependence, reflecting the epidemic process of local transmission [23]. Survival of oysters was lower within the bivalve farming area than outside it. Although the study design was not optimal to investigate survival time of oysters outside the farming areas, it appeared that the disease outbreak started within the bivalve farms. This result agrees well with the fact that aquaculture, which provides high-density populations of susceptible hosts, offers ideal conditions for disease epizootics [25,46].

The disease affecting the oysters likely spread outside of bivalve farms via water currents. Indeed, marine parasites can be transported over considerable distances by seawater movement [25,46,47]. The ability of OsHV-1 to persist in seawater, even briefly, may be what allows the pathogen to spread to the whole lagoon. In a previous study, we showed that mortality of oysters varies between and within farming sites in a way that was consistent with the hydrodynamic regime and connectivity among areas where infected oysters were present [7]. For instance, mortality of oysters maintained in Thau lagoon but outside the farming area coincided with relatively strong currents coming from the farming area where mortality was occurring and OsHV-1 DNA was detected. Recently, the role of physical factors influencing infection prevalence of *Haplosporidium nelsoni*, causative agent of MSX disease in the eastern oyster *Crassostrea virginica*, was investigated by means of high-resolution hydro-dynamical model in Delaware Bay, USA [47]. These authors showed that spatial and temporal dynamics of *H. nelsoni* infection are related to seawater currents: infection prevalence at up bay locations corresponds to periods of enhanced cross-bay and up bay transport. Also, the spread of disease in aquaculture fish species implicate hydrodynamic regime, currents and the proximity of infected farms in the spread of diseases [48−52]. These findings are supported by modelling and have led to the adoption of management practices integrating the tidal excursion [52]. In our study, it is likely that the high survival of oysters observed at the south-eastern extremity of the lagoon reflects the influence of the Sète channel which connects the lagoon to the Mediteranean Sea [53], where no oyster mortality occurs [7].

Disease and related mortalities of oysters spread very rapidly at the lagoon scale, likely reflecting that the sentinel oysters were young, naïve and not selected for resistance. Indeed, susceptibility of oysters to OsHV-1 decreases with age of oysters, with past exposure to the pathogen, and with selection of resistant individuals [7,54−56]. Also, the rapid spread of the disease and related mortalities of oysters is probably typical of this Mediterranean lagoon, where mortality rates of oysters caused by OsHV-1 are the highest compared to other French Atlantic areas [57]. Spatiotemporal studies of disease mortality in oysters are currently underway in other observation areas in France.

Risk of oyster mortality was positively correlated with detection of OsHV-1 DNA in oyster tissues, which agrees well with the idea

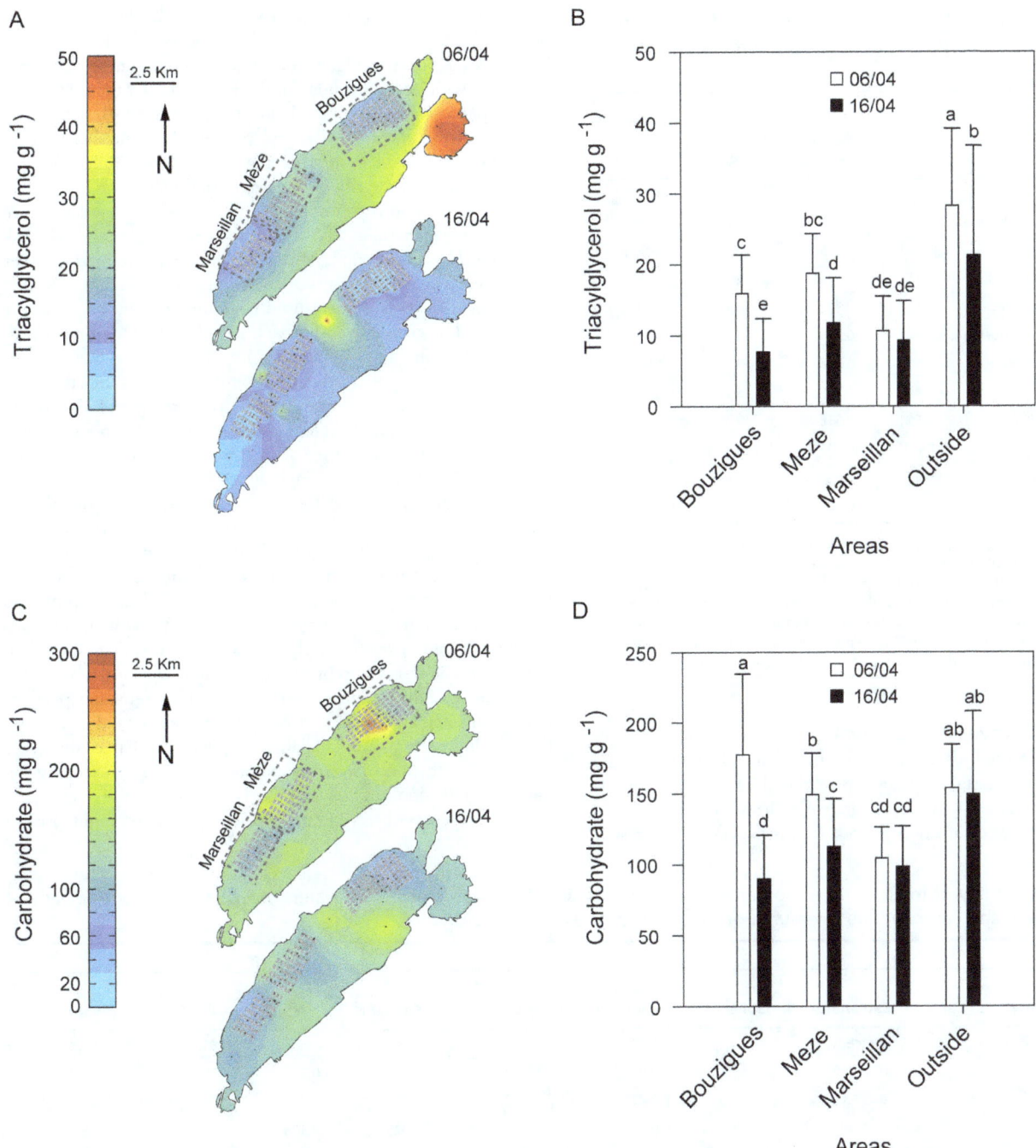

Figure 3. Energetic reserves of oysters in the Thau lagoon. (A, C) Kriged maps of the triacylglycerol and carbohydrate levels in oysters in the Mediterranean Thau lagoon measured on 6 April when seawater temperature reached 16°C, and 16 April at the beginning of the mass mortality phenomenon. Black points represent sampling sites, areas with grey boxes represent individual bivalve farms and dashed rectangles correspond to the three farming areas. (B, D) Triacylglycerol and carbohydrate levels of oysters as a function of time and area in the Thau lagoon. Letters indicate significant differences. Data are means ± SD.

that OsHV-1 is a causal factor of mass mortality [7−9,11,44,56,58−60]. It is, however, noteworthy that detection of OsHV-1 DNA in great amount in oysters tissues (on 16 April, the quantity of the viral DNA had increased up to $9.0 \cdot 10^{8}$ copy number mg^{-1} wet tissue of oysters, regardless of areas) does not necessarily lead to high mortality rate, as observed in oysters maintained free of mortality in the south-easternmost part of the lagoon (present study), in resistant animals [56] or in oysters cemented onto ropes at low density [7]. These previous studies showed that OsHV-1 positive oysters which exhibit no or low mortality clear the viral DNA from their tissue more rapidly than those which suffer high mortality. It seems that the capacity to rapidly eliminate the viral DNA is linked with virus resistance and survival in oysters.

Table 2. Analysis of the effect of covariates on survival of oysters in the Thau lagoon by a Cox proportional hazards model.

Covariate	Hazard ratio	Hazard ratio SE	χ^2	p value
Areas	n.a	n.a	33.2	<0.001
Bouzigues vs Outside	9.671	4.282	26.3	<0.001
Mèze vs outside	14.870	6.975	33.1	<0.001
Marseillan vs outside	13.551	6.876	26.4	<0.001
Marseillan vs Mèze	0.911	0.228	0.1	0.710
Marseillan vs Bouzigues	1.401	0.870	1.9	0.165
Mèze vs Bouzigues	1.537	0.346	3.7	0.056
OsHV-1 DNA	1.405	0.209	5.2	0.022
Triacylglycerol	0.971	0.018	2.4	0.119
Carbohydrate	1.004	0.002	3.3	0.070
Δ Triacylglycerol	0.771	0.082	6.0	0.014
Δ Carbohydrate	0.680	0.108	5.9	0.015

For each covariate, the following elements are provided: its parameter estimate and the corresponding instantaneous hazard ratio and its standard error (SE); the χ^2 statistic (with 1 degree of freedom for all tests except for areas where it was 3) and the resulting p value for the type II test from the complete model. Covariates are the presence or not of OsHV-1 DNA in oysters (on 6 April), their triacylglycerol and carbohydrate reserves (on 6 April), and the relative variation of these energetic reserves (Δ) between 6 and 16 April.

Energetic Reserves and Survival

Our study provides evidence that, under natural conditions, the risk of disease mortality in oysters exposed to the herpesvirus OsHV-1 decreases with increasing mobilization of energetic reserves during the infection period. It was previously suggested that oysters with high energy reserves may be less affected by pathogen infections than those with low levels [7]. For instance, higher levels of triacylglycerol in oysters cemented to ropes coincided with lower mortality and limited proliferation of OsHV-1 compared with oysters held in Australian baskets [7]. Interestingly, the present study reveals that mobilization of energetic reserves in oysters during the period of infection is a better predictor of the risk of disease mortality than the initial level of reserves. It is, however, noteworthy that these two variables are positively correlated, so that their effects on oyster mortality risk are somewhat confounded. Also, it cannot be ascertained whether the decrease in energy reserves contributed to the mass mortality as the effect of other unmeasured confounding factors cannot be ruled out. The correlative approach used in this study prevents us from establishing causal relationship.

The use of energetic reserves during the period of infection may fuel oyster immune response. Therefore, higher mobilization of energetic reserves may lead to a stronger immune response, which increases the survival time of oysters. In bivalves, immunity consists of innate processes, including various serologically active molecules and of phagocytosis accompanied by production of oxygen metabolites and the release of lysosomal enzymes [61]. Recent studies show that bivalves exposed to pathogenic bacteria allocate a part of their energy from feeding and energetic reserves to sustain immune, antioxidant and cytoprotection processes [62,63]. For example, oyster larvae exposed to pathogenic *Vibrio coralliilyticus* had lower triacylglycerol and protein content, the two main energetic reserves in bivalve larvae, and higher abundance of transcripts of antioxidant enzymes and immune-related proteins than uninfected larvae [62]. However, these authors also reported that feeding activity in challenged oysters decreased markedly compared with that of control animals, which is symptomatic of pathogen infection in bivalves [62,64]. Therefore, the decrease in energetic reserves observed during disease outbreaks in the present

Table 3. Summary of multiple regression analyses using fatty acid food web markers in oysters as explanatory variables and triacylglycerol and carbohydrate (measured on 6 April and Δ) as response variables.

Response variables	Explanatory variables	Parameter estimates	SE	Partial r^2	$\sum r^2$	AICC	p value
$\sqrt{\text{Triacylglycerol}}$	Intercept	3.902	0.077			81.5	<0.001
	16:1/16:0	2.851	0.650	0.229	0.229	58.1	<0.001
	20:5n-3	0.228	0.079	0.054	0.283	53.1	0.005
	18:2+18:3	0.384	0.155	0.025	0.308	51.8	0.015
	20:5n-3×(18:2+18:3)	−0.119	0.063	0.024	0.332	50.7	0.062
	PUFA/SFA	−1.131	0.638	0.022	0.355	49.7	0.080
$\sqrt{\text{Carbohydrate}}$	Intercept	11.738	0.198			235.4	<0.001
	20:5n-3	0.981	0.125	0.289	0.289	204.2	<0.001
	PUFA/SFA	−3.030	1.106	0.099	0.388	191.6	0.007
	20:5n-3^2	0.070	0.035	0.025	0.413	189.7	0.050
Log Δ triacylglycerol	Intercept	0.523	0.061			13.0	<0.001
	20:5n-3	0.142	0.035	0.158	0.157	−0.7	<0.001
Log Δ carbohydrate	Intercept	0.340	0.041			−52.7	<0.001
	20:5n-3	0.110	0.025	0.184	0.184	−69.5	<0.001
	18:2+18:3	−0.155	0.064	0.047	0.231	−72.8	0.018
	16:1/16:0	−0.649	0.336	0.031	0.262	−74.4	0.057

All variables were measured in oysters sampled in Thau Mediterranean lagoon.

study may reflect not only the cost of immune response, but also a concomitant reduction in food acquisition.

Energetic Reserves and Food Quality

The local environment causes marked changes in energetic reserves of oysters and their utilisation during the infection period. The observed changes in energetic condition of oysters partly reflect natural variation in food quality. For instance, energetic reserves of oysters were positively correlated with 16-1n7/16:0 and 20:5n-3, two diatom markers commonly used in trophic ecology [34,35]. This result agrees well with previous studies showing that diatoms are the main food source sustaining oyster growth and development in Thau lagoon, whereas terrestrial, animal (small zooplankton such as ciliate protists and tintinnids) and bacterial organic matter are secondary food sources that mostly contribute to the diet of oysters during non-bloom periods [17,65]. Therefore, food quality explains a significant part of the variance in energetic reserves of oysters and their utilisations, which in turn is negatively correlated with the risk of disease mortality.

Although we specifically tested for the effect of food quality on energetic reserves of oysters, we cannot rule out the effect of other factors such as food availability. Several studies have found that shellfish farms have a large impact on plankton communities and biomass. For example, a study conducted in 1991–92 reported that the presence of shellfish farms led to a ~40% deficit in chlorophyll *a* in the western part of the Thau lagoon [66]. It is therefore likely that oysters held outside of the farming area benefit from more food than those within it, thus enhancing their energetic reserves and their disease resistance.

Conclusion

This study shows that local differences in the use of energetic reserves of oysters, partly driven by variation in food quality, play a significant role in the spatial and temporal dynamics of disease mortality. In particular, the relative contribution of diatoms to the diet of oysters correlates with their energetic reserves and their utilisation during the infection period which decreases the risk of disease mortality. Therefore, energetic status and food quality

could have major implications for host-pathogen dynamics in marine ecosystems. Our study provides a better understanding of the factors that contribute to transmission of disease mortalities in the aquatic environment, which is necessary to build realistic predictive modelling of disease mortalities. Such models are needed as tools to test disease control scenarios that could mitigate the impact of disease mortalities and could potentially be adapted to other epizootic events that may occur in the near future.

Supporting Information

File S1 Laboratory analyses.

File S2 Spatial and temporal dynamics of oyster mortality (kriged maps and variogram).

File S3 Spatial distribution of OsHV-1 DNA in oysters.

File S4 Fatty acid food web markers in oysters (kriged maps and result description).

Acknowledgments

The authors thank Marine Miguet and Julien Vicario for their careful lab work, the staff involved in oyster production (Ifremer hatchery and nursery teams) and Nicolas Keck and Hélène Boulet from Laboratoire départemental vétérinaire de l'Hérault for OsHV-1 analyses. The authors express their gratitude to Helen McCombie for revising the English version of this manuscript, Coralie Lupo for providing advice on statistics, and 5 anonymous referees and David Bushek for their constructive reviews.

Author Contributions

Conceived and designed the experiments: FP FL NJ. Performed the experiments: FP FL JB PLG CQ. Analyzed the data: FP FL NJ GD. Contributed reagents/materials/analysis tools: FP ER. Wrote the paper: FP FL ER.

References

1. Harvell CD, Mitchell CE, Ward JR, Altizer S, Dobson AP, et al. (2002) Ecology - Climate warming and disease risks for terrestrial and marine biota. Science 296: 2158–2162. doi: 10.1126/science.1063699.

2. Lafferty KD, Porter JW, Ford SE (2004) Are diseases increasing in the ocean? Annu Rev Ecol Evol Syst 35: 31–54. doi: 10.1146/annurev.ecolsys.35.021103.105704.

3. Mydlarz LD, Jones LE, Harvell CD (2006) Innate immunity environmental drivers and disease ecology of marine and freshwater invertebrates. Annu Rev Ecol Evol Syst 37: 251–288. doi: 10.1146/annurev.ecolsys.37.091305.110103.

4. Comps M, Bonami JR (1977) Virus-infection associated with mortality of oyster *Crassostrea gigas* Thunberg. C R Hebd Séanc Acad Sc D 285: 1139–1140. doi: not found.

5. Samain J-F and McCombie H (2008) Summer mortality of Pacific oyster *Crassostrea gigas*, the Morest project.

6. Jolivel A, Fleury E (2012) Analyse statistique des données de mortalité d'huîtres acquises par l'Observatoire National Conchylicole (RESCO). Rennes 2 Agrocampus. Ouest. pp. 67.

7. Pernet F, Barret J, Gall PL, Corporeau C, Dégremont L, et al. (2012) Mass mortalities of Pacific oysters *Crassostrea gigas* reflect infectious diseases and vary with farming practises in the Thau lagoon. Aquaculture Env Interact 2: 215–237. doi: 10.3354/aei00041.

8. EFSA (2010) Scientific Opinion of the Panel on Animal Health and Welfare on a request from the European Commission on the increased mortality events in Pacific oysters *Crassostrea gigas*. EFSA 8: 1894–1953. doi: 10.2903/j.efsa.2010.

9. Renault T (2011) A review of mortality outbreaks in the Pacific oyster, *Crassostrea gigas*, reported since 2008 in various European Union Member States and the related implementation of Council Directive 2008/88/EC. Bulletin OIE 4: 51–52.

10. Pernet F, Barret J, Marty C, Moal J, Le Gall P, et al. (2010) Environmental anomalies, energetic reserves and fatty acid modifications in oysters coincide

with an exceptional mortality event. Mar Ecol Prog Ser 401: 129–146. doi: 10.3354/meps08407.

11. Segarra A, Pepin JF, Arzul I, Morga B, Faury N, et al. (2010) Detection and description of a particular Ostreid herpesvirus 1 genotype associated with massive mortality outbreaks of Pacific oysters, *Crassostrea gigas*, in France in 2008. Virus Res 153: 92–99. doi: 10.1016/j.virusres.2010.07.011.

12. Beldomenico PM, Telfer S, Gebert S, Lukomski L, Bennett M, et al. (2008) Poor condition and infection: a vicious circle in natural populations. Proc R Soc Lond Ser B-Biol Sci 275: 1753–1759. doi: 10.1098/rspb.2008.0147.

13. Beldomenico PM, Telfer S, Gebert S, Lukomski L, Bennett M, et al. (2009) The vicious circle and infection intensity: The case of *Trypanosoma microti* in field vole populations. Epidemics 1: 162–167. doi: 10.1016/j.epidem.2009.05.002.

14. Townsend AK, Clark AB, McGowan KJ, Miller AD, Buckles EL (2012) Condition, innate immunity and disease mortality of inbred crows. Proc R Soc Lond Ser B-Biol Sci 277: 2875–2883. doi: 10.1098/rspb.2010.0480.

15. Moller AP, Christe P, Erritzoe J, Mavarez J (1998) Condition, disease and immune defence. Oikos 83: 301–306. doi: 10.2307/3546841.

16. Kang CK, Lee YW, Choy EJ, Shin JK, Seo IS, et al. (2006) Microphytobenthos seasonality determines growth and reproduction in intertidal bivalves. Mar Ecol Prog Ser 315: 113–127. doi: 10.3354/meps315113.

17. Pernet F, Malet N, Pastoureaud A, Vaquer A, Quéré C, et al. (2012) Marine diatoms sustain growth of bivalves in a Mediterranean lagoon. J Sea Res 68: 20–32. doi: 10.1016/j.seares.2011.11.004.

18. Delaporte M, Soudant P, Lambert C, Moal J, Pouvreau S, et al. (2006) Impact of food availability on energy storage and defense related hemocyte parameters of the Pacific oyster *Crassostrea gigas* during an experimental reproductive cycle. Aquaculture 254: 571–582. doi: 10.1016/j.aquaculture.2005.10.006.

19. Soudant P, Chu F-LE, Volety A (2013) Host-parasite interactions: Marine bivalve molluscs and protozoan parasites, Perkinsus species. J Invertebr Pathol 114: 196–216. doi: 10.1016/j.jip.2013.06.001.

20. Delaporte M, Soudant P, Lambert C, Jegaden M, Moal J, et al. (2007) Characterisation of physiological and immunological differences between Pacific oysters (*Crassostrea gigas*) genetically selected for high or low survival to summer mortalities and fed different rations under controlled conditions. J Exp Mar Biol Ecol 353: 45–57. doi: 10.1016/j.jembe.2007.09.003.

21. Delaporte M, Soudant P, Moal J, Giudicelli E, Lambert C, et al. (2006) Impact of 20: 4n-6 supplementation on the fatty acid composition and hemocyte parameters of the Pacific oyster *Crassostrea gigas*. Lipids 41: 567–576. doi: 10.1007/s11745-006-5006-9.

22. Delaporte M, Soudant P, Moal J, Lambert C, Quere C, et al. (2003) Effect of a mono-specific algal diet on immune functions in two bivalve species - *Crassostrea gigas* and *Ruditapes philippinarum*. J Exp Biol 206: 3053–3064. doi: 10.1242/jeb.00518.

23. Osnas EE, Heisey DM, Rolley RE, Samuel MD (2009) Spatial and temporal patterns of chronic wasting disease: fine-scale mapping of a wildlife epidemic in Wisconsin. Ecol Appl 19: 1311–1322. doi: 10.1890/08–0578.1.

24. Ostfeld RS, Glass GE, Keesing F (2005) Spatial epidemiology: an emerging (or re-emerging) discipline. Trends Ecol Evol 20: 328–336. doi: 10.1016/j.tree.2005.03.009.

25. McCallum HI, Kuris A, Harvell CD, Lafferty KD, Smith GW, et al. (2004) Does terrestrial epidemiology apply to marine systems? Trends Ecol Evol 19: 585–591. doi: 10.1016/j.tree.2004.08.009.

26. Thieltges D, Reise K (2007) Spatial heterogeneity in parasite infections at different spatial scales in an intertidal bivalve. Oecologia 150: 569–581. doi: 10.1007/s00442-006-0557-2.

27. Binias C, Tu Do V, Jude-Lemeilleur F, Plus M, Froidefond J-M, et al. (2013) Environmental factors contributing to the development of brown muscle disease and perkinsosis in Manila clams (*Ruditapes philippinarum*) and trematodiasis in cockles (*Cerastoderma edule*) of Arcachon Bay. Marine Ecology. doi: 10.1111/maec.12087.

28. de Montaudouin X, Binias C, Lassalle G (2012) Assessing parasite community structure in cockles *Cerastoderma edule* at various spatio-temporal scales. Estuarine, Coastal and Shelf Science 110: 54–60. doi: 10.1016/j.ecss.2012.02.005.

29. Ford SE, Scarpa E, Bushek D (2012) Spatial and temporal variability of disease refuges in an estuary: Implications for the development of resistance. J Mar Res 70: 2–3. doi: 10.1357/002224012802851850.

30. Oliver LM, Fisher WS, Ford SE, Calvo LMR, Burreson EM, et al. (1998) Perkinsus marinus tissue distribution and seasonal variation in oysters *Crassostrea virginica* from Florida, Virginia and New York. Dis Aquat Org 34: 51–61. doi: 10.3354/dao034051.

31. Ford SE (1996) Range extension by the oyster parasite *Perkinsus marinus* into the northeastern United States: Response to climate change? J Shellfish Res 15: 45–56.

32. White DL, Bushek D, Porter DE, Edwards D (1998) Geographic information systems (GIS) and kriging: Analysis of the spatial and temporal distributions of the oyster pathogen *Perkinsus marinus* in a developed and an undeveloped estuary. J Shellfish Res 17: 1473–1476.

33. Gray BR, Bushek D, Drane JW, Porter D (2009) Associations between land use and *Perkinsus marinus* infection of eastern oysters in a high salinity, partially urbanized estuary. Ecotoxicology 18: 259–269. doi: 10.1007/s10646-008-0279-9.

34. Dalsgaard J, St. John M, Kattner G, Muller-Navarra D, Hagen W (2003) Fatty acid trophic markers in the pelagic marine environment. Adv Mar Biol 46: 225–340. doi: 10.1016/S0065-2881(03)46005-7.

35. Kelly JR, Scheibling RE (2012) Fatty acids as dietary tracers in benthic food webs. Mar Ecol Prog Ser 446: 1–22. doi: 10.3354/meps09559.

36. Petton B, Pernet F, Robert R, Boudry P (2013) Experimental investigations of temperature influence on pathogens transmission and subsequent mortalities in juvenile Pacific oysters *Crassostrea gigas* Aquaculture Env Interact 3: 257–273. doi: 10.3354/aei00070.

37. Pepin JF, Riou A, Renault T (2008) Rapid and sensitive detection of ostreid herpesvirus 1 in oyster samples by real-time PCR. J Virol Methods 149: 269–276. doi: 10.1016/j.jviromet.2008.01.022.

38. Kaplan EL, Meier P (1958) Nonparametric Estimation from Incomplete Observations. Journal of the American Statistical Association 53: 457–481. doi: 10.1080/01621459.1958.10501452.

39. Cox DR (1972) Regression Models and Life Tables. Journal of the Royal Statistical Society: Series B (Statistical Methodology) 20: 187–220. doi: 10.1007/978-1-4612-4380-9_37.

40. Lin DY, Wei LJ (1989) The Robust Inference for the Proportional Hazards Model. Journal of the American Statistical Association 84: 1074–1078. doi: 10.2307/2290085.

41. Box GEP, Cox DR (1964) An analysis of transformations. Journal of the Royal Statistical Society: Series B (Statistical Methodology) 26: 211–252. doi: 10.2307/2287791.

42. Legendre P, Legendre L (1998) Numerical ecology. Amsterdam: Elsevier. 853 p.

43. Géovariances (2008) ISATIS Technical References, 9th Edition, Geovariances. Fontainebleau.

44. Schikorski D, Faury N, Pepin JF, Saulnier D, Tourbiez D, et al. (2011) Experimental ostreid herpesvirus 1 infection of the Pacific oyster *Crassostrea gigas*:

45. Sauvage C, Pépin JF, Lapègue S, Boudry P, Renault T (2009) Ostreid herpes virus 1 infection in families of the Pacific oyster, *Crassostrea gigas*, during a summer mortality outbreak: Differences in viral DNA detection and quantification using real-time PCR. Virus Res. doi: 10.1016/j.virusres.2009.02.013.

46. Krkošek M (2010) Review: Host density thresholds and disease control for fisheries and aquaculture. Aquacult Environ Interact 1: 21–32. doi: 10.3354/aei0004.

47. Wang Z, Haidvogel DB, Bushek D, Ford SE, Hofmann EE, et al. (2012) Circulation and water properties and their relationship to the oyster disease MSX in Delaware Bay. J Mar Res 70: 279–308. doi: 10.1357/002224012802851931.

48. Viljugrein H, Staalstrom A, Molvaelr J, Urke HA, Jansen PA (2009) Integration of hydrodynamics into a statistical model on the spread of pancreas disease (PD) in salmon farming. Dis Aquat Org 88: 35–44. doi: 10.3354/dao02151.

49. Gustafson LL, Ellis SK, Beattie MJ, Chang BD, Dickey DA, et al. (2007) Hydrographics and the timing of infectious salmon anemia outbreaks among Atlantic salmon (*Salmo salar* L.) farms in the Quoddy region of Maine, USA and New Brunswick, Canada. Prev Vet Med 78: 35–56. doi: 10.1016/j.prevetmed.2006.09.006.

50. Kristoffersen AB, Viljugrein H, Kongtorp RT, Brun E, Jansen PA (2009) Risk factors for pancreas disease (PD) outbreaks in farmed Atlantic salmon and rainbow trout in Norway during 2003–2007. Prev Vet Med 90: 127–136. doi: 10.1016/j.prevetmed.2009.04.003.

51. Salama NKG, Collins CM, Fraser JG, Dunn J, Pert CC, et al. (2013) Development and assessment of a biophysical dispersal model for sea lice. J Fish Dis 36: 323–337. doi: 10.1111/jfd.12065.

52. Salama NKG, Murray AG (2011) Farm size as a factor in hydrodynamic transmission of pathogens in aquaculture fish production. Aquaculture Env Interact 2: 61–74. doi: 10.3354/aei00030.

53. Lazure P (1992) Etude de la dynamique de l'étang de Thau par modèle numérique tridimensionnel. Vie et Milieu 42: 137–145.

54. Dégremont L (2013) Size and genotype affect resistance to mortality caused by OsHV-1 in *Crassostrea gigas*. Aquaculture 416–417: 129–134. doi: 10.1016/j.aquaculture.2013.09.011.

55. Paul-Pont I, Dhand NK, Whittington RJ (2013) Spatial distribution of mortality in Pacific oysters *Crassostrea gigas*: reflection on mechanisms of OsHV-1 transmission. Dis Aquat Org 105: 127–138. doi: 10.3354/dao02615.

56. Dégremont L (2011) Evidence of herpesvirus (OsHV-1) resistance in juvenile *Crassostrea gigas* selected for high resistance to the summer mortality phenomenon. Aquaculture 317: 94–98. doi: 10.1016/j.aquaculture.2011.04.029.

57. Fleury E, Bedier E (2013) RESCO - REseau d'Observations Conchylicoles : Campagne 2012. 98. Available: http://archimer.ifremer.fr/doc/00142/25346/. Accessed 12 November 2013.

58. Schikorski D, Renault T, Saulnier D, Faury N, Moreau P, et al. (2011) Experimental infection of Pacific oyster *Crassostrea gigas* spat by ostreid herpesvirus 1: demonstration of oyster spat susceptibility. Vet Res 42. doi: 2710.1186/1297-9716-42-27.

59. Martenot C, Oden E, Travaille E, Malas J-P and Houssin M (2011) Detection of different variants of Ostreid Herpesvirus 1 in the Pacific oyster, *Crassostrea gigas* between 2008 and 2010. Virus Res 160: 25–31. doi: 10.1016/j.virusres.2011.04.012.

60. Clegg TA, Morrissey T, Geoghegan F, Martin SW, Lyons K, et al. (2014) Risk factors associated with increased mortality of farmed Pacific oysters in Ireland during 2011. Prev Vet Med 113: 257–267. doi: 10.1016/j.prevetmed.2013.10.023.

61. Gestal C, Roch P, Renault T, Pallavicini A, Paillard C, et al. (2008) Study of Diseases and the Immune System of Bivalves Using Molecular Biology and Genomics. Rev Fish Sci 16: 133–156. doi: 10.1080/10641260802325518.

62. Genard B, Miner P, Nicolas J-L, Moraga D, Boudry P, et al. (2013) Integrative study of physiological changes associated with bacterial infection in pacific oyster larvae. PLoS ONE 8: e64534–e64534. doi: 10.1371/journal.pone.0064534.

63. Wang X, Wang L, Zhang H, Ji Q, Song L, et al. (2012) Immune response and energy metabolism of *Chlamys farreri* under *Vibrio anguillarum* challenge and high temperature exposure. Fish Shellfish Immunol 33: 1016–1026. doi: 10.1016/j.fsi.2012.08.026.

64. Flye-Sainte-Marie J, Pouvreau S, Paillard C, Jean F (2007) Impact of Brown Ring Disease on the energy budget of the Manila clam *Ruditapes philippinarum*. J Exp Mar Biol Ecol 349: 378–389. doi: 10.1016/j.jembe.2007.05.029.

65. Dupuy C, Vaquer A, Lam-Hoai T, Rougier C, Mazouni N, et al. (2000) Feeding rate of the oyster *Crassostrea gigas* in a natural planktonic community of the Mediterranean Thau Lagoon. Mar Ecol Prog Ser 205: 171–184. doi: 10.3354/meps205171.

66. Souchu P, Vaquer A, Collos Y, Landrein S, Deslous-Paoli JM, et al. (2001) Influence of shellfish farming activities on the biogeochemical composition of the water column in Thau lagoon. Mar Ecol Prog Ser 218: 141–152. doi: 10.3354/meps218141.

Kinetics of virus DNA detection by q-PCR in seawater and in oyster samples. Virus Res 155: 28–34. doi: 10.1016/j.virusres.2010.07.031.

Permissions

LIST OF CONTRIBUTORS

Tai Chong Toh, Chin Soon Lionel Ng, Jia Wei Kassler Peh, Kok Ben Toh and Loke Ming Chou
Reef Ecology Laboratory, Department of Biological Sciences, National University of Singapore, Singapore, Singapore

Grace S. Chiu
CSIRO Mathematics, Informatics and Statistics, Commonwealth Scientific and Industrial Research Organisation (CSIRO), Canberra, Australian Capital Territory, Australia

Margaret A. Wu
Business Methods Survey Division, Statistics Canada, Ottawa, Ontario, Canada

Lin Lu
McGregor GeoScience, Bedford, Nova Scotia, Canada

Stine M. Ulven and Mari C. W. Myhrstad
Department of Health, Nutrition and Management, Faculty of Health Sciences, Oslo and Akershus University College of Applied Sciences, Oslo, Norway

Kirsten B. Holven
Department of Nutrition, Institute for Basic Medical Sciences, University of Oslo, Oslo, Norway

Inger Ottestad
Department of Health, Nutrition and Management, Faculty of Health Sciences, Oslo and Akershus University College of Applied Sciences, Oslo, Norway, Department of Nutrition, Institute for Basic Medical Sciences, University of Oslo, Oslo, Norway,

Grethe I. Borge and Gjermund Vogt
Nofima, Norwegian Institute of Food, Fisheries and Aquaculture Research, Ås, Norway

Sahar Hassani
Nofima, Norwegian Institute of Food, Fisheries and Aquaculture Research, Ås, Norway
Centre for Integrative Genetics (CIGENE), Department of Mathematical Sciences and Technology, Norwegian University of Life Science, Ås, Norway

Achim Kohler
Centre for Integrative Genetics (CIGENE), Department of Mathematical Sciences and Technology, Norwegian University of Life Science, Ås, Norway
Nofima, Norwegian Institute of Food, Fisheries and Aquaculture Research, Ås, Norway

Tuulia Hyötyläinen and Matej Orešič
VTT Technical Research Centre of Finland, Espoo, Finland

Kirsti W. Brønner
TINE SA, Centre for Research and Development, Kalbakken, Oslo, Norway

Emily W. Grason and Benjamin G. Miner
Western Washington University, Biology Department, Bellingham, Washington, United States of America
Shannon Point Marine Center, Anacortes, Washington, United States of America

Natacha S. Agudo
Natural Resources Management, The WorldFish Center, Penang, Malaysia

Steven W. Purcell
National Marine Science Centre, Southern Cross University, Coffs Harbour, New South Wales, Australia
Natural Resources Management, The WorldFish Center, Penang, Malaysia

Jeffrey Levinton and Michael Doall
Department of Ecology and Evolution, Stony Brook University, Stony Brook, New York, United States of America

David Ralston
Woods Hole Oceanographic Institution, Woods Hole, Massachusetts, United States of America

Adam Starke and Bassem Allam
School of Marine and Atmospheric Sciences, Stony Brook University, Stony Brook, New York, United States of America

Paulo M. Brito
Department of Zoology, Universidade Estadual do Rio de Janeiro, Rio de Janeiro, Brazil

Dayse A. Silva and Elizeu F. Carvalho
Department of Ecology, Universidade Estadual do Rio de Janeiro, Rio de Janeiro, Brazil

Cesar R. L. Amaral
Department of Zoology, Universidade Estadual do Rio de Janeiro, Rio de Janeiro, Brazil
Department of Ecology, Universidade Estadual do Rio de Janeiro, Rio de Janeiro, Brazil

Verity Nye, Jonathan T. Copley and Paul A. Tyler
Ocean and Earth Science, National Oceanography Centre, University of Southampton, Southampton, United Kingdom
Laetitia Helene Marie Schmitt
Centre for Health Economics/Department of Economics and Related Studies, University of York, Heslington, York, United Kingdom

Cecile Brugere
Stockholm Environment Institute, University of York, Heslington, York, United Kingdom

Martin Ubertini and Karine Grangeré
Université de Caen Basse-Normandie, FRE3484 BioMEA, Caen, France
CNRS INEE, FRE3484 BioMEA, Caen, France

Francis Orvain
Université de Caen Basse-Normandie, FRE3484 BioMEA, Caen, France
CNRS INEE, FRE3484 BioMEA, Caen, France
CNRS, UMR 7208 BOREA, Muséum d'histoire naturelle, CRESCO, Dinard, France

Aline Gangnery and Romain Le Gendre
IFREMER, LERN, Port en Bessin, France

Sébastien Lefebvre
Université de Lille1, UMR CNRS 8187 LOG "Laboratoire d'océanologie et geosciences", Station Marine de Wimereux, Wimereux, France

Catherine Longo, Courtney Scarborough and Benjamin D. Best
National Center for Ecological Analysis and Synthesis, Santa Barbara, California, United States of America

Benjamin S. Halpern
National Center for Ecological Analysis and Synthesis, Santa Barbara, California, United States of America
Bren School of Environmental Science and Management, University of California, Santa Barbara, California, United States of America
Imperial College London, Silwood Park Campus, Ascot, United Kingdom

Darren Hardy
National Center for Ecological Analysis and Synthesis, Santa Barbara, California, United States of America
Digital Library Systems & Services, Stanford University, Stanford, California, United States of America

Scott C. Doney
Marine Chemistry & Geochemistry Department, Woods Hole Oceanographic Institution, Woods Hole, Massachusetts, United States of America

Steven K. Katona
Conservation International, Arlington, Virginia, United States of America

Karen L. McLeod
COMPASS, Oregon State University, Department of Zoology, Corvallis, Oregon, United States of America

Andrew A. Rosenberg
Union of Concerned Scientists, Cambridge, Massachusetts, United States of America

Jameal F. Samhouri
Conservation Biology Division, Northwest Fisheries Science Center, National Marine Fisheries Service, National Oceanic and Atmospheric Administration, Seattle, Washington United States of America

Fernando Norambuena, Michael Lewis, Karen Hermon and Giovanni M. Turchini
School of Life and Environmental Sciences, Deakin University, Warrnambool, Victoria, Australia

Noor Khalidah Abdul Hamid and John A. Donald
School of Life and Environmental Sciences, Deakin University, Waurn Ponds, Geelong, Victoria, Australia

Wei Xin and Xiaoying Li
First Department of Geriatric Cardiology, Chinese PLA General Hospital, Beijing, PR China

Wei Wei
Medical College of Nankai University, Tianjin, PR China

Julien de Lorgeril, Reda Zenagui and Evelyne Bachère
Institut Franc̦ais de Recherche pour l'Exploitation de la Mer, Centre National de la Recherche Scientifique, Montpellier, France

Rafael D. Rosa
Institut Français de Recherche pour l'Exploitation de la Mer, Centre National de la Recherche Scientifique, Montpellier, France
Université Montpellier 2, and Institut de Recherche pour le Développement, UMR 5119 "E´cologie des Systémes Marins Côtiers", Montpellier, France

David Piquemal
Skuld-Tech, Cap Delta, ZAC Euromedecine II, Grabels, France

Sharyn J. Goldstien and David R. Schiel
Marine Ecology Research Group, School of Biological Sciences, University of Canterbury, Christchurch, New Zealand

Graeme J. Inglis
National Institute of Water and Atmospheric Research, Aquatic Biodiversity and Biosecurity, Christchurch, New Zealand

Neil J. Gemmell
Centre for Reproduction and Genomics, Department of Anatomy, University of Otago, Dunedin, New Zealand
Allan Wilson Centre for Molecular Ecology and Evolution, University of Otago, Dunedin, New Zealand

Junfang Zhou, Wenhong Fang , Shuai Zhou, Linlin Hu, Xincang Li, Hang Su and Layue Xie
Key Laboratory of Marine and Estuarine Fisheries Resources and Ecology, East China Sea Fisheries Research Institute, Chinese Academy of Fisheries Science, Shanghai, China

Xianle Yang
Aquatic Pathogen Collection Center of Ministry, Shanghai, China

Xinyong Qi
Shanghai Animal Disease Control Center, Shanghai, China

Yusuke Tani,   Michiteru Kitazaki and Shigeki Nakauchi
Department of Computer Science and Engineering, Toyohashi University of Technology, Toyohashi, Aichi, Japan

Takehiro Nagai
Graduate school of Science and Engineering, Yamagata University, Yonezawa, Yamagata, Japan,

Kowa Koida
Electronics Inspired-Interdisciplinary Research Institute, Toyohashi University of Technology, Toyohashi, Aichi, Japan

Yoann Thomas and Serge Andréfouët
Institut de Recherche pour le Développement, unite de recherche CoRéUs, Nouméa, New Caledonia

Franck Dumas
Institut Français de Recherche pour l'Exploitation de la Mer, unite DYNECO, Plouzané, France

Cristiane T. Elfes
Department of Ecology, Evolution and Marine Biology, University of California Santa Barbara, Santa Barbara, California, United States of America

Catherine Longo, Courtney Scarborough and Benjamin D. Best
National Center for Ecological Analysis and Synthesis, Santa Barbara, California, United States of America

Benjamin S. Halpern
National Center for Ecological Analysis and Synthesis, Santa Barbara, California, United States of America
Bren School of Environmental Science and Management, University of California Santa Barbara, Santa Barbara, California, United States of America
Imperial College London, Silwood Park Campus, Berkshire, London, United Kingdom

Darren Hardy
National Center for Ecological Analysis and Synthesis, Santa Barbara, California, United States of America
Digital Library Systems, Stanford University, Stanford, California, United States of America

Tiago Pinheiro
Atlantic Forest Program, Conservation International Brazil, Belo Horizonte, Minas Gerais, Brazil

Guilherme F. Dutra
Marine Program, Conservation International Brazil, Rio de Janeiro, Rio de Janeiro, Brazil

Marc Suquet, Christian Mingant , Myrina Boulais, Isabelle Queau and Dominique Ratiskol
Ifremer, UMR 6539, PFOM Department, Station Expérimentale d'Argenton, Argenton, France

Catherine Labbé
INRA, UR 1037, LPGP, Rennes, France

Sophie Puyo, Benjamin Quittet and Pierrick Haffray
SYSAAF, LPGP, Rennes, France

Blandine Diss
Satmar, Barfleur, France

Franck Lagarde, Jean Barret, Patrik Le Gall and Emmanuelle Roque D'orbcastel
Ifremer, Laboratoire Environnement Ressource du Languedoc Roussillon, Bd Jean Monnet, Sète, France

Claudie Quere
UMR LEMAR Ifremer/CNRS/UBO/IRD, Technopole de Brest-Iroise, Plouzanè, France

Fabrice Pernet
Ifremer, Laboratoire Environnement Ressource du Languedoc Roussillon, Bd Jean Monnet, Sète, France

Gaetan Daigle
Université Laval, Département de mathématiques et de statistique, Pavillon Alexandre-Vachon, Québec, Québec, Canada

Nicolas Jeannée
Geovariances, Avon, France

Index

A

Actin Cytosqueleton, 150, 155

Antimicrobials, 150, 155

Aquaculture Development, 87-88, 95-96

Aquatic Organisms, 56, 150

Artificial Seeding, 1

Artisanal Fishing, 116, 118-119, 121, 124-128, 200, 207

Autophagy, 150, 154-156, 159

B

Bacterial Isolation, 171

Bathymetry, 53, 99-100, 102, 104-105, 107-108, 111, 186, 196

Benthic-pelagic Coupling, 99-100, 104-105, 108, 113

Biodiversity Conservation, 116, 130, 208

Biotic Metrics, 10, 14, 18-20

Blood Pressure, 22, 140-141, 143, 146, 148

C

Cancer Productus, 33-34

Carbon Storage, 118-119, 124-125, 127-128, 200-201, 203, 208

Cell Adhesion, 150, 154-156

Cell Differentiation, 156, 160

Cell Membranogenesis, 132

Cell Trafficking, 150

Climate Change, 49-50, 53-54, 56, 127-129, 197, 201, 224

Coastal Aquaculture, 87-88, 90, 97

Coastal Management, 1, 97, 101

Coastal Protection, 87, 90-93, 118-119, 121, 124-125, 199-201, 203

Communication, 88, 92, 94, 150, 152, 154-155

Coral Mariculture, 1-2

Cytoskeleton, 156, 159

D

Degraded Reefs, 1-2

Diastolic Blood, 140-141, 143, 146, 148

Dietary Cholesterol, 132-133, 135-139

Disease, 22-23, 29, 31-32, 49-51, 53-56, 92-93, 96, 136, 139, 141, 143, 145, 149, 155, 160, 171-176, 184, 216-220, 222-224

Disease Prevalence, 49-51

Disease Refuge, 49, 54

Donor Colonies, 1-2, 5

E

Endothelial Dysfunction, 140, 145, 149

Endothelial Function, 140-141, 143-146, 148-149

Estuarine Ecosystems, 49, 107, 109, 114

Etiologic Pathogen, 171, 173

F

Filter Feeding, 55, 99

Fine-mesh Enclosures, 39-42, 44-47

Fish Oil Consumption, 22, 140

Fish Oil Supplementation, 22-24, 28, 30-31, 140-141, 144-148

Fishing Pressure, 126, 201

Flow-mediated Dilation, 140, 143, 145-149

Food Provision, 118-119, 121, 125-126, 128, 199-200, 203

G

Growth Rates, 1-6, 8, 39, 44-47, 49-50, 218

H

Habitat Destruction, 49, 207

Heterogeneous Pattern, 72, 84-85

Hlb0905 Strain, 171

Homeostasis, 32, 136, 150, 155-156

Hydrothermal Vents, 72, 76, 84-86

I

Immune-related Genes, 152

Immunocompetent Cells, 150, 154

Iteroparous Reproduction, 72, 84

J

Juvenile Corals, 1-3, 5-9

Juvenile Oysters, 33-36, 50, 56, 211

M

Mariculture, 1-2, 5, 7-8, 35, 39-40, 47, 98, 119, 121, 123-126, 128, 139, 199-201, 203, 206-208

Marine Ecosystems, 116, 208, 223

Marine Invertebrates, 150, 197

Microphytobenthos, 99-101, 107, 110, 113-115, 223

Mortality Rate, 1, 6, 221

Multimetric Health, 10, 18

Muscle Fibers, 172

Muscle Necrosis, 171-173, 176

N

Natural Products, 117-121, 200-201, 203
Neighbor-joining Method, 174, 176
Normoglycemic, 140, 142, 144-145, 147, 149
Nursery Systems, 39, 47-48

O

Ocean Health, 116-117, 119, 127-130, 199-201, 203, 206, 208
Ocean Health Index, 116-117, 119, 128-130, 199-201, 203, 206
Overfishing, 49-50, 55, 190
Oyster Crassostrea Gigas, 150, 159, 214, 220, 223-224
Oyster Survival, 49-50, 52-54, 151, 155-156, 211, 214

P

Parametrisation, 93
Pathogenesis, 150, 156, 159
Pathogenic Bacteria, 150, 159, 176, 222
Pelagic Compartments, 99, 101, 104, 110, 113
Pelagic Maps, 99, 101, 110
Plasma Lipidomic, 22-23, 26, 28, 31
Pollution, 21, 34, 49-50, 87-88, 93, 95, 97, 120, 126, 197-198, 200-201, 203, 207
Post-transplantation, 1-4, 6-8
Predation, 1-2, 9, 33-38, 48, 184, 196
Predators, 33-38, 43, 46-47, 49-50, 53, 111, 193

R

Reef Restoration, 1-2, 8-9
Respiratory Chain, 154, 156
Richibucto Estuary, 10-13, 18, 21
River Discharge, 50, 52-55
Rna Extraction, 134, 151, 157

S

Salinity Variation, 52
Scleractinian Corals, 1-2, 5-6, 9
Shrimp Aquaculture System, 171
Social-ecological Systems, 87, 117
Species Biodiversity, 125
Suspension-feeders, 99, 101, 103, 108, 111, 113-114
Sustainable Development, 97, 171, 206, 218

T

Teleost Fish, 132, 138-139
Tetraodontidae, 57, 59, 65, 70-71
Transplantation, 1-8
Triglycerides, 22-26, 32, 149
Tropical Sea Cucumbers, 39-40, 47

V

Vent Taxa, 72, 84-85
Viral Diseases, 87, 171
Volumetric Growth Rates, 1, 3-6, 8

www.ingramcontent.com/pod-product-compliance
Lightning Source LLC
Chambersburg PA
CBHW080412190526
45161CB00003B/220